Advances in Direction-of-Arrival Estimation

For a list of recent titles in the *Artech House Radar Library*, turn to the back of this book.

Advances in Direction-of-Arrival Estimation

Sathish Chandran
Editor

ARTECH
HOUSE

BOSTON | LONDON
artechhouse.com

Library of Congress Cataloging-in-Publication Data
Advances in direction-of-arrival estimation / [edited by] Sathish Chandran.
   p. cm.—(Artech House radar library)
Includes bibliographical references and index.
ISBN 1-59693-004-7 (alk. paper)
1. Tracking radar.   2. Radio direction finders.   3. Antenna arrays.   I. Chandran, Sathish, 1962–

TK6592.A9A38   2005
621.3848—dc22   2005056277

British Library Cataloguing in Publication Data
Advances in direction-of-arrival estimation.—(Artech House radar library)
1. Radar—Mathematics   2. Algorithms   I. Chandran, Sathish, 1962–
621.3'848'0151

ISBN-10: 1-59693-004-7

Cover design by Cameron, Inc.

© 2006 ARTECH HOUSE, INC.
685 Canton Street
Norwood, MA 02062

All rights reserved. Printed and bound in the United States of America. No part of this book may be reproduced or utilized in any form or by any means, electronic or mechanical, including photocopying, recording, or by any information storage and retrieval system, without permission in writing from the publisher.

All terms mentioned in this book that are known to be trademarks or service marks have been appropriately capitalized. Artech House cannot attest to the accuracy of this information. Use of a term in this book should not be regarded as affecting the validity of any trademark or service mark.

International Standard Book Number: 1-59693-004-7
Library of Congress Catalog Card Number: 2005056277

10 9 8 7 6 5 4 3 2 1

To my wife, Priya, and my children, Uma and Shankar

Contents

Preface xvii

Acknowledgments xix

PART I
Overview 1

CHAPTER 1
Antenna Arrays for Direction-of-Arrival Estimation 3

1.1 Introduction 3
1.2 Spinning DF Antenna 4
1.3 Common DF Antennas 6
1.4 Antenna Arrays 7
    1.4.1 Linear Arrays 8
    1.4.2 Adcock Array 9
    1.4.3 Array Element Spacing 10
    1.4.4 Butler Matrix 12
    1.4.5 Monopulse 15
    1.4.6 Wullenweber Array 15
    1.4.7 Planar Array 17
1.5 DF Algorithms and Arrays 18
    References 19

CHAPTER 2
Detection-Estimation of Gaussian Sources by Nonuniform Linear Antenna Arrays: An Overview of the Generalized Likelihood-Ratio Test Approach 21

2.1 Introduction 21
2.2 GLRT Philosophy for the Detection-Estimation Problem 24
2.3 GLRT Detection-Estimation Results for Uniform and Sparse Antenna Arrays 30
    2.3.1 Superior Case 30
    2.3.2 Conventional Case 33
2.4 Conclusions 39
    References 42

PART II
DOA Methods — 47

CHAPTER 3
DOA Estimation of Wideband Signals — 49
3.1 Incoherent Wideband DOA Estimation — 52
3.2 Coherent Wideband DOA Estimation — 52
    3.2.1 Coherent Signal Subspace Method — 53
    3.2.2 BICSSM — 54
    3.2.3 WAVES — 55
3.3 TOPS — 56
    3.3.1 Theory — 57
    3.3.2 Signal Subspace Projection — 62
    3.3.3 Performance Assessment — 63
    3.3.4 Modification of TOPS for Localization — 66
3.4 Conclusions — 67
    Acknowledgments — 68
    References — 68

CHAPTER 4
Coherent Broadband DOA Estimation Using Modal Space Processing — 71
4.1 Introduction — 71
4.2 System Model — 72
4.3 Focusing Matrices for Coherent Wideband Processing — 73
4.4 Spatial Resampling Method — 74
4.5 Modal Decomposition — 75
    4.5.1 Linear Arrays — 75
    4.5.2 Modal Truncation — 76
4.6 Modal Space Processing — 77
    4.6.1 Focusing Matrices — 77
    4.6.2 Spatial Resampling Matrices — 78
4.7 Modal Space Algorithm — 79
4.8 Simulation — 80
    4.8.1 A Group of Two Sources — 80
    4.8.2 Three Groups of Five Sources — 82
4.9 Conclusions — 85
    References — 85

CHAPTER 5
Target Localization in Three Dimensions — 87
5.1 Direction-of-Arrival Estimation Method — 87
    5.1.1 Extraction of y_i — 90
    5.1.2 Extraction of z_i — 91
    5.1.3 Extraction of d_i — 91
    5.1.4 Pairing y_i, z_i, and d_i — 92
    5.1.5 Complex Amplitude Estimation — 92
    5.1.6 Noisy Signals — 92

5.2	Distance Estimation Method	93
5.3	Target Location Estimation and Simulation Results	96
5.4	Conclusions	101
	References	101

CHAPTER 6
Polarimetric Time-Frequency MUSIC for Direction Finding of Moving Sources with Time-Varying Polarizations — 103

6.1	Introduction	103
6.2	Signal Model	105
	6.2.1 Time-Frequency Distributions	105
	6.2.2 Spatial Time-Frequency Distributions	105
6.3	Spatial Polarimetric Time-Frequency Distributions	106
	6.3.1 Polarimetric Modeling	106
	6.3.2 Polarimetric Time-Frequency Distributions	108
	6.3.3 Spatial Polarimetric Time-Frequency Distributions	108
6.4	Polarimetric Time-Frequency MUSIC	110
6.5	Spatio-Polarimetric Correlations	111
6.6	Moving Sources with Time-Varying Polarizations	112
	6.6.1 Polarization Diversity and Selection of Data Samples	113
	6.6.2 Polarization Diversity Consideration in PTF-MUSIC	114
6.7	Simulation Results	117
6.8	Conclusions	118
	References	120

CHAPTER 7
A One-Dimensional Tree Structure–Based Algorithm for DOA-Delay Joint Estimation — 123

7.1	Introduction	123
7.2	System Model	124
7.3	The Proposed Algorithm	127
	7.3.1 S-ESPRIT	127
	7.3.2 T-ESPRIT	128
	7.3.3 TST-ESPRIT	131
7.4	Simulations and Discussions	138
7.5	Conclusions	141
	References	141
Appendix 7.A Noise Property in (7.24)		142

CHAPTER 8
Direction-of-Arrival Estimation Approach Using Switched Parasitic Arrays — 145

8.1	Introduction	145
8.2	Switched Parasitic Arrays	146
8.3	Wideband Direction-Finding Method for a Two-Path Model	148
	8.3.1 Wideband Signaling in Multipath Environments	149
	8.3.2 Method Description	149
	8.3.3 Method Performance Analysis	151

	8.4	A Modified MUSIC Algorithm for Switched Parasitic Arrays	153
		8.4.1 Method Performance Analysis	155
	8.5	Conclusions	157
		References	159

CHAPTER 9
DOA Estimation for Noncircular Signals: Performance Bounds and Algorithms — 161

9.1	Introduction	161
9.2	Array Signal Model	162
9.3	MUSIC-Like Algorithms	163
	9.3.1 MUSIC-Like Algorithms Built Only from $\mathbf{R}'_{y,T}$	163
	9.3.2 MUSIC-Like Algorithms Built from Both $\mathbf{R}_{y,T}$ and $\mathbf{R}'_{y,T}$	164
9.4	Asymptotically Minimum Variance Estimation	170
9.5	Stochastic Cramér-Rao Bound for Noncircular Gaussian Signals	173
9.6	Stochastic Cramér-Rao Bound for BPSK Signals	175
9.7	Illustrative Examples	177
9.8	Conclusions	187
	References	188

CHAPTER 10
Localization of Scattered Sources — 191

10.1	Introduction	191
10.2	Array Model	192
	10.2.1 Coherently Distributed Source Model	193
	10.2.2 Incoherently Distributed Source Model	195
10.3	Parameter Estimation Techniques	197
	10.3.1 Estimation Techniques for CD Sources	197
	10.3.2 Estimation Techniques for ID Sources	200
10.4	Conclusions	209
	References	209

PART III
Source Localization Problems — 211

CHAPTER 11
Direct Position Determination of Multiple Radio Transmitters — 213

11.1	Introduction	213
11.2	Mathematical Preliminaries	215
11.3	Position Determination	217
	11.3.1 Unknown Waveforms Signals	217
	11.3.2 Known Waveform Signals	219
11.4	Numerical Examples	222
	11.4.1 Example 1	222
	11.4.2 Example 2	223

	11.4.3 Example 3	223
	11.4.4 Example 4	225
	11.4.5 Example 5	225
11.5	Conclusions	225
	References	227

Appendix 11.A Cramér-Rao Bound for Known Waveforms — 228
Appendix 11.B CRB for Unknown Gaussian Waveforms — 230
Appendix 11.C DPD Performance Analysis for Unknown Waveforms — 233

	11.C.1 Expressions for Ω	234
	11.C.2 Expressions for Ψ	235
	11.C.3 Special Case: Uniform Linear Array	237

Appendix 11.D The Frequency Domain Representation of Finite Length Observations — 238

CHAPTER 12

Direction-of-Arrival Estimation Under Uncertainty — 241

12.1	Introduction	241
12.2	Basic DOA Estimation Signal Model	242
12.3	Uncertainties in Signal Model and System Parameters	244
	12.3.1 Sources of Uncertainty	245
	12.3.2 Improving the Robustness	245
12.4	Uncertainty in Noise Model	247
	12.4.1 Qualitative and Quantitative Robustness	248
	12.4.2 Robust Procedures	250
12.5	Array Processing Examples	254
12.6	Conclusions	255
	References	256

CHAPTER 13

Coupling Model in DOA Antenna Arrays — 259

13.1	Introduction	259
13.2	Antenna Steering Vector Model	260
13.3	Steering Vector Errors and Their Influence in DOA Estimation	264
13.4	Coupling Model	266
13.5	Coupling Influence in DOA Estimation	272
13.6	Error Compensation	273
13.7	Conclusions	278
	References	280

PART IV

Specific Applications of DOA Estimation — 283

CHAPTER 14

Maximum Likelihood Estimate of Target Angular Coordinates Under Main Beam Interference: Application to Recorded Live Data — 285

14.1	Introduction	285

14.2	MLE Algorithm for Estimation of Target Angular Coordinates	286
14.3	CRLB of Estimation of Target Angular Coordinates	288
14.4	MLE Algorithm Simulation Results	292
14.5	Test on Recorded Live Data	294
	14.5.1 Description of the Radar for Data Capture	295
	14.5.2 Data Capture Experimental Setup	296
	14.5.3 Comparison Between the Theoretical and Real Radar Beams	297
	14.5.4 MLE on Recorded Live Data	301
14.6	Conclusions	301
	References	302

CHAPTER 15
Direction Estimation of Broadband Sources for Auditory Localization and Spatially Selective Listening   305

15.1	Introduction	305
15.2	Scenario and Model	306
15.3	Characteristics of Natural Sound Sources	307
15.4	Decomposing into Spectral Components	308
15.5	DOA Estimation Based on Time Delays and Relative Attenuation	308
	15.5.1 General Approach	309
	15.5.2 DOA Estimation Using Time Delay Between the Sensors	309
	15.5.3 Subspace Methods in a Reverberant Room	310
	15.5.4 DOA Estimation Using Relative Attenuation Between the Sensors	311
15.6	Spectrally Weighted DOA Estimate	313
15.7	Source Separation Algorithm	314
15.8	Experimental Results	316
	15.8.1 Parameter Values Used in the Implementation	316
	15.8.2 Performance of the Algorithm When the Sources Are Sinusoidal	318
	15.8.3 Performance of the Algorithm with Increasing Number of Sources	318
	15.8.4 Performance of the Algorithm When the Sensor Spacing Is Varied	319
	15.8.5 Resolvability of the Algorithm When Sources Are Closely Spaced	320
	15.8.6 Performance of the Sound Source Separation from Real Two Sensor Array Recordings in Multitalker Scenario	320
	15.8.7 Performance Measure for Listening Tests	321
	15.8.8 Performance of the Algorithm with Headlike Structure	322
15.9	Conclusions	323
	References	323
	Selected Bibliography	324

CHAPTER 16
Multiple Target Estimation in a Surveillance Radar System with a Mechanically Rotating Antenna   327

16.1	Introduction	327

16.2	Data Model and Problem Statement	328
16.3	The Estimation Algorithm	329
	16.3.1 Asymptotic Maximum Likelihood Estimation	329
	16.3.2 The AML-RELAX Estimator	332
16.4	Numerical Performance Analysis	334
16.5	Conclusions	342
	References	342

CHAPTER 17

Joint DOA/DOD/DTOA Estimation System for UWB Double Directional Channel Modeling — 345

17.1	Introduction	345
17.2	UWB Double Directional Channel Sounding System	346
17.3	Parametric Multipath Modeling for UWB	347
	17.3.1 Multipath Model	347
	17.3.2 Subband Model	349
	17.3.3 Spherical Wave Model	349
	17.3.4 Two Single Directional Measurements, Pairing, and Clusterization	350
17.4	ML-Based Parameter Estimation for UWB Directional Channel	351
	17.4.1 Maximum Likelihood Parameter Estimation	352
	17.4.2 Expectation Maximization Algorithm	352
	17.4.3 SAGE Algorithm	353
	17.4.4 Successive Interference Cancellation-Type Procedure	354
	17.4.5 Consideration About the Subband	356
	17.4.6 Deconvolution of Antenna Directivity	356
17.5	Experiments	356
	17.5.1 Specification of Measurement Equipment	356
	17.5.2 Measurement Environment	357
	17.5.3 Results	357
17.6	Conclusions	358
	References	360

CHAPTER 18

Conventional Method Improvement by Signal Property Exploitation — 363

18.1	Cyclostationarity	363
18.2	Data Model	364
18.3	High-Resolution Techniques	365
	18.3.1 Cyclic MUSIC Algorithm	365
	18.3.2 Extended Cyclic MUSIC Algorithm	365
	18.3.3 Root Cyclic MUSIC Algorithm	369
18.4	Nonselective Beamforming Techniques	372
	18.4.1 Data Model	373
	18.4.2 Beamforming for Cyclostationary Signals	374

	18.4.3 Minimum Variance Beamforming for Cyclostationary Signals	374
	18.4.4 Discussion	375
	18.4.5 Simulations	375
18.5	Conclusions	376
	References	377

PART V
Experimental Setup and Results 379

CHAPTER 19
DOA Antenna Array Measurement and System Calibration 381

19.1	Introduction	381
19.2	Steering Vector Model	382
19.3	Calibration Process	386
	19.3.1 Coupling Through Terminals	387
	19.3.2 Radiation Pattern Measurements	388
	19.3.3 Calibration Process for the RF Subsystem	389
	19.3.4 Application Example	391
19.4	Conclusions	392
	References	393

CHAPTER 20
ESPAR Antenna Signal Processing for DOA Estimation 395

20.1	Introduction	395
20.2	ESPAR Antenna Fundamentals	397
	20.2.1 Structure and Formulation	397
	20.2.2 Signal Model	400
20.3	High-Precision DOA Estimation	401
	20.3.1 Correlation Matrix for Single-Port Antennas	402
	20.3.2 Reactance-Domain Signal Processing	403
	20.3.3 Computer Simulations	405
	20.3.4 Experimental Verification	408
20.4	Practical System Applications	412
	20.4.1 Satellite Attitude and Direction Control	412
	20.4.2 Wireless Locator and Homing System	413
	20.4.3 Wireless Ad Hoc Network	414
20.5	Conclusions	414
	Acknowledgments	415
	References	415

CHAPTER 21
A New DOA Estimation Method for Pulsed Doppler Radar 419

21.1	Introduction	419
21.2	Review of MUSIC and BS-MUSIC	421
21.3	Model for HF Radar	423

21.4	Two-Dimensional Prefiltering-Based MUSIC	423
21.5	Algorithm Summary for 2DP-MUSIC	427
21.6	Performance Analysis	427
	21.6.1 Formulation of Sample Null Spectra for MUSIC, BS-MUSIC, and 2DP-MUSIC	428
	21.6.2 Expectation of 2DP-MUSIC's Null Spectrum	429
	21.6.3 Resolution Threshold	430
21.7	Simulations	431
21.8	Experimental Result	433
21.9	Conclusions	435
	References	435
Appendix 21.A Derivation of (21.24)		437
Appendix 21.B Resolution of Spectral MUSIC		438

List of Symbols	441
Acronyms	443
About the Authors	445
Index	459

Preface

Determining the direction-of-arrival (DOA) of any signal of interest has long been of great interest to the research community. It is often required to identify the DOA of signals emanating from every possible direction. This is important for military as well as for civilian applications. The directions to targets are usually expressed by the DOA of the transmitted signals from the sources. Several models based DOA estimation are given in the literature.

If the DOA of the desired signal is known, then adaptive beamforming algorithms can be performed to minimize the signal power of the interference, which will in turn maximize the power of the signal of interest. DOA estimation is one of the main functional requirements for direction-finding smart antennas in future generation mobile and stealth communication systems.

In this book, a reflection of the recent works regarding the DOA estimation employing several related algorithms and their related advantages and disadvantages are illustrated.

This book, *Advances in Direction-of-Arrival Estimation*, is an assembly of the research findings and insights of several scientists and engineers. This book is a detailed reference book for engineers, researchers, and final-year undergraduate and postgraduate students.

The book is organized as follows:

- Part I: Overview;
- Part II: DOA Estimation Methods;
- Part III: Source Localization Problems;
- Part IV: Specific Applications of DOA Estimation;
- Part V: Experimental Setup and Results.

Acknowledgments

I thank deeply all the authors of this book for contributing the chapters about their works. I strongly believe that this book will serve the purpose with which it has been prepared.

Sathish Chandran
Editor
Malaysia
December 2005

PART I
Overview

CHAPTER 1
Antenna Arrays for Direction-of-Arrival Estimation

Randy Haupt and Rodney Martin

1.1 Introduction

Direction finding (DF) is a procedure for determining the bearing of a transmitting source by observing the direction of arrival, along with other properties of the signal [1]. A direction finding system consists of the components shown in Figure 1.1. The first component of the DF system, the antenna arrays, is described in this chapter. Any errors introduced by the antenna propagate through the rest of the system. Errors introduced by the antenna correspond to unwanted amplitude and phase components in the signals, which cause the phase center of the antenna to move. The antenna phase center is the point at which the radiation from the antenna appears to emanate, if it were transmitting. A change in the phase center corresponds to a change in the coherent summation of the signals from the antenna elements. When processed, these components cause an error in the display output. Errors introduced by the antenna include geometrical placement error, cross-polarization, noise, mutual coupling, multipath, and manufacturing tolerances. Important DF antenna properties include gain, sidelobe level, frequency of operation, polarization, bandwidth, and size.

Directivity of the antenna is an important indicator of the antenna's ability to isolate the location of a transmitting signal. Directivity is inversely proportional to the antenna's beamwidth, and the beamwidth is a measure of the uncertainty of the location of the signal. Using the antenna beam to locate a signal is very basic, and this concept is the starting point for this chapter. Mechanically moving the antenna to locate sources limits DF performance. Using a null in conjunction with a main beam improves the DF accuracy. Tremendous improvements are possible by using a phased array antenna. Thus, the majority of this chapter is devoted to antenna arrays.

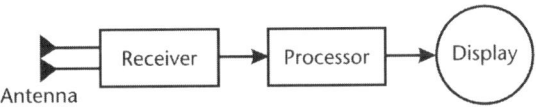

Figure 1.1 Block diagram of a DF system.

1.2 Spinning DF Antenna

Rotating an antenna in azimuth and correlating the antenna position with high output power is the most elementary way of finding the location of an electromagnetic source. If an antenna with a 26.2° beamwidth rotates in azimuth, then the output of the antenna as a function of angle appears in Figure 1.2 when there are two 0-dB sources. If both sources are at 90°, then the output as a function of angle is proportional to the antenna pattern. Both sources appear at the peak of the main beam. Leaving one source at 90° while moving the other source to 100° produces the dashed line pattern. Since both sources are very close together, they cannot be distinguished. In fact, the output indicates that there is a single source at 95°. When the two sources are one full 3-dB beamwidth apart, then the dot-dash line with two peaks and a 3-dB minimum in between occurs. Increasing the separation between the sources to 130° results in two very distinct peaks, as shown by the dotted line.

Sources arriving at the antenna with different power levels further complicate the determination of the directions of the sources. If the source at 90° is at 0 dB and the other source is at 10 dB, then a dramatically different output occurs, as shown in Figure 1.3. Now, only the second source can be easily detected. Even when the second source is at 130°, the first source is only a few decibels above the far field pattern sidelobe level, so it probably would not be detected. Using a low sidelobe antenna helps with this problem, as shown in Figure 1.4. Here, when the second source is at 130°, the first source is significantly higher than the sidelobe level of the array, so it could be distinguished. When the second source is at 100° and 116.2°, however, the increased width of the main beam due to the low sidelobe amplitude taper does not allow a 3 dB dip between the two sources, so they would

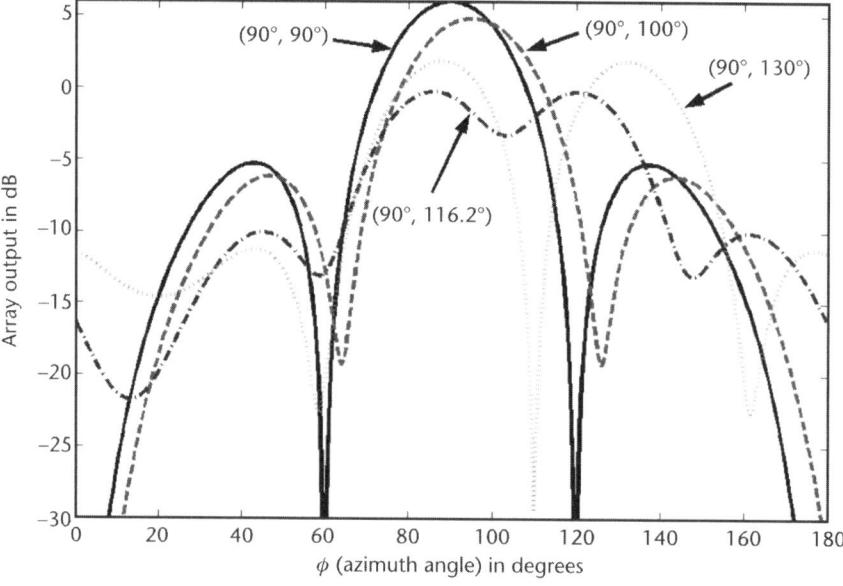

Figure 1.2 The array output for two 0 dB sources, when one source is at 90° and the other source is at 90°, 100°, 116.2°, and 130°.

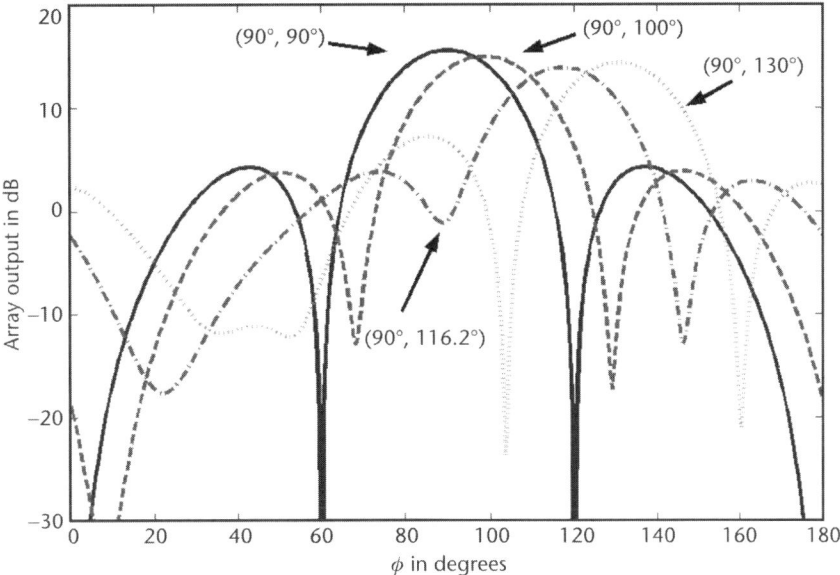

Figure 1.3 Output power when the source at 90° is 0 dB, and the other source at 90°, 100°, 116.2°, and 130° is 10 dB.

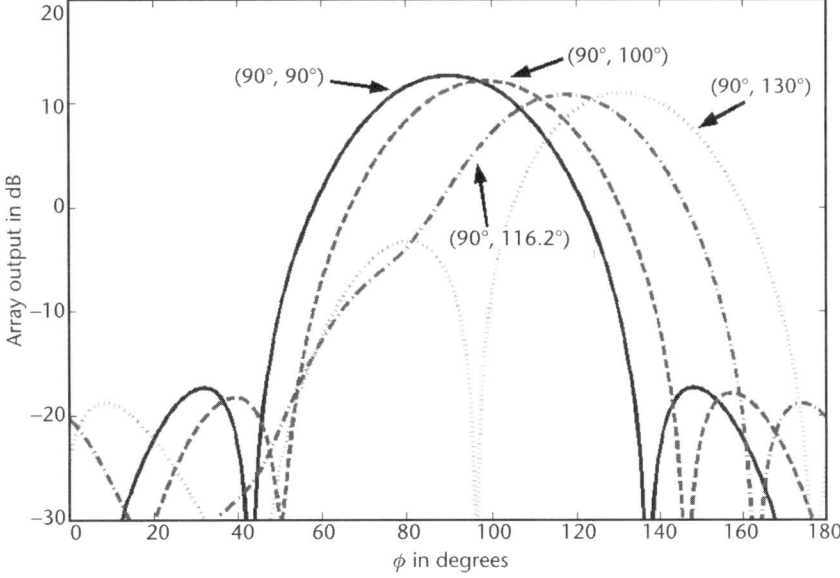

Figure 1.4 Output power when the source at 90° is 0 dB, and the other source at 90°, 100°, 116.2°, and 130° is 10 dB, and the antenna has a maximum sidelobe level of −30 dB.

not be distinguished. Closely spaced targets can be distinguished by making the antenna beamwidth smaller. The beamwidth of the antenna is inversely proportional to the antenna width in the plane of the beamwidth.

Figure 1.5 shows the 3-dB beamwidth for three different aperture widths. Increasing the aperture size improves the antenna's ability to separate multiple sources.

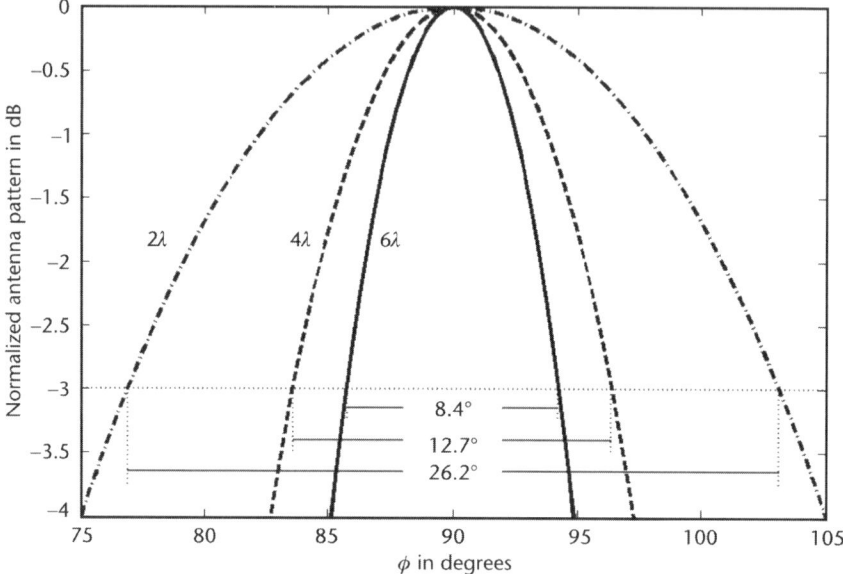

Figure 1.5 Antenna 3-dB beamwidth as a function of aperture size.

A source can often be more accurately pinpointed using a null rather than a main beam. When a source falls in the null, then no output power is received. Nulls tend to have steep sides, so the output power changes very quickly near the null, in contrast to the peak of the main beam. The accuracy of locating a source depends upon the null depth. Figure 1.6 is an example of an antenna with a difference pattern. The difference pattern has a null in place of the main beam peak. Factors, such as noise, tolerances, and so forth, can limit the null depth. A deep null has a narrow width. The steepness of the pattern near the null is higher for a larger aperture. Figure 1.6 shows the null width when the nulls are −10, −20, and −30 dB below the pattern peak. Resolution using the null width is analogous to resolution using the beamwidth.

Peak detection with a spinning DF is usually done with a high gain (narrow beamwidth) antenna, such as a dish. DF using an antenna null can be done with low-gain antennas, such as a dipole or loop. A half-wavelength dipole along the x-axis has nulls when $\phi = 0°$ and $180°$. Sources at $\phi = 0°$ and $180°$ are ambiguous. A loop in the y-z plane also has nulls when $\phi = 0°$ and $180°$. Both types of antennas have been used to find the direction of sources using nulls. DF using a main beam peak is preferred, when the source is weak, the content of the received signal is important, low sidelobes are required, and a large antenna (in terms of wavelengths) is needed. Otherwise, the null locating techniques can more accurately locate a source.

1.3 Common DF Antennas

Besides dipoles, monopoles, and loops, other common antenna types used for DF include reflectors, horns, and spirals [2]. A reflector consists of a large conducting

1.4 Antenna Arrays

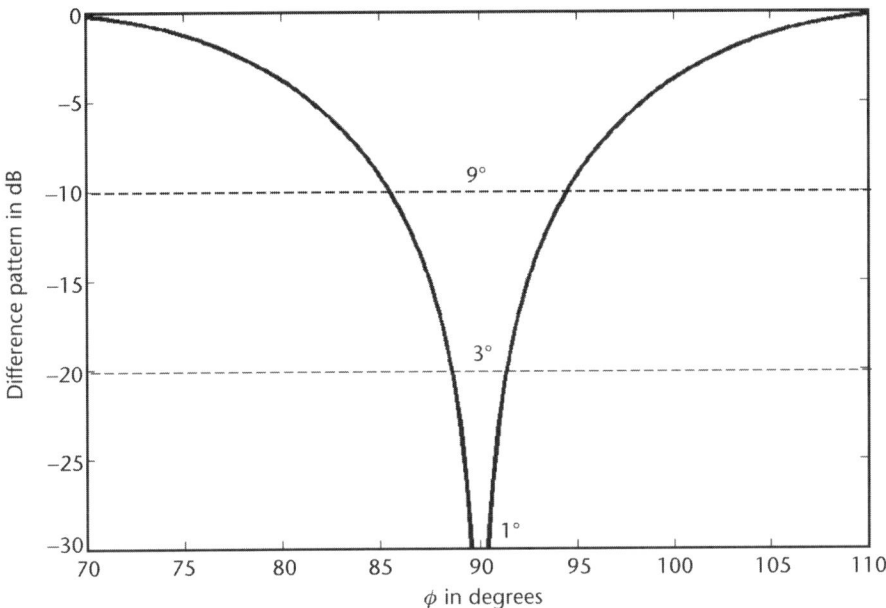

Figure 1.6 Null width as a function of null depth for an antenna pattern.

surface that reflects or scatters the radiation from a feed element. Its simple construction and high gain make it very attractive, especially for spinning DF applications. The reflector can be made broadband by using a broadband feed element, such as a log periodic antenna. Horn antennas have high gain and can be made to be broadband. In addition, multimodal horns have ports corresponding to sum and difference patterns. Spiral antennas are called frequency independent antennas, and can receive both vertical and horizontal polarized signals. Planar spirals radiate equally from each face. In order to direct the beam in only one direction, a cavity filled with an absorber is often placed on one side. Another way to force the spiral to radiate in only one direction is to wrap it around a cone. The antenna then radiates only in the direction of the cone apex. Spirals with four or more arms can be made to radiate modes that include a sum pattern and a difference pattern.

1.4 Antenna Arrays

Arrays have two or more elements or antennas, which have their outputs weighted and added together. They bring an added level of sophistication to DF by electronic beam steering, combined sum and difference patterns, multiple beams, and the ability to separate multiple sources. The expense and complexity of the array offset these advantages. The placement of the elements is critical in determining how the array receives a signal from any angle. The array performance independent of the element is known as the array factor. If all the elements radiate equally in all directions (i.e., isotropic point sources), then it is known as the array response. The array factor for an arbitrary array is given by

$$AF = \sum_{n=1}^{N} w_n e^{jk(x_n \sin\theta \cos\phi + y_n \sin\theta \sin\phi + z_n \cos\theta)} \quad (1.1)$$

where

N = number of elements
$w_n = a_n e^{j\delta_n}$ = complex weight for element n
$k = 2\pi$/wavelength
(x_n, y_n, z_n) = location of element n
(θ, ϕ) = direction in space

1.4.1 Linear Arrays

We'll assume that the ground is in the x-y plane, and distance above the ground is in the z-direction. Thus, azimuth is measured by ϕ from the x-axis, and elevation is measured by θ from the z-axis. A linear array has elements along one of the three axes. In these cases, the array factor becomes greatly simplified.

$$AF_x(\phi) = \sum_{n=1}^{N} w_n e^{jkx_n \cos\phi} \quad (1.2)$$

$$AF_y(\phi) = \sum_{n=1}^{N} w_n e^{jky_n \sin\phi} \quad (1.3)$$

$$AF_z(\phi) = \sum_{n=1}^{N} w_n e^{jkz_n \cos\theta} \quad (1.4)$$

Locating sources in azimuth requires either AF_x or AF_y, since they are parallel to the ground and a function of only ϕ. On the other hand, locating sources in elevation requires an array stacked in the vertical direction, or AF_z. If the elements are equally spaced, then

$$\begin{aligned} AF_x &= 1 + w_1 e^{jkd_x \cos\phi} + w_2 e^{j2kd_x \cos\phi} + \ldots + w_{(N_x-1)} e^{j(N_x-1)kd_x \cos\phi} \\ &= 1 + w_1 e^{j\psi_x} + w_2 e^{j2\psi_x} + \ldots \\ &= 1 + w_1 z_x + w_2 z_x^2 + \ldots \end{aligned} \quad (1.5)$$

When the amplitude weights are all equal to one and the spacing between elements is equal, then the array is known as a uniform linear array (ULA), and its array factor is given by

$$AF_x = 1 + w_1 e^{j\psi_x} + w_2 e^{j2\psi_x} + \ldots + w_{(N_x-1)} e^{j(N_x-1)\psi_x} = \frac{\sin\left(\frac{N\psi_x}{2}\right)}{\sin\left(\frac{\psi_x}{2}\right)} \quad (1.6)$$

The beam of an array with equally spaced elements can be electronically steered by adding the following phase shift

$$\psi_x = kd \cos \phi + \beta = kd(\cos \phi - \cos \phi_0) \tag{1.7}$$

where ϕ_0 is the desired steering direction.

1.4.2 Adcock Array

In 1907, Bellini and Tosi used two orthogonal loop antennas to determine the direction of a source without rotating the antenna [3]. Loop antennas receive both horizontal and vertical polarization, so they are susceptible to unwanted skywave noise. Adcock [4] addressed this problem by using four monopole antennas placed on the edges of a square (see Figure 1.7). If the north (N) and south (S) antennas on the y-axis are out of phase, then they have a pattern similar to that of a loop antenna in the x-z plane. Similarly, if the east (E) and west (W) antennas along the x-axis are out of phase, then they have a pattern similar to that of a loop antenna in the y-z plane. The Adcock antenna receives only vertical polarization, since it uses monopoles.

Sir Watson-Watt developed the principle for finding the elevation and azimuth of a source incident on an Adcock array. The N-S array has an array factor given by [5]

$$AF_{NS} = 2j \sin\left(k\frac{d}{2} \sin \theta \cos \phi\right) \tag{1.8}$$

Likewise, the array factor for the E-W array is given by

$$AF_{EW} = 2j \sin\left(k\frac{d}{2} \sin \theta \sin \phi\right) \tag{1.9}$$

An estimate of the tangent of the azimuth angle is the ratio of the output from the E-W array and the output from the N-S array. Thus,

$$\tan \hat{\phi} = \frac{AF_{EW}}{AF_{NS}} = \frac{\sin\left(k\frac{d}{2} \sin \theta \sin \phi\right)}{\sin\left(k\frac{d}{2} \sin \theta \cos \phi\right)} \tag{1.10}$$

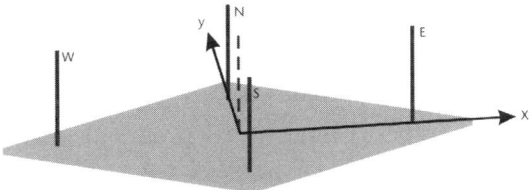

Figure 1.7 Diagram of a four-element Adcock antenna with an optional fifth element in the center.

An estimate of the elevation angle is given by

$$\cos\hat{\theta} = \frac{1}{kd} \sqrt[4]{AF_{EW}^2 + AF_{NS}^2} \qquad (1.11)$$

The azimuth angle of the Adcock array is ambiguous unless another element is inserted in the center of the array (see dashed line in Figure 1.7), or unless the outputs of all the elements are added together to establish a phase reference point [6]. More accurate estimates of the arrival angles are possible by adding element pairs to the four-element Adcock antenna on opposite sides of a circle, with a center at the origin of the x-y axes.

Loop antennas are generally only used when the DF antenna must be extremely compact [2]. Loop antennas have lower sensitivity than Adcock arrays. Even a small horizontal component in the received signal causes significant bearing errors. This error increases with increasing elevation of the transmitting source.

1.4.3 Array Element Spacing

The array factor of a ULA is highly dependent on the number of elements and their spacing. A large array can collect more electromagnetic fields than a small array can. The beamwidth or resolution of the array is a function of the width of the array. For arrays with $d = 0.5\lambda$, increasing the number of elements decreases the beamwidth, thus increasing the array's ability to resolve two signals at different angles. According to the Nyquist sampling theorem, two samples are required for every period of the highest frequency Fourier component of the signal. In this case, two spatial samples are needed for every wavelength, making the element spacing $d = 0.5\lambda$.

Figure 1.8 demonstrates what happens when the sampling theorem is violated. Undersampling results in aliasing or grating lobes. A signal received by an antenna with grating lobes cannot be uniquely located in space, since it is not known through which main lobe the signal entered. If the element spacing exceeds the Nyquist sampling rate of $d = 0.5\lambda$, then grating lobes/nulls result in location ambiguities. It has been shown that a spacing that exceeds one-half wavelength produces ambiguity errors in direction-of-arrival (DOA) estimation algorithms (e.g., MUSIC, Min-Norm, JoDeG, ESPIRIT, and SAGE) [7].

Radio astronomers often need high-resolution apertures that are many miles wide to receive signals from space. Completely filled apertures are prohibitively expensive, so they use sparse arrays, called minimum redundancy arrays [8]. A minimum redundancy array maximizes the distance between grating lobes, while keeping the sidelobes in between the grating lobes at a relatively constant level. Thus, the narrow beamwidth can scan a distant section of space while keeping the grating lobes out of the section being scanned. Figure 1.9 shows the optimum arrangement, where the grating lobes stay just outside of the scan region of the main beam. Spacing between elements occurs at integer multiples of the smallest spacing. Figure 1.10 shows a four-element minimum redundancy array and its array factor, when $d = 1.5\lambda$ and the first grating lobe occurs at 48.2°. Looking at the separation distances between all possible combinations of two elements

1.4 Antenna Arrays

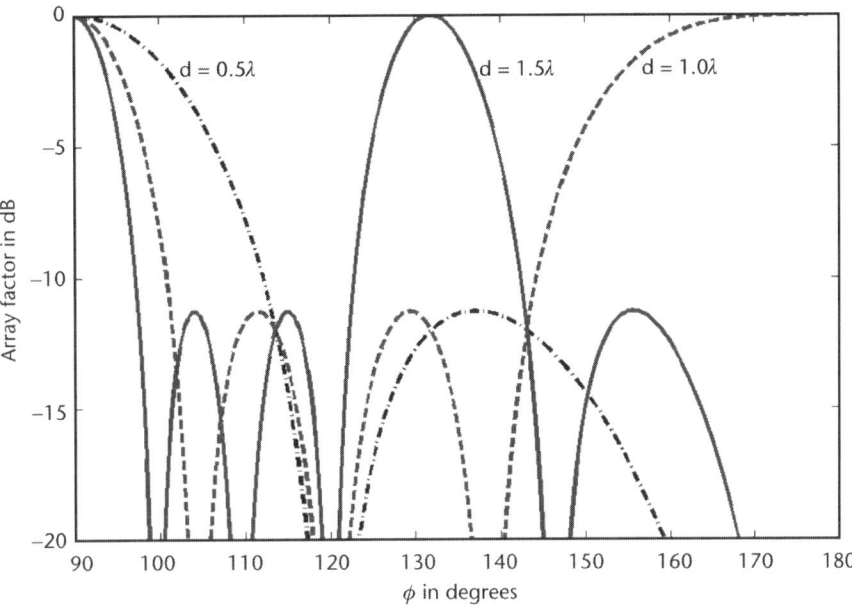

Figure 1.8 Array factors of four-element arrays with different element spacings.

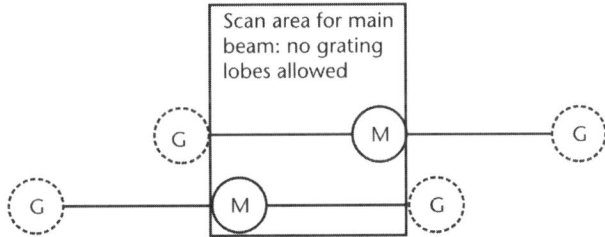

Figure 1.9 The array scans its main beam over the rectangular box. Grating lobes are placed such that they remain just outside the box.

in the array reveals that each integer multiple of d up to the length of the array $d, 2d, \ldots, 6d$, occurs once and only once. Arrays with more than four elements will have more than one sample at some spatial frequencies.

The accuracy of using a single main beam to locate a signal is limited by the 3-dB beamwidth of the array. Using multiple beams significantly increases the array's ability to pinpoint the direction of a signal. A Rotman lens [9], Butler matrix [10], or neural network beamformer [11] allows the array to generate beams at regularly spaced intervals in ϕ, as shown in Figure 1.11. The exact location of the signal is found by interpolating between the angular responses of adjacent beams. As an example, consider an antenna with multiple beams as shown in Figure 1.12. A signal incident at $\phi = 95°$ intercepts beam A at −1.34 dB below its peak and beam B at −5.45 dB below its peak. Beam C is far below the levels of the other two beams, so its response is not considered. Given the responses of the two beams, the direction of arrival can be accurately estimated for a single signal. This process is known as amplitude comparison.

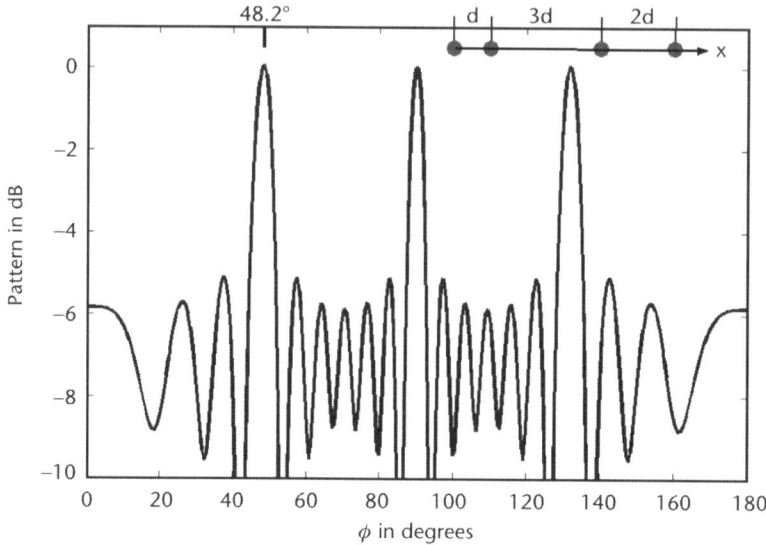

Figure 1.10 Array factor for a four-element minimum redundancy array. The array spacing is shown in the upper right corner.

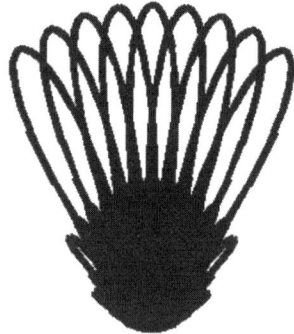

Figure 1.11 An array can have simultaneous multiple overlapping beams.

1.4.4 Butler Matrix

An example of a multiple-beam array is a Butler matrix beamformer [9]. The goal of the Butler matrix is to place predefined, linear phase progressions between elements of a linear array. The Butler matrix is a series of 3-dB quadrature hybrid couplers and phase shifting/delay elements. The 3-dB quadrature hybrid coupler is a device with two inputs (I1, I2) and two outputs (O1, O2). A signal incident to I1 will be split equally between the outputs, O1 and O2. However, the output signals will have a 90° phase shift between them. The phase shifting/delay may be an actual phase shifting device, or, if the matrix is on a printed circuit board, it may be a section of line that has an electrical length of 45° in phase.

To illustrate the performance, a schematic of a four-element Butler matrix is given in Figure 1.13. The circled numbers in the figure will be used to "trace" the phase progression from each of the four inputs—1R, 2L, 2R, and 1L. A source of

1.4 Antenna Arrays

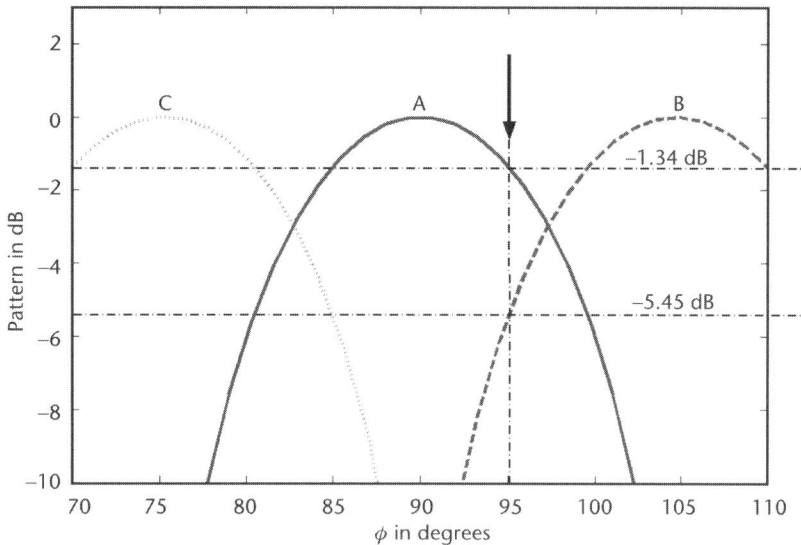

Figure 1.12 Response to a signal of adjacent beams in a multibeam antenna.

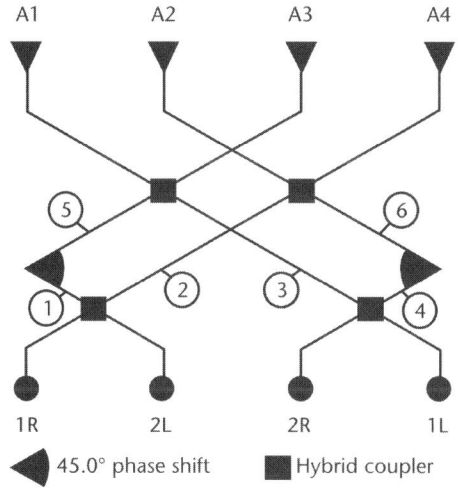

Figure 1.13 Diagram of a four-element Butler matrix.

1V at 0° phase is applied to input 1R, and the trace points 1 and 2 on the output of the first hybrid coupler will be 0.5V at 0° and −90° phase, respectively. Trace points 3 and 4 will have no signal. The signal from trace point 1 is then delayed by 45°, such that trace point 5 will have a signal of 0.5V at −45° phase. This signal is then input to another hybrid coupler, resulting in signals of 0.25V and −45° and −135° phase at antenna elements A1 and A3, respectively. The signal from trace point 2 inputs to a separate coupler, resulting in signals of 0.25V at −90° and −180° phase at elements A2 and A4, respectively. This results in a uniform-amplitude array with a linear phase progression of −45° between successive elements. Each of the trace points will have a voltage level of 0.5V, and the antenna elements will have a voltage level of 0.25V. Note that the phase will have been altered to keep their values between −180° and +180°.

As it is shown, the Butler matrix allows for fixed values of linear phase shift, β. The fixed values are given by

$$\beta_i = 2\pi i/N \text{ for } i = \pm 1/2, \pm 3/2, \ldots, (N-1)/2 \qquad (1.12)$$

Therefore, the beam peaks occur when

$$\psi_x = 0 \text{ or } \phi_i = \cos^{-1}\left(-\frac{\beta_i}{kd}\right) = \cos^{-1}\left(-\frac{2i}{N}\right) \text{ for } i = \pm 1/2, \pm 3/2, \ldots, (N-1)/2 \qquad (1.13)$$

The locations of the four beams are shown in Table 1.1. The four possible array factors are graphed in Figure 1.14.

The structure of the Butler matrix requires 2^n elements. As the number of elements increases, the complexity of the matrix also increases. Figure 1.15 shows a schematic for an eight-element (2^3) Butler matrix. Linear phase progressions of ±22.5°, ±67.5°, ±112.5°, and ±157.5° may be obtained, forming eight distinct beams from the array.

Table 1.1 Phase Shifts and Resulting Beam Locations for the Four-Element Butler Matrix

i	Beam	β_i	Beam Location
1/2	2	45°	75.5°
−1/2	3	−45°	104.5°
3/2	1	135°	41.4°
−3/2	4	−135°	138.6°

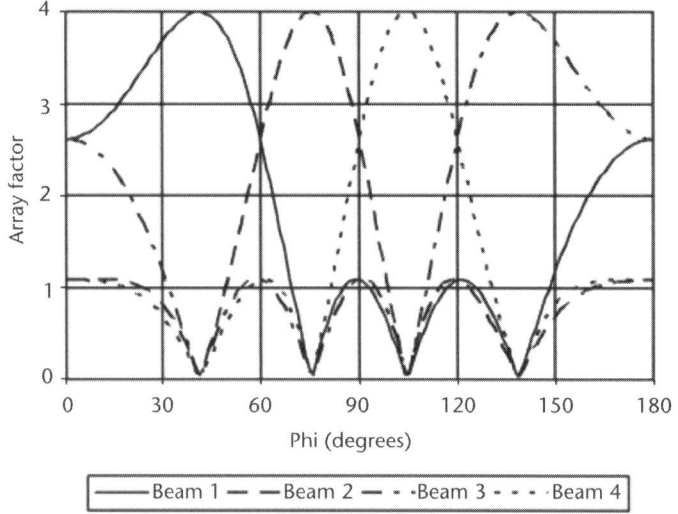

Figure 1.14 Beam locations from the Butler matrix fed array.

1.4 Antenna Arrays

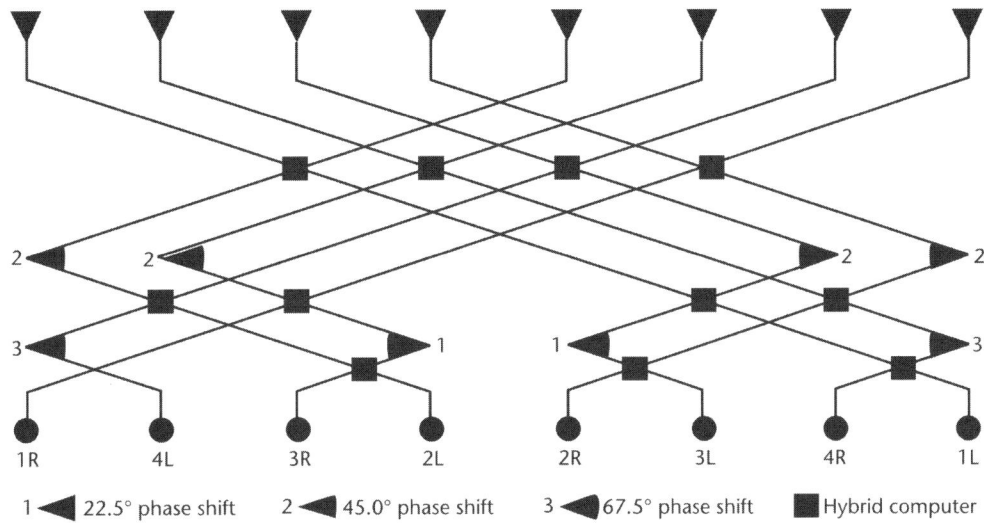

Figure 1.15 Butler matrix for an eight-element array.

1.4.5 Monopulse

A very popular multiple beam angle estimation approach is called monopulse. This approach uses simultaneous sum beams and difference beams. The difference beam has one-half the elements in the aperture receiving a 180° phase shift before adding to the other half.

Amplitude monopulse forms a ratio of the difference output to the sum output. This ratio relates to the angular distance from broadside. The signal incident on the monopulse array in Figure 1.16 has an amplitude ratio given by −6.65 − (−1.35) = −5.30 dB. Since the left difference main beam is 180° out of phase with the right difference main beam and the sum main beam, the receiver can sense which lobe the signal enters. Knowing the amplitude ratio and the relative phase of the received signal provides an accurate estimation of the angle of arrival.

1.4.6 Wullenweber Array

A Wullenweber antenna is a circular array that has only a subset of adjacent elements active at a time (see Figure 1.17) [12]. Phase differentials are compensated on either side of the active elements, so that all signals received have zero phase when the received signal is parallel to the radial line going through the center of the active elements. A commutating feed connects the elements active to the output. As the commutating feed rotates, the main beam points in equally spaced azimuth directions. A circular array has an array factor given by

$$AF_{cir} = \sum_{n=1}^{N} w_n e^{jk\rho \sin\theta \cos(\phi - \phi_n)} \qquad (1.14)$$

where ρ is the radius of the array, and ϕ_n is the angular location of element n. A Wullenweber array only has M elements active at a time, where $M < N$. In order

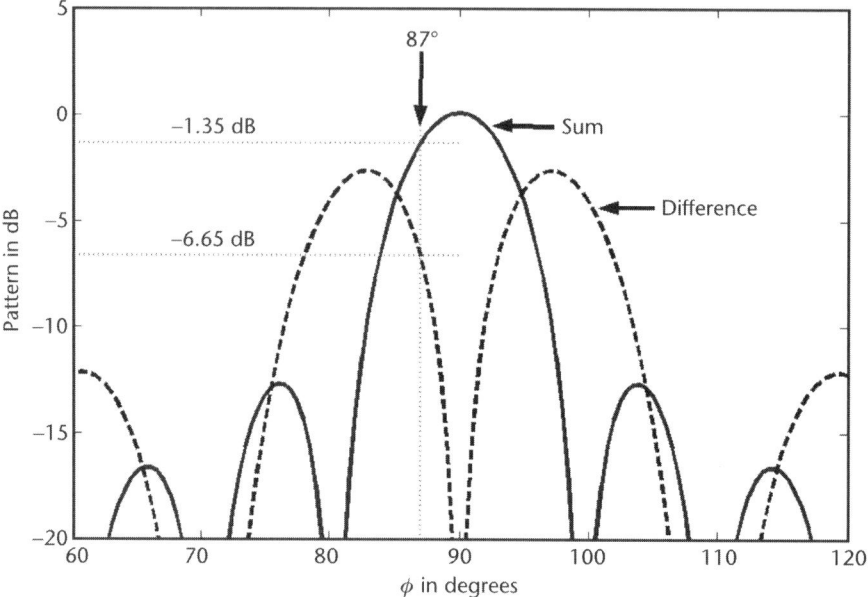

Figure 1.16 A monopulse system with a sum and difference beam.

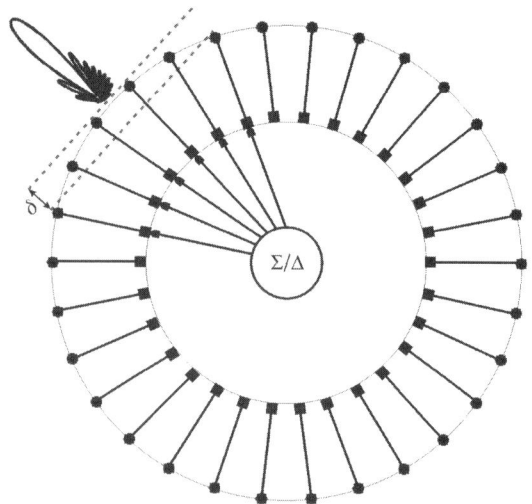

Figure 1.17 Diagram of a Wullenweber array.

to form a coherent beam in the desired direction ϕ_0, the element weights must be given by

$$w_n = e^{-k\rho\cos(\phi_0 - \phi_n)} \tag{1.15}$$

Figure 1.18 shows a 60-element Wullenweber array with 0.5λ spacing between elements, and with its array factor, when the indicated 15 elements (large dark dots on the center circle) are turned on.

1.4 Antenna Arrays

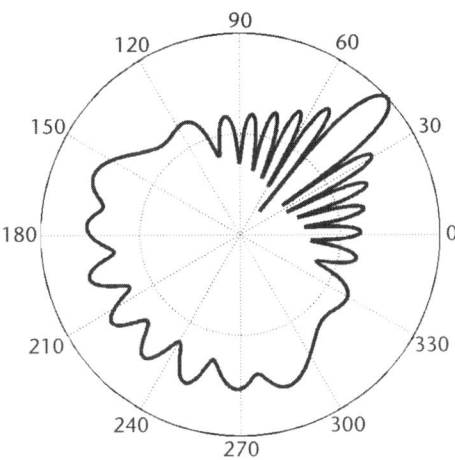

Figure 1.18 The array factor of a 60-element Wullenweber array with 15 elements turned on.

A linear array and a Wullenweber array can distinguish sources in azimuth but not elevation. If the array has N elements, then $N - 1$ nulls can be placed to locate sources. Signal processing techniques derive the null (source) locations from the array weights used to place the nulls. The accuracy in locating a source for a linear array is not a constant at all angles, but the accuracy is constant for the Wullenweber array.

1.4.7 Planar Array

A planar array is needed to locate sources in both azimuth and elevation. Adding more elements to an array increases the cost and computational complexity of the signal processing algorithms. As a result, DOA arrays have only enough elements to locate a desired number of sources. In general, fully populated arrays are not used. Instead, some configuration that covers the extent of the planar array but not the density of the planar array is used [13–17].

The array factor for the linear array along the y-axis also can be written as a polynomial.

$$AF_y = 1 + w_{y1} z_y + w_{y2} z_y^2 + \ldots \tag{1.16}$$

where $z_y = e^{j\psi_y}$. The polynomials of an N_x element array along the x-axis and an N_y element array along the y-axis can be factored to find the zeros

$$(z_x - z_{x1})(z_x - z_{x1}) \ldots (z_x - z_{x(N_x-1)}) = 0 \tag{1.17}$$

$$(z_y - z_{y1})(z_y - z_{y1}) \ldots (z_y - z_{y(N_y-1)}) = 0 \tag{1.18}$$

Signal processing techniques find the zeros of these polynomials. Assuming that each source has a zero, then the location of the sources are found from

$$\frac{\psi_{x1}}{\psi_{y1}} = \frac{kd_x \sin \theta_1 \cos \phi_1}{kd_y \sin \theta_1 \sin \phi_1} = \frac{d_x}{d_y} \tan \phi_1 \qquad (1.19)$$

or

$$\phi_1 = \tan^{-1}\left(\frac{d_y \psi_{x1}}{d_x \psi_{y1}}\right) \qquad (1.20)$$

The elevation angle is found from

$$\begin{aligned}\psi_{x1}^2 + \psi_{y1}^2 &= (kd_x \sin \theta_1 \cos \phi_1)^2 + (kd_y \sin \theta_1 \sin \phi_1)^2 \\ &= (k \sin \theta_1)^2 (d_x^2 \cos^2 \phi_1 + d_y^2 \sin^2 \phi) \\ &= (k \sin \theta_1)^2 c_1^2 \end{aligned} \qquad (1.21)$$

Solving for θ yields

$$\theta_1 = \sin^{-1}\left(\frac{\sqrt{\psi_{x1}^2 + \psi_{y1}^2}}{kc_1}\right) \qquad (1.22)$$

Thus, once the zeros of the array factor polynomial are known, then the source locations in (θ, ϕ) can be found.

DF arrays are sometimes mounted on moving platforms. The element positions are critical to the establishment of the phase relationships necessary for the signal processing. In [18], the authors model arrays whose elements move during the observation interval in an arbitrary but known way. They found these arrays to be more robust to ambiguity errors and to the presence of correlation between the sources.

1.5 DF Algorithms and Arrays

Most of this book is devoted to describing various DF algorithms that use signals obtained from antenna arrays. The chapters are dedicated to various directions of arrival estimation methods, source localization problems, specific application of direction of arrival estimation, and experimental setups and results.

The goal of DF algorithms is to place nulls in the directions of signals by appropriately weighting the array elements. Signal directions are then extracted from the element weights. These signal processing algorithms are usually sensitive to the signal-to-noise ratio (SNR), dynamic range of the received signal powers, location of the antenna, propagation channel effects, and antenna errors. Accurate calibration of the receivers at each element is essential for successful algorithm performance. As noted in this chapter, locating signals in both azimuth and elevation requires a nonlinear array. The associated DF algorithms have a higher level of sophistication. More detailed information on antenna arrays can be found in [9].

References

[1] "IRE Standards on Navigation Aids: Direction Finder Measurements," 1959 IEEE Std. 59 IRE 12, S1, August 8, 1959.

[2] Lipsky, S. E., *Microwave Passive Direction Finding*, New York: John Wiley & Sons, 1987.

[3] "Introduction into Theory of Direction Finding," http://www.rohde-schwarz.com.

[4] Adcock, F., "Improvement in Means for Determining the Direction of a Distant Source of Electromagnetic Radiation," British Patent 1304901919, 1917.

[5] Baghdady, E. J., "New Developments in Direction-of-Arrival Measurement Based on Adcock Antenna Clusters," *Proc. of the IEEE Aerospace and Electronics Conference*, Dayton, OH, May 22–26, 1989, pp. 1873–1879.

[6] "Basics of the Watson-Watt Radio Direction Finding Technique," WN-002, Web Note, RF Products, 1998.

[7] Tan, C. M., M. A. Beach, and A. R. Nix, "Problems with Direction Finding Using Linear Array with Element Spacing More Than Half Wavelength," *First Annual COST 273 Workshop*, Expoo, Finland, May 29–30, 2002, pp. 1–6.

[8] Moffet, A. T., "Minimum-Redundancy Linear Arrays," *IEEE Trans. on Antennas and Propagation Society*, Vol. 16, No. 2, March 1968, pp. 172–175.

[9] Mailloux, R. J., *Phased Array Antenna Handbook*, 2nd ed., Norwood, MA: Artech House, 2005.

[10] Chan, Y. T., et al., "Direction Finding with a Four-Element Adcock-Butler Matrix Antenna Array," *IEEE Trans. on Aerospace and Electronic Systems*, Vol. 37, No. 4, October 2001, pp. 1155–1162.

[11] Southall, H. L., J. A. Simmers, and T. H. O'Donnell, "Direction Finding in Phased Arrays with a Neural Network Beamformer," *IEEE Trans. on Antennas and Propagation Society*, Vol. 43, No. 12, December 1995, pp. 1369–1374.

[12] Frater, M. R., and M. Ryan, *Electronic Warfare for the Digitized Battlefield*, Norwood, MA: Artech House, 2001.

[13] DuFort, E. C., "High-Resolution Emitter Direction Finding Using a Phased Array Antenna," *IEEE Trans. on Acoustics, Speech, and Signal Processing Newsletter*, Vol. 31, No. 6, December 1983, pp. 1409–1416.

[14] Manikas, A., A. Alexiou, and H. R. Karimi, "Comparison of the Ultimate Direction-Finding Capabilities of a Number of Planar Array Geometries," *IEE Proc. Radar, Sonar Navig.*, Vol. 144, No. 6, December 1997, pp. 321–329.

[15] Hua, Y., T. K. Sarkar, and D. D. Weiner, "An L-Shaped Array for Estimating 2-D Directions of Wave Arrival," *IEEE Trans. on Antennas and Propagation Society*, Vol. 39, No. 2, February 1991, pp. 143–146.

[16] Swindlehurst, A., and T. Kailath, "Azimuth/Elevation Direction Finding Using Regular Array Geometries," *IEEE Trans. on Aerospace and Electronic Systems*, Vol. 29, No. 1, January 1993, pp. 145–156.

[17] Liu, T. H., and J. M. Mendel, "Azimuth and Elevation Direction Finding Using Arbitrary Array Geometries," *IEEE Trans. on Antennas and Propagation Society*, Vol. 46, No. 7, July 1998, pp. 2061–2065.

[18] Zeira, A., and B. Friedlander, "Direction Finding with Time-Varying Arrays," *IEEE Trans. on Antennas and Propagation Society*, Vol. 43, No. 4, July 1995, pp. 927–937.

CHAPTER 2

Detection-Estimation of Gaussian Sources by Nonuniform Linear Antenna Arrays: An Overview of the Generalized Likelihood-Ratio Test Approach

Yuri I. Abramovich, Nicholas K. Spencer, and Alexei Y. Gorokhov

2.1 Introduction

Nonuniform linear arrays (NLAs) were introduced many years ago [1–4] and continue to attract interest, mainly due to the significantly larger aperture that such arrays span for a given number of sensors, compared with the corresponding uniform linear array (ULA). Moreover, under certain conditions, NLAs can resolve a *superior* number of sources m, that is,

$$m \geq M \qquad (2.1)$$

Indeed, the ability to unambiguously estimate the parameters of uncorrelated sources, when the number of such sources approaches the number of *coarray* elements, was demonstrated by Pillai et al. [5] for one type of minimum-redundancy sparse antenna array. (The *coarray* is loosely defined as the set of all distinct intersensor distances.) After that publication, considerable attention was attracted to different aspects of DOA estimation in such arrays [6, 7].

Initially, the problem of DOA estimation for a *specified* superior number of uncorrelated sources was addressed [8–11], with the intention of embracing different NLA geometries (e.g., "partially augmentable," noninteger-positioned, and so forth) by a DOA estimation methodology, and to approach the Cramér-Rao bound (CRB) in estimation accuracy for a superior number of sources, since the direct augmentation approach (DAA) introduced in [5] failed to meet these requirements [12, 13].

These studies therefore assumed that the number m of uncorrelated sources was either known or somehow accurately estimated. Naturally, in practical applications where nothing is known about the signal environment a priori, this step is nontrivial, both for *conventional* scenarios

$$m < M \qquad (2.2)$$

and especially for superior scenarios.

It is well known [14] that optimal approaches to sensor array processing involve simultaneously solving the detection and estimation problem, with the standard technique in the statistical estimation literature known as the generalized likelihood-ratio test (GLRT) [15, 16]. This test is based on the maximum-likelihood (ML) estimator [17], and since ML estimation usually requires multidimensional nonlinear (nonconvex) optimization, ML estimation and, consequently, the GLRT technique are usually replaced by some suboptimal method.

In the conventional case, the detection-estimation problem can be addressed in the traditional (suboptimal) way by separating the problem into detection (which is the estimation of the number of sources), followed by the standard DOA estimation problem for that number of sources. The detection problem is typically addressed by traditional information-theoretic criteria (ITC) such as AIC or MDL [5], or their later variants [18]. These techniques address detection without referring to any particular antenna array geometry, or parameters of the sources, because they solve the related problem of testing the number of smallest equal eigenvalues in the covariance matrix of the input data. Subspace-based techniques are then usually applied to solve the DOA estimation problem. One example of this technique is the multiple signal classification (MUSIC), described in [19–21].

Both approaches are applicable to antenna arrays with arbitrary geometries, and this powerful property of these techniques demonstrates its weakness when applied to a nonuniform (sparse) array. This weakness reveals itself when we consider the correspondence between the detection-estimation performance provided by this conventional ITC+MUSIC geometry-invariant approach, compared with the optimum (in the GLRT sense) joint detection-estimation solution that incorporates all available a priori information, including signal model and array geometry. The NLA-specific aspect stems from the fact that, for a special set of DOAs, both conventional ITC and MUSIC techniques fail to provide the correct (unambiguous) solution, even for arbitrarily large sample sizes. This is due to the *manifold ambiguity* property of every NLA [22], whereby there always exists a set of DOAs with linearly dependent array-signal manifold ("steering") vectors. Obviously, the fundamental idea behind the ITC methods [i.e., the equality of the $(M < m)$ smallest eigenvalues of the covariance matrix R] is invalid if the m sources have linearly dependent steering vectors. For the same reason, MUSIC fails to produce the true $m_1 < m$ number of peaks if there exist $(m - m_1)$ DOAs with steering vectors linearly dependent on the steering vectors of the m_1 sources. It could be argued that absolute linear dependence among the array-signal manifold vectors only occurs in a very specific set of DOAs that measure zero, and so the probability of encountering such a scenario in practice is small. Nevertheless, we have demonstrated in [23, 24] that for any finite sample support, there exists a (surprisingly large) continuous DOA region in the vicinity of the true ambiguous set, which results in ambiguous DOA estimates being generated by subspace methods such as MUSIC. We wanted to see whether these scenarios could be properly treated by other techniques. Specifically, we investigated the nonidentifiability conditions for any given NLA geometry, and compared these with the manifold ambiguity conditions that make ITC+MUSIC fail. This study [23, 25, 26] defined nonidentifiability conditions for Gaussian sources for estimation-only (where m is known), and the obviously more relaxed conditions for detection-

2.1 Introduction

estimation (where m is unknown, and the model could be ambiguously presented by a mixture of the wrong number of sources). The derived nonidentifiability conditions cover both the conventional and superior cases. For the former, we showed that most manifoldly ambiguous scenarios are, in fact, identifiable for detection-estimation.

A more fundamental aspect of this comparison, which applies to any antenna geometry, concerns the preasymptotic properties of the traditional techniques, such as ITC and MUSIC. The asymptotic efficiency of MUSIC was proven by Stoica et al. [27], which means that as the sample size N and/or SNR approach infinity, the DOA estimation accuracy will be close to the CRB. However, it has long been known that subspace techniques such as MUSIC suffer a rapid deterioration in performance as the number of snapshots N and/or SNR drop below certain threshold values [28–31]. This is due to a discontinuity in parameter estimates that are specific to the method in question. The sole apparent discontinuity, which is typical for all subspace methods, is induced by the interchange of eigenvectors between the estimated signal and noise subspaces ("subspace swap") [31]. The important issue of whether truly optimal solutions to the detection-estimation problem can suffer performance breakdown had not been addressed. Clearly, a positive answer would define the ultimate limit (in N and/or SNR), beyond which accuracy is impossible by that method. However, if such conditions are significantly different from the MUSIC performance breakdown conditions, then this limit defines a region where a hypothetical optimal detection-estimation technique could overcome the limitations of ITC+MUSIC. These important issues have been addressed [32, 33].

Turning now from the conventional case ($m < M$) to the superior case ($m \geq M$), the difficulties were even greater, since the literature lacked any detection or detection-estimation technique that could handle a superior number of independent Gaussian sources using sparse linear arrays. However, we were able to successfully address this issue [34, 35].

This chapter presents an overview of our recent results that mainly deal with the detection-estimation of Gaussian sources in nonuniform (sparse) linear antenna arrays. Section 2.2 introduces our GLRT technique that can address any parametric Gaussian model in arbitrary antenna arrays. For practically all important cases, implementation of the GLRT-based technique leads to the nonconvex optimization of a likelihood ratio (LR) function, as expected. This generic approach was previously considered impractical because of the inability to identify inappropriate local extrema. To support our GLRT detection-estimation methodology, we introduced [32, 33, 36] scenario-independent, nonasymptotic statistical lower bounds for the optimized LR, which allows a quite efficient "quality assessment" of any solution that is supposed to be optimal in the maximum LR sense. This lower-bound analysis also is discussed in Section 2.2 for uniform and, especially, for nonuniform linear antenna arrays. Section 2.3 introduces several examples that demonstrate our GLRT method, supported by the lower-bound analysis. Section 2.4 presents a summary and conclusions, and refers to some related problems that are also successfully addressed by our generic GLRT detection-estimation technique [36–38].

2.2 GLRT Philosophy for the Detection-Estimation Problem

For any identifiable scenario comprising μ sources, the set of parameters Ψ_μ uniquely specifies the covariance matrix R_μ. Moreover, under the Gaussian model, R_μ exhaustively describes the statistical properties of the observed set of input data $y(t), t = 1, \ldots, N$. For the simple model of independent point (perfectly coherent) Gaussian sources, each source is parameterized by its DOA θ_j and power p_j. Hence, the covariance matrix is specified by the 2μ source parameters, plus the single parameter for the additive white-noise power p_0. For more complicated models, such as spread (distributed, scattered) sources [36, 39], the number of parameters will increase, but the important feature here is that there is a one-to-one correspondence between the set of parameters Ψ_μ and the covariance matrix model R_μ. This implies that the detection-estimation problem for Gaussian sources can be transformed into the problem of selecting in some sense the best covariance matrix model R_μ for the observed training data.

We create a test hypothesis for R_μ using the sufficient statistics of the direct data covariance (DDC) matrix:

$$\hat{R} \equiv \sum_{t=1}^{N} y(t) y^H(t) \qquad (2.3)$$

For example, the modified LR test [40] is

$$H_0: \mathcal{E}\{R_\mu^{-1/2} \hat{R} R_\mu^{-1/2}\} = I_M N \quad \text{against} \qquad (2.4)$$
$$H_1: \mathcal{E}\{R_\mu^{-1/2} \hat{R} R_\mu^{-1/2}\} \neq I_M N$$

which tests whether or not R_μ is equal to the exact (true) covariance matrix of the input data. In [40], this LR test is implemented as

$$\gamma(R_\mu) \underset{H_0}{\overset{H_1}{\lessgtr}} \beta_0 \qquad (2.5)$$

where

$$\gamma(R_\mu) \equiv \left(\frac{e}{N}\right)^{MN} \left[\det(R_\mu^{-1} \hat{R})\right]^N \exp\left[-\text{tr}(R_\mu^{-1} \hat{R})\right] \qquad (2.6)$$

Note that here the LR $\gamma(R_\mu)$ is just a normalized likelihood function

$$\delta(R_\mu) \equiv \frac{1}{(\det R_\mu)^N} \exp\left[-\text{tr}(R_\mu^{-1} \hat{R})\right] \qquad (2.7)$$

and therefore the ML estimates for a given number of sources μ

2.2 GLRT Philosophy for the Detection-Estimation Problem

$$\hat{\Psi}_\mu^{\text{ML}} = \arg \max_{\Psi_\mu} \delta(R_\mu) \qquad (2.8)$$

are the same as for the maximum LR

$$\hat{\Psi}_\mu^{\text{ML}} = \arg \max_{\Psi_\mu} \gamma(R_\mu) \qquad (2.9)$$

The problem of detection-estimation, namely, the simultaneous estimation of the number of sources $\hat{m}$ and their parameters $\hat{\Psi}_m^{\text{ML}}$, could not be solved simply by LR maximization over μ and Ψ_μ because of the well-known "nested" property of the LR, whereby

$$\gamma(R_1^{\text{ML}}) < \ldots \leq \gamma(R_{m_{\max}}^{\text{ML}}) \qquad (2.10)$$

where $\gamma(R_\mu^{\text{ML}})$ is the covariance matrix model constructed with the ML estimates $\hat{\Psi}_\mu^{\text{ML}}$ for μ sources. This nested property means that we may test R_μ^{ML} in the order $\mu = 0, \ldots, m_{\max}$, and, as in the traditional detection philosophy [14], the smallest value of μ that passes the test (i.e., that exceeds the LR threshold) is taken to be the estimate $\hat{m}$. ITC or Bayesian criteria typically address the usual problem of arbitrarily selecting the threshold level (i.e., the probability of false alarm). This defines the optimum model for R_μ as

$$\hat{m} = \arg \min_\mu \left\{ -\ln LR\left[R(\hat{\Psi}_\mu^{\text{ML}})\right] + \nu_\mu \right\} \qquad (2.11)$$

where ν_μ is a penalty function such that

$$\nu_\mu > \nu'_\mu \text{ for } \mu > \mu' \qquad (2.12)$$

and is defined by three different ITC as [18]

$$\nu_\mu = \begin{cases} \nu_{\text{AIC}} \equiv d_\mu & \text{Akaike information criterion} \\ \nu_{\text{MDL}} \equiv \dfrac{1}{2} d_\mu \log N & \text{minimum description length} \\ \nu_{\text{MAP}} \equiv \dfrac{5}{6} d_\mu \log N & \text{maximum a posteriori probability} \end{cases} \qquad (2.13)$$

Here, d_μ is the number of real-valued parameters that fully specify the covariance matrix R_μ for the simple point-source model

$$d_\mu = 2\mu \qquad (2.14)$$

Since the possibly unknown white-noise power p_0 participates in all tested models, it does not affect the penalty function.

We can now see that the general GLRT framework suggests a straightforward approach. For every admissible number of sources μ, we need to find the set of corresponding ML estimates for all the parameters required for the specification of the covariance matrix R_μ^{ML}. From the set of models R_μ^{ML} ($\mu = 0, \ldots, m_{max}$), we then have to choose the optimum one using an ITC (2.11). This framework naturally includes all identifiable Gaussian parametric scenarios, including superior or manifoldly ambiguous ones. This approach requires a careful computational implementation, with an accurate LR optimization. If we were to obtain nonoptimal estimates that do not satisfy the nested property (2.10), then the selection rule for the optimum number of sources does not make sense. Since accurate LR optimization in important cases is associated with nonconvex optimization, it is evident that successful implementation of this detection-estimation methodology strongly depends on use of an efficient optimization routine, and, most importantly, on some tools that can be used to identify and disregard inappropriate solutions.

The problem of efficient optimization has different aspects, only one of which is purely computational speed with a view to real-time implementation. In [34, 35], we have been primarily concerned with designing routines with the highest "success rate," in terms of proximity of the optimization results to the optimal solution. In order to assess this success rate, Monte Carlo simulations can be used with their known exact scenario, but even here it is nontrivial to establish the proximity of any given solution to the true ML one. The traditional approach that compares the accuracy obtained with the CRB does not allow us to identify any particular optimization outlier, and more importantly, cannot be used for optimization efficiency analysis in the "threshold region" where the CRB are inaccurate due to insufficient sample support and/or SNR. In practical applications, where the exact scenario is unknown, the necessary "quality assessment" of a given solution, in terms of its proximity to the optimum one, is even more difficult.

These two problems, which are important for successful GLRT implementation, have been addressed in [32, 33] with the introduction of a statistical maximum LR lower bound. This lower-bound technique arises from the observation that for any admissible model $R_{\mu \geq m}$, the LR $\gamma(R_\mu^{ML})$ can never be smaller than that generated for the same sufficient statistics $\hat{R}$ by the exact covariance matrix R_m:

$$\gamma(R_\mu^{ML}) \geq \gamma(R_m) \text{ for } \mu \geq m \qquad (2.15)$$

This is true because the admissible set of possible covariance matrices R_μ for $\mu \geq m$ obviously includes the exact ones R_m. This observation is also true for the likelihood function $\delta(R_\mu)$ (2.7), but the crucial difference is that the statistical properties of the LR $\gamma(\hat{R}m)$ do not depend on the scenario R_m. Indeed, the LR $\gamma(R_\mu)$ (2.6) is fully specified by the properties of the matrix

$$\hat{C} \equiv R_\mu^{-1/2} \hat{R} R_\mu^{-1/2} \qquad (2.16)$$

where $\hat{R} \sim \mathcal{CW}(N \geq M, M, R_m)$, and for $R_\mu = R_m$, we get [41]

$$\hat{C} \sim \mathcal{CW}(N \geq M, M, I_M) \qquad (2.17)$$

2.2 GLRT Philosophy for the Detection-Estimation Problem

Here $\mathcal{CW}(N, M, R)$ is the central complex Wishart distribution, which for $R = I_M$ is fully specified by the number of sensors M and the sample size N ($N > M$). Based on these scenario-free properties of $\gamma(R_m)$, the following propositions have been made [32, 33].

Proposition 1

For arbitrary $N \geq M$ and the sufficient statistic $\hat{R}$, any set of parameters $\hat{\Psi}_\mu$ that uniquely define the covariance matrix R_μ can be classified as being non-ML-optimal if

$$\gamma(R_\mu) < \alpha \tag{2.18}$$

where

$$\int_\alpha^\infty f[\gamma(R_\mu)]\, d\gamma = P \tag{2.19}$$

and $(1 - P)$ is the desired probability of incorrect identification or false alarm, where the true ML estimate is classified as nonoptimal.

Unlike the CRB that only can be calculated for known parameters, it is important that this bound (2.18) is applicable to the quality assessment of estimates in practical applications.

Note that $\hat{\Psi}_\mu$ could be an arbitrary (complete) set of parameters that specify the covariance matrix R_μ. For example, if $\hat{\Theta}_\mu$ is a set of DOA estimates produced by MUSIC, then we need to find an appropriate set of power estimates $\hat{p}_j$ ($j = 0, \ldots, \mu$) in order to complete the parametric set, and then to test the LR $\gamma(R_\mu)$ by the inequality (2.18).

While this first proposition addresses the problem of "quality assessment" of a given solution to practical detection-estimation, our next proposition concerns the "success rate" assessment of a given optimization method that is applied to LR maximization in (2.9).

Proposition 2

For Monte Carlo simulations, where the set of true parameters Ψ_m is known, an assessment of the LR optimization efficiency for any set of estimates $\hat{\Psi}_\mu$ can be made by a direct comparison of LRs. If

$$\gamma(R_\mu) \leq \gamma(R_m) \tag{2.20}$$

then the set $\hat{\Psi}_\mu$ is not the ML set.

Again, no assumptions on the nature of the parameters estimates $\hat{\Psi}_\mu$ have been made here, nor are the bounds merely asymptotic.

While this inequality means strict non-ML-optimality for $\hat{\Psi}_\mu$, the reverse situation

$$\gamma(R_\mu) > \gamma(R_\mu^1) \geq \gamma(R_m) \tag{2.21}$$

is less straightforward. Indeed, if the LR of any given set $\hat{\Psi}_\mu^1$ (that corresponds to R_μ^1) already exceeds the LR generated by the true parameters Ψ_m, then it is unclear whether or not a further search for the globally optimal solution R_μ^{ML}, such that

$$\gamma(R_\mu^{ML}) > \gamma(R_\mu^1) \geq \gamma(R_m) \qquad (2.22)$$

will lead to better accuracy in the parameter space. In other words, it is unclear that (2.21) implies better estimation accuracy

$$\|\hat{\Psi}\mu^{ML} - \Psi_m\| < \|\hat{\Psi}\mu^1 - \Psi_m\| \qquad (2.23)$$

Note that the existing asymptotic theory of ML estimation typically ignores such an issue, and assumes the existence of a single global likelihood function extremum in the vicinity of the true solution Ψ_m. In practice, under limited sample support and/or SNR conditions, there are solutions that meet the first inequality (2.21), being not globally optimal. Most importantly, from the parameter estimation accuracy viewpoint, all these solutions seem to be statistically equivalent. This implies that any LR optimization routine that generally gives solutions meeting (2.21) can be treated as appropriate.

Another important issue that arises from a consideration of the inequalities (2.21) is the prediction of the "ML performance breakdown" phenomenon [42, 43]. We must consider the possible existence of a solution $\hat{\Psi}_\mu$ that is "better than the true one" in terms of LR [i.e., (2.21) holds], but at the same time is extremely far from the true solution Ψ_m according to the metric $\|\hat{\Psi}_\mu - \Psi_m\|$. Obviously, such a "breakdown" solution could be neither identified nor rectified (i.e., neither "predicted" nor "cured") within the ML paradigm. The sample support and/or SNR conditions, for which the probability of a breakdown solution occurring exceeds some given level, determine the ultimate limit for any detection-estimation scheme within the GLRT paradigm. However, these ultimate ML-breakdown conditions must be less stringent than, say, MUSIC performance breakdown conditions for the same scenario. Therefore, if any particular technique, such as MUSIC, generates a solution that is identified by (2.18) as an outlier, then by applying an "appropriate" optimization routine for $\mu \geq m$, we must get a solution that exceeds the lower bound in (2.18). Such a "rectified" solution can indeed be of the expected high accuracy, or, if ML breakdown occurs, we can transform this "MUSIC outlier" into an "ML outlier" with its now optimally high LR.

Finally, since the scenario-free probability distribution function (PDF) for $\gamma(R_m)$ is available, the ITC (2.11) with the penalty function, which is usually justified by asymptotic arguments, now can be replaced by the more accurate detection rule

$$\hat{m} = \arg \min_\mu \left[LR(R_\mu^{ML}) \geq \alpha \right] \qquad (2.24)$$

where, as in (2.19), the threshold α is defined by

2.2 GLRT Philosophy for the Detection-Estimation Problem

$$\int_{\alpha}^{\infty} f[LR(R_m)] \, dLR = P \tag{2.25}$$

and $(1 - P)$ is the upper bound for the probability of overestimating the number of sources. Given the exact (nonasymptotic) PDF for the "lower bound" $LR(R_m)$, this approach should be more suitable for practical applications.

Instead of the generic test that involves the strict equality of the covariance matrices in (2.4), we can use the sphericity test:

$$H_0: \mathcal{E}\{R_\mu^{-1/2} \hat{R} R_\mu^{-1/2}\} = c_0 I_M \quad \text{against} \tag{2.26}$$

$$H_1: \mathcal{E}\{R_\mu^{-1/2} \hat{R} R_\mu^{-1/2}\} \neq c_0 I_M \quad \text{for } c_0 > 0$$

that leads to the sphericity LR

$$\gamma_{sph}(R_\mu) = \left\{ \frac{\det(R_\mu^{-1} \hat{R})}{\left[\frac{1}{M} \operatorname{tr}(R_\mu^{-1} \hat{R})\right]^M} \right\}^N \tag{2.27}$$

For applications that involve persymmetric Hermitian matrices R_μ (such as uniform linear antenna arrays), we may use the persymmetric sphericity test

$$\gamma_{persph}(R_\mu) = \left\{ \frac{\det(R_\mu^{-1} \hat{R}_p)}{\left[\frac{1}{M} \operatorname{tr}(R_\mu^{-1} \hat{R}_p)\right]^M} \right\}^N \tag{2.28}$$

where

$$\hat{R}_p \equiv \frac{1}{2}(\hat{R} + J\bar{\hat{R}}J) \tag{2.29}$$

and

$$J = \begin{bmatrix} & & 1 \\ & \iddots & \\ 1 & & \end{bmatrix} \tag{2.30}$$

is the exchange matrix, and the bar indicates complex conjugation.

Exact (nonasymptotic) distributions for $\gamma_{sph}(R_m)$ and $\gamma_{persph}(R_m)$ have been derived in [33, 36, 38], and involve quite complicated Meijer's G-functions [44]. Yet, since all LRs [(2.6), (2.27), and (2.28)] are scenario-free and are specified by the properties of the matrix $\hat{C} \sim \mathcal{CW}(N \geq M, M, I_M)$, Monte Carlo simulations can be used to precalculate the required PDFs with any desired accuracy that is sufficient for accurate threshold calculations.

Therefore, the maximum LR lower-bound analysis provides an efficient tool for selecting an appropriate LR optimization routine, and, most importantly, for "quality assessment" of practical solutions, which together make the generic GLRT-based detection-estimation technique practical.

2.3 GLRT Detection-Estimation Results for Uniform and Sparse Antenna Arrays

2.3.1 Superior Case

The new problem of GLRT-based detection-estimation for $m \geq M$ uncorrelated Gaussian sources using "fully or partially augmentable" NLAs has been addressed in detail in [34, 35]. Those papers contain a full description of the quite complicated LR maximization routines that have been proposed. The main idea is to gradually proceed from the DDC matrix $\hat{R}$ with its maximum possible LR of unity into the admissible set generated by the set of augmented M_α-variate positive definite Toeplitz matrices. Such matrices are the covariance matrices for the corresponding virtual ULA that is specified by the coarray of the M-element NLA, and are related by the linear "compression" transformation from the M-variate NLA covariance matrix R_μ by

$$R_\mu = L T_\mu L^T \tag{2.31}$$

where L is the $M \times M_\alpha$ binary selection (or incidence) matrix, whose element L_{jk} is unity if $k = d_j$ and zero otherwise (d_j is the position of the jth sensor in the NLA that ranges over the values $0, \ldots, M_\alpha$).

"Fully augmentable" NLAs have the property that the set of all intersensor distances

$$\mathcal{D} \equiv \{d_j - d_k \mid j, k = 1, \ldots, M; j \geq k\} \tag{2.32}$$

is complete; that is, $\mathcal{D} \equiv \{0, \ldots, d_M\}$. For fully augmentable NLAs, the initial (inadmissible) approximation for the augmented M_α-variate Toeplitz matrix T_μ is generated by the DAA technique of Pillai et al. [5]. Gradual transformations of this (typically nonpositive-definite) Toeplitz matrix into a positive definite Toeplitz matrix whose $(M_\alpha - \mu)$ smallest eigenvalues are equal, take place using an iterative linear programming (ILP) technique, starting from the model with the maximum possible number of sources, $\mu_{\max} = M_\alpha - 1$. For this model, the solution that exceeds the lower-bound threshold (2.18) must exist. Apart from this practical threshold, for Monte Carlo simulations the "strict" inequality (2.21) also should be checked at each iteration in order to assess the computational efficiency of our optimization technique. Given this more appropriate $\hat{T}_{M_\alpha-1}$ matrix, it is then iteratively modified to equalize an increasing number of eigenvalues

$$\hat{R} \to \hat{T} \equiv \hat{T}_{M_\alpha-1} > 0 \to \hat{T}_{M_\alpha-2} > 0 \to \hat{T}_{M_\alpha-3} \to \ldots \quad (2.33)$$

each time necessarily introducing some additional degradation in LR. In this way, the "nested" property (2.10) of the solutions T_μ is maintained.

For "partially augmentable" NLAs, we need to use an additional step that involves the completion of a partially specified Toeplitz matrix [35]. Examples of simulation results that illustrate the high probability of correct identification (that is, detection and estimation without outliers) are introduced in [34, 35].

We now concentrate on simulation results on the maximum LR lower-bound analysis, since this does not depend on any particular LR optimization routine, and so can be applied to various other problems.

Consider the fully augmentable five-sensor NLA

$$d_9 = [0, 2, 5, 8, 9] \quad (2.34)$$

that has an $M_\alpha = 10$-element uniform coarray, and a five-source superior scenario equally separated in terms of $w \equiv \sin\theta$

$$w_5 = [-0.90, -0.68, -0.46, -0.24, -0.02] \quad (2.35)$$

with common SNR = 20 dB and $N = 100$ snapshots. We use the sphericity test (2.27), with

$$\gamma_{sph}(R_m) = [\gamma_0(\hat{C})]^N \quad (2.36)$$

Figure 2.1 presents an analysis of the LR distributions over 1,000 Monte Carlo trials comprising the exact M_α-variate Toeplitz matrix T, via its "contraction" R. Note that this is a sample distribution for the scenario-free distribution $\gamma_0(\hat{C})$, which is shown as a dotted line; and the local ML completion T^{ML}, as described in [34, 35], which is shown as a solid line.

These results show that the lower bound for this scenario is quite tight. Indeed, according to (28) in [33]:

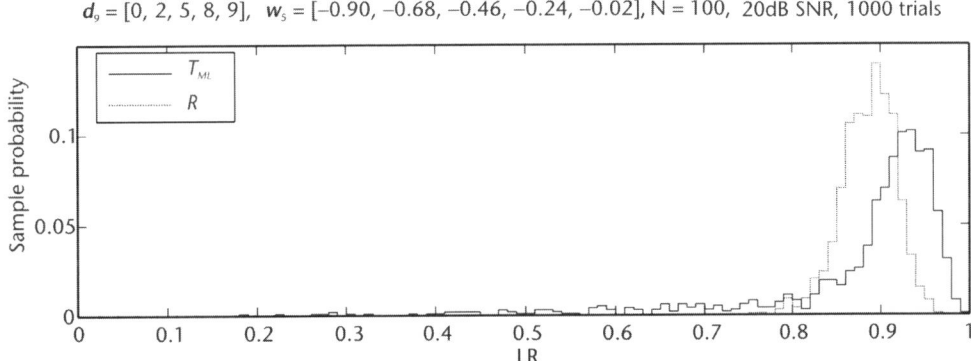

Figure 2.1 Sample LR distributions for a fully augmentable array.

$$\mathcal{E}\{\gamma_0(\hat{C})\} = \{0.7819, 0.8856, 0.9414, 0.9606, 0.9762, 0.9881, 0.9988\} \quad (2.37)$$

respectively, for $N = \{50, 100, 200, 300, 500, 1,000, 10,000\}$. Despite such tight bounds, our optimization resulted in 74.9% of trials exceeding the lower bound, thus satisfying a necessary condition for optimality. While we cannot claim global optimality for any of these trials, this high success rate is statistical evidence for close-to-optimal results.

Our second example result involves the five-sensor partially augmentable array

$$d_{11} = [0, 1, 4, 9, 11] \quad (2.38)$$

that has (only) the sixth lag missing in its coarray. Consider the two seven-source scenarios

$$w_7 = [-0.9, -0.7, -0.5, -0.3, -0.1, 0.1, w_7] \quad (2.39)$$

where we investigate changing the separation of the seventh source as $w_7 = \{0.12, 0.20\}$, with common SNR = 20 dB and $N = 300$ snapshots. Figure 2.2 illustrates the resulting LR distributions for T^{ML} and R, calculated according to [35]. Comparison of these distributions leads to some quite important conclusions.

First, the success rate of LR maximization is quite high in both scenarios: 88.2% for the smaller separation, and 98.0% for the larger. Despite the high performance (in terms of LR), these separation scenarios demonstrate quite different detection-estimation performance. For the larger separation, a high sample probability of correct detection (more than 0.9) was found, with all DOA estimates being extremely close to the true DOAs in terms of CRB. On the contrary, for the smaller separation, not a single trial was successfully identified. Even for those trials where the number of sources ($m = 7$) was correctly estimated (approximately 50%), the seventh estimated DOA was distributed across the entire interval of possible values. The remaining trials underestimated the number of sources, despite the extremely high LR, exceeding that of the true covariance matrix.

This LR analysis demonstrates the anticipated existence of dramatically incorrect solutions (outliers), which nevertheless are "better" than the true solution in

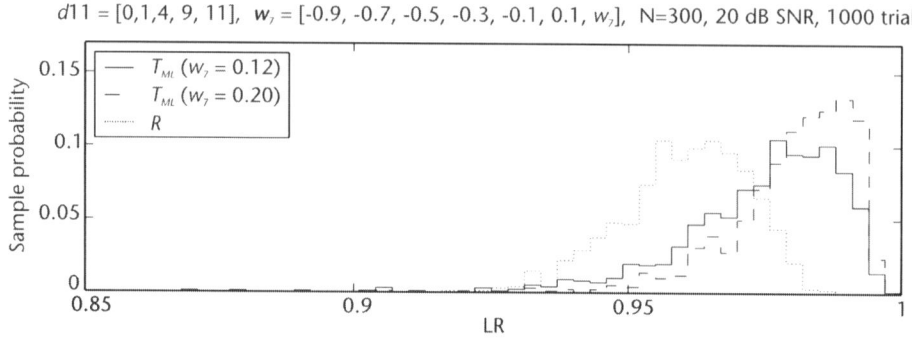

Figure 2.2 Locally optimal ML and maximum LR lower bound distributions for a partially augmentable array. (*From:* [33]. © 2004 IEEE. Reprinted with permission.)

terms of LR. In other words, the discontinuous nature of ML estimation has been demonstrated. The similar "performance breakdown" phenomenon has been described for subspace-based estimation techniques [28, 29], where it has been attributed to a specific "subspace swap" event that is caused by a splitting of the measurement space into an estimated signal subspace and an orthogonal subspace. We have now demonstrated performance breakdown for successful LR maximization in superior scenarios, where there is no noise subspace in R.

Note that in conventional scenarios, threshold conditions for ML estimation are usually specified in terms of SNR that are sufficient to avoid discontinuity for a given scenario and sample support; whereas in superior scenarios, the threshold conditions should be defined in terms of the sample support required for a given scenario and SNR to avoid discontinuity.

2.3.2 Conventional Case

We previously mentioned two separate problems that arise for sparse arrays with a conventional ($m < M$) number of uncorrelated Gaussian sources. The first is NLA-specific and concerns scenarios that are too close to a manifoldly ambiguous scenario for standard detection-estimation techniques (e.g., ITC and MUSIC) to be successful. While the superior-specific methods in [34, 35] that operate over the set of augmented M_α-variate Toeplitz matrices can address this problem, there is another "shortcut" approach that was suggested in [23]. For manifoldly ambiguous cases, the ITC that tests the hypothesis regarding the equality of the smallest $(M - \mu)$ eigenvalues of the covariance matrix of the input data always tends to underestimate the number of sources. On the other hand, subspace techniques such as MUSIC tend to expose all directions within the ambiguous set, regardless of the actual subset of DOAs present in the data. If MUSIC operates in its efficient regime (i.e., close to the CRB), having sufficiently large N and/or SNR, then the set of MUSIC-generated DOA estimates is accurate, although ambiguous. By matching power estimates to the DOA estimates to fit the covariance matrix model, we can once again use our GLRT rule, (2.11) or (2.24), to find the minimal DOA set that generates a sufficiently high LR. Details on this matching routine and its stability are introduced in [24].

The second problem is more generic, and involves detection-estimation behavior in the "threshold" region. While the "performance breakdown" phenomenon for subspace-based techniques such as MUSIC is well documented and understood, the potential discontinuity in ML estimation has been predicted and already demonstrated above for superior scenarios. The most important question that we address here for conventional scenarios is whether the threshold conditions for ML estimation performance breakdown are somewhat different from those for subspace-based techniques. An affirmative answer to this question means that subspace-specific breakdown can be "cured" by accurate LR maximization, up to the point that the ML threshold conditions block all such attempts.

While particular threshold SNR and sample support N values for ML estimation depend on the DOA scenario, they are defined by conditions where outliers generate LRs exceeding that of the true covariance matrix. Therefore, if there exists a substantial "gap" between subspace-specific and ML-specific threshold conditions, then it should be reflected by a substantial gap between the LRs generated by subspace-specific outliers and the optimal LR, (statistically) presented by the maximum LR lower bound (2.18). In turn, that would allow the use of maximum LR lower bound analysis as a simple data-based indicator to determine whether or not "subspace swap" has actually occurred. That is, it would be able to predict performance breakdown for subspace-based DOA estimation techniques.

To demonstrate these capabilities, we introduce a (conventional) three-source scenario with the five-sensor uniform array, so that the lower bounds distributions are exactly the same as in the sparse array case d_9 of the previous section (see Figure 2.1). We only consider here the relatively high sample support $N = 100$, and high (20 dB) common SNR case, where we vary the separation of the third DOA:

$$w_3 = [-0.40, 0, w_3] \quad (2.40)$$

where $w_3 = \{0.03, 0.04, 0.05, 0.06, 0.08, 0.10\}$ varies from extremely close to amply separated. This DOA scenario was chosen specifically to span the range of MUSIC performance breakdown: from 981 outliers out of 1,000 Monte Carlo trials for $w_3 = 0.03$ to only two outliers for $w_3 = 0.10$. Here, we concentrate only on DOA estimation breakdown, so LR maximization was conducted for the true number of sources $m = 3$, with the smallest two eigenvalues in the 5×5 Toeplitz covariance matrix kept equal. An iterated Gohberg–Semencul routine was used to initialize our ML optimization process, with each iteration described in Step 4 of the MMEE Algorithm in [8].

For each separation scenario, we calculate

1. The LR of the MUSIC-derived covariance matrix, computing the source power estimates $\hat{p}$ and the white-noise power estimate $\hat{p}_0$ from the traditional DOA estimates $\hat{\theta}$;
2. $\gamma_0(R)$, where R is the exact covariance matrix;
3. $\gamma_0(R^{ML})$, where R^{ML} is the (local) ML estimate of the M-variate positive definite Toeplitz covariance matrix, with the $(M - m) = 2$ smallest eigenvalues being equal (the computational details for LR maximization appear in [34, 45]).

Additionally, in each of the 1,000 Monte Carlo trials, these two estimators are each determined to be:

1. A normal DOA estimate or an outlier, according to whether or not any particular DOA estimate $\hat{w}_j$ lies outside the region bounded by

$$w_j \pm \frac{1}{2} \min(w_j - w_{j-1}, w_{j+1} - w_j)$$

("correct identification");
2. ML-optimization successful or not.

Since our LR maximization algorithm does not guarantee reaching the global extremum and depends on successful initialization, success for ML purposes is defined according to the condition $\gamma_0(R^{\mathrm{ML}}) \geq \gamma_0(R)$. If this condition is not met, then the estimate cannot be an ML estimate. While these trials should be excluded from the experimental statistics, in practical situations, the particular value $\gamma_0(R)$ is obviously unknown, so we report on the overall success rate of ML optimization. For unsuccessful ML optimization, we also identify those trials that incorrectly identify the DOAs based on R^{ML}. Finally, we also determine the error of the worst DOA estimate, in order to compare this with the CRB.

Figures 2.3 to 2.5 show these LR and error distributions for a representative subset of the above source separations. Figure 2.3 begins with the benign widely separated scenario $w_3 = 0.10$. We see that the LRs for R_{MUSIC}, R, and R^{ML} are statistically similar, although it is clear that the LR-maximized solutions R^{ML} are statistically "better" than the true covariance matrix (in terms of LR). Note that this "improvement" does not involve DOA estimation accuracy improvement, as previously discussed. The two MUSIC estimates with an extremely low LR are classified as outliers. Note again that the distribution of $\gamma_0(R)$ is scenario-free.

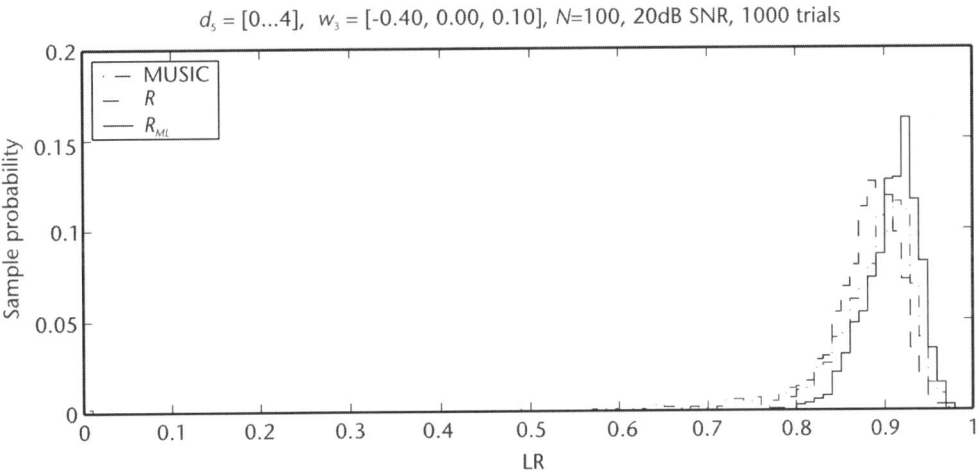

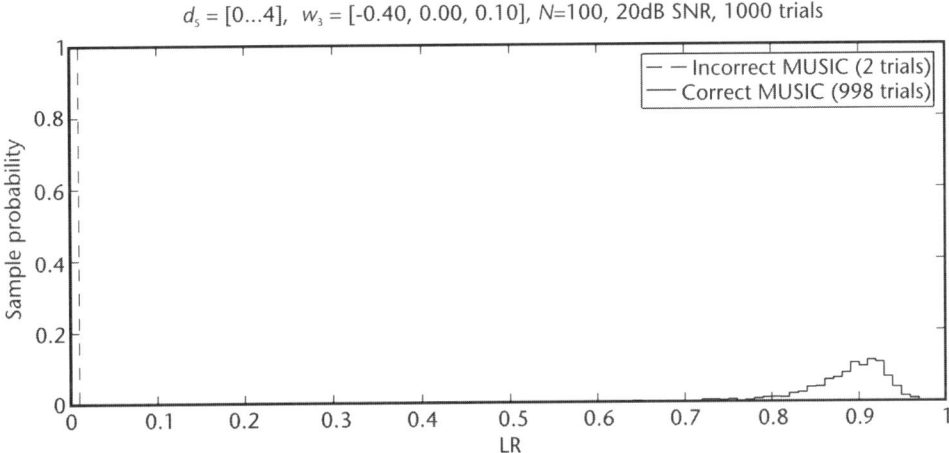

Figure 2.3 LR and accuracy analysis results for $w_3 = 0.10$. (*From:* [33]. © 2004 IEEE. Reprinted with permission.)

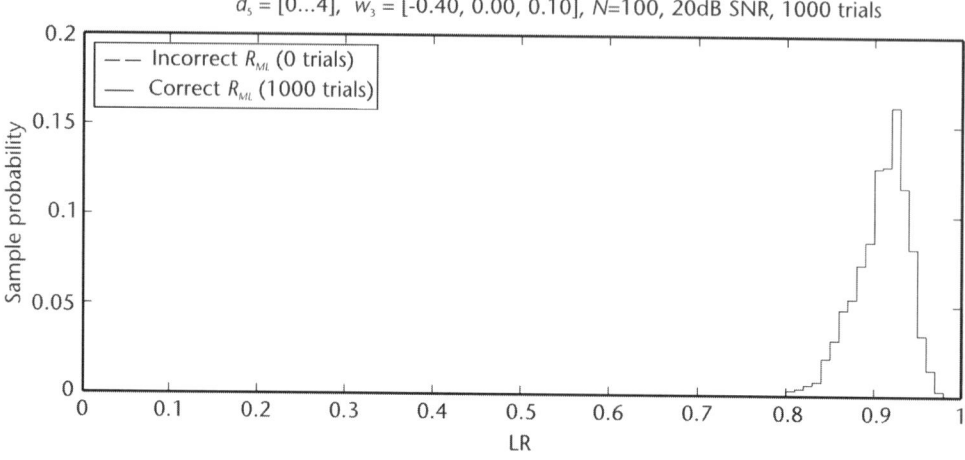

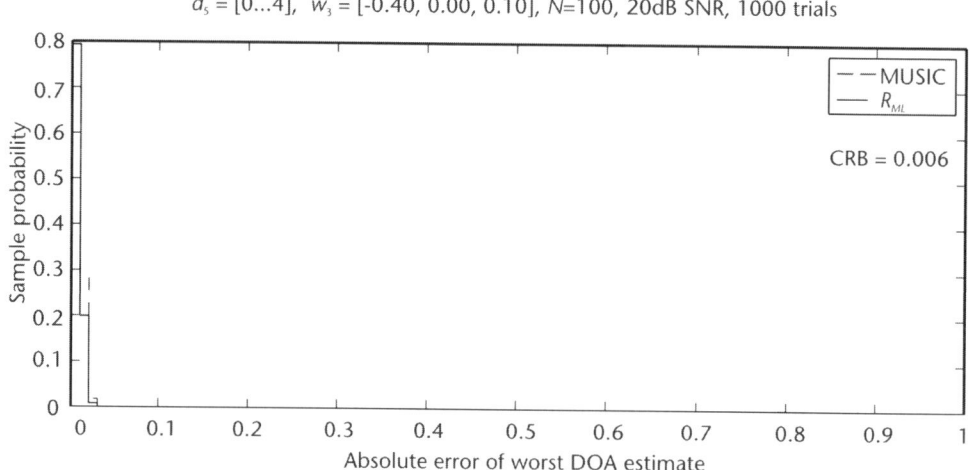

Figure 2.3 (Continued.)

Figure 2.4 shows results for the separation $w_3 = 0.06$, chosen to demonstrate an example where the probability of correct identification by MUSIC is about one-half. In fact, the sample MUSIC performance breakdown rate is 0.541. As expected, the LR distribution for R_{MUSIC} has two widely separated peaks. The peak with extremely low LRs is entirely due to outliers, while the peak with optimally high LRs is due to the 459 correctly identified trials. For ML estimation, R^{ML} still has mainly high LR values, above $\gamma_0(R)$. Specifically, $\gamma_0(R^{\text{ML}})$ was smaller than $\gamma_0(R)$ only in 45 trials. Interestingly, as seen in the third figure, only seven outliers gave very low LRs. The remaining 57 outliers have relatively high LR, and therefore could not be determined to be inappropriate estimates in practice, using (2.18). In fact, only 30 of the 64 outliers were among the unsuccessfully optimized solutions, while in the other 34 cases, these outliers had LRs exceeding those of the true covariance matrix. This example demonstrates the ability of our technique to reliably determine subspace-specific outliers and to efficiently rectify incorrect

2.3 GLRT Detection-Estimation Results for Uniform and Sparse Antenna Arrays

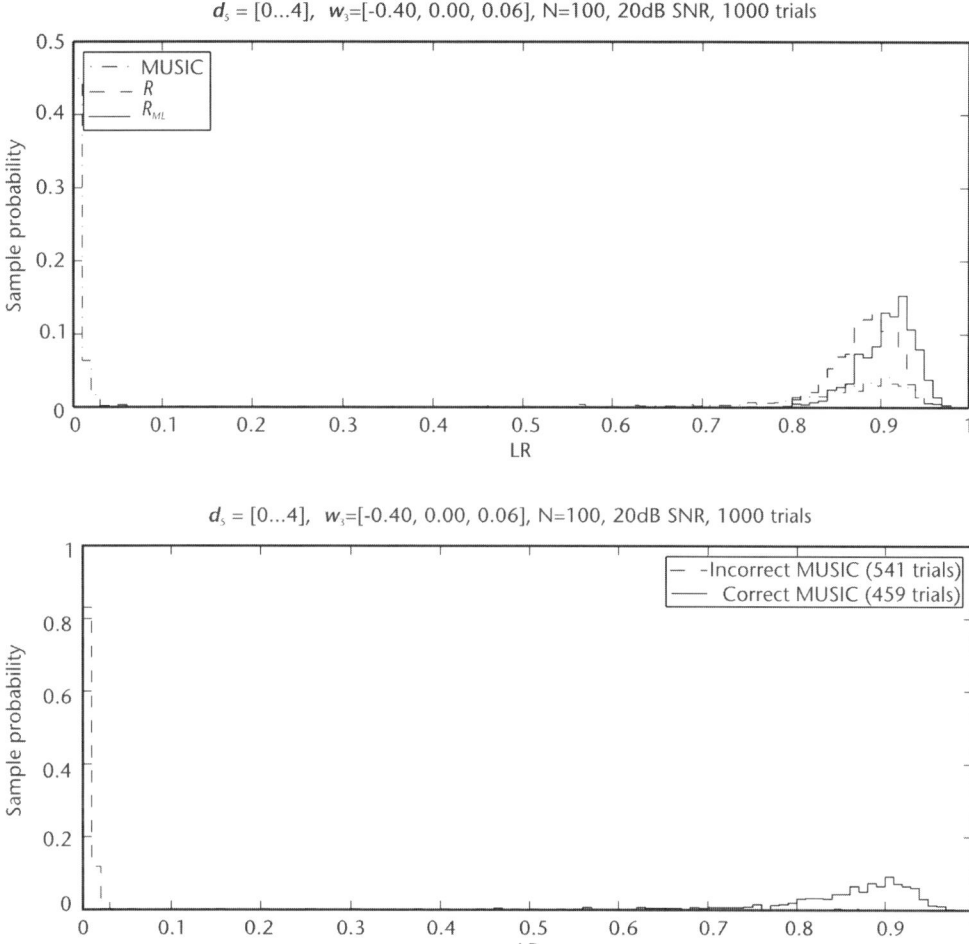

Figure 2.4 LR and accuracy analysis results for $w_3 = 0.06$. (*From:* [33]. © 2004 IEEE. Reprinted with permission.)

estimates by accurate ML optimization. This is possible because there is a predicted "gap" between the threshold conditions for MUSIC and those for ML estimation. DOA estimation accuracy is significantly improved by LR maximization compared with MUSIC, since the outliers have been rectified.

Our final example with $w_3 = 0.03$ (see Figure 2.5) was chosen to represent threshold conditions for ML estimation. Here, MUSIC fails completely, with a sample probability of correct identification of only 0.005. The second figure indicates that most of the incorrectly identified trials have a low LR, and so could be reliably determined as improper estimates. Our R^{ML} solution has a probability of correct identification of 0.408. We emphasize that the success rate of ML optimization is significantly higher at 0.893, with only 17.6% of the 592 incorrectly identified R^{ML} trials also having a LR less than that of R (i.e., unsuccessfully optimized). The vast majority of outliers still had LRs greater than that of R. The third figure leaves no doubt that these outliers, which are "better than" the exact DOAs, could

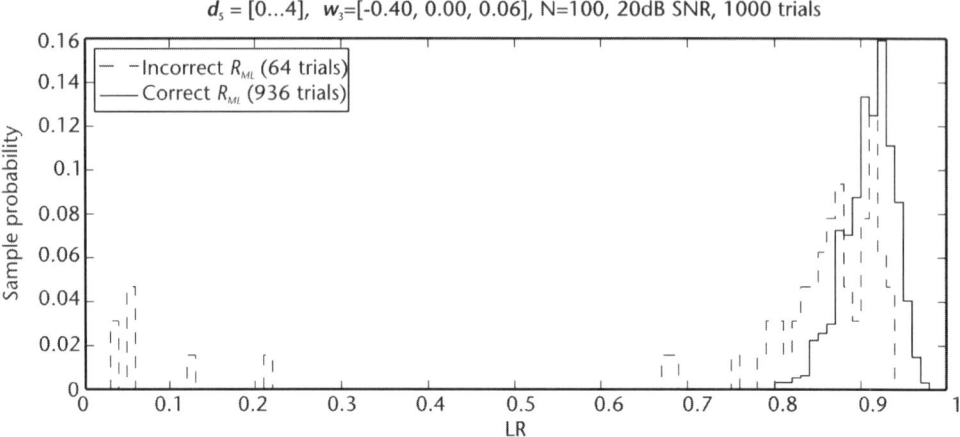

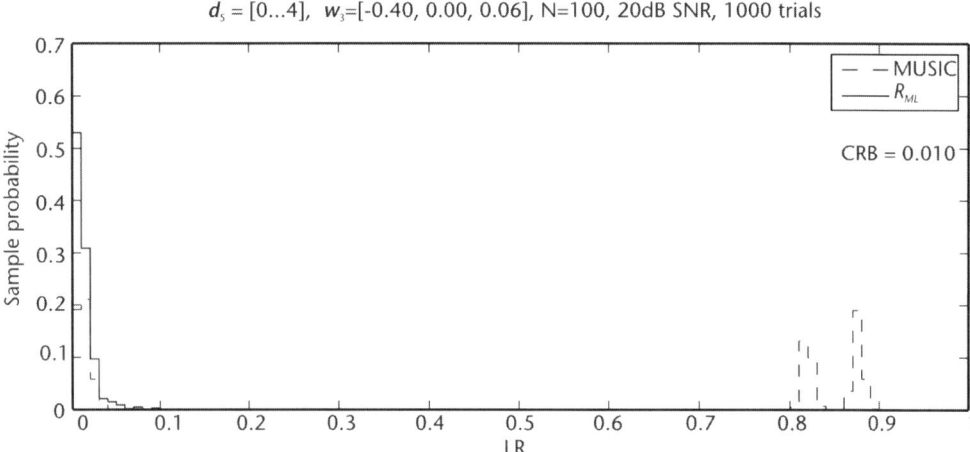

Figure 2.4 (Continued.)

not be determined nor rectified (i.e., "neither predicted nor cured"). We have approached the continuity threshold for ML parameter estimation. Nevertheless, most MUSIC outliers could be determined, and more than one-half of them rectified.

It is reasonable to predict that as $w_3 \to 0$, we would approach the situation where the probability of ML-estimated outliers approaches unity, despite $\gamma_0(T^{ML})$ being properly located above the lower bound $\gamma_0(T)$. This is physically understandable, since LR maximization on a three-source scenario with a two-source model results in essentially the same LR distribution as for the correct three-source model. It is worth mentioning that the minimum source separation of 0.03 is close to the CRB for that scenario (0.021). The ML criterion "does not see" this third source as separate, and uses the additional degree of freedom to "equalize an imaginary source" that is created by random noise data.

First, we have demonstrated the important ability to "predict and cure" performance breakdown of subspace-based DOA estimation by LR analysis and maximization. Second, we have demonstrated the existence of performance breakdown intrinsic to ML optimization, something that could not be overcome.

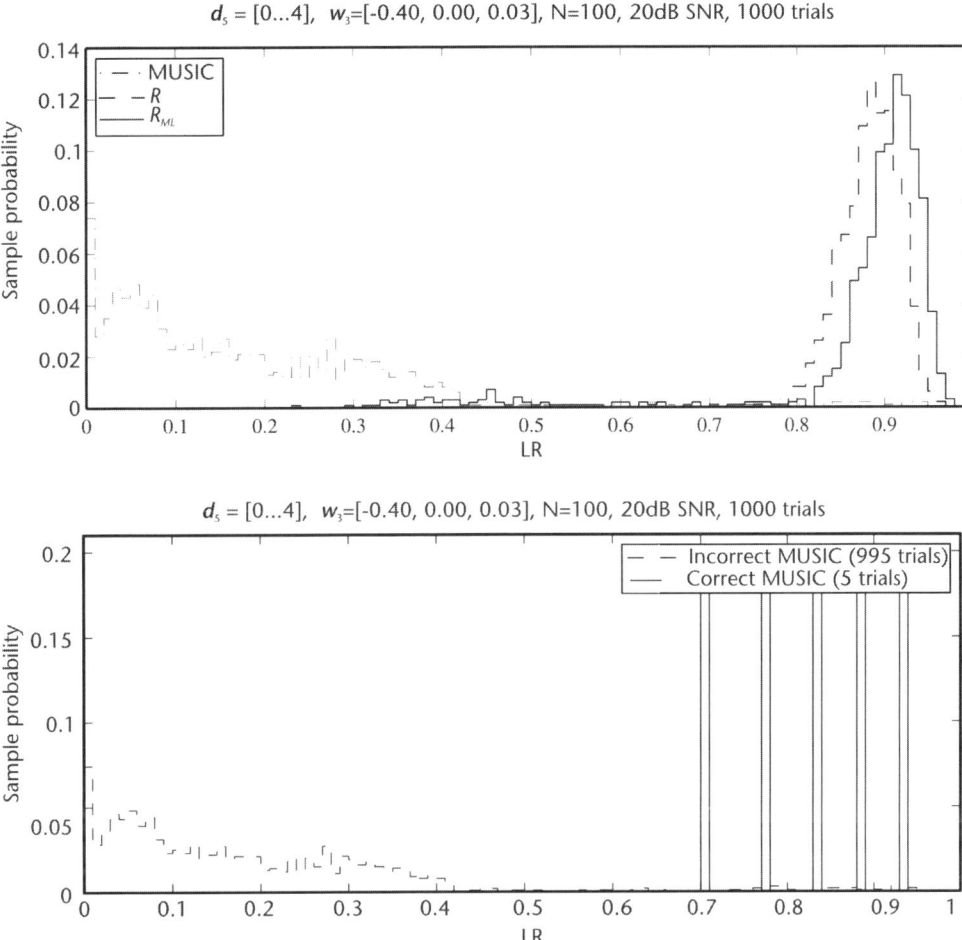

Figure 2.5 LR and accuracy analysis results for $w_3 = 0.03$. (*From:* [33]. © 2004 IEEE. Reprinted with permission.)

2.4 Conclusions

Recent developments in the area of detection-estimation of uncorrelated (independent) Gaussian sources by nonuniform (sparse) linear antenna arrays have addressed a number of theoretically and practically important problems that arose after the first possibility of estimating "more sources than sensors" was raised in 1985 [5].

First, we defined in [23, 25, 26] the necessary and sufficient conditions for a given scenario with $m \leq L$ uncorrelated Gaussian sources to be nonidentifiable for detection-estimation purposes by a NLA whose coarray has L sensors. Nonidentifiability occurs when there exists two or more DOA sets of an admissible number of uncorrelated Gaussian sources that generate identical covariance matrices. These conditions naturally include, as a special case, the nonidentifiability conditions for estimation alone, which occurs for an equal number of sources in two or more ambiguous sets [26]. By testing these conditions against "manifoldly ambiguous" scenarios, with a conventional number of sources $m < M$, that have linearly depend-

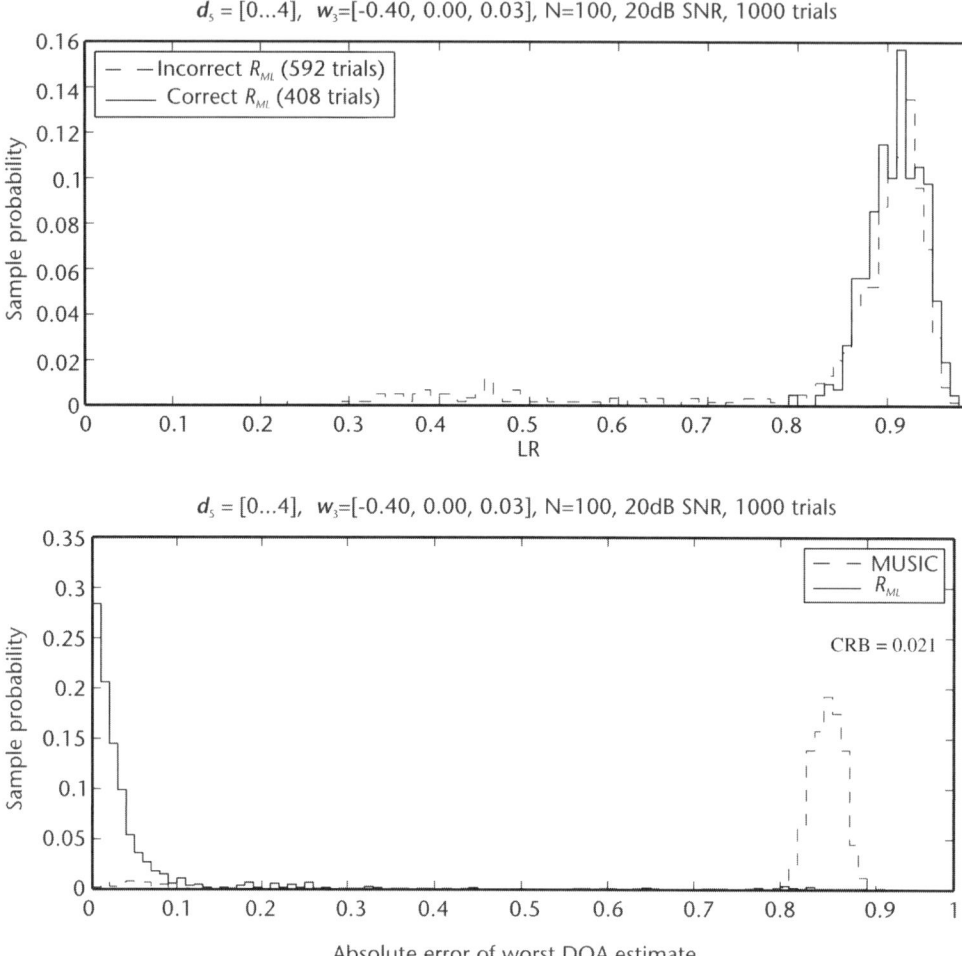

Figure 2.5 (Continued.)

ent array-signal manifold ("steering") vectors, we have been able to demonstrate that these scenarios are identifiable in most cases. While traditional subspace-based techniques (e.g., testing the multiplicity of the minimum eigenvalue by ITC and MUSIC) fail to unambiguously identify such scenarios, the proven identifiability cleared the way for appropriate techniques to be developed. On the other hand, these conditions allow us to find truly nonidentifiable scenarios in some, partially augmentable, arrays, by using the AGS methods proposed in [22] for the identification of manifold ambiguities. These conditions include superior scenarios ($m \geq M$), where little was known about identifiability, mainly because a reliable detection-estimation technique for the superior case did not then exist.

In [34, 35], we suggested the use of the generic GLRT framework for detecting and estimating an admissible number of identifiable sources using sparse arrays. Our approach is to find the complete set of ML estimates that (uniquely) specifies the covariance matrix model R_μ^{ML} for each possible number of sources $\mu = 0, \ldots,$ $m_{\max}$, and then to compare the LR values generated by these matrices for the given

2.4 Conclusions

sufficient statistics $\hat{R}$, selecting the optimum one. By this selection, the scenario is identified. We demonstrated that in a few cases, such as the manifoldly ambiguous conventional scenario, this approach can be used directly, when sufficient sample support and/or SNR lead to the ML-accurate but ambiguous set of DOA estimates produced by MUSIC. In this case, LR analysis of different subsets of the ambiguous DOA set allows us to select the true (identifiable) scenario.

In the more general case of $m \leq M$ and insufficient N and/or SNR, it is necessary to use computationally accurate LR [or likelihood function (LF)] optimization, rather than surrogate methods that rely on asymptotic justification. Since LR (or LF) maximization is generally a nonconvex problem with multiple local extrema, this idea was not previously considered practical. In order to facilitate this GLRT-based approach, we sought a way to assess the quality of any optimization solution that was considered sufficiently close to the ML solution. One such avenue is based on the fact that, for the hypothesis that the number of sources $\mu \geq m$, the optimized LR must exceed the LR generated by the true (exact) covariance matrix. Since the PDF for the latter does not depend on the scenario, but is fully specified by M and N, we were able to propose a statistical test as a "quality assessment" of the model $R_\mu (\mu > m)$, even when the true covariance matrix is unknown. Starting from the hypothesis with the maximum admissible number of sources $m_{\max}$, we proceed with LR optimization routines, using different routines if necessary, until a solution is found that is statistically "as good as" the true solution, in terms of LR. From this solution, and with optimally high LR, we proceed to models with fewer and fewer sources, each time necessarily introducing some additional degradation in LR. In this way, we retain the "nested" property of the detection-estimation problem, even if the particular set of solutions is not the set of globally ML optimal ones.

Apart from its vital application to the statistical lower bound, a direct comparison of the optimized LR with that generated by the true covariance matrix R_0 in Monte Carlo simulations provided us an important insight into the nature of ML estimation in the threshold region that concerns insufficient sample support and/or SNR. This analysis demonstrates that all solutions whose LR exceeds $LR(R_0)$ are equivalent in terms of parameter estimation accuracy, with the global LR maximum being sometimes worse than a local LR maximum, in terms of estimation accuracy, although still being above the LR threshold set by R_0. This observation makes the search for the true global ML solution immaterial if any particular optimization routine steadily generates solutions that are "better than the true one" in terms of LR. If a particular LR optimization routine yields a sufficiently high "success rate" under Monte Carlo trials for a sufficient variety of scenarios of interest, then it can be considered suitable and appropriate for practical applications.

A more fundamental insight given by this comparison is the prediction of an intrinsic discontinuity in ML estimation, irrespective of LR optimization routine. Indeed, if the existence of a completely erroneous solution (in the parameter space) with LR greater than $LR(R_0)$ cannot be excluded, then the conditions under which this phenomenon can repeatedly occur is the region of ML-performance breakdown for the given scenario. Obviously, if a wrong solution "is better than" the true one, then the ML criterion is no longer appropriate. We demonstrated that the onset of this breakdown sets the limit for the entire ML paradigm, including the GLRT methodology. On the positive side, we showed that these conditions are much more stringent than the MUSIC-performance breakdown conditions. We demonstrated that MUSIC, or any other outliers, could be reliably found, when their LR falls below the threshold $LR(R_0)$. We could then propose practical techniques that can "predict" an inappropriate solution, and "cure" it by increasing its LR up to an admissible value.

Supported by the introduced maximum LR lower bound, our GLRT detection-estimation technique demonstrated high performance for superior scenarios, plus the ability to significantly improve upon the performance of standard techniques for conventional scenarios in the threshold region.

While our technique was primarily developed for superior scenarios in sparse arrays, where the covariance matrices do not have a "noise subspace," it is able to address an arbitrary parametric Gaussian model. A typical example that has not been properly addressed by existing techniques is the detection-estimation of spread (i.e., scattered, distributed) Gaussian sources in arbitrary antenna arrays. Spread sources also produce a full-rank covariance matrix, which makes standard subspace techniques inadequate. In [36, 39], we showed high detection-estimation performance.

MUSIC performance breakdown can occur in any array under certain conditions. Not only sparse linear arrays have the ability to extract more independent Gaussian sources than sensors; this potential has been predicted for uniform circular antenna arrays as well. In [38, 43], we adjusted our GLRT technique for such arrays, using a special optimization scheme that takes into account the circular geometry.

Finally, our method has been expanded to address the problem of simultaneous testing of several covariance matrices. A typical example is the testing of the content of some "primary" and "training" (secondary) data to extract the common sources present in both sets, and to find new ones, if any. This problem was successfully addressed in [37, 45].

It is clear that this GLRT technique can be broadly used for the "quality assessment" of existing methods, advancing them where required into the threshold region, and to be used as a framework for detection-estimation for any parametric Gaussian model that is currently poorly addressed by existing methods. While our approach involves optimization and so is computationally demanding, the benefits have now been explored and demonstrated for many practical applications.

References

[1] Barber, N., "Optimum Array for Direction Finding," *NZ J. Science*, Vol. 1, 1958, pp. 35–51.

[2] Moffet, A., "Minimum-Redundancy Linear Arrays," *IEEE Trans. on Antennas and Propagation*, Vol. 16, No. 2, 1968, pp. 172–175.

[3] Bracewell, R., "The Stanford Five-Element Radio Telescope," *IEEE Proc.*, Vol. 61, No. 9, 1973, pp. 1249–1257.

[4] Lang, S., G. Duckworth, and J. McClellan, "Array Design for MEM and MLM Array Processing," in *Proc. ICASSP-81*, 1981, pp. 145–148.

[5] Pillai, S., Y. Bar-Ness, and F. Haber, "A New Approach to Array Geometry for Improved Spatial Spectrum Estimation," *IEEE Proc.*, Vol. 73, No. 10, 1985, pp. 1522–1524.

[6] Haykin, S., et al., "Some Aspects of Array Signal Processing," *IEE Proc. Part F: Radar, Sonar Navig.*, Vol. 139, No. 1, 1992, pp. 1–26.

[7] Chambers, C., et al., "Temporal and Spatial Sampling Influence on the Estimates of Superimposed Narrowband Signals: When Less Can Mean More," *IEEE Trans. on Signal Processing*, Vol. 44, No. 12, December 1996, pp. 3085–3098.

[8] Abramovich, Y., et al., "Positive-Definite Toeplitz Completion in DOA Estimation for Nonuniformlinear Antenna Arrays—Part I: Fully Augmentable Arrays," *IEEE Trans. on Signal Processing*, Vol. 46, No. 9, September 1998, pp. 2458–2471.

[9] Abramovich, Y., N. Spencer, and A. Gorokhov, "Positive-Definite Toeplitz Completion in DOA Estimation for Nonuniform Linear Antenna Arrays—Part II: Partially Augmentable Arrays," *IEEE Trans. on Signal Processing*, Vol. 47, No. 6, June 1999, pp. 1502–1521.

[10] Abramovich, Y., N. Spencer, and A. Gorokhov, "DOA Estimation for Noninteger Linear Antenna Arrays with More Uncorrelated Sources Than Sensors," *IEEE Trans. on Signal Processing*, Vol. 48, No. 4, April 2000, pp. 943–955.

[11] Abramovich Y., and N. Spencer, "Design of Nonuniform Linear Antenna Array Geometry and Signal Processing Algorithm for DOA Estimation of Gaussian Sources," *Digital Signal Processing*, Vol. 10, No. 4, 2000, pp. 340–354.

[12] Abramovich Y., and A. Gorokhov, "Improved Analysis of High Resolution Spatial Spectrum Estimators in Minimum Redundancy Linear Arrays," *Proc. RADAR-94*, Paris, 1994, pp. 127–132.

[13] Gorokhov, A., Y. Abramovich, and J. Böhme, "Unified Analysis of DOA Estimation Algorithms for Covariance Matrix Transforms," *Signal Processing*, Vol. 55, No. 1, 1996, pp. 107–115.

[14] Ottersten, B., et al., "Exact and Large Sample Maximum Likelihood Techniques for Parameter Estimation and Detection in Array Processing," in *Radar Array Processing*, Ch. 4, Springer Series in Information Sciences, Vol. 25, S. Haykin, J. Litva, and T. Shepherd, (eds.), Berlin, Germany: Springer-Verlag, 1993, pp. 99–151.

[15] Silvey, S., *Statistical Inference*, London, England: Penguin, 1970.

[16] Kay, S., *Fundamentals of Statistical Signal Processing, Detection Theory, Volume II*, Upper Saddle River, NJ: Prentice Hall, 1998.

[17] Anderson, T., *An Introduction to Multivariate Statistical Analysis*, New York: John Wiley & Sons, 1958.

[18] Djuric, P., "A Model Selection Rule for Sinusoids in White Gaussian Noise," *IEEE Trans. on Signal Processing*, Vol. 44, No. 7, July 1996, pp. 1744–1757.

[19] Schmidt, R., "A Signal Subspace Approach to Multiple Emitter Location and Spectral Estimation," Ph.D. dissertation, Stanford University, CA, Dept. of Electrical Engineering, 1981.

[20] Bienvenu, G., and L. Kopp, "Optimality of High Resolution Array Processing," *IEEE Trans. on Acoustics, Speech, and Signal Processing*, Vol. 31, 1983, pp. 1235–1248.

[21] Viberg, M., and B. Ottersten, "Sensor Array Processing Based on Subspace Fitting," *IEEE Trans. on Signal Processing*, Vol. 39, 1991, pp. 1110–1121.

[22] Manikas, A., and C. Proukakis, "Modeling and Estimation of Ambiguities in Linear Arrays," *IEEE Trans. on Signal Processing*, Vol. 46, No. 8, August 1998, pp. 2166–2179.

[23] Abramovich, Y., N. Spencer, and A. Gorokhov, "Resolving Manifold Ambiguities in Direction-of-Arrival Estimation for Nonuniform Linear Antenna Arrays," *IEEE Trans. on Signal Processing*, Vol. 47, No. 10, October 1999, pp. 2629–2643.

[24] Abramovich, Y., V. Gaitsgory, and N. Spencer, "Stability of Manifold Ambiguity Resolution in DOA Estimation with Nonuniform Antenna Arrays," *AE U Int. J. Electron. Commun.*, Vol. 53, No. 6, 1999, pp. 364–370.

[25] Abramovich, Y., and N. Spencer, "Detection Nonidentifiability for Independent Gaussian Sources in Nonuniform Linear Antenna Arrays," *Proc. ISSPA-2001*, Vol. 1, Kuala Lumpur, Malaysia, 2001, pp. 116–119.

[26] Abramovich, Y., N. Spencer, and A. Gorokhov, "Detection-Estimation of More Uncorrelated Gaussian Sources than Sensors in Nonuniform Linear Antenna Arrays—Part III: Detection Nonidentifiability," *IEEE Trans. on Signal Processing*, Vol. 51, No. 10, October 2003, pp. 2483–2494.

[27] Stoica, P., and A. Nehorai, "MUSIC, Maximum Likelihood and Cramér–Rao Bound," *IEEE Trans. on Acoustics, Speech, and Signal Processing*, Vol. 37, No. 5, 1989, pp. 720–741.

[28] Tufts, D., A. Kot, and R. Vaccaro, "The Threshold Effect in Signal Processing Algorithms Which Use an Estimated Subspace," in *SVD and Signal Processing II: Algorithms, Analysis and Applications*, R. Vaccaro, (ed.), New York: Elsevier, 1991, pp. 301–320.

[29] Thomas, J., L. Scharf, and D. Tufts, "The Probability of a Subspace Swap in the SVD," *IEEE Trans. on Signal Processing*, Vol. 43, No. 3, March 1995, pp. 730–736.

[30] Stoica, P., V. Simonyté, and T. Söderström, "On the Resolution Performance of Spectral Analysis," *Signal Processing*, Vol. 44, No. 6, 1995, pp. 153–161.

[31] Hawkes, M., A. Nehorai, and P. Stoica, "Performance Breakdown of Subspace-Based Methods: Prediction and Cure," *Proc. ICASSP-2001*, Vol. 6, Salt Lake City, UT, 2001, pp. 405–408.

[32] Abramovich, Y., N. Spencer, and A. Gorokhov, "A Lower Bound for the Maximum Likelihood Ratio: Sparse Antenna Array Applications," *Proc. ICASSP-2002*, Vol. 2, Orlando, FL, 2002, pp. 1145–1148.

[33] Abramovich, Y., N. Spencer, and A. Gorokhov, "Bounds on Maximum Likelihood Ratio—Part I: Application to Antenna Array Detection-Estimation with Perfect Wavefront Coherence," *IEEE Trans. on Signal Processing*, Vol. 52, No. 6, June 2004, pp. 1524–1536.

[34] Abramovich, Y., N. Spencer, and A. Gorokhov, "Detection-Estimation of More Uncorrelated Gaussian Sources Than Sensors in Nonuniform Linear Antenna Arrays—Part I: Fully Augmentable Arrays," *IEEE Trans. on Signal Processing*, Vol. 49, No. 5, May 2001, pp. 959–971.

[35] Abramovich, Y., N. Spencer, and A. Gorokhov, "Detection-Estimation of More Uncorrelated Gaussian Sources Than Sensors in Nonuniform Linear Antenna Arrays—Part II: Partially Augmentable Arrays," *IEEE Trans. on Signal Processing*, Vol. 51, No. 6, June 2003, pp. 1492–1507.

[36] Abramovich, Y., N. Spencer, and A. Gorokhov, "Bounds on Maximum Likelihood Ratio—Part II: Application to Antenna Array Detection-Estimation with Imperfect Wavefront Coherence," *IEEE Trans. on Signal Processing*, Vol. 53, No. 6, June 2005, pp. 2046–2058.

[37] Abramovich, Y., N. Spencer, and P. Turcaj, "Two-Set Adaptive Detection-Estimation of Gaussian Sources in Gaussian Noise," *Signal Processing*, Vol. 84, No. 9, 2004, pp. 1537–1560.

[38] Abramovich, Y., N. Spencer, and A. Gorokhov, "GLRT-Based Detection-Estimation of Independent Gaussian Sources Using Uniform Circular Antenna Arrays," *IEEE Trans. Sig. Proc.*, submitted March 19, 2004.

[39] Abramovich, Y., N. Spencer, and A. Gorokhov, "Detection-Estimation of Distributed Gaussian Sources," *Proc. SAM-2002*, Washington D.C., 2002, pp. 513–517.

[40] Muirhead, R., *Aspects of Multivariate Statistical Theory*, New York: John Wiley & Sons, 1982.

[41] Siotani, M., T. Hayakawa, and Y. Fujikoshi, *Modern Multivariate Statistical Analysis*, Columbus, OH: American Sciences Press, 1985.

[42] Abramovich, Y., and N. Spencer, "Performance Breakdown of Subspace-Based Methods in Arbitrary Antenna Arrays: GLRT-Based Prediction and Cure," *Proc. ICASSP-2004*, Vol. 2, Montreal, Canada, 2004, pp. 117–120.

[43] Spencer, N., and Y. Abramovich, "Performance Analysis of DOA Estimation Using Uniform Circular Antenna Arrays in the Threshold Region," *Proc. ICASSP-2004*, Vol. 2, Montreal, Canada, 2004, pp. 233–236.

[44] Gradshteyn, I., and I. Ryzhik, *Tables of Integrals, Series, and Products*, 6th ed., New York: Academic Press, 2000.

[45] Abramovich, Y., N. Spencer, and P. Turcaj, "GLRT-Based Adaptive Detection-Estimation of Gaussian Sources in Coloured Noise Fields," *Proc. ISSPIT-2002*, CD, Darmstadt, Germany, 2003.

PART II
DOA Methods

CHAPTER 3
DOA Estimation of Wideband Signals
Yeo-Sun Yoon, Lance M. Kaplan, and James H. McClellan

For many applications, wideband signals are exploited for localization. A wideband signal is any signal whose energy is distributed over a bandwidth that is large in comparison to the signal's center frequency. For example, ultrawideband (UWB) noise radars use wideband offering low probability of detection (LPD), while achieving good target detection and high resolution [1]. In the acoustic vehicle tracking scenario, the target emits a set of narrowband harmonics [2].

Direct exploitation of an array of raw wideband signals for localization using traditional narrowband techniques will fail miserably. The shortcoming is that the narrowband methods exploit the fact that time delays directly translate to a phase shift in the frequency domain [3].

$$s(t - \tau) \leftrightarrow S(f)e^{-j2\pi f \tau} \tag{3.1}$$

For narrowband signals, where the bandwidth is small relative to the center frequency f_c, the phase-shift is approximately constant over the bandwidth. The delayed signal in the time domain is

$$S(f)e^{-j2\pi f \tau} \approx S(f)e^{-j2\pi f_c \tau} \leftrightarrow s(t)e^{-j2\pi f_c \tau} \tag{3.2}$$

The phase shift is now independent of time, where the time delay τ is a function of the location of the source relative to the array elements. When the signal is mixed with a pure tone at the center frequency (i.e., baseband conversion), the outputs along a linear array due to P far-field sources are viewed as an approximately constant signal. A well-accepted model for the linear array outputs is given by a weighted sum of steering vectors embedded in noise

$$\mathbf{x}(t) = [\mathbf{a}(\theta_0) \ldots \mathbf{a}(\theta_{P-1})]\mathbf{s}(t) + \mathbf{n}(t) \tag{3.3}$$

where $\mathbf{a}(\theta_i)$ is the $M \times 1$ steering vector, or array manifold, for the ith source,

$$\mathbf{a}(\theta_i) = \begin{bmatrix} 1 & e^{-j2\pi f_c(d_1/c)\sin\theta_i} & \ldots & e^{-j2\pi f_c(d_{M-1}/c)\sin\theta_i} \end{bmatrix}^T \tag{3.4}$$

The array outputs, signal amplitudes, and noise residuals are represented by the $M \times 1$ vector $\mathbf{x}$, the $P \times 1$ vector $\mathbf{s}$, and the $M \times 1$ vector $\mathbf{n}$, respectively. The number of sensors, M, is usually assumed to be larger than the number of sources, P. In the expression for the steering vector, d_m is the sensor displacement of the mth element with respect to the first element, c is the speed of propagation, and θ_i is the azimuth angle pointing to the ith source. For linear array geometry, the elevation angle ϕ_i is completely ambiguous.

On the other hand, for a wideband source, the array output after baseband conversion is no longer constant. For the wideband case, the array outputs are modeled as

$$\mathbf{x}(t) = \int \mathbf{X}(f) e^{-j2\pi(f-f_c)t} \, df \tag{3.5}$$

where

$$\mathbf{X}(f) = [\mathbf{a}(f, \theta_0) \ldots \mathbf{a}(f, \theta_{P-1})] \mathbf{S}(f) + \mathbf{N}(f) \tag{3.6}$$

and

$$\mathbf{a}(f, \theta_i) = \begin{bmatrix} 1 & e^{-j2\pi f(d_1/c)\sin\theta_i} & \ldots & e^{-j2\pi f(d_{M-1}/c)\sin\theta_i} \end{bmatrix}^T \tag{3.7}$$

In a sense, narrowband sources represent a special case of (3.5), where $S(f) = S(f)\delta(f - f_c)$. For the special case of a ULA, the steering vector takes on the Vandermonde form

$$\mathbf{a}(f, \theta_i) = \begin{bmatrix} 1 & e^{-2\pi f v \sin\theta_i} & \ldots & e^{-2\pi f v (M-1) \sin\theta_i} \end{bmatrix}^T \tag{3.8}$$

where $v = d/c$, and d is the spacing between array elements. Figure 3.1 shows the support of the ULA output due to two sources, after taking a Fourier transform in both the temporal and spatial domains. For the narrowband case shown in

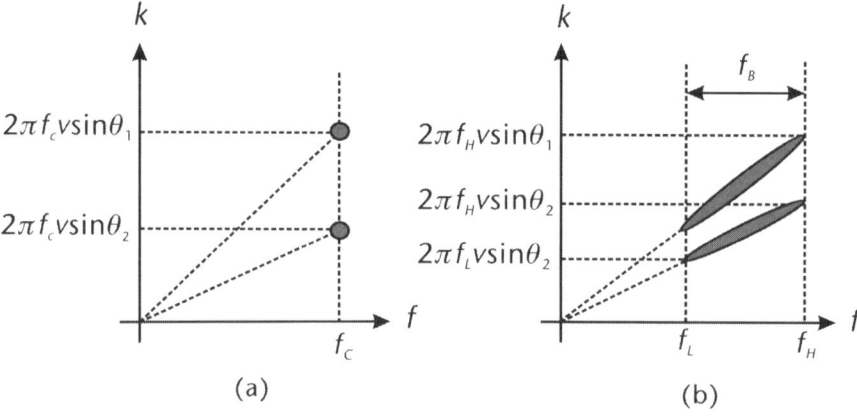

Figure 3.1 Frequency support for the array output, where the horizontal and vertical axes represent temporal and spatial frequencies, respectively, for: (a) two narrowband sources, and (b) two wideband sources.

Figure 3.1(a), the sources are well-separated via the spatial frequency. For the wideband case without bandpass filtering shown in Figure 3.1(b), the spatial bandwidths of the two sources overlap, and it is difficult to separate the two targets.

Usually, the array outputs are collected in a sequence of snapshots, where the time interval between snapshots is large enough for the signal amplitudes $S(f)$ to decorrelate over time. As a result, the covariance for the array output at frequency f; that is, $\mathbf{X}(f)\exp\{-j2\pi(f-f_c)t\}$, is modeled as

$$\mathbf{R}(f) = \mathbf{A}(f, \boldsymbol{\theta})\mathbf{R}_s(f)\mathbf{A}^H(f, \boldsymbol{\theta}) + \sigma^2(f)\mathbf{I} \tag{3.9}$$

where

$$\mathbf{A}(f_i, \boldsymbol{\theta}) = [\mathbf{a}(f, \theta_0) \quad \ldots \quad \mathbf{a}(f, \theta_{P-1})] \tag{3.10}$$

$\mathbf{R}_s(f)$ is the covariance matrix for $\mathbf{S}(f)$, and $\sigma^2(f)$ is the noise variance. The composite covariance matrix for array output $\mathbf{X}(f)$ is

$$\mathbf{R} = \int \mathbf{R}(f)\, df \tag{3.11}$$

For narrowband sources, $\mathbf{R} = \mathbf{R}(f_c)$. The matrix $\mathbf{R}(f)$ has P dominant eigenvalues whose corresponding eigenvectors represent a signal subspace. The P steering vectors $\mathbf{a}(f, \theta_i)$ also span the signal subspace. The remaining M-P eigenvectors span the noise subspace. Similar statements are made about the composite covariance matrix for narrowband signals. A large number of the DOA estimation techniques (e.g., MUSIC [4] and ESPRIT [5]) exploit the fact that the DOAs define the signal subspace. However, when the sources are wideband, the number of significant eigenvalues for the covariance matrix is larger than P, due to the mixing of different frequency components [6]. As a result, when the bandwidth of the source increases, it becomes more difficult to distinguish the signal and noise subspaces from the composite covariance matrix.

It is possible to pass the output of each array element through a narrowband bandpass filter, so that the composite matrix follows (3.9) for $f = f_i$. Then, a narrowband DOA estimator is applied. However, this methodology ignores valuable information in other frequency bands that can enhance estimation. Researchers have proposed a number of methods to exploit the richer structure of the array output available with wideband sources [7–9]. For example, Su and Morf modeled the wideband sources as a rational transfer function driven by white noise, and used modal decomposition to estimate the delays on each mode [7]. Agrawal and Prasad used an array manifold vector for wideband sources instead of the conventional narrowband array manifold, assuming that the power spectral density of the signal is flat [9].

Other methods decompose the array outputs into multiple narrowband frequency components using filter banks, usually a fast Fourier transform (FFT). This chapter focuses on these methods. Initially, the output at each array is segmented temporally into J snapshots. The temporal FFT is applied to each snapshot to determine K frequency components. The array output for the jth snapshot and the ith frequency component is represented as $\mathbf{x}_{j,i}$. Then, the sample covariance matrix,

$$\hat{\mathbf{R}}_i = \frac{1}{J} \sum_{j=0}^{J-1} \mathbf{x}_{j,i} \mathbf{x}_{j,i}^H \qquad (3.12)$$

is computed for $i = 0, \ldots, K - 1$, as an estimate of the covariance matrix given by (3.9) for frequency $f = f_i$. Throughout this chapter, the number of array elements, sources, and frequency bins will be denoted an M, P, and K, respectively.

The wideband DOA estimators discussed here are divided into two broad categories, *incoherent* and *coherent*, according to how information from the covariance matrices is used. In the next two sections, each category is discussed. Then, a new wideband method, Test of Orthogonality of Projected Signal (TOPS) subspaces [10], is introduced. The subsequent section compares the performance of the different wideband methods through simulations. Then, application of TOPS to a more general localization problem is briefly covered. Finally, some concluding remarks are provided.

3.1 Incoherent Wideband DOA Estimation

The incoherent methods process each frequency bin independently and average the DOA estimates over all the bins. Since each decomposed signal is approximated as a narrowband signal, any narrowband DOA estimation method is applicable. For example, when MUSIC [4] is used as the narrowband method, K frequency bins are summed incoherently

$$\hat{\theta} = \arg \min_{\theta} \sum_{i=0}^{K-1} \mathbf{a}^H(f_i, \theta) \mathbf{W}_i \mathbf{W}_i^H \mathbf{a}(f_i, \theta) \qquad (3.13)$$

where $\mathbf{W}_i$ is the noise subspace, and $\mathbf{a}(f_i, \theta)$ is the array manifold at frequency f_i. The noise subspace is usually unavailable. Therefore, it is estimated from the estimated form as the sample covariance matrix given in (3.12).

Examples of incoherent methods appear in [11, 12]. The incoherent combination of information in the different frequency bands provides good DOA estimates in favorable situations, such as high SNR and well-separated source locations. However, when the SNR is low or the noise level is not uniform over frequency bins, the output of the incoherent methods is worse than processing only a single frequency bin. Furthermore, when two signals are close to each other, the averaging process could make the estimates worse [13]. In this case, high SNR is not always a favorable condition. Since the width of the peaks for the likelihood surface of MUSIC are narrow, multiple peaks around the source location can appear when the likelihood surfaces for different frequency bins are averaged together.

3.2 Coherent Wideband DOA Estimation

The coherent methods attempt to align the signal subspaces that are associated with the DOA of the sources among all frequency bins. This alignment is accomplished by

3.2 Coherent Wideband DOA Estimation

a transformation of the covariance matrices associated with each bin. After the transformation, the signal and noise subspaces are coherent over the frequency bins, so that the formation of the composite covariance matrix is meaningful. This section reviews the coherent signal subspace method (CSSM) [13], and the weighted average of signal subspace (WAVES) method [14].

3.2.1 Coherent Signal Subspace Method

CSSM was introduced by Wang and Kaveh [13]. In this method, an estimate of the composite covariance matrix is formed via

$$\mathbf{R}_{com} = \sum_{i=0}^{K-1} \alpha_i \mathbf{T}_i \mathbf{R}_i \mathbf{T}_i^H \qquad (3.14)$$

where α_i is a weighting, and $\mathbf{T}_i$ is the focusing matrix for frequency f_i, such that

$$\min_{\mathbf{T}_i} \| \mathbf{A}(f_r, \boldsymbol{\theta}_f) - \mathbf{T}_i \mathbf{A}(f_i, \boldsymbol{\theta}_f) \|_F \qquad (3.15)$$

where f_r is the reference frequency, $\boldsymbol{\theta}_f$ is the set of *focusing angles*, and $\|\cdot\|_F$ denotes the Frobenius norm. The reference frequency may be within the signal's frequency band. Once the composite matrix is formed, any narrowband method is applied to the matrix $\mathbf{R}_{com}$ to generate DOA estimates.

When $\alpha_i = 1$,

$$\begin{aligned}\mathbf{R}_{com} &= \sum_{i=0}^{K-1} \mathbf{T}_i \mathbf{R}_i \mathbf{T}_i^H \\ &= \sum_{i=0}^{K-1} \mathbf{T}_i \{ \mathbf{A}(f_i, \boldsymbol{\theta}_f) \mathbf{R}_s(f_i) \mathbf{A}^H(f_i, \boldsymbol{\theta}_f) + \sigma^2(f_i) \mathbf{I} \} \mathbf{T}_i^H \\ &\approx \mathbf{A}(f_r, \boldsymbol{\theta}_f) \sum_{i=0}^{K-1} \mathbf{R}_s(f_i) \mathbf{A}^H(f_i, \boldsymbol{\theta}_f) + \sum_{i=0}^{K-1} \sigma^2(f_i) \mathbf{T}_i \mathbf{T}_i^H \end{aligned} \qquad (3.16)$$

In other words, CSSM tries to generate the covariance matrix associated with frequency f_r by averaging the transformed covariance matrices of all frequency bins. The method assumes that the array response matrix for frequency f_i is formed by the array manifolds associated with the focusing angles [i.e., $\mathbf{A}(f_i, \boldsymbol{\theta}_f)$]. Therefore, the validity of the general covariance matrix depends on how closely $\mathbf{A}(f_i, \boldsymbol{\theta}_f)$ matches the array response matrix of the sensor outputs i.e., $[\mathbf{A}(f_i, \boldsymbol{\theta}_f)]$. Equivalently, the success of CSSM depends on how close the focusing angles, $\boldsymbol{\theta}_f$, match the true DOAs. However, the DOAs are the parameters that are being investigated. One can view CSSM as a refinement approach where low-resolution DOA estimates are usually used as focusing angles. As shown in [13], the CSSM estimator that determines focusing angles in this manner provides satisfactory estimates.

Focusing matrices $\mathbf{T}_i$, as well as focusing angles, are other important factors that determine the performance of CSSM estimators. One well-known focusing matrix is the signal subspace transformation (SST) matrix [15]. The SST matrix is optimal in the sense of the focusing loss, which is defined as the ratio of SNR after transformation, divided by the SNR before focusing. Specifically, the focusing loss g is [16]

$$g = \frac{tr\left\{\mathbf{R}_n^{-1} \sum_{i=0}^{K-1} \mathbf{T}_i \mathbf{A}(f_i, \boldsymbol{\theta}) \mathbf{R}_s(f_i) \mathbf{A}^H(f_i, \boldsymbol{\theta}) \mathbf{T}_i^H \right\}}{tr\left\{\sum_{i=0}^{K-1} \mathbf{A}(f_i, \boldsymbol{\theta}) \mathbf{R}_s(f_i) \mathbf{A}^H(f_i, \boldsymbol{\theta})\right\}} \quad (3.17)$$

where $tr\{\cdot\}$ denotes the trace, and $\mathbf{R}_n$ is related to the noise covariance matrix after focusing via

$$\mathbf{R}_n = \frac{\sum_{i=0}^{K-1} \sigma^2(f_i) \mathbf{T}_i \mathbf{T}_i^H}{\sum_{i=0}^{K-1} \sigma^2(f_i)} \quad (3.18)$$

It is easy to show that the maximum value $g = 1$ is obtained when $\mathbf{T}_i \mathbf{T}_i^H$ is independent of the ith frequency bin [15]. The rotational signal subspace (RSS) focusing matrix, which was proposed in [16], is one special case of the SST matrices, because $\mathbf{T}_i \mathbf{T}_i^H = \mathbf{I}$ [15]. Therefore, once the focusing angles $\boldsymbol{\theta}_f$ are determined, the RSS focusing matrix $\mathbf{T}_i$ is found via

$$\min_{\mathbf{T}_i} \|\mathbf{A}(f_r, \boldsymbol{\theta}_f) - \mathbf{T}_i \mathbf{A}(f_i, \boldsymbol{\theta}_f)\|_F \quad (3.19)$$

for $i = 0, \ldots, K-1$ subject to $\mathbf{T}_i \mathbf{T}_i^H = \mathbf{I}$. The solution to the above optimization problem is

$$\mathbf{T}_i = \mathbf{V}_i \mathbf{U}_i^H \quad (3.20)$$

where the columns of $\mathbf{U}_i$ and $\mathbf{V}_i$ are the left and right singular vectors of $\mathbf{A}(f_r, \boldsymbol{\theta}_f) \mathbf{A}^H(f_i, \boldsymbol{\theta}_f)$ [16].

CSSM is robust against low SNR, and exhibits higher resolution than the incoherent methods [13]. However, the overall performance seems dominated by the quality of the focusing angles [10]. Furthermore, errors in the focusing angles can bias the CSSM estimates [17].

3.2.2 BICSSM

The beamforming-invariant coherent signal subspace method (BICSSM) was introduced in [18]. Like CSSM, it uses a transformation matrix to form the composite covariance matrix and applies a subspace method over the composite matrix. Unlike CSSM, the transformation matrices in BICSSM are actually a bank of beamformers tuned to certain look directions. Let us consider $\hat{M}$ look directions. For the $\hat{m}$th look direction, the differences between the beam responses in different frequency bins are minimized. Let $\mathbf{t}_{r,\hat{m}}$ be an $M \times 1$ vector, representing the beamforming weights associated with the reference frequency bin for $\hat{m}$th look direction. Then, the beamforming weights for other frequency bins in the $\hat{m}$th look direction are determined by

$$\mathbf{t}_{i,\hat{m}} = \arg \min_{\mathbf{c}} \int_{\Omega} \rho(\theta) |\mathbf{t}_{r,\hat{m}}^H \mathbf{a}(f_r, \theta) - \mathbf{c}^H \mathbf{a}(f_i, \theta)|^2 d\theta \qquad (3.21)$$

where Ω is the field of view (FOV), $\vec{\beta}$ is a vector representing DOA, and $\rho(\theta)$ is a weighting function. Usually, $\rho(\theta) = 1$. The reference beamforming weights $\mathbf{t}_{r,\hat{m}}$ should have a beam pattern that exhibits low sidelobes outside of the FOV. Furthermore, the beam patterns for $\mathbf{t}_{r,\hat{m}}$ should be invariant between the frequency components. Finally, the transformation for the ith frequency bin is

$$\mathbf{T}_i = [\mathbf{t}_{i,0} \quad \cdots \quad \mathbf{t}_{i,\hat{M}-1}]^H \qquad (3.22)$$

BICSSM removes the need to estimate focusing angles by attempting to cohere the transformed steering matrices within a FOV. Once all the $\mathbf{T}_i$'s are found, it generates the composite correlation matrix via (3.14). Since $\mathbf{T}_r \mathbf{a}(f_r, \theta) \approx \mathbf{T}_i \mathbf{a}(f_i, \theta)$ in the FOV, the composite correlation matrix is

$$\mathbf{R}_{gen} \approx \mathbf{B}_r \left(\sum_{i=0}^{K-1} w_i \mathbf{R}_s(f_i) \right) \mathbf{B}_r^H \qquad (3.23)$$

where $\mathbf{B}_r = \mathbf{T}_r \mathbf{A}(f_r, \boldsymbol{\theta})$. Narrowband MUSIC is then applied to this general correlation matrix, with the array manifold being $\mathbf{T}_r \mathbf{a}(f_r, \theta)$.

The design of the reference beamforming weights $\mathbf{T}_r$ is crucial to obtain good performance for BICSSM. The reference beamforming weights are determined by a more sophisticated way to reduce focusing error [14]. Although BICSSM is able to avoid the need for initial estimates of the DOA, reference beamformers that exhibit low sidelobes outside the FOV and consistent beam responses over different frequency bins may not exist for certain array geometries. Furthermore, the processor may not know the FOV.

3.2.3 WAVES

Recently, the WAVES method was proposed in [14]. It uses a pseudodata matrix that is inspired by the weighted subspace fitting (WSF) method [19]. Specifically, WAVES estimates the noise subspaces at a reference frequency by taking a singular value decomposition of the pseudodata matrix $\mathbf{Z}$,

$$\mathbf{Z} = [\mathbf{T}_0 \mathbf{F}_0 \mathbf{P}_0 \quad \mathbf{T}_1 \mathbf{F}_1 \mathbf{P}_1 \quad \cdots \quad \mathbf{T}_{K-1} \mathbf{F}_{K-1} \mathbf{P}_{K-1}] \qquad (3.24)$$

where $\mathbf{T}_i$ is the focusing matrix, and $\mathbf{F}_i$ is the signal subspace at frequency f_i. The focusing matrix $\mathbf{T}_i$ is obtained using focusing angles or via BICSSM [18]. The matrix $\mathbf{P}_i$ is a diagonal weighting matrix. The kth diagonal element of the weighting matrix $\mathbf{P}_i$ is

$$[\mathbf{P}_i]_{(k,k)} = \frac{\lambda_{i,k} - \sigma_n^2}{\sqrt{\lambda_{i,k} \sigma_n^2}} \qquad (3.25)$$

where $\lambda_{i,k}$ is the kth largest eigenvalue of $\mathbf{R}_i$, and σ_n^2 is the noise power, which is assumed to be a constant over all the frequency bins. As shown in [19], the weighting matrix is optimal when the signal and noise follow the Gaussian distribution. WAVES determines the noise subspace $\mathbf{U}_N$, which is $M \times (K-1)P$ matrix via the singular value decomposition (SVD) of the $M \times KP$ pseudodata matrix

$$\mathbf{Z} = [\mathbf{U}_S \ \mathbf{U}_N] \begin{bmatrix} \mathbf{\Sigma}_S & 0 \\ 0 & \mathbf{\Sigma}_N \end{bmatrix} \begin{bmatrix} \mathbf{V}_S^H \\ \mathbf{V}_N^H \end{bmatrix} \qquad (3.26)$$

Finally, the MUSIC algorithm is applied with the noise subspace $\mathbf{U}_N$.

Conceptually, WAVES is similar to CSSM because it exploits the noise subspace of the composite-like correlation matrix

$$\mathbf{R}_{gen} = \sum_{i=0}^{K-1} \mathbf{T}_i \left(\mathbf{F}_i \mathbf{P}_i \mathbf{P}_i^H \mathbf{F}_i^H \right) \mathbf{T}_i^H \qquad (3.27)$$

However, in WAVES, the covariance matrices for each frequency bin are filtered before forming the composite matrix. For instance, the eigenvalues associated with the noise subspaces are set to zero, and the weighting matrix $\mathbf{P}_i$ transforms the signal eigenvalues to emphasize the subspaces whose energy is significantly higher than the noise floor. The nonlinear transformation of the eigenvalues makes WAVES more robust than CSSM against noise.

A significant drawback to WAVES is that it still requires the use of transformation matrices. If the transformation matrices are formed as in CSSM, then the quality of the DOA estimates depends on initial DOA estimates for the focusing angles. On the other hand, the transformation matrices are generated via the BICSSM method without needing any initial DOA estimates. However, the estimation performance is not as good as when using transformation matrices via the CSSM approach, especially in low SNR [18].

3.3 TOPS

Recently, the new wideband DOA estimation method TOPS was proposed [10, 20]. Unlike the coherent methods, the new method does not require initial focusing angles that can induce errors on the estimates. Unlike the incoherent methods,

3.3 TOPS

TOPS integrates information for all frequency bins before forming the estimates of the DOAs. This section describes the theory behind TOPS, and then shows the numerical simulation results.

3.3.1 Theory

This section assumes an arbitrary array geometry that is either one-dimensional or two-dimensional. For a two-dimensional array, the DOA includes both the azimuth angle θ and the elevation angle ϕ. For a linear array, the elevation angle is completely ambiguous and is ignored. For a two-dimensional array that lies in the x-z plane, the angles (θ, ϕ) and $(\theta, \pi - \phi)$ are ambiguous [21]. The following theory assumes that for an M-element array, any set of array manifolds associated with M unambiguous DOAs are linearly independent. This assumption is certainly valid for the ULA because of the Vandermonde structure of the array manifold [21]. However, it may not be valid for other array geometries. For these cases, all subspace methods will suffer from false peaks, due to unanticipated ambiguous DOAs.

The mth element of the array manifold is

$$[\mathbf{a}(f, \vec{\alpha})]_m = \exp\{-j2\pi f(\vec{r}_m \cdot \vec{\alpha})\} \tag{3.28}$$

where $\vec{r}_m = (x_m, y_m, z_m)$ is the mth sensor position vector, and the DOAs are represented by the slowness vector $\vec{\alpha}$ [21]

$$\vec{\alpha} = \frac{1}{c}(\sin\theta\sin\phi, \cos\theta\sin\phi, \cos\phi) \tag{3.29}$$

Note that the magnitude of the slowness vector is the reciprocal of the propagation speed c. If the mth elements of two array manifolds at different frequencies are multiplied, then we obtain

$$[\mathbf{a}(f_1, \vec{\alpha}_1)]_m [\mathbf{a}(f_2, \vec{\alpha}_2)]_m = \exp\{-2\pi(f_1+f_2)(\xi\vec{r}_m \cdot \vec{\alpha}_1 + (1-\xi)\vec{r}_m \cdot \vec{\alpha}_2)\} \tag{3.30}$$

where $\xi = f_1/(f_1 + f_2)$. If the slowness vectors of the two array manifolds are the same, $(\vec{\alpha}_1 = \vec{\alpha}_2)$, then (3.30) becomes

$$[\mathbf{a}(f_1, \vec{\alpha}_1)]_m [\mathbf{a}(f_2, \vec{\alpha}_1)]_m = \exp\{-2\pi(f_1+f_2)(\vec{r}_m \cdot \vec{\alpha}_1)\} \tag{3.31}$$
$$= [\mathbf{a}(f_1+f_2, \vec{\alpha}_1)]_m$$

which is the array manifold at frequency $(f_1 + f_2)$ corresponding to the slowness vector $\vec{\alpha}_1$. The following theorem shows that (3.30) is also a valid array manifold, as long as the array elements lie in a plane or along a line.

Lemma 3.1
Let $\mathbf{a}(f_i, \vec{\alpha})$ be the array manifold vector of an arbitrary one-dimensional or two-dimensional array, and define a transformation matrix Φ as

$$\Phi(f_i, \vec{\alpha}) = diag\{\mathbf{a}(f, \vec{\alpha})\} \tag{3.32}$$

Then,

$$\Phi(f_2, \vec{\alpha}_2)\mathbf{a}(f_1, \vec{\alpha}_1) = \mathbf{a}(f_3, \vec{\alpha}_3) \tag{3.33}$$

where $f_3 = f_1 + f_2$, and $\vec{\alpha}_3$ is a valid slowness vector.

Proof: The phase of the mth elements of the left-hand side of (3.33) is

$$2\pi f_1(\vec{r}_m \cdot \vec{\alpha}_1) + 2\pi f_2(\vec{r}_m \cdot \vec{\alpha}_2) = 2\pi f_3[\vec{r}_m \cdot \{\xi\vec{\alpha}_1 + (1-\xi)\vec{\alpha}_2\}] \tag{3.34}$$

Define a vector $\vec{\beta}$,

$$\vec{\beta} = \xi\vec{\alpha}_1 + (1-\xi)\vec{\alpha}_2 \tag{3.35}$$

If $\vec{\alpha}_1 = \vec{\alpha}_2$, then $\vec{\beta} = \vec{\alpha}_1$ and the theorem is true. When $\vec{\alpha}_1 \neq \vec{\alpha}_2$, the $\vec{\beta}$ is not a valid slowness vector, since the magnitude is less than $1/c$ by the triangle inequality,

$$|\vec{\beta}| < \xi|\vec{\alpha}_1| + (1-\xi)|\vec{\alpha}_2| = \frac{1}{c} \tag{3.36}$$

However, if there exists a valid slowness vector $\vec{\alpha}_3$ for any $\vec{\beta}$, such that

$$\vec{r}_m \cdot \vec{\alpha}_3 = \vec{r}_m \cdot \vec{\beta} \tag{3.37}$$

$\mathbf{a}(f_3, \vec{\beta})$ is an array manifold vector. Define $\mathbf{x}$ to be the vector whose elements are the x-component of the sensor locations,

$$\mathbf{x} = [x_0 \quad \ldots \quad x_{M-1}]^T \tag{3.38}$$

Define $\mathbf{y}$ and $\mathbf{z}$ in a similar manner. Let $\vec{\alpha}_3 = (\alpha_{3,x}, \alpha_{3,y}, \alpha_{3,z})$ and $\vec{\beta} = (\beta_x, \beta_y, \beta_z)$. Then, (3.37) can be rewritten as

$$\alpha_{3,x}\mathbf{x} + \alpha_{3,y}\mathbf{y} + \alpha_{3,z}\mathbf{z} = \beta_x\mathbf{x} + \beta_y\mathbf{y} + \beta_z\mathbf{z} \tag{3.39}$$

If the array geometry is not collapsed into a lower dimension, then $\mathbf{x}$, $\mathbf{y}$, and $\mathbf{z}$ are independent, and $\vec{\alpha}_3$ and $\vec{\beta}$ must be the same, element by element. When the array is one-dimensional or two-dimensional, there always exists a $\vec{\alpha}_3$ that satisfies (3.37) and slowness vector constraint; that is, $|\vec{\alpha}_3| = 1/c$. In fact, multiple slowness vectors within the standard DOA ambiguity space are valid [21].

The lemma states that the transformation matrix given by (3.32) can transform an array manifold corresponding to one frequency into another array manifold corresponding to another frequency. Furthermore, if the DOA in the transformation matrix matches the DOA of the original array manifold, then the transformation preserves the DOA. Note the similarity between the transformation matrix (3.32) and the focusing matrix in the steered covariance method (STCM) in [22]. In fact,

$\mathbf{\Phi}(f, \vec{\alpha})$ is the conjugate transpose of the STCM matrix. Although these two matrices are very similar, they do completely different jobs. While STCM steers the sensor output toward $\vec{\alpha}$, the transformation matrix (3.32) preserves $\vec{\alpha}$ in the array manifold, resulting in different methods. For the remainder of the discussion, the slowness vector $\vec{\alpha}$ is replaced by the azimuth angle θ, assuming a one-dimensional array. Extension to the general case for a two-dimensional array is straightforward. Note that the signal subspace spans the same space as the steering vectors that comprise the array response matrix,

$$\mathbf{F}_i = \mathbf{A}(f_i, \hat{\boldsymbol{\theta}}_i)\mathbf{G}_i \tag{3.40}$$

where $\mathbf{G}_i$ is a $P \times P$ nonsingular matrix. Lemma 3.1 leads to the next theorem that serves as the basis for the TOPS algorithm. The idea is that the array manifolds at a reference frequency f_0 are transformed by $\mathbf{\Phi}(\Delta f_i, \tilde{\theta})$ into array manifolds for other frequencies f_i. Note that the reference frequency is now f_0 instead of f_r, to emphasize that the reference frequency is one of the decomposed frequency bins, unlike other methods in the previous section. If the parameter $\tilde{\theta}$ corresponds to an actual DOA, then the transformed array manifold for that DOA is orthogonal to the noise subspace at frequencies f_i. A test for this orthogonality can lead to a DOA estimation technique via the next theorem.

Theorem 3.1
Assume that $2P < M + 1$ and $K > P$. Let $\mathbf{U}_i(\tilde{\theta})$ be an $M \times P$ matrix for $i = 1, \ldots, K - 1$, such that

$$\mathbf{U}_i(\tilde{\theta}) = \mathbf{\Phi}(\Delta f_i, \tilde{\theta})\mathbf{F}_0 \tag{3.41}$$

where $\Delta f_i = f_i - f_0$. Define the $P \times (K - 1)(M - P)$ matrix

$$\mathbf{D}(\tilde{\theta}) = \begin{bmatrix} \mathbf{U}_1^H(\tilde{\theta})\mathbf{W}_1 & \mathbf{U}_2^H(\tilde{\theta})\mathbf{W}_2 & \ldots & \mathbf{U}_{K-1}^H(\tilde{\theta})\mathbf{W}_{K-1} \end{bmatrix} \tag{3.42}$$

where $\mathbf{W}_i$ is the noise subspace at frequency f_i. Then,

1. If $\tilde{\theta} = \theta_l$ for some l, then $\mathbf{D}(\tilde{\theta})$ loses its rank (rank-deficient).
2. If $\mathbf{D}(\tilde{\theta})$ is rank-deficient, then $\tilde{\theta} = \theta_l$ for some l.

Proof: The theorem states that $\tilde{\theta} = \theta_l$ for some l, if and only if $\mathbf{D}(\tilde{\theta})$ is rank-deficient. For notational convenience, we will use $\mathbf{A}_i(\boldsymbol{\theta})$ and $\mathbf{a}_i(\theta)$ in place of $\mathbf{A}(f_i, \boldsymbol{\theta})$ and $\mathbf{a}(f_i, \theta)$, respectively.

Proof of 1: Let $\tilde{\theta}$ be equal to the lth DOA, θ_l. Since $\mathbf{W}_i$ is the noise subspace $\mathbf{A}_i^H \mathbf{W}_i = 0$, we can write

$$\mathbf{a}_i^H(\theta_l)\mathbf{W}_i = \mathbf{0}^T \tag{3.43}$$

for all $l = 0, 1, \ldots, P - 1$, and $i = 0, 1, \ldots, K - 1$. We know that

$$\mathbf{U}_i(\tilde{\theta}) = \mathbf{\Phi}(\Delta f_i, \tilde{\theta})\mathbf{A}_0(\boldsymbol{\theta})\mathbf{G}_0 = \mathbf{A}_i(\hat{\boldsymbol{\theta}}_i)\mathbf{G}_0 \qquad (3.44)$$

where $\mathbf{G}_0$ is a nonsingular $P \times P$ matrix, and

$$\sin\left[\hat{\boldsymbol{\theta}}_i\right]_l = \frac{f_0}{f_i} \sin \theta_l + \frac{\Delta f_i}{f_i} \sin \tilde{\theta} \qquad (3.45)$$

for $l = 0, 1, \ldots, P - 1$. Since $\tilde{\theta} = \theta_l$,

$$\left[\hat{\boldsymbol{\theta}}_0\right]_l = \ldots = \left[\hat{\boldsymbol{\theta}}_{K-1}\right]_l = \theta_l \qquad (3.46)$$

Therefore,

$$\mathbf{U}_i^H(\tilde{\theta})\mathbf{W}_i = \mathbf{G}_0^H \begin{bmatrix} \mathbf{a}_i^H([\hat{\boldsymbol{\theta}}_i]_0)\mathbf{W}_i \\ \vdots \\ \mathbf{a}_i^H(\theta_l)\mathbf{W}_i \\ \vdots \\ \mathbf{a}_i^H([\hat{\boldsymbol{\theta}}_i]_{P-1})\mathbf{W}_i \end{bmatrix} \qquad (3.47)$$

$$= \mathbf{G}_0^H \begin{bmatrix} * \\ \mathbf{0}^T \\ * \end{bmatrix} \leftarrow l\text{th row}$$

Then, the $\mathbf{D}(\tilde{\theta})$ matrix becomes

$$\mathbf{D}(\tilde{\theta}) = \begin{bmatrix} \mathbf{U}_1^H \mathbf{W}_1 & \ldots & \mathbf{U}_{K-1}^H \mathbf{W}_{K-1} \end{bmatrix} = \mathbf{G}_0^H \begin{bmatrix} * & * & * \\ \mathbf{0}^T & \mathbf{0}^T & \mathbf{0}^T \\ * & * & * \end{bmatrix} \qquad (3.48)$$

and it loses rank. This proves that $\tilde{\theta} = \theta_l$ is a sufficient condition for $\mathbf{D}(\tilde{\theta})$ to lose rank.

Proof of 2: Assume that none of the DOAs are equal to $\tilde{\theta}$. We need to show that $\mathbf{D}(\tilde{\theta})$ is full rank. Because $\mathbf{G}_0$ is a nonsingular square matrix, the matrix $\mathbf{D}(\tilde{\theta})$ is rank-deficient, if and only if the following matrix is rank-deficient

$$\mathbf{L} = \begin{bmatrix} \mathbf{L}_1 & \ldots & \mathbf{L}_{K-1} \end{bmatrix} \qquad (3.49)$$

where

$$\mathbf{L}_i = \mathbf{A}_i^H(\hat{\boldsymbol{\theta}}_i)\mathbf{W}_i \qquad (3.50)$$

Since **L** has more columns than rows, the matrix is rank-deficient, if and only if the left null space is not empty; that is, $\mathscr{LN}\{\mathbf{L}\} \neq \varnothing$ where $\mathscr{LN}$ denotes the left null space. Equivalently,

$$\bigcap_{i=1}^{K-1} \mathscr{LN}\{\mathbf{L}_i\} \neq \varnothing \qquad (3.51)$$

The size of the left null space of $\mathbf{L}_i$ depends on how many columns of the transformed array response matrix $\mathbf{A}_i(\hat{\boldsymbol{\theta}}_i)$ correspond to true DOAs. Let s represent the number of such columns. With slight abuse of the notation, let $\hat{\mathbf{A}}_i = \mathbf{A}(\hat{\boldsymbol{\theta}}_i)$ and $\mathbf{A}_i = \mathbf{A}(\boldsymbol{\theta}_i)$. Define **B** and **C** as

$$\mathbf{B} = \begin{bmatrix} \mathbf{A}_i & \hat{\mathbf{A}}_i \end{bmatrix} \qquad (3.52)$$

and

$$\mathbf{C} = \begin{bmatrix} \mathbf{A}_i & \mathbf{W}_i \end{bmatrix} \qquad (3.53)$$

Because $2P < M + 1$, the rank of the $M \times 2P$ matrix **B** is determined by the number of linearly independent columns of **B**. Since s columns of $\hat{\mathbf{A}}_i$ are equal to s corresponding columns of $\mathbf{A}_i$, and the columns represent array manifolds, the rank of **B** is $2P - s$. If we multiply $\mathbf{B}^H$ by **C**, then

$$\mathbf{B}^H \mathbf{C} = \begin{bmatrix} \mathbf{A}_i^H \\ \hat{\mathbf{A}}_i^H \end{bmatrix} \begin{bmatrix} \mathbf{A}_i & \mathbf{W}_i \end{bmatrix} = \begin{bmatrix} \mathbf{A}_i^H \mathbf{A}_i & \mathbf{0} \\ \hat{\mathbf{A}}_i^H \mathbf{A}_i & \mathbf{L}_i \end{bmatrix} \qquad (3.54)$$

Sylvester's inequality states that the rank of $\mathbf{B}^H \mathbf{C}$ is bounded by [23]

$$\gamma(\mathbf{B}) + \gamma(\mathbf{C}) - M \leq \gamma(\mathbf{B}^H \mathbf{C}) \leq \min\{\gamma(\mathbf{B}), \gamma(\mathbf{C})\} \qquad (3.55)$$

where γ denotes the rank of a matrix. Therefore,

$$2P - s + M - M \leq \gamma(\mathbf{B}^H \mathbf{C}) \leq \min(2P - s, M) \qquad (3.56)$$

$$\Rightarrow \gamma(\mathbf{B}^H \mathbf{C}) = 2P - s$$

Since the upper right block is a zero matrix and the upper left block is full rank, the rank of $\mathbf{L}_i$ is $2P - s$. In other words, the dimension of the left null space of $\mathbf{L}_i$ is s. The s columns of $\hat{\mathbf{A}}_i$ that correspond to true DOAs are orthogonal to all columns of $\mathbf{W}_i$. As a result, the matrix $\mathbf{L}_i$ has s rows of zeros. This means that s elementary vectors span the entire left null space of $\mathbf{L}_i$. An elementary vector is a vector where one element is one, and the remaining elements are zeros. For the condition in (3.51) to be true, one or more elementary vectors must form the left null space of $\mathbf{L}_i$ for $i = 1, \ldots, K - 1$. For this to be the case, the same columns of $\hat{\mathbf{A}}_i$ must correspond to a true DOA for $i = 1, \ldots, K - 1$. If we label one such column as l, then the transformed angle is given by (3.45). As f_i increases, the transformed angle is either monotonically increasing or decreasing, depending on whether $\tilde{\theta} > \theta_l$ or $\tilde{\theta} < \theta_l$. At most, the transformed angle can correspond to a true DOA for $P - 1$ frequency bins. In other words, the corresponding elementary vector is in the left null space of $\mathbf{L}_i$ for, at most, $P - 1$ bins. Therefore, for (3.51) to be true, $K - 1 < P$. Once we know that $K > P$, then $\mathbf{L}$ and $\mathbf{D}(\tilde{\theta})$ are full rank. This proves that if $\tilde{\theta} \neq \theta_l$ for $l = 0, \ldots, P - 1$, then $\mathbf{D}(\tilde{\theta})$ is full rank. Thus, Statement 2 is true.

The properties of the matrix $\mathbf{D}(\tilde{\theta})$ as given in the above theorem motivate the TOPS algorithm. In short, $\mathbf{D}(\tilde{\theta})$ is rank-deficient if and only if $\tilde{\theta}$ is a true DOA.

3.3.2 Signal Subspace Projection

In practice, estimated covariance matrices have to be used in place of real covariances. Using the sample covariance as computed in (3.12), one can determine estimates of $\hat{\mathbf{F}}_1$ and $\hat{\mathbf{W}}_i$ for $i = 0, 1, \ldots, K - 1$ via an eigenvalue decomposition of $\hat{\mathbf{R}}_i$. The estimation performance depends on the accuracy of the estimated covariance matrix. The quality of the estimated covariance matrix is fully determined by the number of snapshots and the SNR. It is possible to reduce some error terms in the matrix $\hat{\mathbf{D}}(\tilde{\theta})$ by using subspace projections. To temper the estimation error, define a projection matrix

$$\mathbf{P}_i(\tilde{\theta}) = \mathbf{I} - \mathbf{a}_i(\tilde{\theta})\mathbf{a}_i^H(\tilde{\theta})\left\{\mathbf{a}_i^H(\tilde{\theta})\mathbf{a}_i(\tilde{\theta})\right\}^{-1} \quad (3.57)$$

that projects onto the null space of $\mathbf{a}_i(\tilde{\theta})$. A more robust $\hat{\mathbf{D}}(\tilde{\theta})$ matrix is formed by replacing the term $\mathbf{U}_i(\tilde{\theta})$ by

$$\mathbf{U}_i'(\tilde{\theta}) = \mathbf{P}_i(\tilde{\theta})\mathbf{U}_i(\tilde{\theta}) \quad (3.58)$$

By this modification, one of the estimation error terms that could affect the rank of $\mathbf{D}(\tilde{\theta})$ is removed. It is reasonable to assume that the estimation error for the target subspace at f_0 is uncorrelated with the estimation error for the noise subspace at the other frequency bins [24]. As a result, the overall noise is reduced, after including the projection matrix. As a result, we expect that the introduction of the projection methods will improve the noise robustness of the TOPS algorithm, which exploits the rank deficiency of the $\mathbf{D}(\tilde{\theta})$ matrix when $\tilde{\theta}$ corresponds to a true DOA. See [10, 20] for proof.

Since the signal and noise subspaces have to be estimated from sample covariance matrices, $\mathbf{D}(\tilde{\theta})$ is rarely rank-deficient. However, we expect that when $\tilde{\theta}$ corresponds to a DOA, then the minimum singular value of $\mathbf{D}(\tilde{\theta})$ will be near zero, and the condition number $\mathbf{D}(\tilde{\theta})$ will be high [25]. Empirical evidence indicates that the smallest singular value is a better indicator for approximate rank-deficiency. Therefore, we use this to indicate a likelihood that the source DOA corresponds to $\tilde{\theta}$. The likelihood function is

$$f(\tilde{\theta}) = \frac{1}{\sigma_{\min}(\tilde{\theta})} \quad (3.59)$$

where $\sigma_{\min}(\tilde{\theta})$ is the smallest singular value of $\mathbf{D}(\tilde{\theta})$. The DOA estimates are $\tilde{\theta}$ at peaks of (3.59).

3.3.3 Performance Assessment

This section evaluates the performance of four wideband DOA methods through computer simulation. Among those tested, CSSM and WAVES are coherent methods, IMUSIC is an incoherent method, and the fourth method is TOPS. Additional simulation results are available in [20]. The test signal is a wideband signal generated as a sum of complex exponentials

$$s_i(t) = b_i(t) \sum_{n=0}^{N_f-1} \exp\{j2\pi f_n t + \mu_n\} \qquad (3.60)$$

where $b_i(t)$ is a complex circular Gaussian random variable, and μ_n is a uniformly distributed random variable over $[0, 2\pi)$. The signal $s_i(t)$ represents a sum of sinusoids with random phase, representing N_f different frequencies. The magnitude $b_i(t)$ is assumed to vary slowly in time, so that the sampled composite covariance matrix formed by the DFT and (3.12) is valid.

Figure 3.2 shows examples of the TOPS likelihood surfaces for one-dimensional and two-dimensional arrays. For the one-dimensional case shown in Figure 3.2(a), a 10-sensor ULA is searching over azimuth for target sources. Three sources are located at 8°, 33°, and 37°. Clearly, TOPS can resolve the two closely spaced sources. For the two-dimensional case, a 7-sensor circular array is searching over azimuth θ and elevation ϕ. The DOA is the pair (θ, ϕ), and the likelihood surface shown in Figure 3.2(b) is parameterized by that pair. Two sources are impinging the two-dimensional array at DOAs of (10°, 20°) and (19°, 15°). It is clear from Figure 3.2(b) that TOPS is able to accurately estimate both the azimuth and elevation angles.

A further comparison of the three wideband methods and TOPS was conducted for the one-dimensional case using the 10-element ULA. Again, three sources are located at 8°, 33°, and 37°. Two hundred Monte Carlo runs were performed to obtain meaningful performance statistics (see Figure 3.3). For the coherent methods, RSS focusing matrices [16] were used. The focusing angles were determined by the strategy in [16], where Capon's method was employed as the initial estimator [21]. Five frequency bins are used for IMUSIC and TOPS. The coherent methods processed 13 frequency bins. The reference frequency bin f_1 for TOPS was selected as the bin where the difference between the smallest signal eigenvalue and the largest noise eigenvalue is maximized. CSSM and WAVES use the largest power frequency bin for Capon's method in the focusing process.

Figure 3.3 summarizes the results for the simulations. Figure 3.3(a) plots the sum of the root mean square (RMS) errors of the three sources versus SNR, and Figure 3.3(b) shows the probability of resolution versus SNR. The probability of resolution denotes the probability that the two closest sources (i.e., 33° and 37° in these simulations), are resolved. Note that the RMS error was calculated from the runs in which all three signals were resolved. This means that the number of

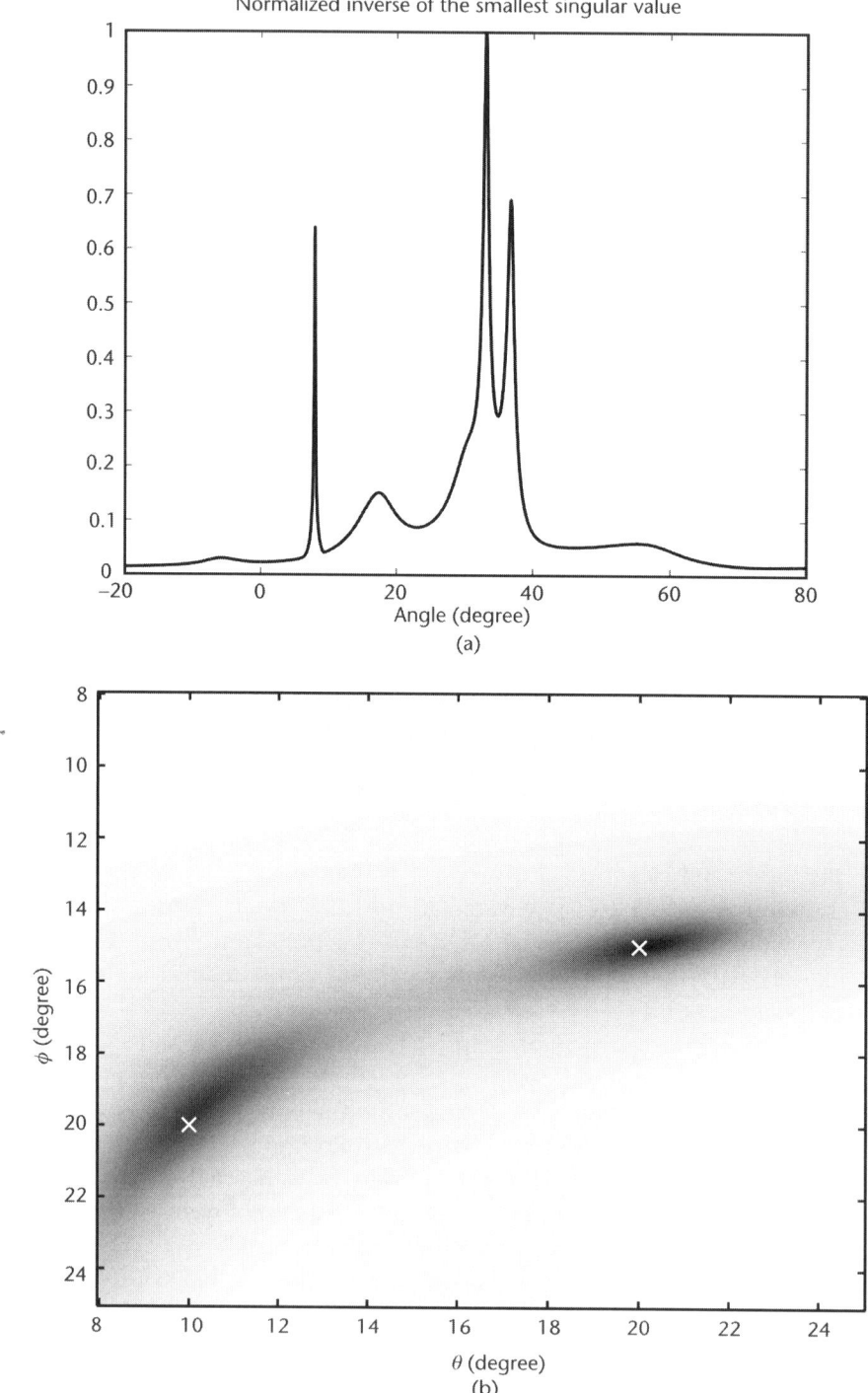

Figure 3.2 The likelihood surface of the TOPS estimator for: (a) 10-sensor ULA with three signals present at 8°, 33°, and 37°; and (b) 7-sensor two-dimensional circular array with two signals at [10°, 20°] and [19°, 15°]. Note that two-dimensional arrays can estimate both the azimuth (θ) and elevation (ϕ).

3.3 TOPS

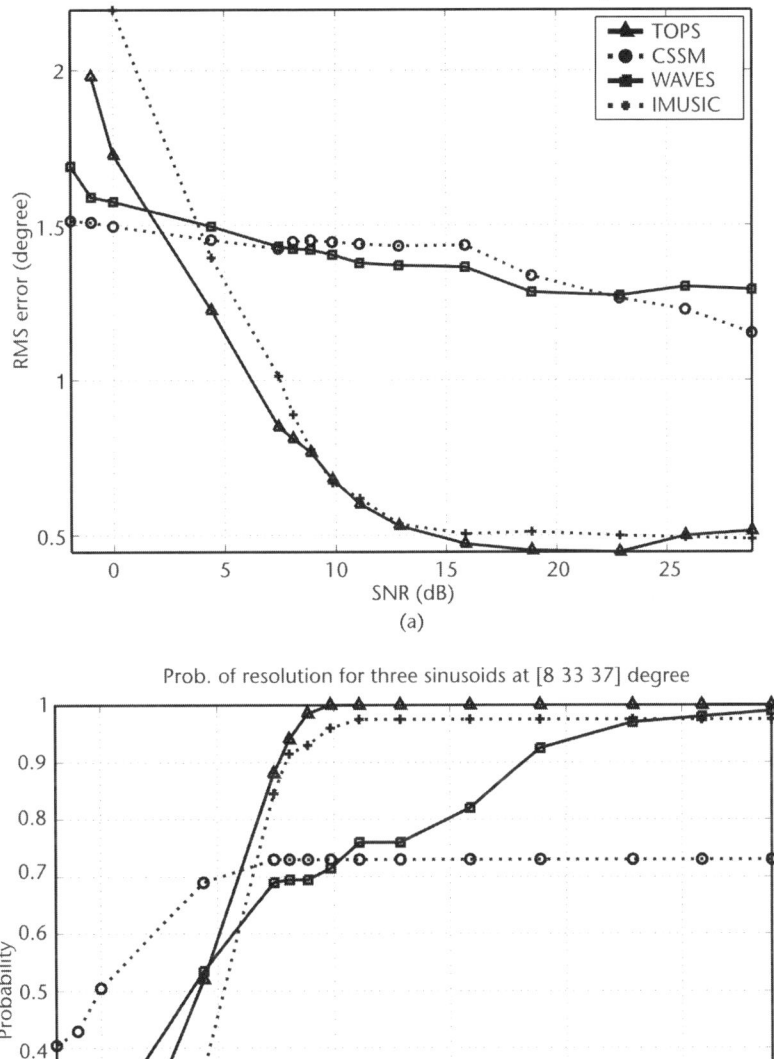

Figure 3.3 Performance comparison using a Monte Carlo simulation for a 10-element ULA, with three targets at 8°, 33°, and 37°. (a) RMS error versus SNR, and (b) probability of resolution versus SNR.

results used in RMS error calculation is different for each of the methods, due to the difference in the probability of resolution. The performance of the two coherent methods is very similar. For high SNR, IMUSIC and TOPS outperformed the coherent methods in terms of both RMS error and probability of resolution. However, as the SNR decreases, the probability of resolution for TOPS and IMUSIC drops dramatically and the RMS error rises exponentially. When the SNR is lower than 0 dB, the two coherent methods work better than TOPS and IMUSIC. While the performance of TOPS and IMUSIC are similar at high SNR, as the SNR degrades, the performance degradation of IMUSIC is faster than that of TOPS. As a result, in the middle range of SNRs (approximately from 3 to 10 dB), TOPS has the best performance.

3.3.4 Modification of TOPS for Localization

Wideband DOA estimation methods are modified for other problems, such as localization. In this section, we discuss the application of seismic sensors for landmine detection and localization [3, 25]. The seismic sensors are configured as an array that acts as both a transmitter to excite the ground and as a receiver to measure the seismic waves that reflect from objects such as landmines. In this application, the location of the landmines is estimated using subspace DOA methods because the received signal is modeled as an array manifold that depends on the landmine locations. In the experimental setup considered below, the transmitters generate a differentiated Gaussian signal, which is the wideband signal of interest. In previous work, only narrowband estimation techniques are applied over a single frequency bin of data [3, 25]. Thus, there is an opportunity to employ wideband localization techniques in an effort to improve the estimation of landmine position. In this section, we describe how TOPS and IMUSIC are applied to this problem.

Both MUSIC and TOPS depend on knowing the exact form of the array manifold vector. For localization, the array manifold vector for seismic signal at frequency f is

$$\mathbf{g}(f, \vec{r}) = [G(f, z_0) \quad \ldots \quad G(f, z_{M-1})]^T \qquad (3.61)$$

where $\vec{r} = (x, y)$ is the target's location, z_i is the distance between $\vec{r}$ and the ith receiver, and $G(\cdot)$ is the Green's function of the medium. The simplified array manifold approximation is

$$\mathbf{g}(f, \vec{r}) \cong \left[\frac{1}{k(f)z_0} e^{jk(f)z_0} \quad \ldots \quad \frac{1}{k(f)z_{M-1}} e^{jk(f)z_{M-1}} \right]^T \qquad (3.62)$$

where $k(f)$ is the frequency-dependent wavenumber. This form still contains an exponent, so TOPS and IMUSIC are modified to work with this type of array manifold. More details about the localization form of TOPS are found in [14].

Figure 3.4 shows the calculated likelihood surfaces (versus x-y location) for TOPS and IMUSIC, when the scene contains a single 9-cm diameter landmine buried at a depth of 2 cm. The raw data was collected using 15 receivers and

3.4 Conclusions

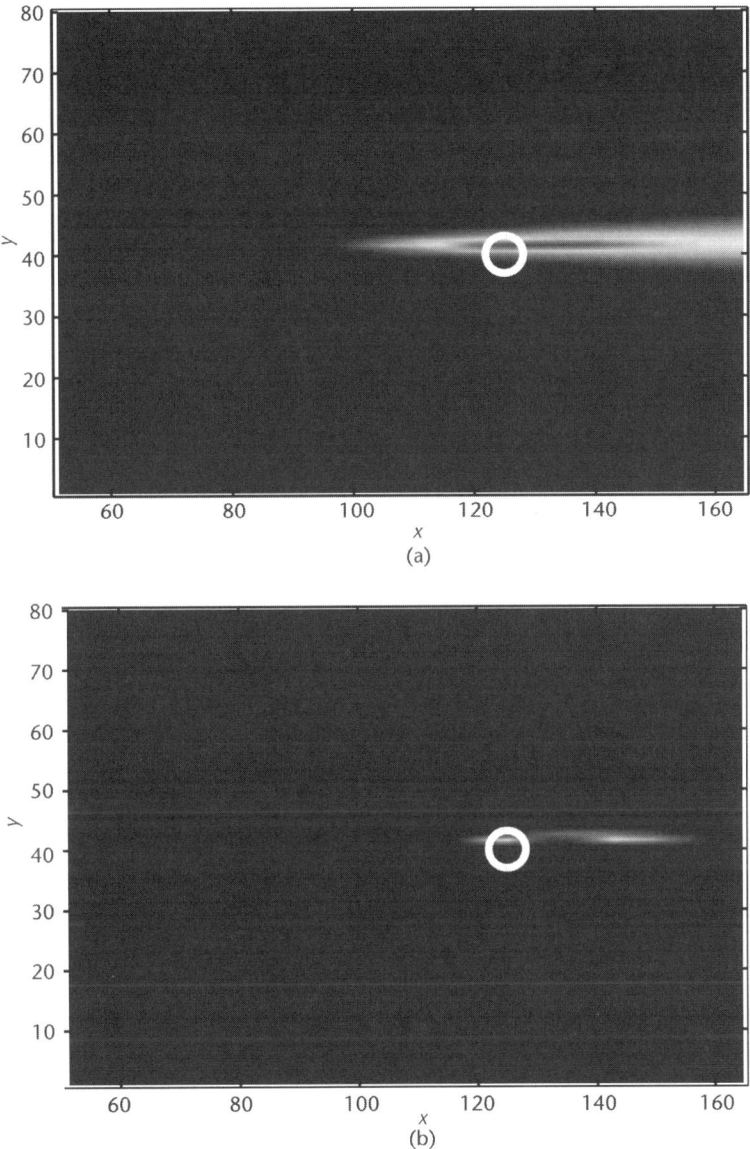

Figure 3.4 Likelihood surfaces versus (x, y) for two wideband DOA methods for landmine detection: (a) IMUSIC, and (b) TOPS. The white circles indicate the true location and size of the buried mines.

6 transmitters. When comparing TOPS to IMUSIC, both the downrange resolution and the cross range resolution are higher for TOPS.

3.4 Conclusions

In this chapter, DOA estimation methods for wideband sources are discussed. Wideband signals differ from narrowband signals in the sense that the time-delay is not approximated as a phase-shift. This complicates the DOA estimation methods, since most of the narrowband techniques exploit the phase-shift to estimate DOAs. Many wideband methods perform narrowband decomposition of the wideband sources. Methods that use those decomposed signals independently are called the incoherent methods. Since the decomposed signal is regarded as narrowband, any narrowband technique is applicable. On the other hand, there are methods that find coherent statistics out of all the decomposed signals. Although these coherent methods outperform the incoherent methods in low SNR, they require initial DOAs that are called focusing angles.

We propose a new method that might be classified as a *noncoherent* method [10, 20, 26, 27]. The new method simultaneously uses all the decomposed signals. Unlike incoherent methods, the new method is not a simple average of narrowband estimates from each band. Unlike the coherent methods, the new method does not attempt to align sampled covariance matrices over each frequency bin. The new method does not require a priori knowledge to initialize the focusing matrix. It looks for consistency between the signal subspace and noise subspaces for different frequency bands. Computer simulations showed the high performance of the new method over previous methods in the mid-to-high SNR range. Seismic data processing has shown one possible application of the new method.

Acknowledgments

This chapter was prepared through collaborative participation in the Advanced Sensors and the Communications and Networks Collaborative Technology Alliances sponsored by the U.S. Army Research Laboratory under Cooperative Agreements DAAD19-01-2-008 and DAAD19-01-2-0011, respectively. The U.S. government is authorized to reproduce and distribute reprints for government purposes notwithstanding any copyright notation thereon. The views and conclusions contained in this chapter are those of the authors and should not be interpreted as presenting the official policies either express or implied of the Army Research Laboratory or the U.S. government.

References

[1] Garmatyuk, D., and R. Narayanan, "ECCM Capabilities of An Ultrawide-Band Bandlimited Random Noise Imaging Radar," *IEEE Trans. Aerosp. Electron. Syst.*, Vol. 38, No. 4, October 2002, pp. 1243–1255.

[2] Lake, D., "Harmonic Phase Coupling for Battlefield Acoustic Target Identification," *Proc. IEEE Acoustics, Speech, and Signal Processsing (ICASSP'98)*, Seattle, WA, May 1998.

[3] Stoica, P., and R. Moses, *Introduction to Spectral Analysis*, Upper Saddle River, NJ: Prentice Hall, 1997.

[4] Schmidt, R., "Multiple Emitter Location and Signal Parameter Estimation," *IEEE Trans. on Antennas and Propagation*, Vol. AP-34, No. 3, March 1986, pp. 276–280.

[5] Roy, R., and T. Kailath, "ESPRIT-Estimation of Signal Parameters Via Rotational Invariance Techniques," *IEEE Trans. on Signal Processing*, Vol. 37, No. 7, July 1989, pp. 984–995.

[6] Zatman, M., "How Narrow Is Narrowband?" *IEE Proc.-Radar, Sonar Navig.*, Vol. 145, No. 2, April 1998, pp. 85–91.

[7] Su, G., and M. Morf, "The Signal Subspace Approach for Multiple Wideband Emitter Location," *IEEE Trans. on Acoustics, Speech, and Signal Processing*, Vol. ASSP-31, No. 6, December 1983, pp. 1502–1522.

[8] Buckley, K., and L. Griffiths, "Broad-Band Signal-Subspace Spatial-Spectrum (BASS-ALE) Estimation," *IEEE Trans. on Acoustics, Speech, and Signal Processing*, Vol. 36, No. 7, July 1988, pp. 953–964.

[9] Agrawal, M., and S. Prasad, "Broadband DOA Estimation Using Spatial-Only Modeling of Array Data," *IEEE Trans. on Signal Processing*, Vol. 48, No. 3, March 2000, pp. 663–670.

[10] Yoon, Y.-S., L. Kaplan, and J. McClellan, "TOPS: A New DOA Estimation Method for Wideband Sources," *IEEE Trans. on Signal Processing*, submitted.

[11] Chandran, S., and M. K. Ibrahim, "DOA Estimation of Wide-Band Signals Based on Time-Frequency Analysis," *IEEE J. of Oceanic Engineering*, Vol. 24, No. 1, January 1999, pp. 116–121.

[12] Wax, M., and T. Kailath, "Spatio-Temporal Spectral Analysis by Eigen-Structure Methods," *IEEE Trans. on Acoustics, Speech, and Signal Processing*, Vol. ASSP-32, No. 4, August 1984, pp. 817–827.

[13] Wang, H., and M. Kaveh, "Coherent Signal-Subspace Processing for the Detection and Estimation of Angles of Arrival of Multiple Wide-Band Sources," *IEEE Trans. on Acoustics, Speech, and Signal Processing*, Vol. ASSP-33, August 1985, pp. 823–831.

[14] Di Claudio, E. D., and R. Parisi, "WAVES: Weighted Average of Signal Subspaces for Robust Wideband Direction Finding," *IEEE Trans. on Signal Processing*, Vol. 49, No. 10, October 2001, pp. 2179–2190.

[15] Doron, M., and A. Weiss, "On Focusing Matrices for Wide-Band Array Processing," *IEEE Trans. on Signal Processing*, Vol. 40, No. 6, June 1992, pp. 1295–1302.

[16] Hung, H., and M. Kaveh, "Focusing Matrices for Coherent Signal-Subspace Processing," *IEEE Trans. on Acoustics, Speech, and Signal Processing*, Vol. ASSP-36, No. 8, August 1988, pp. 1272–1282.

[17] Swingler, D. N, and J. Krolik, "Source Location Bias in the Coherently Focused High-Resolution Broad-Band Beamformer," *IEEE Trans. on Acoustics, Speech, and Signal Processing*, Vol. 37, No. 1, January 1989, pp. 143–145.

[18] Lee, T. -S., "Efficient Wide-Band Source Localization Using Beamforming Invariance Technique," *IEEE Trans. on Signal Processing*, Vol. 42, June 1994, pp. 1376–1387.

[19] Viberg, M., B. Ottersten, and T. Kailath, "Detection and Estimation in Sensor Arrays Using Weighted Subspace Fitting," *IEEE Trans. on Signal Processing*, Vol. 39, November 1991, pp. 2436–2449.

[20] Yoon, Y.-S., "Direction-of-Arrival Estimation for Wideband Sources Using Sensor Arrays," Ph.D. thesis, Georgia Institute of Technology, Atlanta, GA, 2004.

[21] Johnson, D., and D. Dudgeon, *Array Signal Processing: Concepts and Techniques*, Englewood Cliffs, NJ: Prentice Hall, 1993.

[22] Krolik, J., "Focused Wideband Array Processing for Spatial Spectral Estimation," Ch. 6 in *Advances in Spectrum Analysis and Processing, Vol. 2*, S. Haykin, (ed.), Upper Saddle River, NJ: Prentice Hall, 1991.

[23] Chen, C.-T., *Linear System Theory and Design*, New York: Holt, Rinehart, and Winston, 1984.

[24] Kaveh, M., and A. Barabell, "The Statistical Performance of the MUSIC and the Minimum-Norm Algorithms in Resolving Plane Waves in Noise," *IEEE Trans. on Acoustics, Speech, and Signal Processing*, Vol. ASSP-34, No. 2, April 1986, pp. 331–341.

[25] Golub, G., and C. Van Loan, *Matrix Computations*, Baltimore, MD: Johns Hopkins University Press, 1996.

[26] Alam, M., et al., "Time-Reverse Imaging for the Detection of Landmines," *Proc. of SPIE Defense and Security Symposium*, Orlando, FL, April 2004.

[27] Yoon, Y. -S., L. Kaplan, and J. McClellan, "Direction-of-Arrival Estimation of Wideband Signal Sources Using Arbitrary Shaped Multidimensional Arrays," *IEEE Int. Conf. Acoustics, Speech, and Signal Processing*, Montreal, Canada, May 2004.

CHAPTER 4
Coherent Broadband DOA Estimation Using Modal Space Processing

Thushara D. Abhayapala

4.1 Introduction

The problem of estimation of DOA of coherent broadband sources has applications in wireless communication systems, especially systems with smart antennas [1]. DOA estimation and adaptive beamforming are the main issues in smart antenna systems. However, in a complex multipath environment, received signals from different directions may be correlated, which prevents the application of narrowband DOA estimation techniques to estimate DOA of broadband sources. This chapter introduces a novel coherent broadband DOA estimation technique based on *modal decomposition* of wavefields.

Wang and Kaveh [2] introduced the use of focusing matrices for the purpose of CSS processing for DOA estimation of far-field wideband sources. In this method, the wideband array data are first decomposed into several narrowband components. The focusing matrices are used for the alignment of the signal subspaces of narrowband components within the bandwidth of the signals, followed by the averaging of narrowband array data covariance matrices into a single covariance matrix. Then, any signal subspace direction finding procedure, such as MUSIC [3] or its variants, maximum likelihood (ML), or minimum variance (MV), can be applied to this averaged covariance matrix to obtain the desired parameter estimates. The design of focusing matrices in CSS method requires preliminary estimates of the direction of arrivals. Further, this method is applicable to only a pair of sources. In later years, the CSS technique was further developed and refined [4, 5] to account for multiple sources, but the problem of prior information about the DOA still remained. In this chapter, we use *modal decomposition* of wavefields to propose novel focusing matrices that do not require preliminary DOA estimates and are completely independent of the signal environment.

The spatial resampling method is a technique that does not require preliminary knowledge of DOA in order to localize wideband sources. It was first introduced by Krolik and Swingler [6] and is motivated by treating the output of a discrete array as being the result of spatially sampling a continuous linear array. The same concept also is known as an interpolated array approach used in [7]. Krolik and Swingler [6] used digital interpolation methods to resample the array data. The spatial resampling method requires more than one resampling matrix to be constructed for different field-of-view or sectors. An alternative technique is suggested in this chapter using modal decomposition of wavefields. Under this technique a set of resampling matrices has been proposed, which is the same for the full field of view of the array data, unlike in the case of [7].

The application of modal decomposition of wavefields enables insights into the structure of focusing and spatial resampling matrices. It can be observed that the computational complexity of CSS and spatial resampling methods can be reduced by combining the focusing and spatial resampling matrices to form modal covariance matrices. Use 2 of modal decomposition of wavefields in any array signal processing application is termed as *modal space processing* (MSP) since it converts measured array data into *modal space*. The modal space can be viewed as a vector space with a predefined orthogonal basis set, which is the natural basis set for wavefields.

4.2 System Model

Let us consider a double-sided linear array of $2Q + 1$ sensors, located at distances x_q, $q = -Q, \ldots, 0, 1, \ldots, Q$ from the array origin, which receives signals from V sources in space. Let $\Theta = [\theta_1, \theta_2, \ldots, \theta_v, \ldots, \theta_V]$ be a vector containing bearings of each source with reference to the array axis, where θ_v is the direction of the vth source. We assume that the source signal and the noise are confined in a bandwidth of $k \in [k_l; k_u]$, where k_l and k_u are lower and upper band edges, respectively. We use wavenumber $k = 2\pi f/c$, where f is the frequency in hertz, and c is the speed of wave propagation, to represent frequency in this chapter. The signal received at each sensor is discrete Fourier transformed (DFT) into M distinct frequency bins within the design bandwidth. The array output in the mth frequency bin can be represented as:

$$z(k_m) = \sum_{v=1}^{V} a(\theta_v; k_m) s_v(k_m) + n(k_m) \tag{4.1}$$

where $s_v(\cdot)$ is the signal received from the vth source at the origin, $n(\cdot)$ is the uncorrelated noise data, and

$$a(\theta; k) = [e^{-ikx_{-Q}\cos\theta}, \ldots, e^{-ikx_Q \cos\theta}]' \tag{4.2}$$

where $[\cdot]'$ denotes the transpose operator, and $i = \sqrt{-1}$. We write (4.1) in matrix form as

$$z(k_m) = A(\Theta; k_m) s(k_m) + n(k_m) \tag{4.3}$$

for $m = 1, \ldots, M$, where

$$A(\Theta; k) = [a(\theta_1; k), \ldots, a(\theta_V; k)] \tag{4.4}$$

and

$$s(k) = [s_1(k), \ldots, s_V(k)]' \tag{4.5}$$

The objective of any DOA technique is to determine the DOA Θ from the observed data $z(k_m)$, $m = 1, \ldots, M$.

Most DOA techniques use the correlation matrix of the received data in their DOA algorithms. The correlation matrix of the observed data in the mth frequency bin is defined as

$$R_z(k_m) \triangleq E\{z(k_m)z(k_m)^H\} \tag{4.6}$$

where $[\cdot]^H$ denotes conjugate transpose operation, and $E\{\cdot\}$ is the expectation operator. Substituting (4.3) in (4.6), we have

$$R_z(k_m) = A(\Theta; k_m)R_s(k_m)A^H(\Theta; k_m) + E\{n(k_m)n(k_m)^H\} \tag{4.7}$$

where

$$R_s(k_m) \triangleq E\{s(k_m)s(k_m)^H\} \tag{4.8}$$

is the source correlation matrix. Here, we assume that the source signals and noise are uncorrelated.

4.3 Focusing Matrices for Coherent Wideband Processing

In this section, we briefly outline the focusing method. The first step following the frequency decomposition of the array data vector is to align or focus the signal space at all frequency bins into a common bin at a reference frequency by focusing matrices $T(k_m)$ that satisfy

$$T(k_m)A(\Theta; k_m) = A(\Theta; k_0) \quad m = 1, \ldots, M \tag{4.9}$$

where $k_0 \in [k_l; k_u]$ is some reference frequency, and $A(\Theta; k)$ is the direction matrix defined by (4.4). Applying the M focusing matrices to the respective array data vectors (4.3) gives the following focused array data vector,

$$T(k_m)z(k_m) = A(\Theta; k_0)s(k_m) + T(k_m)n(k_m) \quad m = 1, \ldots, M$$

Then the focused and frequency averaged data covariance matrix can be defined by

$$R \triangleq \sum_{m=1}^{M} E\{T(k_m)z(k_m)[T(k_m)z(k_m)]^H\} \tag{4.10}$$

$$= \sum_{m=1}^{M} T(k_m)E\{z(k_m)z^H(k_m)\}T^H(k_m)$$

We use (4.6), (4.7), and (4.9) to get

$$R = A(\Theta; k_0)\overline{R}_s A^H(\Theta; k_0) + R_{\text{noise}} \quad (4.11)$$

where

$$\overline{R}_s \triangleq \sum_{m=1}^{M} R_s(k_m) \quad (4.12)$$

and the transformed noise covariance matrix

$$R_{\text{noise}} \triangleq \sum_{m=1}^{M} T(k_m) E\{n(k_m) n^H(k_m)\} T^H(k_m) \quad (4.13)$$

The focused data covariance matrix (4.11) is now in a form in which almost any narrowband direction finding procedure may be applied. Here, we apply the MV method of spatial spectral estimation [8] to the frequency averaged data covariance matrix R.

Several methods of forming focusing matrices have been suggested in the literature. The focusing methods of [2, 4, 5, 9] require preliminary DOA estimates in order to construct the focusing matrices. This constitutes a severe disadvantage in practical applications, since it leads to biased DOA estimates. In Section 4.5, we show how to design focusing matrices without preliminary DOA estimates, using modal decomposition of wavefields.

4.4 Spatial Resampling Method

Spatial resampling [6] is another method used to focus the wideband array data to a single frequency, so that existing narrowband techniques may be used to estimate the DOA. The basic idea of spatial sampling is outlined below.

Suppose we have a separate uniform array with half wavelength spacing for each frequency bin with the same effective array aperture in terms of wavelength. Thus, for M frequencies, there are M arrays, and the sensor separation of the mth array is $\lambda_m/2$, where $\lambda_m = 2\pi/k_m$. If each array has $2Q + 1$ sensors, then the aperture length is the same for all frequencies in terms of corresponding wavelength. Then the mth array steering vector for far-field sources is given by

$$\begin{aligned} a(\theta; k_m) &= \left[e^{i\pi Q \cos\theta}, \ldots, e^{i\pi \cos\theta}, 1, e^{-i\pi \cos\theta}, \ldots, e^{-i\pi Q \cos\theta} \right] \\ &= a(\theta) \quad m = 1, \ldots, M \end{aligned}$$

The steering vectors of all arrays are equal, and hence, from (4.4), the DOA matrices of all arrays are the same:

$$A^{(m)}(\Theta; k_m) = A(\Theta) \quad m = 1, \ldots, M \quad (4.14)$$

where $A^{(m)}(\Theta; k_m)$ is the DOA matrix of the mth subarray. Hence, if we have M arrays for each frequency bin with the same aperture, then their covariance matrices can be averaged over frequency without losing DOA information. The average covariance matrix can then be used with existing narrowband DOA techniques to estimate DOA angles.

However, it is not actually practical to have a separate array for each frequency. This problem can be overcome by having a single array and using the received array data to form the array data for M virtual arrays by interpolation/extrapolation of the received array data. This is tantamount to constructing a continuous sensor using the received array data and resampling it. There are several methods reported in the literature. In [7] the field of view of the array is divided into several sectors, and a different interpolation matrix is calculated for each sector using a least-squares fit.

4.5 Modal Decomposition

At the physical level, sensor array signal processing is characterized by the classical wave equation. The general solution to the wave equation can be decomposed into *modes* that are orthogonal basis functions of the spatial coordinates. These modes exhibit interesting mathematical properties, and form a useful basis set to analyze and synthesize an arbitrary wavefield, a response of an array of sensors, or a spatial aperture. Modes of a two- and three-dimensional wavefield are called cylindrical and spherical harmonics, respectively. In this section, we show how to decompose any wavefield due to far-field sources.

In the antenna literature, these modes have been used to synthesize antenna shapes [10, 11], to represent electromagnetic fields radiated by circular antennas [12], and to compute antenna couplings [13]. Recently, they have been used for near-field broadband design [14, 15], directional soundfield recording [16] and reproduction [17], spatial wireless channel characterization [18, 19], and capacity calculation of multiple-input, multiple-output (MIMO) channels [20].

4.5.1 Linear Arrays

The term $e^{-ikx\cos\theta}$ represents the phase delay of a signal received at a point located at a distance x from the origin for a signal from the direction θ. This term is an integral part of any array processing algorithm. Here, we separate the distance dependency and angle dependency into two terms. This separation is very useful in any algorithm, since it effectively separates the sensor geometry from the signal direction.

Using the Jacobi-Anger expansion [21], we write

$$e^{-ikx\cos\theta} = \sum_{n=0}^{\infty} i^n(2n+1)j_n(kx)P_n(\cos\theta) \quad (4.15)$$

where n is a nonnegative integer to index modes, $j_n(\cdot)$ is the spherical Bessel function, and $P_n(\cdot)$ is the Legendre function. We have the following comments on this expansion:

1. The series expansion (4.15) gives an insight into the spatial wavefield along a linear array.
2. Equation (4.15) can be viewed as a Fourier series–type expansion of a function, where $P_n(\cos\theta)$, $n = 0, \ldots, \infty$ are the orthogonal basis set.
3. Observe that in each term of the series, the arrival angle θ dependency is separated out from the sensor location x and the frequency k. Therefore, we may use the above expansion to write the array DOA matrix $A(\Theta; k)$ as a product of two matrices, one depending on DOA angles and the other depending on frequency and sensor locations.
4. However, note that (4.15) has an infinite number of terms. Thus, the usefulness of (4.15) depends on the number of significant terms needed in any numerical evaluation.

4.5.2 Modal Truncation

For a finite aperture array with a finite bandwidth signal environment, (4.15) can be safely truncated by finite number of terms (e.g., N) without generating significant modeling errors. We show this in the following. Figure 4.1 shows plots of a few spherical Bessel functions $j_n(\cdot)$ against its argument. We can observe from Figure 4.1 that for a given kx, the function $j_n(kx) \to 0$ as n becomes large. This observation is supported by the following asymptotic form [22]

$$j_n(kx) \approx \frac{(kx)^n}{1 \cdot 3 \cdot 5 \ldots (2n+1)} \quad \text{for } kx \ll n \qquad (4.16)$$

Therefore, we can notice that the factor $(2n+1)j_n(kx_q)$ in (4.15) decays as n grows larger beyond $n = kx_q$. Suppose that the minimum frequency of the signal

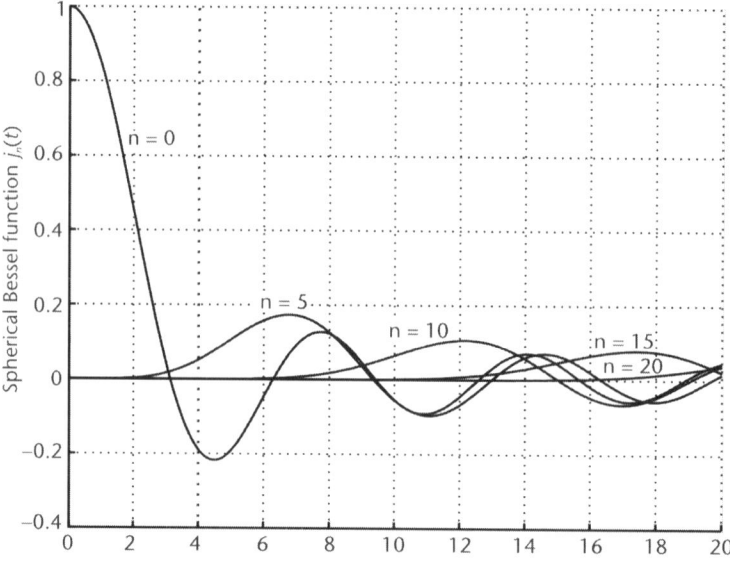

Figure 4.1 Spherical Bessel functions of order $n = 0, 5, 10, 15,$ and 20.

band is k_l. Then, we can truncate (4.15) to N terms if $N > k_l x_Q$, where x_Q is the distance to the Qth sensor (the maximum array dimension). It is difficult to derive an analytical expression for N, but a convenient rule of thumb [23] is $N \sim 2k_l x_Q$. More recent work [18] shows that (4.15) can be truncated by $N = \lceil kex/2 \rceil$ terms with negligible error.

4.6 Modal Space Processing

In this section, we use the modal decomposition developed in the previous section to (1) design focusing matrices, (2) spatial resampling matrices, and (3) introduce a novel modal space processing DOA technique.

4.6.1 Focusing Matrices

We use modal analysis techniques to propose novel focusing matrices that do not require preliminary DOA estimates and are completely independent of the signal environment. Here we only consider a linear (possibly nonuniform) array but it may be generalized to arbitrary array configurations.

We substitute the first $N + 1$ terms of (4.15) into (4.2) and thus write the array steering vector for far-field sources as

$$a(\theta; k) = J(k) \begin{bmatrix} P_0(\cos\theta) \\ \vdots \\ P_N(\cos\theta) \end{bmatrix} \quad (4.17)$$

where

$$J(k) = \begin{bmatrix} i^0(2\cdot 0 + 1)j_0 kx_{-Q} & \cdots & i^N(2N + 1)j_N kx_{-Q} \\ \vdots & \ddots & \vdots \\ i^0(2\cdot 0 + 1)j_0 kx_Q & \cdots & i^N(2N + 1)j_N kx_Q \end{bmatrix} \quad (4.18)$$

We use (4.17) in (4.4) to write the array DOA matrix for far-field signal environment as

$$A(\Theta; k) = J(k)P(\Theta) \quad (4.19)$$

where the $(N + 1) \times V$ matrix

$$P(\Theta) = \begin{bmatrix} P_0(\cos\theta_1) & \cdots & P_0(\cos\theta_V) \\ \vdots & \ddots & \vdots \\ P_N(\cos\theta_1) & \cdots & P_N(\cos\theta_V) \end{bmatrix} \quad (4.20)$$

The $(2Q + 1) \times (N + 1)$ matrix $J(k)$ depends on the frequency k and the sensor locations, and is independent of the DOA of the signals. Suppose $(2Q + 1) > (N + 1)$, and $J(k)$ has full rank $N + 1$ if the sensor locations are chosen appropriately. With this assumption, and using (4.19), we can propose a set of focusing matrices $T(k_m)$ given by

$$T(k_m) = J(k_0)[J^H(k_m)J(k_m)]^{-1}J^H(k_m) \quad m = 1, \ldots, M \quad (4.21)$$

which satisfies the focusing requirement (4.9); recall that k_0 is the reference frequency.

The major advantage of the focusing matrices (4.21) over the existing methods is that these matrices do not need preliminary DOA estimates and accurately focus signal arrivals from all directions. Note that these matrices are fixed for a given array geometry and frequency band of interest. Thus, they can be calculated in advance in time critical applications, such as smart antennas, to save computational time.

4.6.2 Spatial Resampling Matrices

In this section, we show how to use the modal decomposition to find a transformation matrix to calculate array data for M virtual arrays for the spatial resampling method described in Section 4.4. Sensor locations for the real array can be arbitrary on a line—there is no requirement for it to be a uniformly spaced array. From (4.19), the real array DOA matrix in the mth frequency bin is given by

$$A(\Theta; k_m) = J(k_m)P(\Theta) \quad (4.22)$$

and the DOA matrix of the mth virtual array at frequency k_m would be

$$A^{(m)}(\Theta; k_m) = J^{(m)}(k_m)P(\Theta) \quad (4.23)$$

where, from (4.18), and with $k_m x_q = q\pi$,

$$J^{(m)}(k_m) = \begin{bmatrix} i^0(2 \cdot 0 + 1)j_0(-\pi Q) & \ldots & i^N(2N + 1)j_N(-\pi Q) \\ \vdots & \ddots & \vdots \\ i^0(2 \cdot 0 + 1)j_0(\pi Q) & \ldots & i^N(2N + 1)j_N(\pi Q) \end{bmatrix}$$

$$= \bar{J} \quad m = 1, \ldots, M$$

which is a constant matrix, independent of m and k_m. Therefore, we can write

$$A^{(m)}(\Theta; k_m) = \bar{J}P(\Theta) \quad (4.24)$$
$$= A(\Theta) \quad m = 1, \ldots, M$$

which is the same for all frequency bins. We need to design the set of spatial resampling matrices $T(k_m)$, $m = 1, \ldots, M$, such that

$$A^{(m)}(\Theta; k_m) = T(k_m)A(\Theta; k_m) \quad m = 1, \ldots, M \qquad (4.25)$$

By substituting (4.22) and (4.24) into (4.25) and using the pseudo inverse of $J(k_m)$, we obtain the least-square solution

$$T(k_m) = \bar{J}[J^H(k_m)J(k_m)]^{-1}J^H(k_m) \quad m = 1, \ldots, M \qquad (4.26)$$

These spatial resampling matrices, which act as focusing matrices, can be used to align the array data in different frequency bins, so that narrowband DOA techniques can be applied. Similar to the focusing matrices (4.21), these spatial resampling matrices (4.26) do not require preliminary DOA estimation and depend only on the array geometry and the frequency. They are independent of the angle of arrival and fixed for full field of view.

4.7 Modal Space Algorithm

Observe that the proposed focusing matrices (4.21) and the spatial resampling matrices (4.26) have a common (generalized inverse) matrix factor

$$G(k_m) \triangleq [J^H(k_m)J(k_m)]^{-1}J^H(k_m) \quad m = 1, \ldots, M \qquad (4.27)$$

and only differ by the frequency independent factors $J_0(k_0)$ and $\bar{J}$. Note that from (4.19),

$$G(k_m)A(\Theta; k) = P(\Theta) \quad m = 1, \ldots, M \qquad (4.28)$$

$G(k_m)$ transforms the array DOA matrix into a frequency invariant DOA matrix. Therefore, we can use $G(k_m)$ instead of $T(k_m)$ to align the broadband array data to form a frequency averaged covariance matrix. Intuitively, one can say that the matrices $G(k_m)$ transform the $2Q + 1$ array data vector $z(k_m)$ into a $N + 1$ *modal data* vector in *modal space*. Now we can estimate the frequency averaged *modal covariance matrix* as

$$\hat{R} = \sum_{m=1}^{M} G(k_m)z(k_m)z^H(k_m)G^H(k_m) \qquad (4.29)$$

and the MV spectral estimate

$$\hat{Z}(\theta) = \frac{1}{\begin{bmatrix} P_0(\cos\theta) \\ \vdots \\ P_N(\cos\theta) \end{bmatrix} \hat{R}^{-1}[P_0(\cos\theta), \ldots, P_N(\cos\theta)]} \qquad (4.30)$$

Comments:

1. The MSP method involves less computation compared to the other two methods since the modal space has less dimensions ($N + 1$) than the signal subspace ($2Q + 1$).
2. As for the other two methods, the modal space method does not require preliminary DOA estimates.
3. One can consider the modal space method as a superset of focusing matrices and spatial resampling methods.
4. Given the frequency averaged modal covariance matrix (4.29), any other narrowband DOA technique, such as MUSIC or its variants, ML can be used.
5. This method can be extended to find the range and angle of near-field sources by using the modal expansion of a spherical wavefront [21]. Readers are referred to [15] for details.

4.8 Simulation

In this section, the simulation results have been presented in order to demonstrate the effectiveness of the MSP method. A linear array of 19 nonuniformly spaced sensors has been used for the MSP technique. The use of a nonuniformly spaced sensor array for broadband application has been discussed in [14]. The sensor spacing is kept uniform while performing the simulation of examples that follow the algorithms suggested in past literature [2, 4]. These simulations are presented in this section for comparison of results. The source signal and the noise are stationary zero-mean white Gaussian processes. Noise at each sensor is independent of the noise at any other sensor. The signal received at each sensor is discrete Fourier transformed to get 33 uniformly spaced narrowband frequency bins within the desired bandwidth. For each trial, 64 independent snapshots are generated for every frequency bin. The frequency averaged modal covariance matrix is calculated using (4.29). The sources are then localized by using MV direction finding procedure (4.30) as implemented for narrowband source localization.

4.8.1 A Group of Two Sources

The signal environment consists of two completely correlated sources at angles $\Theta = [38° \ 43°]$. Let $s_1(t)$ be the source at 38°, and the source at 43° is a delayed version of $s_1(t)$ and is given by $s_2(t) = s_1(t - t_o)$, with $t_o = 0.125$ seconds, or, equivalently in the frequency domain, $s_2(f) = s_1(f)e^{-jft_o}$. Here, $s_1(f)$ is the Fourier transformed signal of $s_1(t)$. The signal-to-noise ratio is 10 dB. The two signals $s_1(t)$ and $s_2(t)$ can be viewed as multipath signals from a single source.

The signals used lie within a bandwidth of 40 Hz with mid-band frequency at 100 Hz. This gives a lower band edge ($f_l = 80$ Hz) to upper band edge ($f_u = 120$ Hz) ratio of 2:3. All the signal parameters are kept identical to those described in [2]. The signals are captured by a linear array of 19 sources. Figure 4.2 shows the spectral estimate obtained using MSP. The vertical lines indicate the correct direction of arrival of the sources. For comparison, the results obtained using the method

4.8 Simulation

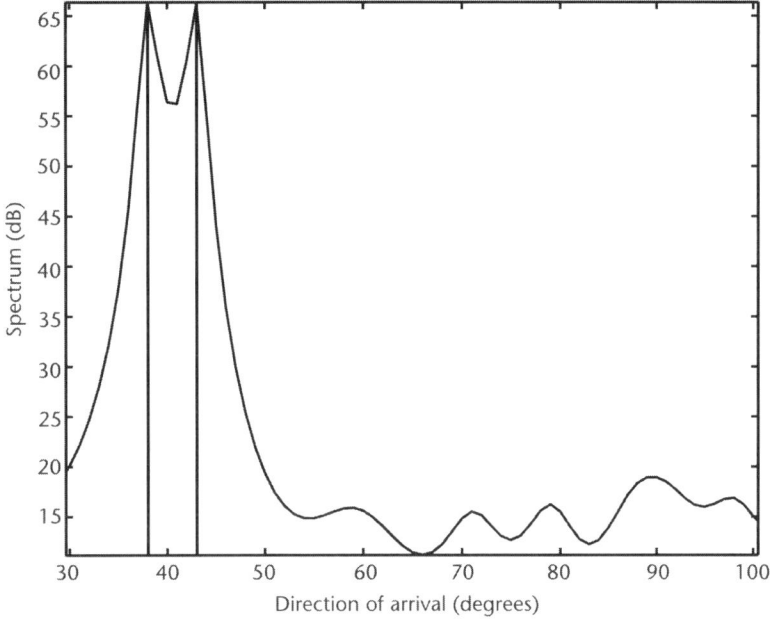

Figure 4.2 The estimated spatial spectrum of the correlated sources using the MSP method.

described in [2] are shown in Figure 4.3. A preliminary angle estimate of 40.4° has been necessary to correctly estimate the direction of arrivals using the later technique, whereas no prior knowledge of angles is required for the MSP technique. The graphs reveal that both processes localize the sources with fine accuracy.

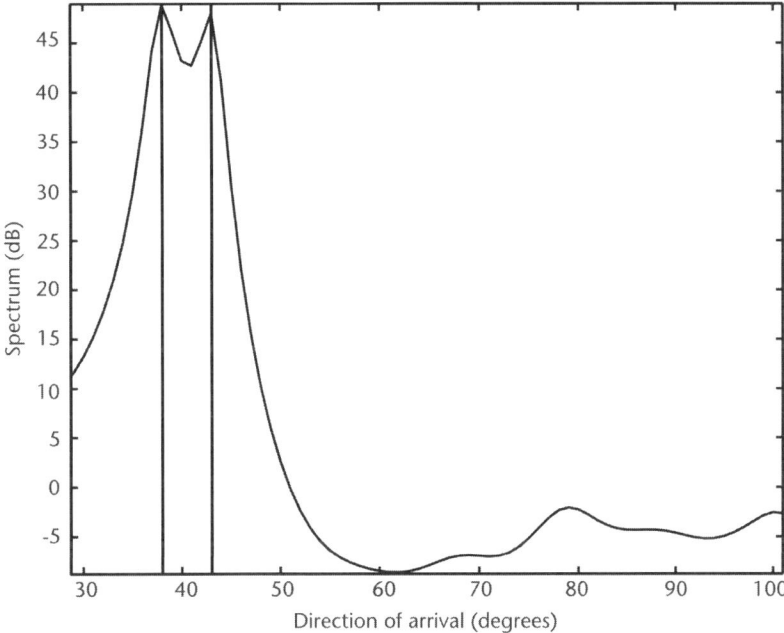

Figure 4.3 The estimated spatial spectrum of the correlated sources using the algorithm of [2].

However, a focusing angle of 53° in the case of [2] will result in Figure 4.4, which cannot resolve the true direction of arrivals.

4.8.2 Three Groups of Five Sources

The number of sources are now increased to five, with bearings Θ = [53°, 58°, 98°, 103°, 145°]. Complete correlation exists between the first and second sources. A frequency band of f = [80:120] Hz is used to compare the results with those obtained using the focusing matrix proposed in [4]. Fifteen independent trials were carried out that showed similar results.

Figure 4.5 shows one realization obtained by using MSP with MV spectral estimate. Here, the number of modes used is N = 15. Reducing the value of N degrades the performance of the procedure, while increasing its value produces no appreciable improvement. The number of sensors is 19, and they are nonuniformly spaced. All the sources are clearly detected without any prior knowledge of source environment. These results can be compared with Figure 4.6, which shows the spatial spectrum of multigroup sources using the technique described in [4]. However, prior knowledge of source directions is required by this technique, and preliminary angle estimates used for this example is β = [53°, 55°, 59°, 96.7°, 100.5°, 104.3°, 144°].

The above simulation is performed for wider bands of frequency of bandwidth [300:3,000] Hz, and the results show that MSP produces better results (see Figure 4.7) as compared to the technique proposed in [4] (see Figure 4.8). A total number of 45 sensors and 55 frequency bins are used in the simulation.

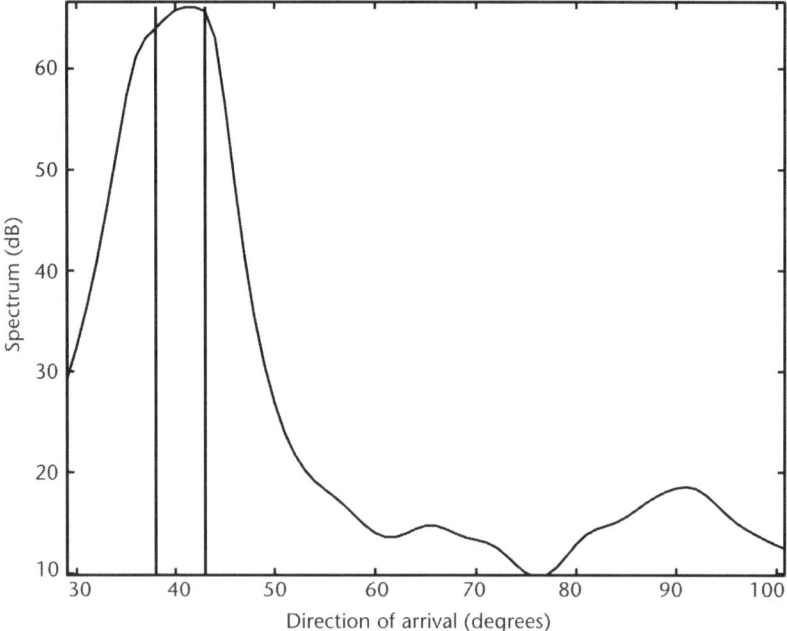

Figure 4.4 The estimated spatial spectrum of the correlated sources using prior angle estimation of 53°.

4.8 Simulation

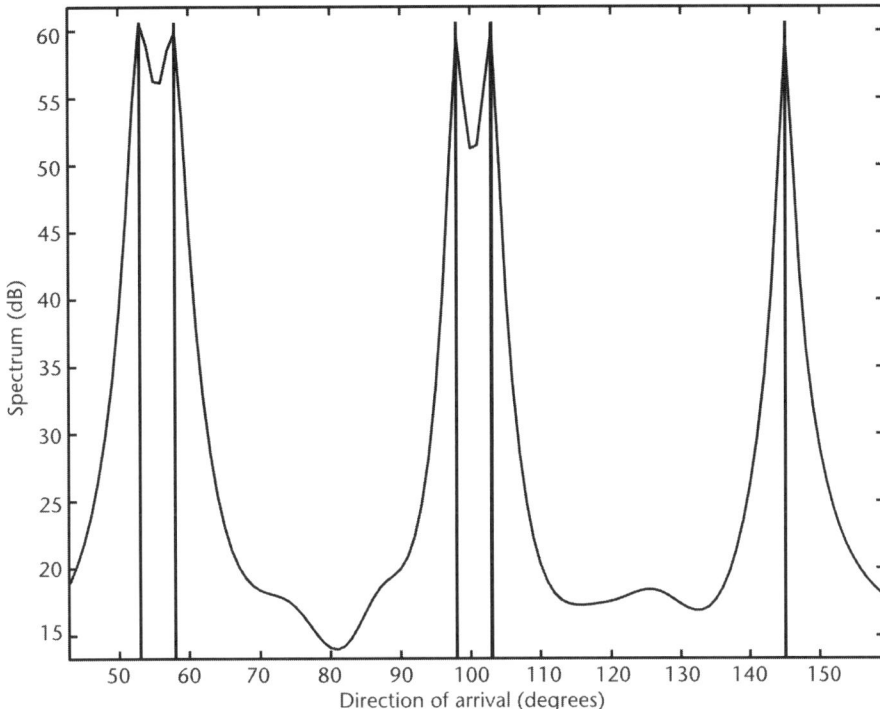

Figure 4.5 The estimated spatial spectrum of the multigroup sources using the MSP method.

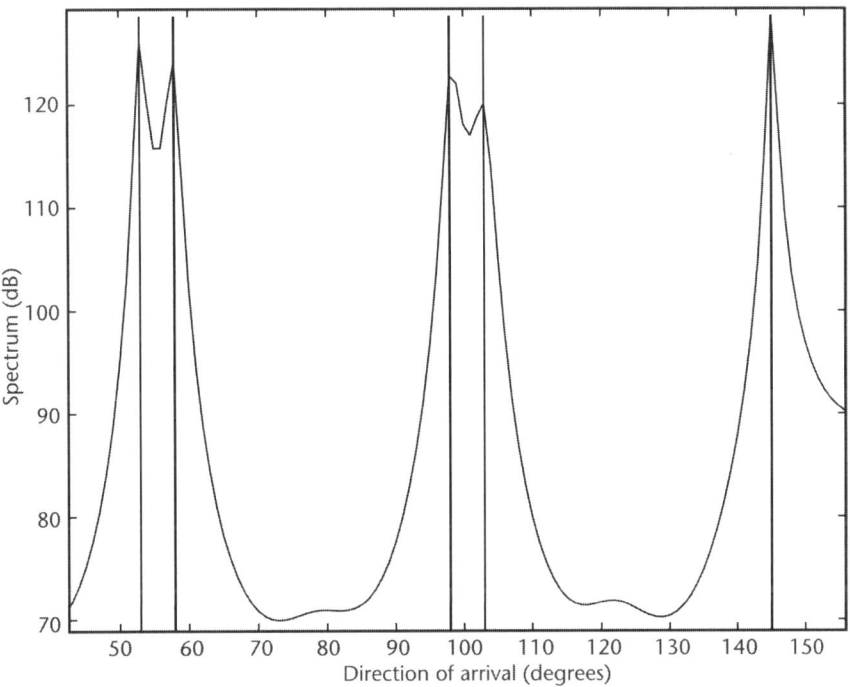

Figure 4.6 The estimated spatial spectrum of the multigroup sources using algorithm of [4].

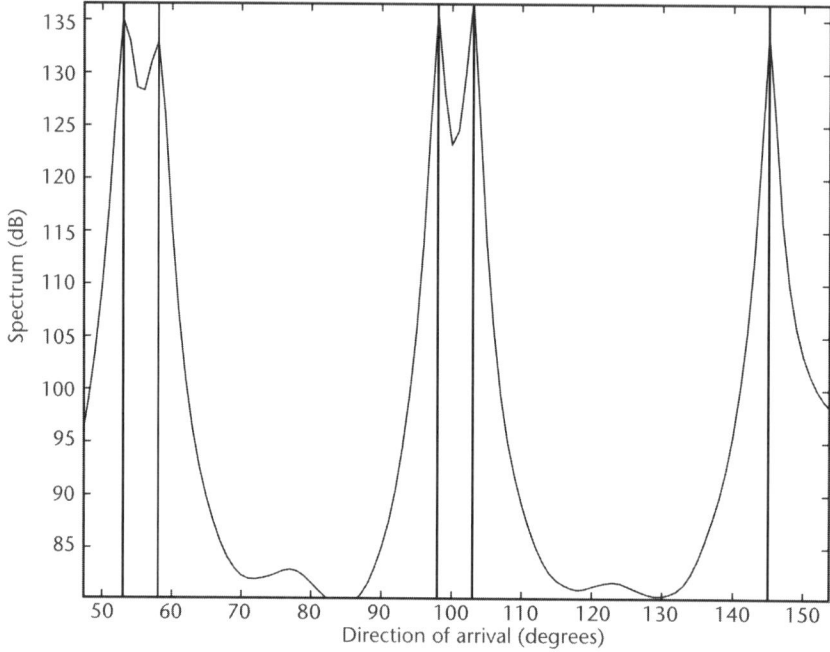

Figure 4.7 The estimated spatial spectrum of the multigroup sources using the MSP method for a wider frequency band [300:3,000] Hz.

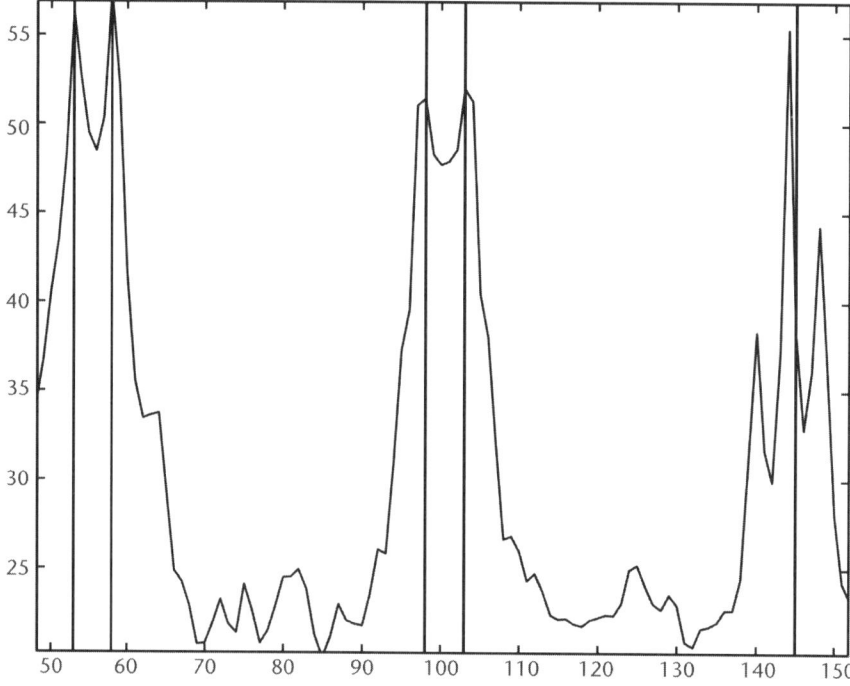

Figure 4.8 The estimated spatial spectrum of the multigroup sources using algorithm of [4] for a wider frequency band [300:3,000] Hz.

4.9 Conclusions

In this chapter, MSP is introduced as a tool to solve coherent broadband source localization problems. MSP direction-of-arrival estimation techniques do not require any preliminary knowledge of DOA angles or the number of sources.

References

[1] Do-Hong, T., and P. Russer, "Signal Processing for Wideband Array Applications," *IEEE Microwave Magazine*, March 2004, pp. 54–67.

[2] Wang, H., and M. Kaveh, "Coherent Signal-Subspace Processing for the Detection and Estimation of Angles of Arrival of Multiple Wide-Band Sources," *IEEE Trans. on Acoustics, Speech, and Signal Processing*, Vol. ASSP-33, No. 4, August 1985, pp. 823–831.

[3] Schmidt, R. O., "Multiple Emitter Location and Signal Parameter Estimation," *IEEE Trans. on Antennas and Propagation*, Vol. AP-34, No. 3, March 1986, pp. 276–280.

[4] Hung, H., and M. Kaveh, "Focusing Matrices for Coherent Signal-Subspace Processing," *IEEE Trans. on Acoustics, Speech, and Signal Processing*, Vol. 36, No. 8, August 1988, pp. 1272–1281.

[5] Sivanand, S., J. Yang, and M. Kaveh, "Focusing Filters for Wideband Direction Finding," *IEEE Trans. on Signal Processing*, Vol. 39, February 1991, pp. 437–445.

[6] Krolik, J., and D. Swingler, "Focused Wide-Band Array Processing by Spatial Resampling," *IEEE Trans. on Acoustics, Speech, and Signal Processing*, Vol. 38, No. 2, February 1990, pp. 356–360.

[7] Friedlander, B., and A. Weiss, "Direction Finding for Wide-Band Signals Using an Interpolated Array," *IEEE Trans. on Signal Processing*, Vol. 41, No. 4, April 1993, pp. 1618–1634.

[8] Owsley, N. L., "Sonar Array Processing," in *Array Signal Processing*, S. Haykin, (ed.), Upper Saddle River, NJ: Prentice Hall, 1985.

[9] Doron, M. A., and A. J. Weiss, "On Focusing Matrices for Wideband Array Processing," *IEEE Trans. on Signal Processing*, Vol. 40, No. 16, June 1992, pp. 1295–1302.

[10] Garbacz, R. J., and D. M. Pozar, "Antenna Shape Synthesis Using Characteristic Modes," *IEEE Trans. on Antennas and Propagation*, Vol. AP-30, May 1982, pp. 340–350.

[11] Harackiewicz, F. J., and D. M. Pozar, "Optimum Shape Synthesis of Maximum Gain Omnidirectional Antennas," *IEEE Trans. on Antennas and Propagation*, Vol. AP-34, February 1986, pp. 254–258.

[12] Liand, L. W., et al., "Exact Solutions of Electromagnetic Fields in Both Near and Far Zones Radiated by Thin Circular-Loop Antennas: A General Representation," *IEEE Trans. on Antennas and Propagation*, Vol. 45, No. 12, December 1997, pp. 1741–1748.

[13] Yaghjian, A. D., "Efficient Computation of Antenna Coupling and Fields Within the Nearfield Region," *IEEE Trans. on Antennas and Propagation*, Vol. AP-30, No. 1, January 1982, pp. 113–128.

[14] Abhayapala, T. D., R. A. Kennedy, and R. C. Williamson, "Nearfield Broadband Array Design Using a Radially Invariant Modal Expansion," *J. Acoust. Soc. Amer.*, Vol. 107, January 2000, pp. 392–403.

[15] Ward, D. B., and T. D. Abhayapala, "Range and Bearing Estimation of Wideband Sources Using an Orthogonal Beamspace Processing Structure," *Proc. IEEE Int. Conf. Acoust., Speech, Signal Processing, ICASSP 2004*, Vol. 2, May 2004, pp. 109–112.

[16] Abhayapala, T. D., and D. B. Ward, "Theory and Design of Higher Order Sound Field Microphones Using Spherical Microphone Array," *Proc. IEEE Int. Conf. Acoust., Speech, Signal Processing, ICASSP 2002*, Vol. 2, May 2002, pp. 1949–1952.

[17] Ward, D. B., and T. D. Abhayapala, "Reproduction of a Plane-Wave Sound Field Using an Array of Loudspeakers," *IEEE Trans. Speech and Audio Proc.*, Vol. 9, No. 6, September 2001, pp. 697–707.

[18] Jones, H. M., R. A. Kennedy, and T. D. Abhayapala, "On Dimensionality of Multipath Fields: Spatial Extent and Richness," *Proc. IEEE Int. Conf. Acoustics, Speech, and Signal Processing, ICASSP 2002*, Vol. 3, Orlando, FL, May 2002, pp. 2837–2840.

[19] Abhayapala, T. D., T. S. Pollock, and R. A. Kennedy, "Spatial Decomposition of MIMO Wireless Channels," *Proc. IEEE 7th International Symposium on Signal Processing and Its Applications*, Vol. 1, July 2003, pp. 309–312.

[20] Pollock, T. S., T. D. Abhayapala, and R. A. Kennedy, "Introducing Space into Space-Time MIMO Capacity Calculation: A New Closed Form Upper Bound," *Proc. Int. Conf. Telecommunicatons, ICT 2003*, Papeete, Tahiti, February 2003, pp. 1536–1541.

[21] Colton, D., and R. Kress, *Inverse Acoustic and Electromagnetic Scattering Theory*, 2nd ed., New York: Springer, 1998.

[22] Skudrzyk, E., *The Foundations of Acoustics*, New York: Springer-Verlag, 1971.

[23] Abhayapala, T., "Modal Analysis and Synthesis of Broadband Nearfield Beamforming Arrays," Ph.D. dissertation, Research School of Information Sciences and Engineering, Australian National University, Canberra ACT, December 1999.

CHAPTER 5
Target Localization in Three Dimensions
Santana Burintramart and Tapan K. Sarkar

In many applications, an adaptive antenna array known as a smart antenna is a promising technology for contemporary developments in both military and civilian applications. One major part of the smart antenna is the estimation of the source directions. This procedure is called DOA estimation. In some applications, such as mobile communication, the statistical-based DOA estimations can be applied. However, in some specific applications, such as radar, application of a statistical-based method has no meaning, since the environment changes rapidly. For example, it is challenging for an airborne radar to determine the direction of a jammer, which would improve the signal-to-noise ratio and improve detection or tracking. Since the platform on which the radar is mounted is moving very rapidly, other methods that are not based on statistical approaches make more sense.

In this section, the simulations are designed mainly for radar application purposes. Therefore, there will be both transmitting and receiving antennas at specific locations. The DOA estimation method, based on a single snapshot of data [1], will be used in determining the directions of returning waves to the receiver. In addition to the DOA estimation, we utilize the same concept to measure the distances between objects in space. Once the DOAs and distances are combined together, the target of interest can be located.

The chapter is organized as follows. In the Section 5.1, the method to be used for the DOA estimation in our numerical simulation is proposed. In Section 5.2, the technique of determining propagation delays is discussed. By knowing the delays, the distance between targets can be evaluated. Section 5.3 describes the DOA and distance estimations with some simulation results. Finally, the conclusion is presented in the last section.

5.1 Direction-of-Arrival Estimation Method

In this section, a two-dimensional DOA estimation method used in the simulations will be introduced. This method is a modified version of the method described in [2], which reduces the computational complexity in the estimation. For noiseless two-dimensional data, voltages induced in a two-dimensional planar array consisting of ideal radiating point sources at element (m, n) lying on an xy-plane can be formulated as:

$$x(m; n) = \sum_{i=1}^{I} A_i \exp(j\gamma_i + j2\pi m \Delta_y \sin \theta_i \sin \phi_i + j2\pi n \Delta_x \sin \theta_i \cos \phi_i) \quad (5.1)$$

where $0 \leq m \leq M - 1$ and $0 \leq n \leq N - 1$. A_i and γ_i are the amplitude and phase of each of the ith signal, and ϕ_i and θ_i are the azimuth and elevation angles of the direction of arrival. I is the number of incoming signals. The array has dimension $M \times N$ with element spacing Δ_x and Δ_y along the x and y directions. The goal is to estimate the number of signals, I, and the DOA, ϕ_i and θ_i. From (5.1), we can write the equation in a compact form by:

$$x(m; n) = \sum_{i=1}^{I} c_i y_i^m z_i^n \quad (5.2)$$

with:

$$c_i = A_i \exp(j\gamma_i) \quad (5.3)$$

$$y_i = \exp(j2\pi \Delta_y \sin \theta_i \sin \phi_i) \quad (5.4)$$

$$z_i = \exp(j2\pi \Delta_x \sin \theta_i \cos \phi_i) \quad (5.5)$$

If the poles y_i and z_i are known, then the angles (ϕ_i, θ_i) can simply be solved from the following [1]:

$$\phi_i = \arctan\left[\left(\frac{\ln y_i}{\ln z_i}\right) \frac{\Delta_x}{\Delta_y}\right] \quad (5.6)$$

$$\theta_i = \arcsin\left[\left(\frac{1}{j2\pi\Delta_x}\right) \sqrt{\left(\frac{\Delta_x}{\Delta_y} \ln y_i\right)^2 + (\ln z_i)^2}\right] \quad (5.7)$$

Next, the received signal $x(m; n)$ will be formed as a matrix in a particular fashion whose rank equals the number of signals. First, a $(B \times L)$ block Hankel matrix is defined as:

$$X_{m,n} = \begin{bmatrix} x(m; n) & x(m; n+1) & \cdots & x(m; n+L-1) \\ x(m; n+1) & x(m; n+2) & \cdots & x(m; n+L) \\ \vdots & \vdots & & \vdots \\ x(m; n+B-1) & x(m; n+B) & \cdots & x(m; n+B+L-2) \end{bmatrix}_{(B \times L)} \quad (5.8)$$

This is similar to the matrix in [2], except that the first element of the block is allowed to be any element from the original two-dimensional array data. One can verify that the matrix $X_{m,n}$ will have rank I, which is the number of the DOAs if B and L are greater than I. However, if the signal is contaminated by noise, then the rank may not be easily determined. We then want to estimate the rank by introducing a new matrix with a partition-and-stacking process. Therefore, one can determine the rank quite easily in this new approach. The criteria to determine the rank will be discussed later in this section. By partitioning and stacking the data, B and L are not necessarily greater than I. However, to ensure that the rank of a new matrix is equal to the number of the DOAs, some constraints must be satisfied. We will consider these conditions later. Using the block matrix in (5.8), we construct a new matrix X_{diag} as follows:

$$X_{diag} = \begin{bmatrix} X_{0,0} & X_{0,1} & \cdots & X_{0,Q} \\ X_{1,0} & X_{1,1} & \cdots & X_{1,Q} \\ \vdots & \vdots & & \vdots \\ X_{P,0} & X_{P,1} & \cdots & X_{P,Q} \end{bmatrix}_{B(P+1) \times L(Q+1)} \quad (5.9)$$

X_{diag} is a $B(P + 1) \times L(Q + 1)$ matrix, where P and Q are the maximum number of block rows and columns, respectively. Once the parameters are chosen, P and Q can be found by:

$$P = M - 1 \quad \text{and} \quad Q = N - L - B + 1 \quad (5.10)$$

We note that $X_{m,n}$ is formed by using the voltages induced by the mth row of the receiving array. Therefore, P is equal to the number of rows in the receiving array.

The poles (y_i, z_i) can be extracted from the X_{diag} matrix. First, we start our decomposition of the matrix $X_{m,n}$ as:

$$X_{m,n} = Z_L C Y_d^m Z_d^n Z_R \quad (5.11)$$

where

$$Z_L = \begin{bmatrix} 1 & 1 & \cdots & 1 \\ z_1 & z_2 & \cdots & z_I \\ \vdots & \vdots & \cdots & \vdots \\ z_1^{B-1} & z_2^{B-1} & \cdots & z_I^{B-1} \end{bmatrix} \quad (5.12)$$

$$Z_R = \begin{bmatrix} 1 & z_1 & \cdots & z_1^{L-1} \\ 1 & z_2 & \cdots & z_2^{L-1} \\ \vdots & \vdots & \cdots & \vdots \\ 1 & z_I & \cdots & z_I^{L-1} \end{bmatrix} \quad (5.13)$$

and $Y_d = diag(y_1, y_2, \ldots, y_I)$, $Z_d = diag(z_1, z_2, \ldots, z_I)$, and $C = diag(c_1, c_2, \ldots, c_I)$. The *diag* in these formulations stands for a diagonal matrix whose nonzero components are on the diagonal. Substituting (5.11) into (5.9), X_{diag} becomes:

$$X_{diag} = E_L C E_R \tag{5.14}$$

with:

$$E_L = \begin{bmatrix} Z_L \\ Z_L Y_d \\ \vdots \\ Z_L Y_d^P \end{bmatrix} \tag{5.15}$$

$$E_R = [Z_R, Z_d Z_R, \ldots, Z_d^Q Z_R] \tag{5.16}$$

The criteria that ensure $rank(X_{diag})$ to be equal to the number of the DOAs will be determined. From (5.14), $rank(X_{diag})$ will be I if and only if $rank(E_L) = rank(E_R) = I$. Therefore, the criteria will be on E_L and E_R. For E_L, it has been shown in [2] that $rank(E_L) = I$ only if:

$$B(P + 1) \geq I \tag{5.17}$$

Similarly, in the case of E_R, we can show that E_R will have $rank(E_R) = I$ if and only if

$$L + Q \geq I \tag{5.18}$$

Combining these two criteria with the fact that $L + B - 1 \leq N$, we get the final conditions for $rank(X_{diag}) = I$. These conditions are:

$$\left. \begin{aligned} N - B + 1 &\geq I \\ B(P + 1) &\geq I \end{aligned} \right\} \tag{5.19}$$

It is important to note that once the matrix pencil (MP) method [1, 2] is applied to X_{diag}, the maximum rank that can be estimated is $rank(X_{diag}) = 1$, because the MP will partition X_{diag} into two submatrices, as will be explained in the following steps. Next, the procedures of extracting the poles (y_i, z_i) will be discussed in detail. To extract the poles, the MP method is applied to the X_{diag} in such a way that poles associated with each dimension can be extracted. Here, the MP method is directly applied to the data, and it can distinguish between coherent signals. ESPRIT cannot do the same unless extra care is taken, because it deals with the covariance matrix of the data.

5.1.1 Extraction of y_i

To extract y_i, two submatrices A_1 and B_1 are formed from X_{diag} by deleting the first B rows and the last B rows from X_{diag}, respectively. Then, the matrix pencil pair is defined by:

5.1 Direction-of-Arrival Estimation Method

$$A_1 - \lambda_1 B_1 = 0 \qquad (5.20)$$

This formulation can be written in another form as:

$$E_{L1} C E_R - \lambda_1 E_{L2} C E_R = 0 \qquad (5.21)$$
$$E_{L1} C (Y_d - \lambda_1 I) E_R = 0$$

where E_{L1} and E_{L2} are E_L, with the first and the last B rows deleted, respectively. Now (5.21) is reduced to a generalized eigenvalue problem [3]. Its associated eigenvalues $\{y_i; i = 1, \ldots, I\}$ can be found from $B_1^\dagger A_1$, where $X^\dagger$ denotes the Moore-Penrose pseudoinverse of matrix X [4].

5.1.2 Extraction of z_i

Similarly, we extract the poles $\{z_i; i = 1, \ldots, I\}$ by generating a pair of matrix pencils:

$$A_2 - \lambda_2 B_2 = 0 \qquad (5.22)$$

where A_2 and B_2 are X_{diag} with the first and the last L columns deleted, respectively. As before, we can write (5.22) as:

$$E_L C E_{R1} - \lambda_2 E_L C E_{R2} = 0 \qquad (5.23)$$
$$E_L C (Z_d - \lambda_2 I) E_{R2} = 0$$

where E_{R1} and E_{R2} are E_R, with the first and the last L columns deleted, respectively. In the same manner as (5.21), the poles z_i can be determined by solving the eigenvalue problem associated with $B_2^\dagger A_2$. Although we have already obtained two sets of poles, the order of the poles in each set is not in the correct pair; in other words, y_1 may not be an associated pole of z_1. If we put the incorrect pair of poles into (5.6) and (5.7), then we will not get a correct estimate for the DOA. Therefore, we need to find a way to rearrange those poles in a correct ordered pair.

5.1.3 Extraction of d_i

After obtaining the two sets of poles y_i and z_i, we define the diagonal matrix pencil pair as:

$$D_1 - \beta D_2 = 0 \qquad (5.24)$$

where $D_1 = X_{diag}$, with the first B rows and the first L columns deleted, and $D_2 = X_{diag}$, with the last B rows and the last L columns deleted. Equation (5.24) also can be decomposed as:

$$E_{L1} C E_{R1} - \beta E_{L2} C E_{R2} = 0 \qquad (5.25)$$
$$C(Y_d Z_d - \beta I) E_{L2} E_{R2} = 0$$

Once solving the eigenvalue problem of $D_2^\dagger D_1$, another set of poles $\{d_i; i = 1, \ldots, I\}$ can be obtained. Since $D_1 - \beta D_2$ is different along the diagonal of the matrix X_{diag}, this method is called the "diagonal matrix pencil method" [5]. Next, we will illustrate how the pairing is done.

5.1.4 Pairing y_i, z_i, and d_i

Once the three sets of poles are obtained, the pairing can be done easily by multiplying each pole pair y_i and z_i. The multiplication for all the possible combinations will be equal to a pole d_i:

$$\{(y_i, z_j) \mid y_i z_j = d_k : i = 1, \ldots, I; j = 1, \ldots, I; k = 1, \ldots, I\} \quad (5.26)$$

This simple pairing step reduces the computation complexity in the pairing step of [2]. However, it is worth noting that care must be taken in this pairing step, since the poles are complex numbers. By substituting the correct pairs of poles (y_i, z_i) into (5.6) and (5.7), the DOAs can be estimated.

5.1.5 Complex Amplitude Estimation

When the poles (y_i, z_i) are correctly paired, the next step is to estimate the complex amplitude c_i of the incoming signal, as formulated in (5.3). One possible way to estimate c_i is to solve the following least-squares problem [1]:

$$\begin{bmatrix} 1 & 1 & \cdots & 1 \\ z_1 & z_2 & \cdots & z_I \\ \vdots & \vdots & \ddots & \vdots \\ z_1^{N-1} & z_2^{N-1} & \cdots & z_I^{N-1} \\ y_1 & y_2 & \cdots & y_I \\ y_1 z_1 & y_2 z_2 & \cdots & y_I z_I \\ \vdots & \vdots & \ddots & \vdots \\ y_1 z_1^{N-1} & y_2 z_2^{N-1} & \cdots & y_I z_I^{N-1} \\ \vdots & \vdots & \vdots & \vdots \\ y_1^{M-1} & y_2^{M-1} & \cdots & y_I^{M-1} \\ y_1^{M-1} z_1 & y_2^{M-1} z_2 & \cdots & y_I^{M-1} z_I \\ \vdots & \vdots & \ddots & \vdots \\ y_1^{M-1} z_1^{N-1} & y_2^{M-1} z_2^{N-1} & \cdots & y_I^{M-1} z_I^{N-1} \end{bmatrix} \begin{bmatrix} c_1 \\ c_2 \\ \vdots \\ c_I \end{bmatrix} = \begin{bmatrix} x(0; 0) \\ x(0; 1) \\ \vdots \\ x(0, N-1) \\ x(1, 0) \\ x(1, 1) \\ \vdots \\ x(1, N-1) \\ \vdots \\ x(M-1, 0) \\ x(M-1, 1) \\ \vdots \\ x(M-1, N-1) \end{bmatrix}$$

(5.27)

5.1.6 Noisy Signals

So far, the number of signals I is assumed to be known from the $rank(X_{diag})$. However, if there is noise in the signals, then $rank(X_{diag})$ will not be equal to I.

Instead, $rank(X_{diag})$ will be greater than I. The DOA estimation will be corrupted by noise in the matrix X_{diag}. Estimation of I and noise reduction can be performed via the SVD [4]. To do this, we first decompose X_{diag} using the SVD, yielding:

$$X_{diag} = U\Sigma V^H \qquad (5.28)$$

where U and V are unitary matrices whose columns are the eigenvectors of $X_{diag}X_{diag}^H$ and $X_{diag}^H X_{diag}$, respectively. X_{diag} denotes the complex transpose of X, and Σ is a diagonal matrix containing the singular values of X_{diag}. It is suggested that the number of dominant singular values in Σ will correspond to the number of signals by I [1]. It also can be estimated by observing the ratio of the nth singular value and the maximum singular value:

$$\frac{\sigma_n}{\sigma_{\max}} \approx 10^{-w} \qquad (5.29)$$

with w as the number of accurate significant decimal digits of the data $x(m; n)$ [6]. From the specified accuracy level, the number of signals I can be obtained. The remaining singular values and their corresponding eigenvectors are assumed to correspond to noise and modeling error in the data. Once I is known, a rank-deficient matrix $\tilde{X}_{diag}$ will be generated as follows:

$$\tilde{X}_{diag} = \tilde{U}\tilde{\Sigma}\tilde{V}^H \qquad (5.30)$$

where $\tilde{U}$ and $\tilde{V}$ are U and V with only the first I columns selected, respectively. $\tilde{\Sigma}$ is a diagonal matrix with I dominant singular values. Finally, the matrix $\tilde{X}_{diag}$ instead of X_{diag} will be used in the processes of extracting the poles. We note here that in the pairing step, we have to relax the equality in (5.26), since the computed poles will estimated instead of exact. Therefore, the pair of poles (y_i, z_i) that gives the closest value to d_i will be chosen to be a correct pair:

$$\{(y_i, z_j) \mid y_i z_j \approx d_k : i = 1, \ldots, I; j = 1, \ldots, I; k = 1, \ldots, I\} \qquad (5.31)$$

Thus far, we have discussed a method to estimate the two-dimensional DOAs and their complex amplitudes by using a planar array. In the next section, the method of estimating distances between targets is discussed.

5.2 Distance Estimation Method

In this section, the method of estimating distances between the targets by using an antenna array based on the MP method is introduced. The method utilizes frequency domain information of the returning signals to estimate the separation distances. Since the phase shifts in frequency relate to delays in time, if we know the phase shifts, we could estimate the time delay of the returning signals. In other words, we could know the distances between targets. Therefore, we will focus on estimation of the phase shifts in the frequency domain. The received signal is in a frequency domain with a frequency step Δ_f Hz. The initial phases at each frequency of the transmitted signal are assumed to be known a priori. The received signal samples in the frequency domain at frequency $(f_0, f_1, \ldots, f_K)$ can be represented by:

$$S(f_k) = \sum_{p=1}^{P} A_p(f_k) \exp\{j2\pi[f_0 + k\Delta_f](\tau_{Sp} + \tau_{Rp}) + j\delta_p\} \qquad (5.32)$$

where P is the number of received signals, A_p is the complex amplitude of the pth signal, and δ_p is the phase shift due to the reflection from the pth target. τ_{Sp} and τ_{Rp} are the propagation delays along the transmitting and receiving paths due to the target p, respectively.

We assume that the received signal can be separated into P independent signals, which will be shown in the next section. For one independent signal, we get:

$$S_p(f_k) = A_p(f_k) \exp\{j2\pi[f_0 + k\Delta_f](\tau_{Sp} + \tau_{Rp}) + j\delta_p\} \qquad (5.33)$$

Next, we define:

$$z_p = \exp\left[j2\pi \frac{\Delta_f}{c}(R_{Sp} + R_{Rp})\right] \qquad (5.34)$$

where R_{Sp} and R_{Rp} are ranges corresponding to the transmitting and receiving paths, respectively, and c is the propagation speed of the signal. By normalizing the amplitude of each frequency component, the relationship between $S_p(f_k)$ and $S_p(f_{k+1})$ can only be due to the phase shift caused by its propagation delay. Thus, we get:

$$S_p(f_{k+1}) = z_p S_p(f_k) \qquad (5.35)$$

Next, the MP method [3] is applied to solve (5.35) for z_p. The normalized frequency domain data is formed as a Hankel matrix as:

$$[S_p] = \begin{bmatrix} S_p(0) & S_p(1) & \cdots & S_p(K-L+1) \\ S_p(1) & S_p(2) & \cdots & S_p(L) \\ \vdots & \vdots & & \vdots \\ S_p(L-1) & S_p(L) & \cdots & S_p(K) \end{bmatrix} \qquad (5.36)$$

where L is called the pencil parameter. Then we solve the generalized eigenvalue problem:

$$[S]_2 - \eta[S]_1 = 0 \qquad (5.37)$$

where $[S]_1$ and $[S]_2$ are $[S_p]$ in (5.36) with the last and the first row deleted, respectively. As mentioned, the signal information in (5.36) contains only one

5.2 Distance Estimation Method

independent signal from (5.32). Therefore, the $rank([S_p])$ will be one, and only one eigenvalue, η, will be the solution. Once z_p is obtained, the distance traveled by the signal can be calculated by:

$$R_{Sp} + R_{Rp} = \frac{c \operatorname{Im}(\ln z_p)}{2\pi \Delta_f} \tag{5.38}$$

where $\operatorname{Im}(x)$ denotes the imaginary part of x. The distance traveled for each of the returning signal is now known. If we combine this information with its corresponding direction, then the target location can be estimated. However, notice that all the information is contained in phase of z_p. Therefore, there will be an ambiguity in the estimation if the phase of z_p, if it is greater than 2π. To prevent this error in the estimation, the limitation of the separation distance can be estimated from

$$0 \leq \frac{\Delta_f}{c}(R_{Sp} + R_{Rp}) \leq 1$$

or:

$$0 \leq R_{Sp} + R_{Rp} \leq \frac{c}{\Delta_f} \tag{5.39}$$

It is seen that the estimated distance depends on the frequency step used, and the smaller the frequency step, the longer the separation distance can be estimated.

As discussed so far, the frequency domain data applied to (5.36) is free from excessive noise, and corresponds to only one signal. However, even though the signal S_p in (5.33) is independent of the other received signal, noise in the signal or receiver can still corrupt and make $rank([S_p])$ larger than one. Therefore, more than one eigenvalue, η, will be obtained when the problem in (5.37) is solved. Those additional eigenvalues will correspond to noise in the estimation. To overcome this, the SVD is applied to (5.36), obtaining:

$$[S_p] = U_p \Sigma_p V_p^H \tag{5.40}$$

Here, again, U_p and V_p are unitary matrices, and Σ_p is a diagonal matrix with one dominant singular value along the diagonal. Because $[S_p]$ contains only information on one signal, only the first column of U_p that corresponds to the signal of interest is selected. The rest belong to the noise space in the received signal. Then, U_1 and U_2 are defined from the first column of U_p, with the last and the first elements deleted. The value of z_p can be found by solving the eigenvalue of $U_1^\dagger U_2$ in the same way as extracting poles in the previous section. We note here that noise in $[S_p]$ has already been filtered out via the SVD. Therefore, we are finding the eigenvalue of the filtered data. However, the estimation of z_p will be more accurate if the number of elements in U_1 and U_2 are larger. Thus, it is recommended to choose the pencil parameter L around two-thirds of the data available.

In the next section, the two-dimensional DOA and distance estimations are combined together to estimate target locations in space. The separation of S_p will be explained when the two-dimensional DOA estimation is done. Some simulation results will be also shown in the next section.

5.3 Target Location Estimation and Simulation Results

In this section, the diagonal matrix pencil method is used for the two-dimensional DOA estimation. The distances between the targets are then estimated by using the technique described in Section 5.2. For a better understanding, the simulation scenario used in our simulations is explained along with the discussion of the estimations. We assume that a transmitting antenna is placed at a coordinate $(x, y, z) = (0, -100, 0)$m in free space, as shown in Figure 5.1. A 10×10 array of radiating point sources with 1m spacing between the elements (Δ_y and Δ_z) with the origin of the coordinate system is located at the center. The array orientation is along the y-z plane.

At each antenna element, the received signal is considered to be in the frequency domain rather than in the time domain. This frequency domain information can be simply obtained by passing the received signal through an FFT, which can be either in the software or the hardware. Once the frequency domain data is available at every antenna element, we then start with the DOA estimation. The data at the same frequency from each antenna element, $x(m; n)$, will be put into the matrix X_{diag} as in (5.9). The matrix X_{diag} will be passed through the processes of obtaining the poles and pairing them. Once the correct pairs of poles are obtained, the DOAs, $(\theta, \phi)_i$, and their corresponding complex amplitudes, c_i, can be determined.

The DOA and complex amplitude estimations will be performed for all frequency samples f_0 to f_K with a step frequency Δ_f Hz. Once the estimations are

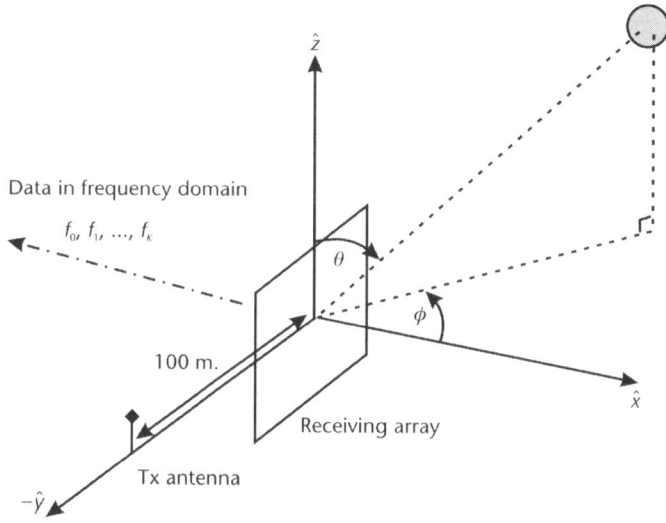

Figure 5.1 Configuration of the transmitting and receiving antennas, receiving signals in the frequency domain, and a sphere target is located in space.

5.3 Target Location Estimation and Simulation Results

finished, the DOA of each target will be averaged to generate the final result. Then, at the next step, the distance is estimated. Since the complex amplitudes of the incoming signals have already been estimated, we will use the amplitudes that correspond to a particular target in one direction $(\theta, \phi)_i$ for all frequencies $(f_0, f_1, \ldots, f_K)$ in estimating the separation distance. The signal in (5.32) will be separated into P independent signals automatically. In other words, we are determining the separation distance of each target one by one, by using their corresponding complex amplitudes from frequencies f_0 to f_K. It is important to note that in order to satisfy (5.35), the complex amplitudes must be normalized by their magnitudes. Thus, the additional term is the phase shift due to the delay in the propagation. The MP method explained in Section 5.2 is now used to find the separation distances of the targets.

Since the distances obtained are total distances traveled by the signal, the correction of the offset between the transmitting and the reference element of the array can be done by:

$$R_i = \frac{T_i^2 - d^2}{2(T_i + d \sin \theta_i \sin \phi_i)} \quad (5.41)$$

where

$$T_i = R_{Si} + R_{Ri} \quad (5.42)$$

and d is the spacing between the transmitter to the reference of the array ($d = 100$ in our case). R_i is the distance of the ith target from the origin. If the geometry of the antennas is different, then (5.41) must be changed accordingly. Finally, the target locations can be plotted by using a spherical coordinate system of $(R, \theta, \phi)_i$. Next, the nature of the simulations is discussed in detail.

To verify the methods outlined above, the electromagnetic modeling software (i.e., WIPL-D [7]) is used in the simulations. All the effects of the scatterers, including mutual couplings between the antenna elements, have already been taken into account. The simulation is done in free space. Therefore, the speed of propagation will be equal to the speed of light, $c \approx 3 \times 10^8$ m/s. For simplicity, the targets are perfect conducting (PEC) spheres placed at arbitrary locations in free space that are to be estimated. The total traveling distances must be satisfied (5.39) with a specific frequency step Δ_f. The sphere size has a radius of 0.5m. Because PEC spheres are used as the targets, phase shift due to the reflection at the target in (5.32) is 180°. The transmitting antenna is a thin dipole, with a length of 1.2m and a radius of 1 mm. The frequency range used in the simulation is from 100 to 102 MHz, with $\Delta_f = 50$ kHz. Twenty DOA estimations are used in obtaining the complex amplitudes of the signal for estimation of the separation, and the DOAs are averaged for the final result. Before the simulation results are shown, we define the error in the estimation along the k direction as:

$$e_k \triangleq \left| \frac{\eta_{\text{exact}} - \eta_{\text{est}}}{\eta_{\text{exact}}} \right| \times 100\% \quad (5.43)$$

where η_{exact} and η_{est} are the exact and estimated values, respectively. Figure 5.2 shows the result of two target locations being estimated without noise, and Table 5.1 includes the details of the results.

It is important to note that, even though noise has not been added to the received signals, there is some disturbance in the signals. This disturbance is caused by mutual coupling between the spheres themselves, and between the spheres and the transmitting antenna. Therefore, it cannot be expected that the rank of X_{diag} will be exactly equal to the number of incoming signals or targets. Care must be taken in selecting the number of targets from the singular values. There is also a direct propagating wave to the array in our simulations, which needs to be removed from the transmitting antenna. This direct wave can be easily suppressed in a real system by using a T-R switch that prevents the direct wave through the receiver.

From Table 5.1, it is seen that the true location of the two spheres are at (250, 700, 300) and (600, 200, 500), respectively. The estimation errors for this simulation were less than 1% for both spheres.

Figures 5.3 and 5.4 and Tables 5.2 and 5.3 show other results for noise-contaminated signals, where a 30-dB SNR noise is applied to each antenna element.

In Figure 5.3 and Table 5.2, two spheres are located at (300, 600, 200) and (700, 500, 200). In Figure 5.4 and Table 5.3, the first sphere is located at (300, 600, 200), the second sphere is at (700, 500, 100), and the third sphere is at (900, 100, 450).

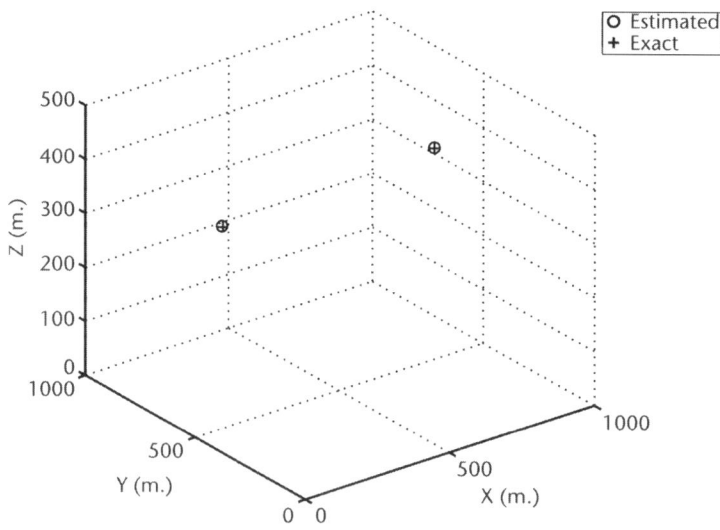

Figure 5.2 Results for two spheres at (250, 700, 300) and (600, 200, 500) without noise but with unmodeled mutual coupling.

Table 5.1 True Locations, Estimated Locations, and Their Errors

Coordinate	Target 1			Target 2		
	x	y	z	x	y	z
True location (m)	250	700	300	600	200	500
Estimated location (m)	247.88	701.46	301.64	600.11	199.26	501.35
Error e_x, e_y, e_z (%)	0.84	0.21	0.55	0.02	0.37	0.27

5.3 Target Location Estimation and Simulation Results

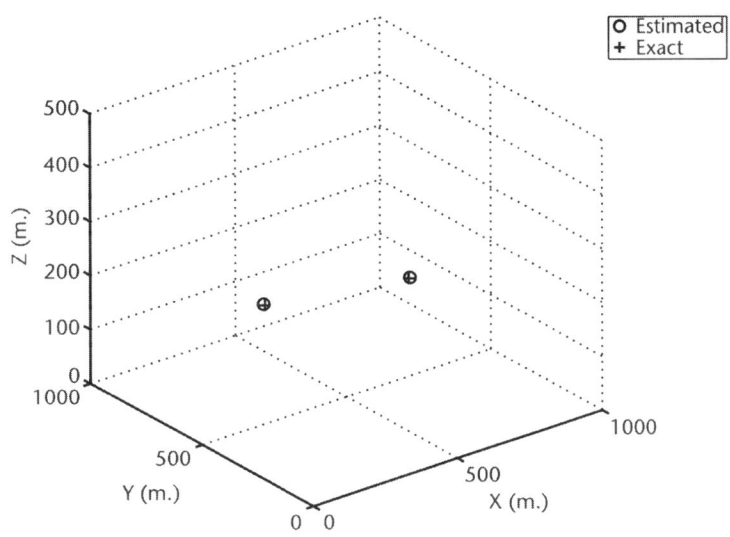

Figure 5.3 Results for two spheres at (300, 600, 200) and (700, 500, 200) with a 30-dB SNR.

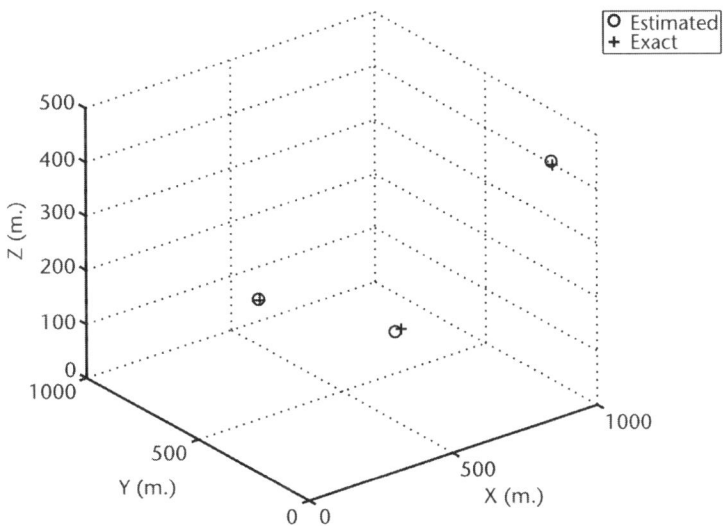

Figure 5.4 Results for three spheres at (300, 600, 200), (700, 500, 100), and (900, 100, 450) with a 30-dB SNR.

Table 5.2 Results for Two Spheres, with a 30-dB SNR

	Target 1			Target 2		
Coordinate	x	y	z	x	y	z
True location (m)	300	600	200	700	500	200
Estimated location (m)	297.82	603.06	201.78	708.49	505.40	198.79
Error e_x, e_y, e_z (%)	0.73	0.51	0.89	1.21	1.08	0.61

Table 5.3 Estimated Results for the Three Spheres, with a 30-dB SNR

Coordinate	Target 1			Target 2			Target 3		
	x	y	z	x	y	z	x	y	z
True (m)	300	600	200	700	500	100	900	100	450
Estimated (m)	297.48	601.84	201.88	685.49	508.37	96.03	896.86	101.04	456.84
e_k (%)	0.84	0.31	0.94	2.07	1.67	3.97	0.35	1.04	1.52

As mentioned in [5], the diagonal matrix pencil method works well for high SNR if there are many incoming signals. Since there are three signals from the targets, and mutual couplings exist between the spheres, the estimation error is quite high, at 30 dB SNR. Increasing the number of antenna elements in the array can reduce this error. In some situations, close targets need to be detected. For example, in surveillance radar systems, the detection of close flying aircraft is one of the required capabilities. Two close targets are simulated in the next simulation, in order to determine whether our method can estimate the targets. The results are shown in Figure 5.5 and Table 5.4, where the spheres are located at (400, 600, 200) and (450, 600, 200). Their estimated locations are (415.70, 613.08, 204.48) and (444.56, 577.76, 199.88), respectively. Even though the error in the estimation is quite high, we can still determine the number of targets and their directions.

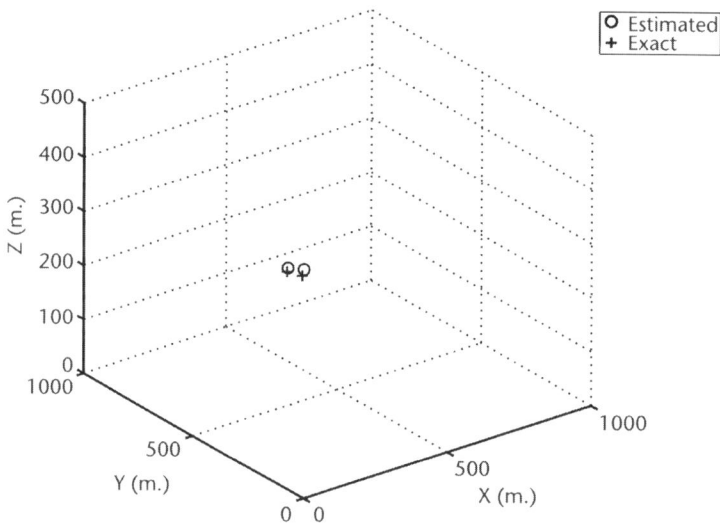

Figure 5.5 Results for two close spheres with 50m spacing at (400, 600, 200) and (450, 600, 200) with 30 dB SNR.

Table 5.4 Estimation Results for Two Spheres Close to Each Other with a 30-dB SNR

Coordinate	Target 1			Target 2		
	x	y	z	x	y	z
True location (m)	400	600	200	450	600	200
Estimated location (m)	415.70	613.08	204.48	444.56	577.76	199.88
Error e_x, e_y, e_z (%)	3.92	2.18	2.24	1.21	3.71	0.06

5.4 Conclusions

So far, we have shown the simulation results for determining target locations by applying the method of DOA estimation. To locate a target in space, the direction that points to the target and the distance to the target need to be estimated. Since the DOA estimation approach used in our simulation also can estimate complex amplitudes of the signal, we utilize that information in determining the distance to the object, or we will determine the distance by estimating the delay of the wave propagation. Combining both DOAs and distances, the target can be located. Since the method is sensitive to noise, the number of array elements has to be increased to reduce the estimation error. The mutual coupling between targets also can be considered as another incoming signal to the antenna. Therefore, care must be taken in selecting the number of targets and other parameters in the methods. Even though the maximum distance, which can be estimated, is limited by phase information of z_p as mentioned in (5.39), this technique could be used as additional information to increase accuracy of any target acquisition system, where the distance information can be obtained by other means.

Finally, in this section, two modifications of the matrix pencil method are utilized in determining target locations in free space. Some simulation results based on the electromagnetic software modeling are presented.

References

[1] Sarkar, T. K., et al., *Smart Antennas*, New York: John Wiley & Sons, 2003.

[2] Hua, Y., "Estimating Two-Dimensional Frequencies by Matrix Enhancement and Matrix Pencil," *IEEE Trans. on Signal Processing*, Vol. 40, No. 9, September 1992, pp. 2267–2280.

[3] Hua, Y., and T. K. Sarkar, "Matrix Pencil Method for Estimating Parameters for Exponentially Damped/Undamped Sinusoids in Noise," *IEEE Trans. on Acoustics, Speech, and Signal Processing*, Vol. 36, No. 5, May 1990, pp. 814–824.

[4] Golub, G. H., and C. F. Van Loan, *Matrix Computations*, 3rd ed., Baltimore, MD: Johns Hopkins University Press, 1996.

[5] Burintramart, S., "Estimation of the Two-Dimensional Direction of Arrival by The Diagonal Matrix Pencil Method," *ACES Conference*, Syracuse, NY, 2004.

[6] Sarkar, T. K., and O. Pereira, "Using the Matrix Pencil Method to Estimate the Parameters of a Sum of Complex Exponentials," *IEEE Antennas and Propagation Magazine*, Vol. 37, No. 1, February 1995.

[7] Kolundzija, B. M., J. S. Ognjanovic, and T. K. Sarkar, *WIPL-D: Electromagnetic Modeling of Composite Metallic and Dielectric Structures*, Norwood, MA: Artech House, 2000, http://wipl-d.com.

CHAPTER 6
Polarimetric Time-Frequency MUSIC for Direction Finding of Moving Sources with Time-Varying Polarizations

Yimin Zhang, Baha A. Obeidat, and Moeness G. Amin

The spatial polarimetric time-frequency distributions (SPTFD) have been introduced as a platform for processing polarized nonstationary signals incident on multiple dual-polarized double-feed antennas. Based on this platform, the polarimetric time-frequency MUSIC (PTF-MUSIC) methods have been developed for DOA estimation of nonstationary sources with distinct polarization characteristics. In this chapter, we examine the feasibility of the PTF-MUSIC methods for tracking of moving sources with time-varying polarization characteristics. We demonstrate the significance of polarization diversity in challenging direction finding problems, where the sources are closely spaced, and discuss important issues relevant to utilization of polarization diversity. The SPTFD has the capability of incorporating both the instantaneous polarization information and time-frequency signatures of the different sources in the field of view. The incorporation of specific time-frequency points or regions, where one or more signals reside, enhances SNR and allows source discrimination and source elimination. This, in turn, leads to DOA performance improvement and reductions in the required number of array sensors. The PTF-MUSIC significantly outperforms the existing time-frequency MUSIC, polarimetric MUSIC, and conventional MUSIC direction finding techniques.

6.1 Introduction

Time-frequency distributions (TFDs) have been used for nonstationary signal analysis and synthesis in various areas, including speech, biomedicine, the automotive industry, and machine monitoring [1–3]. Over the past few years, the spatial dimension has been incorporated, along with the time and frequency variables, into quadratic and higher-order TFDs, and has led to the introduction of spatial time-frequency distributions (STFDs) as a powerful tool for nonstationary array signal processing [4–7]. The steering or mixing matrix defines the relationship between the TFDs of the sensor data to the TFDs of the individual source waveforms. It was found to be similar to that encountered in the traditional data covariance matrix approach to array processing. This similarity has allowed subspace-based estimation methods to be applied to time-frequency signal representations, utilizing the source instantaneous frequency for direction finding. It has been shown that the MUSIC [8, 9], maximum likelihood [10], and ESPRIT [11] techniques based on SPTFDs outperform their counterparts that are based on covariance matrices, when the sources are of nonstationary temporal characteristics.

Polarization and polarization diversities, on the other hand, are commonly used in wireless communications and various types of radar systems [12, 13]. Antenna and target polarization properties are widely employed in remote sensing and synthetic aperture radar (SAR) applications [14–16]. Airborne and spaceborne platforms, as well as meteorological radars, include polarization information [17, 18]. Polarization plays an effective role in identification of targets in the presence of clutter [19, 20], and has been incorporated in antenna arrays to improve signal parameter estimations, including DOA and time-of-arrival (TOA) [21, 22].

To combine time-frequency (t-f) signal representations and polarimetric signal processing, we have introduced the SPTFDs for double-feed dual-polarized arrays, which allow the utilization of both the t-f and polarization signatures of the sources in the field of view [23–25]. We then used the SPTFD to develop the PTF-MUSIC technique, which is formulated based on the source t-f and polarization properties. This new technique is employed for DOA estimation of polarized nonstationary signals, and has been shown to outperform the MUSIC techniques that only incorporate either the t-f or the polarimetric source characteristics [24]. The application of the SPTFDs to an ESPRIT-like method was introduced in [23, 26].

In this chapter, we examine the feasibility of the PTF-MUSIC methods for tracking of moving sources with time-varying polarization characteristics. We demonstrate the significance of polarization diversity in direction finding problems in which the sources are closely spaced, and discuss the necessary treatment of the data samples to effectively utilize the polarization diversity. The SPTFD incorporates the instantaneous polarization information at selected auto-term t-f points where the energy of desired source signals is localized. The selection of t-f regions in constructing the SPTFD matrices not only reduces noise, but also enables source exclusion from post-eigendecomposition and subspace estimation. The reduction of the number of sources reduces the required number of array sensors, and leads to performance enhancement. As a result, PTF-MUSIC outperforms existing polarimetric MUSIC techniques.

This chapter is organized as follows. Section 6.2 discusses the signal model and briefly reviews TFDs and STFDs. Section 6.3 considers dual-polarized antenna arrays, and introduces the concept of SPTFDs. The PTF-MUSIC algorithm is proposed in Section 6.4. Section 6.5 considers the spatio-polarimetric correlation, which is a useful parameter to measure the closeness between two sources in the joint spatial and polarimetric domains. Section 6.6 considers DOA tracking of moving sources with time-varying polarization characteristics. Computer simulations, demonstrating the effectiveness of the proposed methods, are provided in Section 6.7.

Throughout this chapter, lowercase bold letters denote vectors, and uppercase bold letters denote matrices. The operation $E[\cdot]$ denotes expectation, $(\cdot)^*$ denotes complex conjugate, $(\cdot)^T$ denotes transpose, and $(\cdot)^H$ denotes conjugate transpose (Hermitian). We use $(\cdot)^{[i]}$ to denote polarization, $\|\cdot\|$ to denote vector norm, $\otimes$ to

6.2 Signal Model

denote the Kronecker product operator, and $\circ$ to denote the Hadamard product operator.

6.2.1 Time-Frequency Distributions

The Cohen's class of TFDs of a signal $x(t)$ is defined as [1],

$$D_{xx}(t, f) = \iint \phi(t - u, \tau) x\left(u + \frac{\tau}{2}\right) x^*\left(u - \frac{\tau}{2}\right) e^{-j2\pi f \tau} \, du \, d\tau \quad (6.1)$$

where t and f represent the time and frequency indexes, respectively, $\phi(t, \tau)$ is the t-f kernel, and τ is the time-lag variable. In this chapter, all the integrals are from $-\infty$ to ∞.

The cross-term TFD of two signals $x_i(t)$ and $x_k(t)$ is defined by

$$D_{x_i x_k}(t, f) = \iint \phi(t - u, \tau) x_i\left(u + \frac{\tau}{2}\right) x_k^*\left(u - \frac{\tau}{2}\right) e^{-j2\pi f \tau} \, du \, d\tau \quad (6.2)$$

6.2.2 Spatial Time-Frequency Distributions

The STFDs have been developed for single-polarized antenna arrays [4–6]. Consider n narrowband signals impinging on an array of m single-polarized antenna elements. The following linear data model is assumed,

$$\mathbf{x}(t) = \mathbf{y}(t) + \mathbf{n}(t) = \mathbf{A}(\boldsymbol{\Phi})\mathbf{s}(t) + \mathbf{n}(t) \quad (6.3)$$

where the $m \times n$ matrix $\mathbf{A}(\boldsymbol{\Phi}) = [\mathbf{a}(\phi_1), \mathbf{a}(\phi_2), \ldots, \mathbf{a}(\phi_n)]$ is the mixing matrix that holds the spatial information of the n signals, $\boldsymbol{\Phi} = [\phi_1, \phi_2, \ldots, \phi_n]$, and $\mathbf{a}(\phi_1)$ is the spatial signature for source i. Each element of the $n \times 1$ vector $\mathbf{s}(t) = [s_1(t), s_2(t), \ldots, s_n(t)]^T$ is a monocomponent signal. Due to the signal mixing occurring at each sensor, the elements of the $m \times 1$ sensor data vector $\mathbf{x}(t)$ are multicomponent signals. $\mathbf{n}(t)$ is an $m \times 1$ additive noise vector that consists of independent zero-mean, white, and Gaussian distributed processes.

The STFD of a data vector $\mathbf{x}(t)$ is expressed as [4]

$$\mathbf{D}_{\mathbf{xx}}(t, f) = \iint \phi(t - u, \tau) \mathbf{x}\left(u + \frac{\tau}{2}\right) \mathbf{x}^H\left(u - \frac{\tau}{2}\right) e^{-j2\pi f \tau} \, du \, d\tau \quad (6.4)$$

where the (i, k)th element of $\mathbf{D}_{\mathbf{xx}}(t, f)$ is given by (6.2) for $i, k = 1, 2, \ldots, m$. The noise-free STFD is obtained by substituting (6.3) in (6.4), resulting in

$$\mathbf{D}_{\mathbf{xx}}(t, f) = \mathbf{A}(\boldsymbol{\Phi}) \mathbf{D}_{\mathbf{ss}}(t, f) \mathbf{A}^H(\boldsymbol{\Phi}) \quad (6.5)$$

where $\mathbf{D}_{ss}(t, f)$ is the TFD matrix of $\mathbf{s}(t)$, which consists of auto- and cross-source TFDs. With the presence of the noise, which is uncorrelated with the signals, the expected value of $\mathbf{D}_{xx}(t, f)$ yields

$$E[\mathbf{D}_{xx}(t, f)] = \mathbf{A}(\boldsymbol{\Phi}) E[\mathbf{D}_{ss}(t, f)] \mathbf{A}^H(\boldsymbol{\Phi}) + \sigma^2 \mathbf{I} \qquad (6.6)$$

where σ^2 is the noise power and $\mathbf{I}$ is the identity matrix.

Equation (6.6) relates the STFD matrix to the source TFD matrix. It is similar to the formula that is commonly used in narrowband array processing problems, relating the source covariance matrix to the sensor spatial covariance matrix. It is clear, therefore, that the two subspaces spanned by the principal eigenvectors of $\mathbf{D}_{xx}(t, f)$ and the columns of $\mathbf{A}(\boldsymbol{\Phi})$ are identical. By constructing the STFD matrix from the t-f points of highly localized signal energy, the corresponding signal and noise subspace estimates become more robust to noise than their counterparts which were obtained using the data covariance matrix [5, 9, 10]. Further, source elimination, rendered through the selection of specific t-f regions, improves DOA estimations.

6.3 Spatial Polarimetric Time-Frequency Distributions

6.3.1 Polarimetric Modeling

For a far-field transverse electromagnetic (TEM) wave incident on the array, shown in Figure 6.1, the electric field can be described as

$$\begin{aligned}\vec{E}(t) &= E_\theta(t)\hat{\theta} + E_\phi(t)\hat{\phi} \\ &= [E_\theta(t) \cos(\theta) \cos(\phi) - E_\phi(t) \sin(\phi)]\hat{x} \\ &\quad + [E_\theta(t) \cos(\theta) \sin(\phi) - E_\phi(t) \cos(\phi)]\hat{y} + E_\theta(t) \sin(\theta)\hat{z}\end{aligned} \qquad (6.7)$$

where $\hat{\phi}$ and $\hat{\theta}$ are, respectively, the spherical unit vectors along the azimuth and elevation angles ϕ and θ, viewed from the source. The unit vectors $\hat{x}$, $\hat{y}$, and $\hat{z}$ are

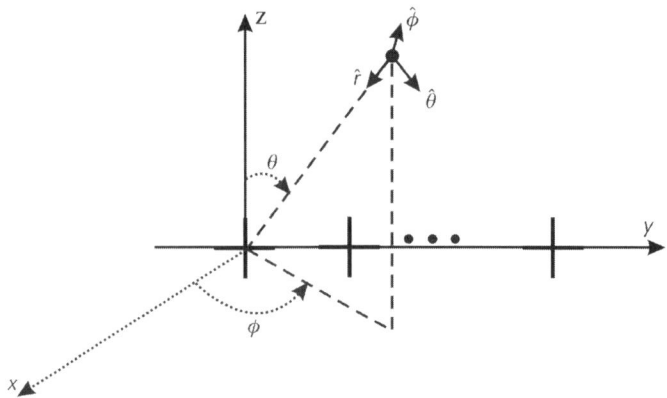

Figure 6.1 Dual-polarized array.

defined along the x, y, and z directions, respectively. For simplicity and without loss of generality, it is assumed that the source signal is in the x-y plane, whereas the array is located in the y-z plane. Accordingly, $\theta = 90°$, $\hat{\theta} = -\hat{z}$, and

$$\vec{E} = -E_\phi(t)\sin(\phi)\hat{x} + E_\phi(t)\cos(\phi)\hat{y} + E_\theta(t)\hat{z} \qquad (6.8)$$

We denote $s(t)$ as the source signal magnitude measured at the receiver reference sensor, with polarization angle $\gamma \in [0°, 90°]$, and polarization phase difference $\eta \in (-180°, 180°]$. The source's vertical and horizontal polarization components, $s^{[v]}(t)$ and $s^{[h]}(t)$, can then be expressed in terms of the respective spherical fields, $E_\theta(t)$ and $E_\phi(t)$, as

$$E_\theta(t) = s^{[v]}(t) = s(t)\cos(\gamma), \; E_\phi(t) = s^{[h]}(t) = s(t)\sin(\gamma)e^{j\eta} \qquad (6.9)$$

A signal is referred to as linearly polarized if $\eta = 0°$ or $\eta = 180°$. Substituting (6.9) into (6.8) results in

$$\vec{E}(t) = s(t)[-\cos(\gamma)\sin(\phi)\hat{x} + \cos(\phi)\sin(\gamma)e^{j\eta}\hat{y} + \cos(\gamma)\hat{z}] \qquad (6.10)$$

Now we consider that n signals impinge on the array, consisting of m dual-polarized antennas. The vertical and horizontal components of the ith source are expressed as

$$s_i^{[v]}(t) = s_i(t)\cos(\gamma_i) = c_{i1}s_i(t), \; s_i^{[h]}(t) = s_i(t)\sin(\gamma_i)e^{j\eta_i} = c_{i2}s_i(t) \qquad (6.11)$$

where the parameters $c_{i1} = \cos(\gamma_i)$ and $c_{i2} = \sin(\gamma_i)e^{j\eta_i}$ denote the vertical and horizontal polarization coefficients, respectively. The signal received at the lth dual-polarization antenna, with vertical and horizontal antennas located in the $\hat{z}$ and $\hat{y}$ directions, is expressed as

$$\begin{aligned}\underline{y}_l(t) &= \left[y_l^{[v]}(t), y_l^{[h]}(t)\right]^T \\ &= \sum_{i=1}^n \left[a_{il}^{[v]}\vec{E}_i(t)\cdot\hat{z}, a_{il}^{[h]}\vec{E}_i(t)\cdot\hat{y}\right]^T \qquad (6.12) \\ &= \sum_{i=1}^n \left[a_{il}^{[v]}s_i^{[v]}(t), a_{il}^{[h]}s_i^{[h]}(t)\cos(\phi_i)\right]^T\end{aligned}$$

where $\vec{E}_i(t)$ is the electric field corresponding to the ith signal, $a_{il}^{[v]}$ and $a_{il}^{[h]}$, respectively, are the lth elements of the vertically and horizontally polarized array vectors, $\mathbf{a}^{[v]}(\phi_i)$ and $\mathbf{a}^{[h]}(\phi_i)$, and "$\cdot$" represents the dot product. It is assumed that the array has been calibrated, and both $\mathbf{a}^{[v]}(\phi_i)$ and $\mathbf{a}^{[h]}(\phi_i)$ are known and normalized, such that $\|\mathbf{a}^{[v]}(\phi_i)\|^2 = \|\mathbf{a}^{[h]}(\phi_i)\|^2 = m$. It is noted that, since the $\cos(\phi_i)$ term is common in all the elements of the horizontally polarized array manifold, it can be absorbed in the array calibration for the region of interest, and removed from further consideration. Then, the above equation is simplified to

$$\underline{y}_l(t) = \left[a_{il}^{[v]} s_i^{[v]}(t), a_{il}^{[h]} s_i^{[h]}(t) \right]^T$$
$$= s_i(t) \left(\left[a_{il}^{[v]}, a_{il}^{[h]} \right]^T \circ [c_{i1}, c_{i2}]^T \right) \quad (6.13)$$
$$= s_i(t) \mathbf{a}_{il} \circ \mathbf{c}_i$$

where the vector

$$\mathbf{c}_i = [c_{i1}, c_{i2}]^T = [\cos(\gamma_i), \sin(\gamma_i) e^{j\eta_i}]^T \quad (6.14)$$

represents the polarization signature of the ith signal.

6.3.2 Polarimetric Time-Frequency Distributions

For the lth dual-polarized sensor, we define the self-polarized TFD as

$$D_{x_l^{[i]} x_l^{[i]}}(t, f) = \iint \phi(t-u, \tau) x_l^{[i]}\left(u+\frac{\tau}{2}\right) \left[x_l^{[i]}\left(u-\frac{\tau}{2}\right)\right]^* e^{-j2\pi f\tau} \, du \, d\tau$$
(6.15)

and the cross-polarized TFD as

$$D_{x_l^{[i]} x_l^{[k]}}(t, f) = \iint \phi(t-u, \tau) x_l^{[i]}\left(u+\frac{\tau}{2}\right) \left[x_l^{[k]}\left(u-\frac{\tau}{2}\right)\right]^* e^{-j2\pi f\tau} \, du \, d\tau$$
(6.16)

where the superscripts i and k denote either v or h. Define

$$\underline{x}_l(t) = \left[x_l^{[v]}, x_l^{[h]} \right]^T \quad (6.17)$$

to include the two polarization components received in a dual-polarized antenna. Then, the self- and cross-polarized TFDs constitute the 2×2 polarimetric TFD (PTFD) matrix

$$\mathbf{D}_{\underline{x}_l \underline{x}_l}(t, f) = \iint \phi(t-u, \tau) \underline{x}_l\left(u+\frac{\tau}{2}\right) \underline{x}_l^H\left(u-\frac{\tau}{2}\right) e^{-j2\pi f\tau} \, du \, d\tau \quad (6.18)$$

The diagonal entries of $\mathbf{D}_{\underline{x}_l \underline{x}_l}(t, f)$ are the self-polarized TFDs $D_{x_l^{[i]} x_l^{[i]}}(t, f)$, whereas the off-diagonal elements are the cross-polarized terms $D_{x_l^{[i]} x_l^{[k]}}(t, f), i \neq k$.

6.3.3 Spatial Polarimetric Time-Frequency Distributions

Equations (6.12) to (6.18) correspond to the case of a single dual-polarization sensor. With an array of m dual-polarized antennas, the data vector, for each polarization i, where $i = v$ or $i = h$, is expressed as

$$\mathbf{x}^{[i]}(t) = \left[x_1^{[i]}(t), x_2^{[i]}(t), \ldots, x_m^{[i]}(t)\right]^T = \mathbf{y}^{[i]}(t) + \mathbf{n}^{[i]}(t) = \mathbf{A}^{[i]}(\mathbf{\Phi})\mathbf{s}^{[i]}(t) + \mathbf{n}^{[i]}(t) \tag{6.19}$$

We generalize the PTFDs to a multisensor receiver. Based on (6.19), the following extended data vector can be constructed for both polarizations,

$$\begin{aligned}
\mathbf{x}(t) &= \begin{bmatrix} \mathbf{x}^{[v]}(t) \\ \mathbf{x}^{[h]}(t) \end{bmatrix} = \begin{bmatrix} \mathbf{A}^{[v]}(\mathbf{\Phi}) & 0 \\ 0 & \mathbf{A}^{[h]}(\mathbf{\Phi}) \end{bmatrix} \begin{bmatrix} \mathbf{s}^{[v]}(t) \\ \mathbf{s}^{[h]}(t) \end{bmatrix} + \begin{bmatrix} \mathbf{n}^{[v]}(t) \\ \mathbf{n}^{[h]}(t) \end{bmatrix} \\
&= \begin{bmatrix} \mathbf{A}^{[v]}(\mathbf{\Phi}) & 0 \\ 0 & \mathbf{A}^{[h]}(\mathbf{\Phi}) \end{bmatrix} \begin{bmatrix} \mathbf{Q}^{[v]} \\ \mathbf{Q}^{[h]} \end{bmatrix} + \begin{bmatrix} \mathbf{n}^{[v]}(t) \\ \mathbf{n}^{[h]}(t) \end{bmatrix} \\
&= \mathbf{B}(\mathbf{\Phi})\mathbf{Q}\mathbf{s}(t) + \mathbf{n}(t)
\end{aligned} \tag{6.20}$$

where

$$\mathbf{B}(\mathbf{\Phi}) = \begin{bmatrix} \mathbf{A}^{[v]}(\mathbf{\Phi}) & 0 \\ 0 & \mathbf{A}^{[h]}(\mathbf{\Phi}) \end{bmatrix} \tag{6.21}$$

is block-diagonal, and

$$\mathbf{Q} = \begin{bmatrix} \mathbf{Q}^{[v]} \\ \mathbf{Q}^{[h]} \end{bmatrix} \tag{6.22}$$

is the polarization signature vector of the sources, where

$$\mathbf{q}^{[v]} = [\cos(\gamma_1), \ldots, \cos(\gamma_n)]^T, \quad \mathbf{Q}^{[v]} = diag(\mathbf{q}^{[v]}) \tag{6.23}$$

$$\mathbf{q}^{[h]} = [\sin(\gamma_1)e^{j\eta_1}, \ldots, \sin(\gamma_n)e^{j\eta_n}]^T, \quad \mathbf{Q}^{[h]} = diag(\mathbf{q}^{[h]}) \tag{6.24}$$

Accordingly,

$$\begin{aligned}
\mathbf{B}(\mathbf{\Phi})\mathbf{Q} &= \begin{bmatrix} \mathbf{a}^{[v]}(\phi_1)\cos(\gamma_1) & \cdots & \mathbf{a}^{[v]}(\phi_n)\cos(\gamma_n) \\ \mathbf{a}^{[h]}(\phi_1)\sin(\gamma_1)e^{j\eta_1} & \cdots & \mathbf{a}^{[h]}(\phi_n)\sin(\gamma_n)e^{j\eta_n} \end{bmatrix} \\
&= [\tilde{\mathbf{a}}(\phi_1) \quad \cdots \quad \tilde{\mathbf{a}}(\phi_n)]
\end{aligned} \tag{6.25}$$

The above matrix can be viewed as the extended mixing matrix, with

$$\tilde{\mathbf{a}}(\phi_k) = \begin{bmatrix} \mathbf{a}^{[v]}(\phi_k)\cos(\gamma_k) \\ \mathbf{a}^{[h]}(\phi_k)\sin(\gamma_k)e^{j\eta_k} \end{bmatrix} \tag{6.26}$$

representing the joint spatio-polarimetric signature of source k.

It is clear that the dual-polarization array, compared to the single-polarization case, doubles the vector space dimensionality. Specifically, when $\mathbf{a}^{[v]}(\phi_k) = \mathbf{a}^{[h]}(\phi_k) = \mathbf{a}(\phi_k)$, the above equation simplifies to

$$\tilde{\mathbf{a}}(\phi_k) = \mathbf{a}^{[h]}(\phi_k) \otimes \mathbf{c}_k \qquad (6.27)$$

It is now possible to combine the polarimetric, spatial, and t-f properties of the source signals incident on the receiver array. The STFD of the dual-polarization data vector, $\mathbf{x}(t)$, can be used in formulating the following SPTFD matrix

$$\mathbf{D}_{\mathbf{xx}}(t, f) = \iint \phi(t - u, \tau) \mathbf{x}\left(u + \frac{\tau}{2}\right) \mathbf{x}^H\left(u - \frac{\tau}{2}\right) e^{-j2\pi f \tau} \, du \, d\tau \qquad (6.28)$$

This SPTFD matrix embodies the information required to address typical problems in array processing, including direction finding, as will be shown in the next section.

When the effect of noise is ignored, the SPTFD matrix is related to the source TFD matrix by the following equation

$$\mathbf{D}_{\mathbf{xx}}(t, f) = \mathbf{B}(\mathbf{\Phi}) \mathbf{Q} \mathbf{D}_{\mathbf{ss}}(t, f) \mathbf{Q}^H \mathbf{B}^H(\mathbf{\Phi}) \qquad (6.29)$$

6.4 Polarimetric Time-Frequency MUSIC

Time-frequency MUSIC (TF-MUSIC) has been recently introduced for improved spatial resolution for signals with clear t-f signatures [8]. It is based on the eigenstructure of the STFD matrix. The PTF-MUSIC is an important generalization of the TF-MUSIC to deal with diversely polarized signals and polarized arrays. It is based on the SPTFD matrix of (6.28).

Consider the following spatial signature matrix

$$\mathbf{F}(\phi) = \frac{1}{\sqrt{m}} \begin{bmatrix} \mathbf{a}^{[v]}(\phi) & 0 \\ 0 & \mathbf{a}^{[h]}(\phi) \end{bmatrix} \qquad (6.30)$$

corresponding to DOA ϕ. Since $\|\mathbf{a}^{[i]}(\phi)\|^2 = m$, $\mathbf{F}^H(\phi)\mathbf{F}(\phi)$ is the 2×2 identity matrix.

To search in the joint spatial and polarimetric domains, we define the following spatio-polarimetric search vector

$$\mathbf{f}(\phi, \mathbf{c}) = \frac{\mathbf{F}(\phi)\mathbf{c}}{\|\mathbf{F}(\phi)\mathbf{c}\|} = \mathbf{F}(\phi)\mathbf{c} \qquad (6.31)$$

where the vector $\mathbf{c} = [c_1, c_2]^T$ is a unit norm vector with unknown polarization coefficients c_1 and c_2. In (6.31), we have used the fact that

$$\|\mathbf{F}(\phi)\mathbf{c}\| = [\mathbf{c}^H \mathbf{F}^H(\phi)\mathbf{F}(\phi)\mathbf{c}]^{1/2} = (\mathbf{c}^H \mathbf{c})^{1/2} = 1$$

The PTF-MUSIC spectrum is given by the following function [23, 24]

$$P(\phi) = \left[\min_{\mathbf{c}} \mathbf{f}^H(\phi, \mathbf{c}) \mathbf{U}_n \mathbf{U}_n^H \mathbf{f}(\phi, \mathbf{c}) \right]^{-1} = \left[\min_{\mathbf{c}} \mathbf{c}^H \mathbf{F}^H(\phi) \mathbf{U}_n \mathbf{U}_n^H \mathbf{F}(\phi) \mathbf{c} \right]^{-1} \tag{6.32}$$

where $\mathbf{U}_n$ is the noise subspace obtained from the SPTFD matrix using the selected t-f points. Time-frequency averaging and joint block-diagonalization are two known techniques that can be used to integrate the different STFD or SPTFD matrices constructed from multiple t-f points [5, 8, 27]. The selection of those points from high energy concentration regions pertaining to all or some of the sources enhances the SNR, and allows the t-f–based MUSIC algorithms to be more robust to noise compared to their conventional MUSIC counterparts [5].

In (6.32), the term in brackets is minimized by finding the minimum eigenvalue of the 2×2 matrix $\mathbf{F}^H(\phi) \mathbf{U}_n \mathbf{U}_n^H \mathbf{F}(\phi)$. Thus, a computationally expensive search in the polarization domain is avoided by performing a simple eigendecomposition on the 2×2 matrix. As a result, the PTF-MUSIC spectrum can be expressed as

$$P(\phi) = \lambda_{\min}^{-1} \left[\mathbf{F}^H(\phi) \mathbf{U}_n \mathbf{U}_n^H \mathbf{F}(\phi) \right] \tag{6.33}$$

where $\lambda_{\min}[\cdot]$ denotes the minimum eigenvalue operator. The DOAs of the sources are estimated as the locations of the highest peaks in the PTF-MUSIC spectrum. For each angle ϕ_k corresponding to one of the n signal arrivals, $k = 1, 2, \ldots, n$, the polarization parameters of the respective source signal can be estimated from

$$\hat{\mathbf{c}}(\phi_k) = \mathbf{v}_{\min} \left[\mathbf{F}^H(\phi_k) \mathbf{U}_n \mathbf{U}_n^H \mathbf{F}(\phi_k) \right] \tag{6.34}$$

where $\mathbf{v}_{\min}[\cdot]$ is the eigenvector corresponding to the minimum eigenvalue $\lambda_{\min}[\cdot]$.

6.5 Spatio-Polarimetric Correlations

In a conventional single-polarized array problem, the spatial resolution capability of an array greatly depends on the correlation between the propagation signatures of the source arrivals [5, 28]. This is determined by the normalized inner product of the respective array manifold vectors. In the underlying problem, in which both the spatial and polarimetric dimensions are involved, the joint spatio-polarimetric correlation coefficient between sources l and k is defined using the extended array manifold $\tilde{\mathbf{a}}(\phi)$, as

$$\beta_{l,k} = \frac{1}{m} \tilde{\mathbf{a}}^H(\phi_k) \tilde{\mathbf{a}}(\phi_l)$$

$$= \frac{1}{m} \left\{ c_{k1}^* c_{l1} \left[\mathbf{a}^{[v]}(\phi_k) \right]^H \mathbf{a}^{[v]}(\phi_l) + c_{k2}^* c_{l2} \left[\mathbf{a}^{[h]}(\phi_k) \right]^H \mathbf{a}^{[h]}(\phi_l) \right\} \tag{6.35}$$

$$= c_{k1}^* c_{l1} \beta_{l,k}^{[v]} + c_{k2}^* c_{l2} \beta_{l,k}^{[h]}$$

where

$$\beta_{l,k}^{[i]} = \frac{1}{m}[\mathbf{a}^{[i]}(\phi_k)]^H \mathbf{a}^{[i]}(\phi_l)$$

is the spatial correlation coefficient between sources l and k for polarization i, with i denoting v or h.

An interesting case arises when the vertically and horizontally polarized array manifolds are identical; that is, when $\mathbf{a}^{[v]}(\phi) = \mathbf{a}^{[h]}(\phi)$. In this case, $\beta_{l,k}^{[v]} = \beta_{l,k}^{[h]}$, and the joint spatio-polarimetric correlation coefficient becomes the product of the individual spatial and polarimetric correlations; that is,

$$\beta_{l,k} = \beta_{l,k}^{[v]} \rho_{l,k} \tag{6.36}$$

with

$$\rho_{l,k} = \mathbf{c}_k^H \mathbf{c}_l = \cos(\gamma_l)\cos(\gamma_k)e^{j(\eta_l - \eta_k)} + \sin(\gamma_l)\sin(\gamma_k) \tag{6.37}$$

representing the polarimetric correlation coefficient. In particular, for linear polarizations, $\eta_l = \eta_k = 0$, and (6.37) reduces to

$$\rho_{l,k} = \cos(\gamma_l - \gamma_k) \tag{6.38}$$

Since $|\rho_{l,k}| \leq 1$, the equality holds only when the two sources have identical polarization states, and the spatio-polarization correlation coefficient is always equal to or smaller than that of the individual spatial correlation coefficient. The reduction in the correlation value due to polarization diversity, through the introduction of $\rho_{l,k}$, translates to improved distinctions of sources. As such, two sources that could be difficult to resolve using the single-polarized spatial array manifold $\mathbf{a}^{[v]}(\phi)$ or $\mathbf{a}^{[h]}(\phi)$ can be easily separated using the extended spatio-polarized array manifold, defined by $\tilde{\mathbf{a}}(\phi)$. This improvement is more evident in the case when the source spatial correlation is high, but the respective polarimetric correlation is low.

6.6 Moving Sources with Time-Varying Polarizations

In this section, we consider the performance of the tracking of moving targets with time-varying polarization signatures. Time-varying polarizations are often observed when active or passive sources move or change orientations [29]. The performance of polarimetric MUSIC (P-MUSIC) and PTF-MUSIC techniques are discussed and compared. The SPTFD maintains the instantaneous polarization information at the autoterm t-f points, whereas the energy of desired source signals is localized. Careful treatment of the data samples is required to effectively utilize the polarization diversity, since the construction of the covariance and SPTFD matrices using all available data samples may compromise the polarization distinctions between sources.

Because the SNR enhancement using multisensor t-f processing is well documented [5, 6, 10], we consider the noise-free case in this section, for notational simplicity. In this case, the received signal vector is

6.6 Moving Sources with Time-Varying Polarizations

$$\mathbf{x}(t) = \begin{bmatrix} \mathbf{A}^{[v]}[\boldsymbol{\Phi}(t)][\mathbf{q}^{[v]}(t) \circ \mathbf{s}(t)] \\ \mathbf{A}^{[h]}[\boldsymbol{\Phi}(t)][\mathbf{q}^{[h]}(t) \circ \mathbf{s}(t)] \end{bmatrix} \qquad (6.39)$$

with $\mathbf{A}^{[i]}[\boldsymbol{\Phi}(t)] = \{\mathbf{a}^{[i]}[\phi_1(t)], \ldots, \mathbf{a}^{[i]}[\phi_n(t)]\}$, $i = v, h$ is the time-varying array response matrix. Note that we used $\boldsymbol{\Phi}(t) = [\phi_1(t)], \ldots, \phi_n(t)]$ to emphasize the fact that the DOAs are now time-varying.

6.6.1 Polarization Diversity and Selection of Data Samples

In order to utilize polarization diversity properly, careful attention should be paid to the selection of data samples. We use the P-MUSIC as the example.

Given the time-varying nature of the source signal polarizations, the covariance matrix of the received signal vector is

$$\mathbf{R}_{\mathbf{xx}}(t) = E[\mathbf{x}(t)\mathbf{x}^H(t)] \qquad (6.40)$$

$$= E\left(\begin{Bmatrix} \mathbf{A}^{[v]}[\boldsymbol{\Phi}(t)][\mathbf{q}^{[v]}(t) \circ \mathbf{s}(t)] \\ \mathbf{A}^{[h]}[\boldsymbol{\Phi}(t)][\mathbf{q}^{[h]}(t) \circ \mathbf{s}(t)] \end{Bmatrix} \begin{Bmatrix} \mathbf{A}^{[v]}[\boldsymbol{\Phi}(t)][\mathbf{q}^{[v]}(t) \circ \mathbf{s}(t)] \\ \mathbf{A}^{[h]}[\boldsymbol{\Phi}(t)][\mathbf{q}^{[h]}(t) \circ \mathbf{s}(t)] \end{Bmatrix}^H\right)$$

When the covariance matrix is replaced by using a time average over the entire data sample, and considering the special case where the spatial, temporal, and polarimetric signatures can be decoupled, it becomes

$$\overline{\mathbf{R}}_{\mathbf{xx}} = \overline{\mathbf{B}} \left\{ \begin{array}{cc} \overline{\mathbf{q}^{[v]}(t)[\mathbf{q}^{[v]}(t)]^H} \circ \overline{\mathbf{R}}_{ss} & \overline{\mathbf{q}^{[v]}(t)[\mathbf{q}^{[h]}(t)]^H} \circ \overline{\mathbf{R}}_{ss} \\ \overline{\mathbf{q}^{[h]}(t)[\mathbf{q}^{[v]}(t)]^H} \circ \overline{\mathbf{R}}_{ss} & \overline{\mathbf{q}^{[h]}(t)[\mathbf{q}^{[h]}(t)]^H} \circ \overline{\mathbf{R}}_{ss} \end{array} \right\} \overline{\mathbf{B}}^H \qquad (6.41)$$

$$= \overline{\mathbf{B}} \left(\left\{ \begin{array}{cc} \overline{\mathbf{q}^{[v]}(t)[\mathbf{q}^{[v]}(t)]^H} & \overline{\mathbf{q}^{[v]}(t)[\mathbf{q}^{[h]}(t)]^H} \\ \overline{\mathbf{q}^{[h]}(t)[\mathbf{q}^{[v]}(t)]^H} & \overline{\mathbf{q}^{[h]}(t)[\mathbf{q}^{[h]}(t)]^H} \end{array} \right\} \circ \left\{ \begin{array}{cc} \overline{\mathbf{R}}_{ss} & \overline{\mathbf{R}}_{ss} \\ \overline{\mathbf{R}}_{ss} & \overline{\mathbf{R}}_{ss} \end{array} \right\} \right) \overline{\mathbf{B}}^H$$

where $\overline{\mathbf{R}}_{ss} = \overline{\mathbf{s}(t)\mathbf{s}^H(t)}$, and

$$\overline{\mathbf{B}} = \begin{bmatrix} \overline{\mathbf{A}^{[v]}[\boldsymbol{\Phi}(t)]} & 0 \\ 0 & \overline{\mathbf{A}^{[h]}[\boldsymbol{\Phi}(t)]} \end{bmatrix}$$

The time-varying source signal polarization vectors are defined, similarly to (6.23) and (6.24), as

$$\mathbf{q}^{[v]}(t) = \{\cos[\gamma_1(t)], \ldots, \cos[\gamma_n(t)]\}^T$$

$$\mathbf{q}^{[h]}(t) = \{\sin[\gamma_1(t)]e^{j\eta_1(t)}, \ldots, \sin[\gamma_n(t)]e^{j\eta_n(t)}\}^T$$

Consider an alternative scenario where the polarimetric signature is time-invariant. Then, the covariance matrix can be expressed as

$$\mathbf{R}'_{xx} = \overline{\mathbf{B}} \begin{bmatrix} \mathbf{q}^{[v]}(\mathbf{q}^{[v]})^H & \mathbf{q}^{[v]}(\mathbf{q}^{[h]})^H \\ \mathbf{q}^{[h]}(\mathbf{q}^{[v]})^H & \mathbf{q}^{[h]}(\mathbf{q}^{[h]})^H \end{bmatrix} \circ \begin{bmatrix} \overline{\mathbf{R}}_{ss} & \overline{\mathbf{R}}_{ss} \\ \overline{\mathbf{R}}_{ss} & \overline{\mathbf{R}}_{ss} \end{bmatrix} \overline{\mathbf{B}}^H \qquad (6.42)$$

The effect of the signal time-varying polarization on the covariance matrix is clear from (6.41) and (6.42). When $\mathbf{q}^{[i]}(t)[\mathbf{q}^{[k]}(t)]^H = \mathbf{q}^{[i]}(t)[\mathbf{q}^{[k]}]^H$ for $i, k = v, h$, the corresponding covariance matrices are identical, and the two cases of sources with time-varying and time-invariant polarizations will lead to the same performance. Consider, for example, two linearly polarized sources, whose polarization angles change linearly over $[0°, 90°]$ for γ_1 and $[90°, 0°]$ for γ_2. The DOAs are assumed to be time-invariant. This case is equivalent to both sources, assuming fixed, time-invariant polarization of $\gamma = 45°$, and thereby, the source polarization diversity cannot be utilized in DOA estimation using P-MUSIC.

To utilize the source signal polarization diversity, the received data covariance matrix in (6.41) should be constructed using a selected set of data samples (e.g., the moving average of the received data samples), such that the polarization distinction between sources is maintained, even if the source DOAs show insignificant changes over the observation period. In general, for moving target tracking problems, the selection of the data window for the moving average covariance matrix should be determined, such that the DOA variation over the window is small.

However, we maintain that the reduction of data samples may compromise the subspace estimation, and, subsequently, DOA estimation performance, particularly when the SNR is relatively low, and/or the sources are closely spaced. The use of t-f representations overcomes this shortcoming, and provides robust subspace estimation in rapidly time-varying polarization environments.

6.6.2 Polarization Diversity Consideration in PTF-MUSIC

We now show that the instantaneous polarization information is incorporated in the SPTFD. Therefore, PTF-MUSIC maintains polarization diversity, while providing SNR enhancement and source discriminations.

In the presence of time-varying polarized sources, the auto- and cross-polarized SPTFD matrix can be expressed as

$$\begin{aligned} D_{\mathbf{x}^{[i]}\mathbf{x}^{[k]}}(t, f) &= \iint \phi(t-u, \tau) \mathbf{x}^{[i]}\left(u+\frac{\tau}{2}\right)\left[\mathbf{x}^{[k]}\left(u-\frac{\tau}{2}\right)\right]^H e^{-j2\pi f\tau} du\, d\tau \\ &= \iint \phi(t-u, \tau) \mathbf{A}^{[i]}\left[\mathbf{\Phi}\left(t+\frac{\tau}{2}\right)\right] \left(\left\{\mathbf{q}^{[i]}\left(u+\frac{\tau}{2}\right)\left[\mathbf{q}^{[k]}\left(u-\frac{\tau}{2}\right)\right]^H\right\}\right. \\ &\quad \left. \circ \left\{\mathbf{s}\left(u+\frac{\tau}{2}\right)\left[\mathbf{s}\left(u-\frac{\tau}{2}\right)\right]^H\right\}\right)\left\{\mathbf{A}^{[k]}\left[\mathbf{\Phi}\left(t-\frac{\tau}{2}\right)\right]\right\}^H e^{-j2\pi f\tau} du\, d\tau \\ &= \iint \phi(t-u, \tau) \mathbf{A}^{[i]}\left[\mathbf{\Phi}\left(t+\frac{\tau}{2}\right)\right]\left[\mathbf{G}^{[ik]}(u, \tau) \circ \mathbf{K}(u, \tau)\right] \\ &\quad \left\{\mathbf{A}^{[k]}\left[\mathbf{\Phi}\left(t-\frac{\tau}{2}\right)\right]\right\}^H e^{-j2\pi f\tau} du\, d\tau \qquad (6.43) \end{aligned}$$

6.6 Moving Sources with Time-Varying Polarizations

where

$$\mathbf{G}^{[ik]}(t, \tau) = \mathbf{q}^{[i]}\left(t + \frac{\tau}{2}\right)\left[\mathbf{q}^{[k]}\left(t - \frac{\tau}{2}\right)\right]^H$$

and

$$\mathbf{K}(t, \tau) = \mathbf{s}\left(t + \frac{\tau}{2}\right)\mathbf{s}^H\left(t - \frac{\tau}{2}\right)$$

When the window length of the applied t-f is small

$$\mathbf{D}_{\mathbf{x}^{[i]}\mathbf{x}^{[k]}}(t, f) \approx \mathbf{A}^{[i]}[\Phi(t)]\mathbf{D}_{\mathbf{s}^{[i]}\mathbf{s}^{[k]}}(t, f)\{\mathbf{A}^{[k]}[\Phi(t)]\}^H \quad (6.44)$$

where

$$\mathbf{D}_{\mathbf{s}^{[i]}\mathbf{s}^{[k]}}(t, f) = \iint \phi(t - u, \tau)\mathbf{G}^{[ik]}(u, \tau) \circ \mathbf{K}(u, \tau) e^{-j2\pi f\tau} \, du \, d\tau \quad (6.45)$$

We assume that the frequency and the polarization signatures of the sources change almost linearly within the temporal span of the t-f kernel. Then, using the first-order Taylor series expansion, the polarization-dependent terms can be approximated as

$$\gamma_i\left(t + \frac{\tau}{2}\right) = \gamma_i(t) + \frac{\tau}{2}\dot{\gamma}_i(t)$$

where

$$\dot{\gamma}_i(t) = \frac{d}{dt}\gamma_i(t)$$

The autoterms of the source polarization information, which reside on the diagonals of $\mathbf{G}^{[vv]}(t, \tau)$, $\mathbf{G}^{[vh]}(t, \tau)$, $\mathbf{G}^{[hv]}(t, \tau)$, and $\mathbf{G}^{[vv]}(t, \tau)$, are given by

$$\left[\mathbf{G}^{[vv]}(t, \tau)\right]_{ii} = \frac{1}{2}\{\cos[2\gamma_i(t)] + \cos[\tau\dot{\gamma}_i(t)]\} \quad (6.46)$$

$$\left[\mathbf{G}^{[vh]}(t, \tau)\right]_{ii} = \frac{1}{2}\{\sin[2\gamma_i(t)] - \sin[\tau\dot{\gamma}_i(t)]\} \quad (6.47)$$

$$\left[\mathbf{G}^{[hv]}(t, \tau)\right]_{ii} = \frac{1}{2}\{\sin[2\gamma_i(t)] + \sin[\tau\dot{\gamma}_i(t)]\} \quad (6.48)$$

$$\left[\mathbf{G}^{[hh]}(t, \tau)\right]_{ii} = \frac{1}{2}\{-\cos[2\gamma_i(t)] + \cos[\tau\dot{\gamma}_i(t)]\} \quad (6.49)$$

respectively. The second sinusoidal terms in (6.47) and (6.48) result in zero values in $\mathbf{D}_{\mathbf{s}_i^{[l]} \mathbf{s}_i^{[k]}}(t, f)$ for symmetric t-f kernels. Therefore, $\mathbf{D}_{\mathbf{s}_i^{[l]} \mathbf{s}_i^{[k]}}(t, f)$ can be expressed at the autoterm points as

$$D_{s_i^{[v]} s_i^{[v]}}(t, f) = \frac{1}{2} \cos[2\gamma_i(t)] D_{s_i s_i}(t, f) + c_{ii}(t, f) \qquad (6.50)$$

$$D_{s_i^{[h]} s_i^{[h]}}(t, f) = -\frac{1}{2} \cos[2\gamma_i(t)] D_{s_i s_i}(t, f) + c_{ii}(t, f) \qquad (6.51)$$

$$D_{s_i^{[v]} s_i^{[h]}}(t, f) = D_{s_i^{[h]} s_i^{[v]}}(t, f) = \frac{1}{2} \sin[2\gamma_i(t)] D_{s_i s_i}(t, f) \qquad (6.52)$$

with

$$c_{ii}(t, f) = \frac{1}{2} \iint \phi(t - u, \tau) \cos[\tau \dot{\gamma}_i(t)] [\mathbf{K}(t, \tau)]_{ii} e^{-j2\pi f \tau} du \, d\tau \qquad (6.53)$$

When different sources are uncorrelated, their t-f signatures have no significant overlap. If the t-f points located in the autoterm region of the ith source are used in constructing the SPTFD matrix, then

$$\mathbf{D_{xx}}(t, f) = \begin{bmatrix} \mathbf{a}^{[v]}[\phi_i(t)] & 0 \\ 0 & \mathbf{a}^{[h]}[\phi_i(t)] \end{bmatrix} \mathbf{M}_i \begin{bmatrix} \mathbf{a}^{[v]}[\phi_i(t)] & 0 \\ 0 & \mathbf{a}^{[h]}[\phi_i(t)] \end{bmatrix}^H \qquad (6.54)$$

where

$$\mathbf{M}_i = \frac{1}{2} D_{s_i s_i}(t, f) \begin{bmatrix} \cos[2\gamma_i(t)] & \sin[2\gamma_i(t)] \\ \sin[2\gamma_i(t)] & -\cos[2\gamma_i(t)] \end{bmatrix} + \begin{bmatrix} c_{ii}(t, f) & 0 \\ 0 & c_{ii}(t, f) \end{bmatrix} \qquad (6.55)$$

In the new structure of the SPTFD matrix of (6.54), the source time-varying polarization has the effect of loading the diagonal elements with $c_{ii}(t, f)$, and, as such, alters the eigenvalues of the above 2×2 matrix. However, the eigenvectors of $\mathbf{M}_i$ remain unchanged. The new eigenvalues are

$$\lambda_{i,1,2} = c_{ii}(t, f) \pm \frac{1}{2} D_{s_i s_i}(t, f)$$

The signal polarization signature (i.e., the eigenvector corresponding to the maximum eigenvalue), is $\mathbf{v}_{i,\max} = \{\cos[\gamma_i(t)], \sin[\gamma_i(t)]\}^T$. Therefore, in the context of PTF-MUSIC, the instantaneous polarization characteristics can be utilized for source discriminations.

6.7 Simulation Results

Two chirp signals are considered, and their parameters are listed in Table 6.1. The two sources impinge on a ULA of five cross-dipoles with an interelement spacing of one-half wavelength. The array responses in both horizontal and vertical polarizations are identical. The input SNR is 5 dB. The source signals' polarization angles $[\gamma_i(t), i = 1, 2]$ change linearly with time over the observation period of $T = 256$ samples (see Figure 6.2). Note that sources with polarization angles of 0° and 180° are both horizontally polarized, whereas a polarization angle difference of 90° implies orthogonal polarizations. Figure 6.3 shows the polarization-averaged pseudo Wigner-Ville distribution (PWVD) of the received data at the reference sensor with a window length of 65. When the sources have low polarization correlation, the cross-terms of the signal components corresponding to the vertical and horizontal polarizations cancel out, resulting in low polarization-averaged cross-term values.

In tracking the moving sources' DOAs, different sets of SPTFD matrices are constructed. Each set uses P consecutive (neighboring) t-f points of the two sources' autoterms, with the middle t-f point being at t_i. The objective is to examine the proposed algorithm performance at different source polarization states, and to

Table 6.1 Signal Parameters of Moving Sources with Time-Varying Polarizations

	Start Frequency	End Frequency	DOA	γ	η
Source 1	0	0.20	−15° to +5°	0° to 180°	0°
Source 2	0.20	0.40	−10° to +10°	180° to 0°	0°

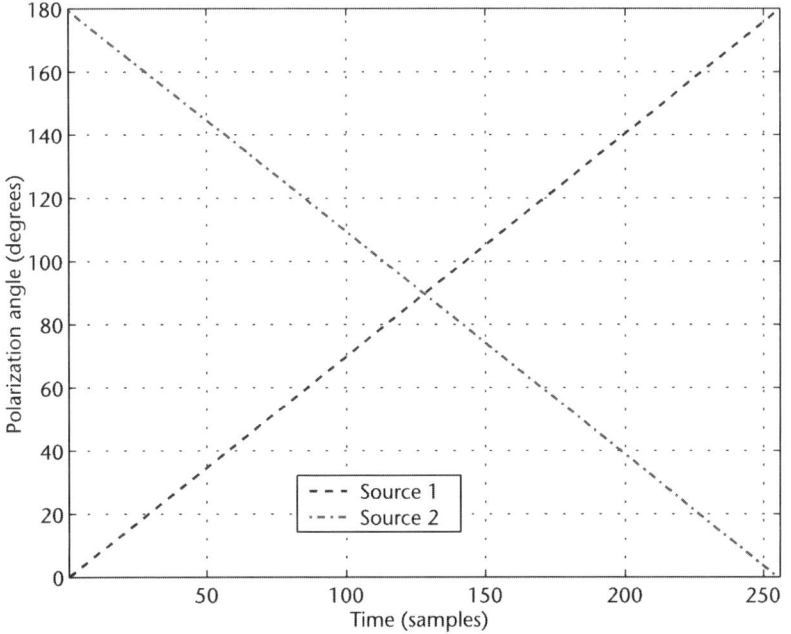

Figure 6.2 Time-varying polarization signatures.

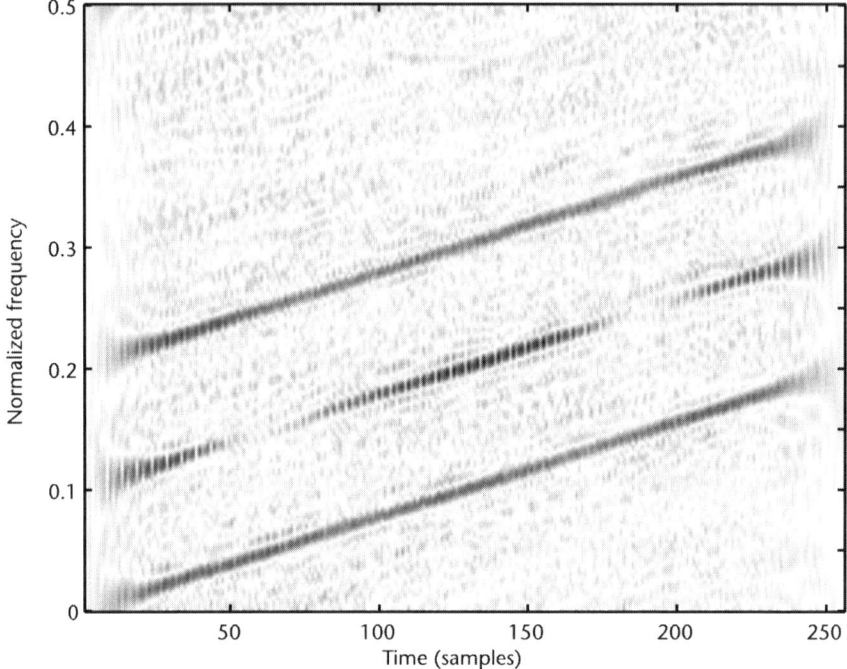

Figure 6.3 Polarization-averaged PWVD of the received data at the reference sensor.

demonstrate the tracking accuracy of the algorithm. This is achieved by separately applying the PTF-MUSIC for each set of SPTFD matrices.

Figure 6.4(a) shows the tracking accuracy of the PTF-MUSIC algorithm for the DOA estimations of the two sources with $P = 31$, and using a PWVD with window length $L = 65$. The performance of P-MUSIC is also shown in Figure 6.4(b) for $P = 31$, and in Figure 6.4(c) for $P = 95$. The tracking accuracy of the algorithms was evaluated at 49 time points that are spaced five samples apart (i.e., $t_1 = 8$ and $t_{49} = 248$). We note that both techniques use the same data samples. The PTF-MUSIC generally outperforms the P-MUSIC, as can be seen in Figure 6.4. The performance of the PTF-MUSIC is particularly impressive in regions where both the signal energy is highly localized in the t-f domain and the polarization distinction between the two sources is high. The importance of polarization diversity is emphasized by the fact that, in the central and edge regions where the two sources have low polarization distinction, the performance of both PTF-MUSIC and P-MUSIC suffers.

6.8 Conclusions

This chapter considers the application of PTF-MUSIC for tracking moving sources with time-varying polarizations. It has been shown that the difference in the instantaneous polarizations of the sources can be uniquely utilized by the proposed approach to maintain polarization diversity. It is shown that the PTF-MUSIC

6.8 Conclusions

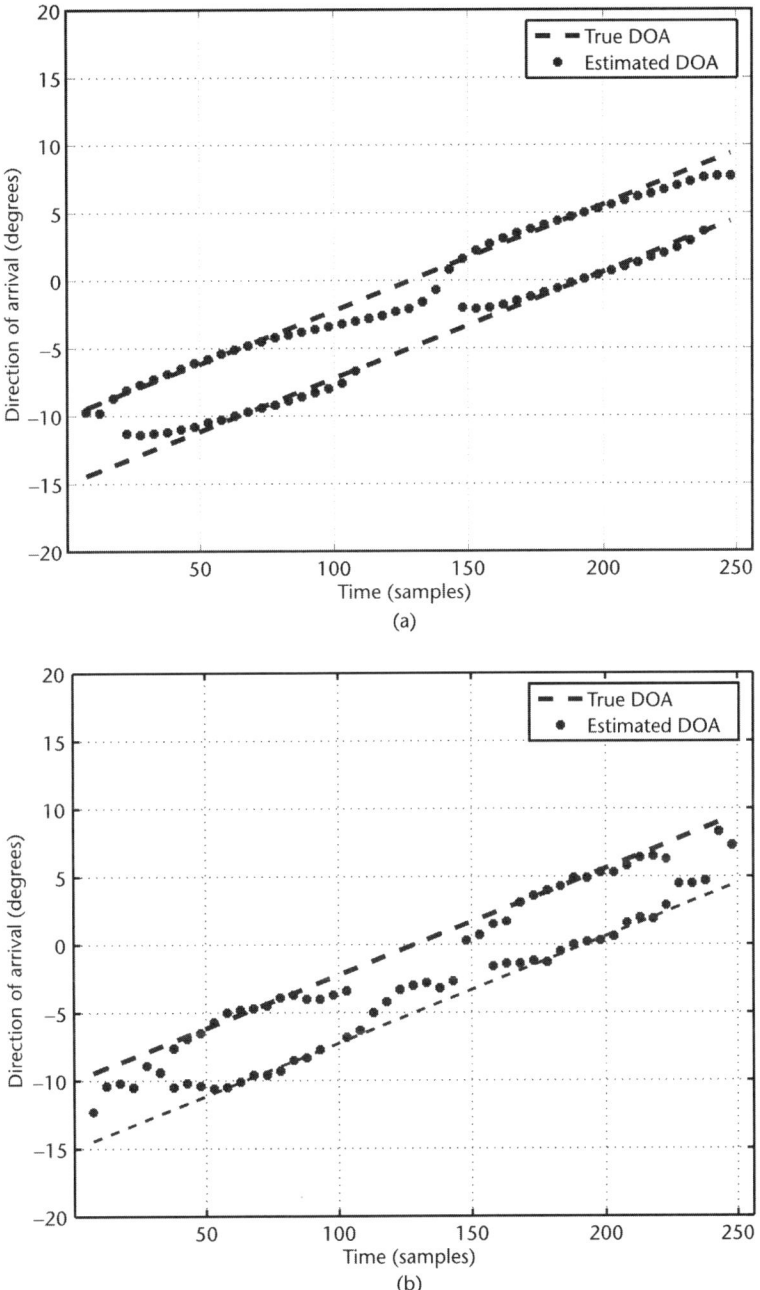

Figure 6.4 PTF-MUSIC and P-MUSIC tracking performance. (a) PTF-MUSIC, using 31 time-frequency points; (b) P-MUSIC, using 31 data samples; and (c) P-MUSIC, using 95 data samples.

approach is superior to a covariance matrix approach that is based on a moving window of the received data. The proposed approach also allows for the discrimination of sources based on their time-frequency signal characteristics.

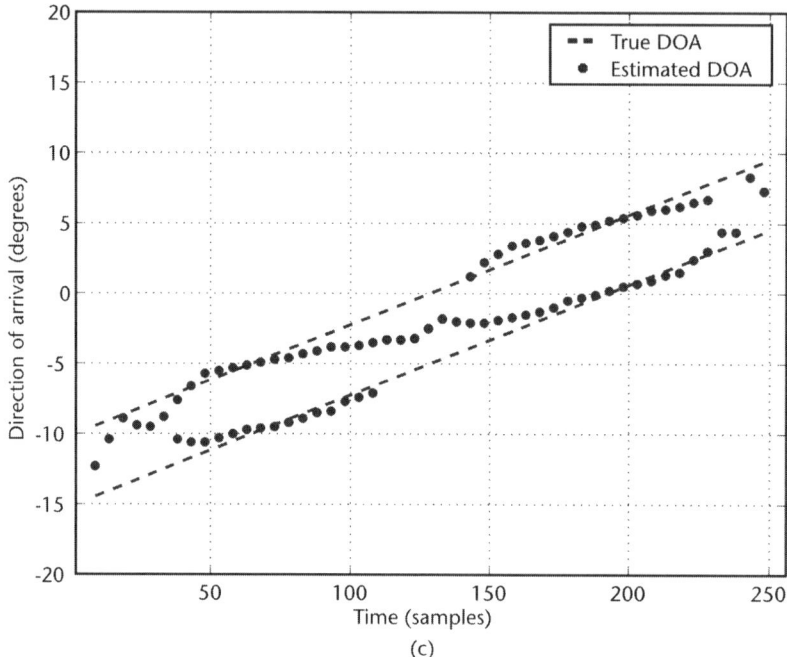

Figure 6.4 (Continued.)

References

[1] Cohen, L., "Time-Frequency Distributions—A Review," *IEEE Proc.*, Vol. 77, No. 7, July 1989, pp. 941–981.

[2] Qian, S., and D. Chen, *Joint Time-Frequency Analysis—Methods and Applications*, Englewood Cliffs, NJ: Prentice Hall, 1996.

[3] Boashash, B., (ed.), *Time-Frequency Signal Analysis and Processing*, New York: Elsevier, 2003.

[4] Belouchrani, A., and M. G. Amin, "Blind Source Separation Based on Time-Frequency Signal Representations," *IEEE Trans. on Signal Processing*, Vol. 46, No. 11, November 1998, pp. 2888–2897.

[5] Zhang, Y., W. Mu, and M. G. Amin, "Subspace Analysis of Spatial Time-Frequency Distribution Matrices," *IEEE Trans. on Signal Processing*, Vol. 49, No. 4, April 2001, pp. 747–759.

[6] Amin, M. G., et al., "Spatial Time-Frequency Distributions: Theory and Applications," in *Wavelets and Signal Processing*, L. Debnath, (ed.), Boston, MA: Birkhauser, 2003.

[7] Mu, W., M. G. Amin, and Y. Zhang, "Bilinear Signal Synthesis in Array Processing," *IEEE Trans. on Signal Processing*, Vol. 51, January 2003, pp. 90–100.

[8] Belouchrani, A., and M. G. Amin, "Time-Frequency MUSIC," *IEEE Signal Processing Letters*, Vol. 6, May 1999, pp. 109–110.

[9] Amin, M. G., and Y. Zhang, "Direction Finding Based on Spatial Time-Frequency Distribution Matrices," *Digital Signal Processing*, Vol. 10, No. 4, October 2000, pp. 325–339.

[10] Zhang, Y., W. Mu, and M. G. Amin, "Time-Frequency Maximum Likelihood Methods for Direction Finding," *J. Franklin Inst.*, Vol. 337, No. 4, July 2000, pp. 483–497.

[11] Hassanien, A., A. B. Gershman, and M. G. Amin, "Time-Frequency ESPRIT for Direction-of-Arrival Estimation of Chirp Signals," *IEEE Sensor Array and Multichannel Signal Processing Workshop*, Rosslyn, VA, August 2002, pp. 337–341.

[12] Lee, W. C. Y., and Y. S. Yeh, "Polarization Diversity for Mobile Radio," *IEEE Trans. on Communications*, Vol. COM-20, May 1972, pp. 912–923.

[13] D. Giuli, "Polarization Diversity in Radars," *Proc. IEEE*, Vol. 74, No. 2, February 1986, pp. 245–269.

[14] Kozlov, A. I., L. P. Ligthart, and A. I. Logvin, *Mathematical and Physical Modelling of Microwave Scattering and Polarimetric Remote Sensing*, Boston, MA: Kluwer Academic, 2001.

[15] McLaughlin, D. J., et al., "Fully Polarimetric Bistatic Radar Scattering Behavior of Forested Hills," *IEEE Trans. on Antennas and Propagation*, Vol. 50, No. 2, February 2002, pp. 101–110.

[16] Sadjadi, F., "Improved Target Classification Using Optimum Polarimetric SAR Signatures," *IEEE Trans. on Aerospace and Electronic Systems*, Vol. 38, No. 1, January 2002, pp. 38–49.

[17] Pazmany, A. L., et al., "An Airborne 95 GHz Dual-Polarized Radar for Cloud Studies," *IEEE Trans. on Geoscience and Remote Sensing*, Vol. 32, No. 4, July 1994, pp. 731–739.

[18] Yueh, S. H., W. J. Wilson, and S. Dinardo, "Polarimetric Radar Remote Sensing of Ocean Surface Wind," *IEEE Trans. on Geoscience and Remote Sensing*, Vol. 40, No. 4, April 2002, pp. 793–800.

[19] Ioannidids, G. A., and D. E. Hammers, "Optimum Antenna Polarizations for Target Discrimination in Clutter," *IEEE Trans. on Antennas and Propagation*, Vol. AP-27, 1979, pp. 357–363.

[20] Garren, D. A., et al., "Full Polarization Matched-Illumination for Target Detection and Identification," *IEEE Trans. on Aerospace and Electronic Systems*, Vol. 38, No. 3, July 2002, pp. 824–837.

[21] Ferrara, E. R., and T. M. Parks, "Direction Finding with an Array of Antennas Having Diverse Polarizations," *IEEE Trans. on Antennas and Propagation*, Vol. 31, March 1983, pp. 231–236.

[22] Li, J., and R. J. Compton, "Angle and Polarization Estimation Using ESPRIT with a Polarization Sensitive Array," *IEEE Trans. on Antennas and Propagation*, Vol. 39, September 1991, pp. 1376–1383.

[23] Zhang, Y., M. G. Amin, and B. A. Obeidat, "Polarimetric Array Processing for Nonstationary Signals," in *Adaptive Antenna Arrays: Trends and Applications*, S. Chandran, (ed.), Berlin, Germany: Springer-Verlag, 2004.

[24] Zhang, Y., B. A. Obeidat, and M. G. Amin, "Spatial Polarimetric Time-Frequency Distributions for Direction-of-Arrival Estimations," *IEEE Trans. on Signal Processing* (in press).

[25] Amin, M. G., and Y. Zhang, "Bilinear Signal Synthesis Using Polarization Diversity," *IEEE Signal Processing Letters*, Vol. 11, No. 3, March 2004, pp. 338–340.

[26] Obeidat, B. A., Y. Zhang, and M. G. Amin, "Polarimetric Time-Frequency ESPRIT," *Annual Asilomar Conf. on Signals, Systems, and Computers*, Pacific Grove, CA, November 2003.

[27] Belouchrani, A., M. G. Amin, and K. Abed-Meraim, "Direction Finding in Correlated Noise Fields Based on Joint Block-Diagonalization of Spatio-Temporal Correlation Matrices," *IEEE Signal Processing Letters*, Vol. 4, No. 9, September 1997, pp. 266–268.

[28] Stoica, P., and A. Nehorai, "MUSIC, Maximum Likelihood and Cramer-Rao Bound," *IEEE Trans. on Acoustics, Speech, and Signal Processing*, Vol. 37, No. 5, May 1989, pp. 720–741.

[29] Zhang, Y., and M. G. Amin, "Joint Doppler and Polarization Characterization of Moving Targets," *IEEE AP-S Int. Symp.*, Monterey, CA, June 2004.

CHAPTER 7
A One-Dimensional Tree Structure–Based Algorithm for DOA-Delay Joint Estimation

Yung-Yi Wang, Jiunn-Tsair Chen, and Wen-Hsien Fang

7.1 Introduction

The DOA-delay estimation is a classical problem encountered in radar, sonar, and geophysics. It also finds applications in source localization, accident reporting, cargo tracking, and intelligent transportation [1]. Furthermore, in a multiray wireless communication system, one can obtain a better channel estimate by jointly exploring the ray DOAs and the ray propagation delays, significantly improving the system performance [2, 3].

Some algorithms for joint estimation of the DOAs and the multiray propagation delays were suggested recently. For example, Swindlehurst et al. [4–6] proposed several computationally efficient algorithms for the estimation of the delays of a multiray channel, and solved the spatial signatures (or DOAs) as a least-square problem. Clark et al. [7] proposed a two-dimensional IQML algorithm that could be extended to jointly estimate the channel parameters. All of the algorithms proposed in [4–7] take advantage of the Vandermonde structure of the estimated channel pulse response in the frequency domain. However, if two or more rays have close time delays, then the data covariance matrix becomes ill-conditioned, and these algorithms may not work properly, even if these rays possess diverse DOAs. In addition, the IQML-based algorithms suffer from the initialization problems. Based on the knowledge of the transmitted signals, Bertaux et al. [8] developed a PML technique that uses the iterative Gauss-Newton procedure to estimate the spatial-temporal parameters of a multipath channel. However, as a consequence of the stacking of the observed data matrix into a high-dimensional vector, the PML technique in [8] calls for enormous computational overhead.

Among other alternatives are the subspace-based algorithms. Ogawa et al. [9] presented a channel sounding method using unmodulated carriers. The parameter pairs were then extracted by invoking a two-dimensional windowed ESPRIT algorithm. Vanderveen et al. proposed the JADE-ESPRIT [10] algorithm, which exploited the properties of the space-time structure by stacking the received data. After a high-dimensional eigendecomposition on the covariance matrix is performed, the channel parameters can be estimated by a two-dimensional searching on the DOA-delay plane. However, the required computations make the JADE-ESPRIT also unfavorable for real-time implementation. To lower the computational overhead, the SI-JADE and JADE-ESPRIT algorithms [11–13] were advocated, which utilized shift invariance property of the estimated channel matrix. By stacking the submatrices of the estimated channel matrix, the SI-JADE transforms the joint estimation problem into a matrix pencil problem. It follows that the rank-reducing numbers of the matrix pencil are the corresponding DOAs/delays. The JADE-ESPRIT algorithm is similar to the SI-JADE, except that the former uses multiple channel estimates, while the latter uses only one channel estimate. As a result, the JADE-ESPRIT performance is usually much better than the SI-JADE performance in terms of accuracy.

In this chapter, we present a low-complexity, yet high-accuracy, ESPRIT-based algorithm [14]—the TST-ESPRIT algorithm, which combines the techniques of temporal filtering and of spatial beamforming with three one-dimensional ESPRITs (i.e., 1 S-ESPRIT and two T-ESPRIT algorithms) to jointly estimate the DOAs-delays of interest, based on the data samples received from an antenna array. The basic idea behind the proposed approach is to group and isolate the signal of each incoming ray using the space-time characteristics of the multiray wireless channel. To achieve this, the T-ESPRIT and the S-ESPRIT algorithms are first employed to estimate the group delays[1] and the DOAs, which are used for temporal filtering right after the first T-ESPRIT, and for spatial beamforming right after the S-ESPRIT, respectively. Thereafter, the other T-ESPRIT algorithm is employed to estimate the ray delays.

The proposed approach possesses some distinctive features. First, compared to the algorithms mentioned in [4–7], the tree-structured TST-ESPRIT algorithm not only inherently resolves incoming rays with either very close DOAs or very close delays, but it also automatic pairs the estimated DOAs and delays. In addition, due to the employment of the temporal filtering process, the number of antennas required by the TST-ESPRIT can be less than that of the incoming rays. Second, in contrast to the JADE-ESPRIT algorithm [10], the TST-ESPRIT algorithm requires eigendecompositions for much smaller covariance matrices. Thus, it calls for substantially lower computational complexity. Third, compared to the SI-JADE and the JADE-ESPRIT algorithms [11, 12], the proposed approach performs better in estimation accuracy, especially in an environment with low SNR.

This chapter is organized as follows. Section 7.2 introduces the system model with multiray fading channels, where we assume the propagation rays to be specular. In Section 7.3, the S-ESPRIT algorithm [14] and the T-ESPRIT algorithm [5] are reviewed, followed by the proposition of the TST-ESPRIT algorithm. The issue of computational complexity is addressed as well. In Section 7.4, simulation results are presented to verify the performance of the proposed approach. Section 7.5 provides a concluding remark to summarize this chapter.

7.2 System Model

In this chapter, we assume a TDMA wireless system, such as the IS-136 [15] and GSM [16], as our target application system. The radio channel in a wireless

---

1. Differing from the ray delay, the group delay is defined as the average delay of a group of rays.

7.2 System Model

communication system is often characterized by a multiray propagation model. In large cells with high base station antenna platforms, the propagation environment is aptly modeled by a few dominant specular rays—typically 2 to 6. In such a case, the baseband signals received at the antenna array during the nth user can be expressed as follows [17]:

$$\mathbf{x}^{(n)}(t) = \sum_{k=1}^{Q} \mathbf{a}(\theta_k) \beta_k^{(n)} \tilde{s}(t - \tau_k) + \mathbf{n}(t) \quad (7.1)$$

where $\mathbf{x}^{(n)}(t)$ is the received baseband signals in the nth time burst, $\mathbf{n}(t)$ is the spatially and temporally white additive Gaussian noises with zero-means and equal variances σ_n^2, $\mathbf{a}(\theta_k)$ is the normalized steering vector $[\|\mathbf{a}(\theta)\| = 1$ for all $\theta]$ of a signal arriving from direction θ_k, $\beta_k^{(n)}$ is the ray amplitude that is a complex Gaussian random process, $\tilde{s}(\cdot)$ is the transmitted complex baseband signal, τ_k is the propagation delay of the kth ray, and Q is the total number of rays present in the system.

In a linear time-invariant system, the transmitted signal $\tilde{s}(t)$ can be represented as a convolution of the data bits $s_l^{(n)}$ and the pulse-shaping function $g(t)$

$$\tilde{s}(t) = \sum_{l} s_l^{(n)} \cdot g(t - lT)$$

where T is the symbol period. Therefore, after sampling at a rate of T/P, the signal received during the nth time burst can be written as

$$\mathbf{X}^{(n)} = \left\{ \mathbf{x}^{(n)}(t_0), \mathbf{x}^{(n)}\left(t_0 - \frac{T}{P}\right), \ldots, \mathbf{x}^{(n)}\left[t_0 - \left(N - \frac{1}{P}\right)T\right] \right\}$$

$$= \underbrace{\mathbf{A}(\boldsymbol{\theta})}_{M \times Q} \underbrace{\mathbf{B}^{(n)}}_{Q \times Q} \underbrace{\mathbf{G}^T(\boldsymbol{\tau})}_{Q \times LP} \underbrace{\mathbf{S}^{(n)}}_{LP \times NP} + \underbrace{\mathbf{N}^{(n)}}_{M \times NP} \quad (7.2)$$

where t_0 is the sampling reference time of the nth data burst, and the superscript $(\cdot)^T$ represents the matrix transpose operation. In (7.2), NP consecutive samples in time are considered, M is the number of antennas, L is the maximum length of the channel pulse response divided by the symbol period T, and P is the oversampling factor. The array response matrix $\mathbf{A}(\boldsymbol{\theta}) = [\mathbf{a}(\theta_1), \mathbf{a}(\theta_2), \ldots, \mathbf{a}(\theta_Q)]$, where $\mathbf{a}(\theta_i)$ denotes the array response vector of the ith ray; $\mathbf{B}^{(n)} = \text{diag}(\beta_1^{(n)}, \ldots, \beta_Q^{(n)})$ with $\beta_i^{(n)}$ being the complex fading amplitudes of the ith ray during the nth burst; and $\mathbf{G}(\boldsymbol{\tau}) = [\mathbf{g}(\tau_1), \mathbf{g}(\tau_2), \ldots, \mathbf{g}(\tau_Q)]$ with

$$\mathbf{g}(\tau_i) = \left[g(t_0 - \tau_i), g\left(t_0 - \frac{T}{P} - \tau_i\right), \ldots, g\left(t_0 - \frac{(LP-1)T}{P} - \tau_i\right) \right]^T$$

where τ_i is the time delay of the ith incoming ray. As a result of the convolution between the data bits and the pulse shaping function, the data matrix $\mathbf{S}^{(n)}$ is a Toeplitz matrix with $\left[s_k^{(n)}, 0, s_{k-1}^{(n)}, 0, \ldots, s_{k-L+1}^{(n)}, 0 \right]^T$ as its first column and

$\left[s_k^{(n)}, 0, s_{k+1}^{(n)}, 0, \ldots, s_{k+N-1}^{(n)}, 0 \right]$ as its first row, where $\{s_i^{(n)}\}$ is the transmitted data bits of the nth burst and 0 is a zero vector of dimension $1 \times (P \times 1)$. The noise matrix of the nth burst $\mathbf{N}^{(n)}$ is assumed to be spatially and temporally white Gaussian noise with noise power σ_n^2. Note that $\mathbf{A}(\boldsymbol{\theta})$ and $\mathbf{G}(\boldsymbol{\tau})$ are assumed constants among data bursts.

In a TDMA system, a known training sequence is usually embedded in each transmitted burst. We denote the data matrix formed by the training sequence as $\mathbf{S}^{(n)} = \mathbf{S}_t$, independent of n. Specifically, $\mathbf{S}_t$ is regarded as prior information in the proposed approach to estimate the channel parameters. Extracting the training portion from (7.2), we have

$$\mathbf{X}_t^{(n)} = \mathbf{A}(\boldsymbol{\theta}) \mathbf{B}^{(n)} \tilde{\mathbf{G}}^T(\boldsymbol{\tau}) + \mathbf{N} \quad (7.3)$$

where $\tilde{\mathbf{G}}^T(\boldsymbol{\tau}) = \mathbf{S}_t^T \cdot \mathbf{G} = [\tilde{\mathbf{g}}(\tau_1), \ldots, \tilde{\mathbf{g}}(\tau_Q)]$, and $\tilde{\mathbf{g}}(\tau_i) = \mathbf{S}_t^T \cdot \mathbf{g}(\tau_i)$ is the convolution between the training sequence and the time-shifted pulse-shaping function. Similar to the normalization of the array steering vector, we assume $\|\mathbf{g}(\tau)\| = 1$ for all τ by adjusting power between $\mathbf{B}^{(n)}$ and $\tilde{\mathbf{G}}(\boldsymbol{\tau})$. We define $\tilde{\mathbf{g}}(\tau)$ as the normalized temporal array vector. The dimension of $\mathbf{X}_t^{(n)}$ is $M \times N_t$, where N_t is the length of the extracted training sequence.

In the applications of wireless communication, the Q fading rays are usually assumed to be mutually uncorrelated, and their fading amplitudes are assumed to be zero-mean complex Gaussian distributed [18]. Hence, the covariance matrix of the fading vector, $\boldsymbol{\beta}^{(n)} = \left[\beta_1^{(n)}, \ldots, \beta_Q^{(n)} \right]^T$, is

$$E\left[\boldsymbol{\beta}^{(n)} \cdot \left(\boldsymbol{\beta}^{(n)} \right)^H \right] = diag(\sigma_1^2, \ldots, \sigma_Q^2) \triangleq \mathbf{P} \quad (7.4)$$

and

$$E\left[\boldsymbol{\beta}^{(n)} \cdot \left(\boldsymbol{\beta}^{(n)} \right)^T \right] = 0 \quad (7.5)$$

where $E[\cdot]$ represents the statistical average operation, the superscript $(\cdot)^H$ denotes the Hermitian operation, and σ_i^2 is the average signal power of ray i. Since $\mathbf{S}_t$ is a known Toeplitz matrix, the only stochastic term in (7.3) is the fading matrix $\mathbf{B}^{(n)}$. We refer to the covariance matrix of the columns of $\mathbf{X}_t^{(n)}$ as the spatial covariance matrix $\mathbf{R}^s$, which is given by

$$\mathbf{R}^s = E\left[\mathbf{X}_t^{(n)} \left(\mathbf{X}_t^{(n)} \right)^H \right] = \mathbf{A}\mathbf{P}\mathbf{A}^H + \sigma_n^2 \cdot \mathbf{I} \quad (7.6)$$

where $\mathbf{A}$ instead of $\mathbf{A}(\boldsymbol{\theta})$ is used to simplify the notation. From (7.6), after the exclusion of the noise subspace, the spatial covariance matrix $\mathbf{R}^s$ and the spatial signature matrix $\mathbf{A}$ share the same column space.

Similarly, the rows of $\mathbf{X}_t^{(n)}$ are the temporal sampling vectors of the received signals. We may express the temporal covariance matrix $\mathbf{R}^t$ as

$$\mathbf{R}^t = E\left[\left(\mathbf{X}_t^{(n)} \right)^T \left(\mathbf{X}_t^{(n)} \right)^* \right] = \tilde{\mathbf{G}}\mathbf{P}\tilde{\mathbf{G}}^H + \sigma_n^2 \cdot \mathbf{I} \quad (7.7)$$

where the superscript $(\cdot)^*$ denotes the complex conjugate operation. From (7.7), after the exclusion of the noise subspace, the temporal covariance matrix $\mathbf{R}^t$ and the temporal signature matrix $\tilde{\mathbf{G}}$ share the same column space.

The number of the incoming rays Q can be obtained by methods of AIC or MLD [19]. Assuming that Q is known a priori, the eigendecomposition of the covariance matrices, (7.6) and (7.7), can be expressed respectively as

$$\mathbf{R}^s = \mathbf{V}_s^s \mathbf{\Lambda}_s^s (\mathbf{V}_s^s)^H + \mathbf{V}_n^s \mathbf{\Lambda}_n^s (\mathbf{V}_n^s)^H \quad (7.8)$$

and

$$\mathbf{R}^t = \mathbf{V}_s^t \mathbf{\Lambda}_s^t (\mathbf{V}_s^t)^H + \mathbf{V}_n^t \mathbf{\Lambda}_n^t (\mathbf{V}_n^t)^H \quad (7.9)$$

The column vectors of $\mathbf{V}_s^s$ and $\mathbf{V}_s^t$ are the eigenvectors that span the signal subspace of $\mathbf{R}^s$ and $\mathbf{R}^t$, respectively, corresponding to the Q largest eigenvalues. The column spaces of $\mathbf{V}_n^s$ and $\mathbf{V}_n^t$, spanned by the rest of the $M - Q$ and $N_t - Q$ eigenvectors of $\mathbf{R}^s$ and $\mathbf{R}^t$, are the orthogonal complement of the column spaces of $\mathbf{V}_s^s$ and $\mathbf{V}_s^t$, respectively. Both $(\mathbf{\Lambda}_s^s, \mathbf{\Lambda}_n^s)$ and $(\mathbf{\Lambda}_s^t, \mathbf{\Lambda}_n^t)$ are diagonal matrix pairs with the associated eigenvalues as their diagonal elements.

Since the covariance matrix of the received signal is usually unavailable, the implementation of the ESPRIT instead employs the sample covariance matrices. For example, if K data bursts are observed, then the sampled spatial covariance matrix $\hat{\mathbf{R}}^s$ and the sampled temporal covariance matrix $\hat{\mathbf{R}}^t$, respectively, are estimated by [20]

$$\hat{\mathbf{R}}^s = \frac{1}{N_t K} \sum_{n=1}^{K} \mathbf{X}_t^{(n)} \cdot (\mathbf{X}_t^{(n)})^H \quad (7.10)$$

and

$$\hat{\mathbf{R}}^t = \frac{1}{MK} \sum_{n=1}^{K} (\mathbf{X}_t^{(n)})^T \cdot (\mathbf{X}_t^{(n)})^* \quad (7.11)$$

To simplify the notation, hereafter we will ignore the burst index superscript $(\cdot)^{(n)}$ of $\mathbf{X}_t^{(n)}$.

7.3 The Proposed Algorithm

In this section, we first briefly review the S-ESPRIT algorithm and the T-ESPRIT algorithm [14], respectively, for DOA estimation and delay estimation. Next, the TST-ESPRIT algorithm is proposed, and a 3-ray scenario is used as an example to illustrate the proposed algorithm.

7.3.1 S-ESPRIT

The S-ESPRIT algorithm is a subspace-based algorithm for DOA estimation. It takes advantage of the geometric symmetry of the antenna array, which possesses a shift invariance property. Since the spatial signal subspace $\mathbf{V}_s^s$ shares the same vector space as the column space of the spatial signature matrix $\mathbf{A}$, it follows that there exists a nonsingular $M \times M$ transformation matrix $\mathbf{T}$, such that

$$\mathbf{A} = \mathbf{V}_s^s \cdot \mathbf{T} \tag{7.12}$$

By defining two selection matrices as

$$\mathbf{J}_1 \triangleq [\mathbf{I}_{M-1} \quad \mathbf{0}] \text{ and } \mathbf{J}_2 \triangleq [\mathbf{0} \quad \mathbf{I}_{M-1}] \tag{7.13}$$

where $\mathbf{0}$ is an $N \times 1$ zero vector. We can obtain two submatrices from the signal subspace matrix as

$$\mathbf{V}_1^s = \mathbf{J}_1 \cdot \mathbf{V}_s^s = \mathbf{A}_1 \cdot \mathbf{T} \tag{7.14}$$

and

$$\mathbf{V}_2^s = \mathbf{J}_2 \cdot \mathbf{V}_s^s = \mathbf{A}_2 \cdot \mathbf{T} \tag{7.15}$$

where both $\mathbf{A}_1 = \mathbf{J}_1 \cdot \mathbf{A}$ and $\mathbf{A}_2 = \mathbf{J}_2 \cdot \mathbf{A}$ are of dimension $(M-1) \times Q$. For example, the shift invariance property of a ULA, with a distance between antenna elements of $d = \lambda/2$, relates $\mathbf{A}_1$ to $\mathbf{A}_2$ via

$$\mathbf{A}_1 = \mathbf{A}_2 \cdot \mathbf{\Phi} \tag{7.16}$$

where

$$\mathbf{\Phi} = diag\{\phi_1, \ldots, \phi_Q\} \tag{7.17}$$

and $\phi_k = e^{-j\pi \sin \theta_k}$. Substituting (7.16) into (7.14) and (7.15), we can obtain

$$\mathbf{V}_1^s = \mathbf{V}_2^s \cdot \mathbf{\Gamma} \tag{7.18}$$

where $\mathbf{\Gamma} = \mathbf{T}^{-1} \mathbf{\Phi} \mathbf{T}$. The DOAs, associated with the diagonal elements of $\mathbf{\Phi}$, then can be extracted by performing an eigendecomposition on $\mathbf{\Gamma}$.

7.3.2 T-ESPRIT

Similar to the S-ESPRIT algorithm, which uses the spatial samples of the receive signal to estimate the DOAs of the incoming rays, the T-ESPRIT algorithm uses the temporal samples of the receive signal to estimate the path delays of the incoming rays. However, since the temporal domain does not possess a shift invariance structure, some signal preprocessing is required, which transforms the temporal sampled signals into the sum of multiple exponential sinusoidal functions with the delay information of each path in the exponent of one function. Figure 7.1 illustrates

7.3 The Proposed Algorithm

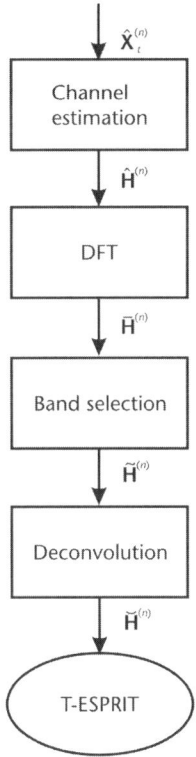

Figure 7.1 The preprocess of the T-ESPRIT algorithm.

the whole preprocess applied to $\mathbf{X}_t$ before the employment of the T-ESPRIT algorithm. The steps are described in details as follows.

Channel Estimation

As shown in Figure 7.1, the pseudoinverse of the training matrix, $\mathbf{S}_t^\dagger$, is post-multiplied to $\mathbf{X}_t$. A channel estimate, $\hat{\mathbf{H}}$, can be obtained in the least-square sense. More specifically, $\hat{\mathbf{H}}$, in a noise-free situation, can be expressed as

$$\hat{\mathbf{H}} = \mathbf{X}_t \mathbf{S}_t^\dagger = \mathbf{A}\mathbf{B}\mathbf{G}^T(\tau) \tag{7.19}$$

Discrete Fourier Transform

Note that each row of $\mathbf{G}^T(\tau)$ is the sampled value of a delayed version of the known waveform $g(t - \tau)$ at a rate of T/P. The kth component of the LP-point DFT of $\mathbf{g}(\tau_i)$, denoted by $\mathfrak{g}_{\tau_i}(k)$, can be expressed as

$$\mathfrak{g}_{\tau_i}(k) = \mathbf{g}^T(\tau_i) \cdot \begin{bmatrix} 1 \\ \phi^{k-1} \\ \vdots \\ \phi^{(k-1)(LP-1)} \end{bmatrix} = \mathfrak{g}_0(k) \cdot e^{-j\frac{2\pi\tau_i}{L}}$$

where $g_0(k)$ is the kth element of the DFT of $\mathbf{g}(0)$ and $\phi \triangleq e^{-j\frac{2\pi}{LP}}$. Hence, after performing the DFT on $\hat{\mathbf{H}}$, we obtain

$$\overline{\mathbf{H}} = \hat{\mathbf{H}} \cdot \mathbf{W}_{LP} \qquad (7.20)$$
$$= \mathbf{ABV}^T(\tau) \cdot diag\{g_0(0), \ldots, g_0(LP-1)\}$$

where $\mathbf{W}_{LP}$ is the DFT matrix given by

$$\mathbf{W}_{LP} = \begin{bmatrix} 1 & 1 & \cdots & 1 \\ 1 & \phi & \cdots & \phi^{LP-1} \\ \vdots & \vdots & \ddots & \vdots \\ 1 & \phi^{LP-1} & \cdots & \phi^{(LP-1)^2} \end{bmatrix} \qquad (7.21)$$

and

$$\mathbf{V}(\tau) = [\mathbf{v}(\tau_1), \ldots, \mathbf{v}(\tau_Q)]$$

with

$$\mathbf{v}(\tau_k) = \begin{cases} \left[1, \varphi_k, \ldots, \varphi_k^{\frac{LP-1}{2}}, \varphi_k^{-\frac{LP-1}{2}}, \ldots, \varphi_k^{-1}\right]^T & \text{if } LP \text{ is odd} \\ \left[1, \varphi_k, \ldots, \varphi_k^{\frac{LP}{2}-1}, \varphi_k^{-\frac{LP}{2}}, \varphi_k^{-\frac{LP}{2}+1}, \ldots, \varphi_k^{-1}\right]^T & \text{if } LP \text{ is even} \end{cases}$$

and $\varphi_k = e^{-j\frac{2\pi}{L}\tau_k}$. It is obvious that $\mathbf{v}(\tau_k)$ is shift invariant if the right-half portion is circularly shifted to the left.

Band Selection and Deconvolution

Signal deconvolution is invoked to remove the $diag\{g_0(0), \ldots, g_0(LP-1)\}$ in (7.20). However, since the employed pulse-shaping function has only finite bandwidth, the trivial components in $\{g_0(0), \ldots, g_0(LP-1)\}$ may contribute to noise enhancement. Therefore, we discard those trivial components before carrying out the deconvolution. With the circular shift operation mentioned above, a band selection matrix $\mathbf{J}_i$ then can be defined as

$$\mathbf{J}_t = \begin{bmatrix} \mathbf{0}_{\frac{LW}{2} \times \frac{LW}{2}} & & \mathbf{I}_{\frac{LW}{2}} \\ \mathbf{0}_{(LP-LW) \times \frac{LW}{2}} & & \mathbf{0}_{(LP-LW) \times \frac{LW}{2}} \\ \mathbf{I}_{\frac{LW}{2}} & & \mathbf{0}_{\frac{LW}{2} \times \frac{LW}{2}} \end{bmatrix}$$

where LW is the number of the selected nontrivial components in $\{g_0(0), \ldots, g_0(LP-1)\}$. The resulting $\tilde{\mathbf{H}} = \bar{\mathbf{H}} \cdot \mathbf{J}_t$ becomes

$$\tilde{\mathbf{H}} = \mathbf{A}\mathbf{B}\tilde{\mathbf{V}}^T(\boldsymbol{\tau})\mathbf{S}_J \qquad (7.22)$$

where $\mathbf{S}_J = diag\{g_0(0), \ldots, g_0(LP-1)\} \cdot \mathbf{J}_t$ is a diagonal matrix with the LW nontrivial components of $\{g_0(0), \ldots, g_0(LP-1)\}$ as its diagonal elements, and $\tilde{\mathbf{V}}_T(\boldsymbol{\tau}) = \mathbf{V}^T(\boldsymbol{\tau}) \cdot \mathbf{J}_i$ is of size $Q \times LW$. Note that the row vectors of $\tilde{\mathbf{V}}^T(\boldsymbol{\tau})$ possesses the shift invariance structure.

Finally, the deconvolution is accomplished by postmultiplying $\mathbf{S}_J^{-1}$ to $\tilde{\mathbf{H}}$, which yields

$$\check{\mathbf{H}} = \tilde{\mathbf{H}}\mathbf{S}_J^{-1} = \mathbf{A}\mathbf{B}\tilde{\mathbf{V}}^T(\boldsymbol{\tau}) \qquad (7.23)$$

The T-ESPRIT algorithm then applies the ESPRIT algorithm to rows of $\check{\mathbf{H}}$ to estimate the path delays. The overall procedures of the T-ESPRIT algorithm are illustrated in Figure 7.2.

7.3.3 TST-ESPRIT

The rationale of the TST-ESPRIT is to incorporate three one-dimensional ESPRIT (S-ESPRIT and T-ESPRIT) algorithms with beamforming and filtering techniques to group, isolate, estimate, and then pair the two-dimensional parameters of a fading channel. To simplify the algorithm description, we first assume that there are only three rays present in the system. The general procedure of the TST-ESPRIT is summarized at the end of this section.

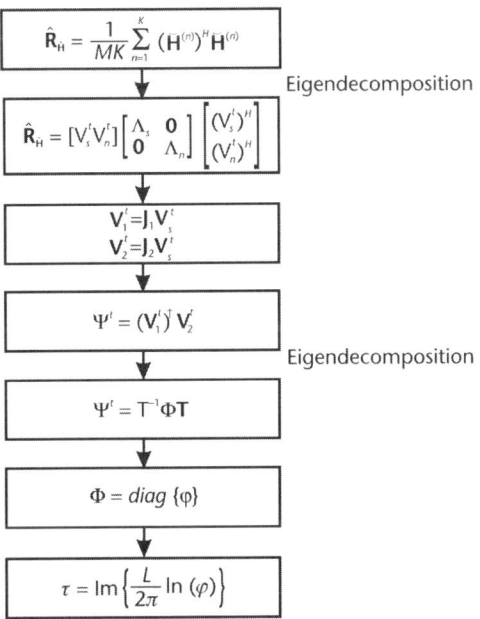

Figure 7.2 The T-ESPRIT algorithm.

As shown in Figure 7.3, three rays are characterized by their temporal-spatial coordinates on the DOA-delay plane. Note that rays 1 and 2 possess close time delays ($\tau_1 \approx \tau_2$), but diverse DOAs ($\theta_1 < \theta_2$), while rays 1 and 3 are close in the DOAs ($\theta_1 \approx \theta_3$), but have far-apart delays ($\tau_1 < \tau_3$). The tree structure of the TST-ESPRIT algorithm for this scenario is illustrated in Figure 7.4. Corresponding to Figure 7.4, Figure 7.3 also shows the evolution of the data contents as the parameters of ray 1 are estimated in the tree structure.

The TST-ESPRIT treats those temporally close rays as a group. Therefore, rays 1 and 2 in Figure 7.3 are regarded as one group, while ray 3 is considered as another group. By applying the T-ESPRIT to the rows of $\mathbf{X}_t$, the resulting group delays are estimated, and denoted by $\hat{t}_1$ and $\hat{t}_2$. Based on these group delay estimates, $\hat{t}_1$ and $\hat{t}_2$, we define the temporal filtering matrices $\mathbf{U}_i^t$ as

$$\mathbf{U}_1^t = \mathbf{I} - \tilde{\mathbf{g}}(t_1)\tilde{\mathbf{g}}^H(t_1), \text{ and } \mathbf{U}_2^t = \mathbf{I} - \tilde{\mathbf{g}}(t_2)\tilde{\mathbf{g}}^H(t_2)$$

Note that $\mathbf{U}_1^t$ (or $\mathbf{U}_2^t$) is also the complement projection matrix of $\tilde{\mathbf{g}}(\hat{t}_1)$ [or $\tilde{\mathbf{g}}(\hat{t}_2)$], with $\tilde{\mathbf{g}}^H(\hat{t}_1) \cdot \mathbf{U}_1^t = \mathbf{0}^T$ [or $\tilde{\mathbf{g}}^H(\hat{t}_2) \cdot \mathbf{U}_2^t = \mathbf{0}^T$]. In the 3-ray scenario shown in Figure 7.3, we have $\tau_1 \approx \tau_2 \approx \hat{t}_1 < \tau_3 \approx \hat{t}_2$, which implies that

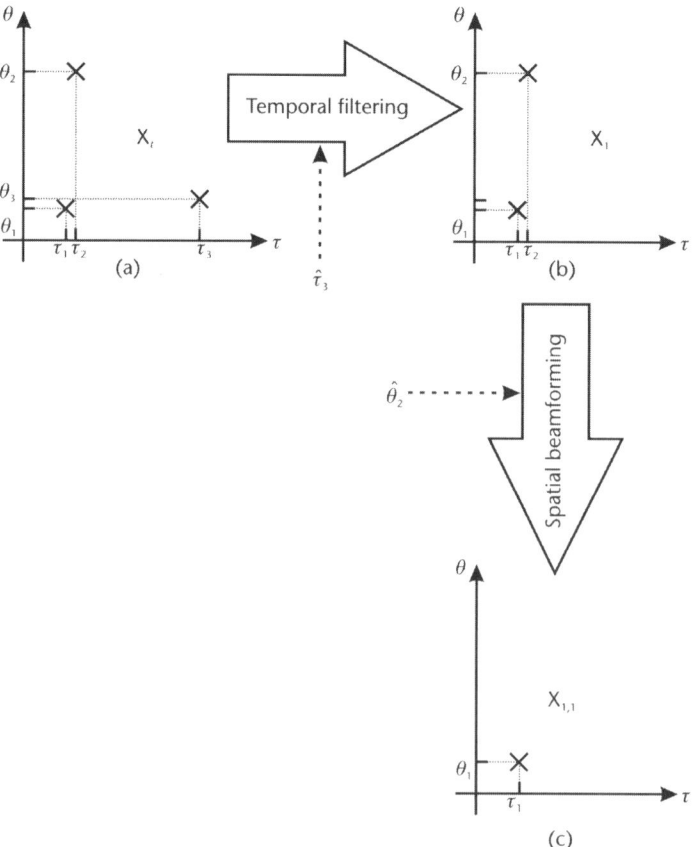

Figure 7.3 (a–c) Evolution of the signal contents for parameter estimation of ray 1.

7.3 The Proposed Algorithm

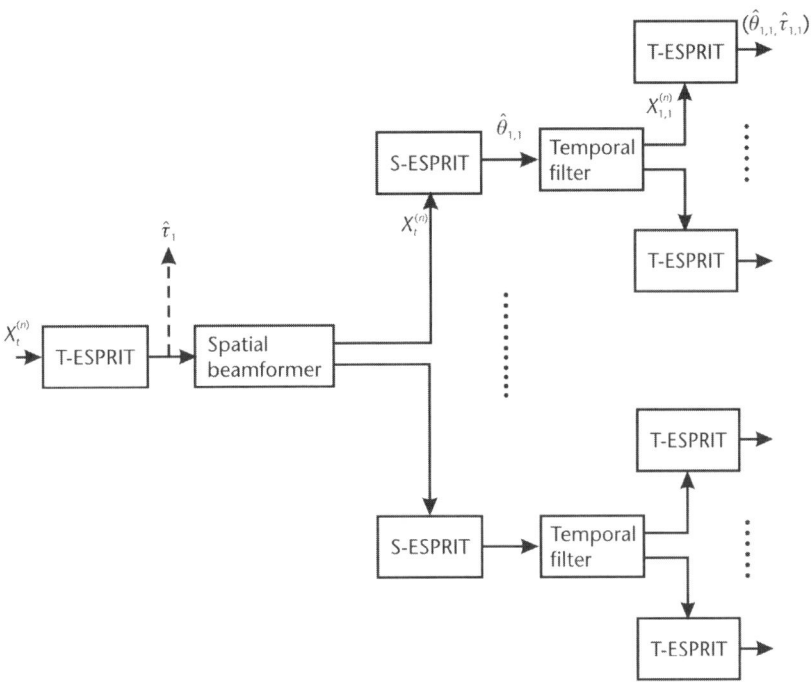

Figure 7.4 The tree structure of the TST-ESPRIT algorithm.

$$\|\tilde{\mathbf{g}}^T(\tau_1) \cdot \mathbf{U}_1^t\| \approx \|\tilde{\mathbf{g}}^T(\tau_2) \cdot \mathbf{U}_1^t\| \approx 0 \ll \|\tilde{\mathbf{g}}^T(\tau_3) \cdot \mathbf{U}_1^t\|$$

and

$$\|\tilde{\mathbf{g}}^T(\tau_1) \cdot \mathbf{U}_2^t\| \approx \|\tilde{\mathbf{g}}^T(\tau_2) \cdot \mathbf{U}_2^t\| \gg \|\tilde{\mathbf{g}}^T(\tau_3) \cdot \mathbf{U}_2^t\| \approx 0$$

where the notation $\|\cdot\|$ denotes the 2-norm of a vector. With these facts, the TST-ESPRIT postmultiplies $\mathbf{U}_i^t$ to $\mathbf{X}_t$, which is referred to as the *temporal filtering process* to separate the rays with delay τ_1 or τ_2 from the ray with delay τ_3. As a result, two group matrices, denoted as $\mathbf{X}_1$ and $\mathbf{X}_2$, are generated as

$$\mathbf{X}_1 = \mathbf{X}_t \cdot \mathbf{U}_2^t \tag{7.24}$$

$$\approx [\mathbf{a}(\theta_1), \ \mathbf{a}(\theta_2)] \begin{bmatrix} \beta_1 & 0 \\ 0 & \beta_2 \end{bmatrix} \begin{bmatrix} \tilde{\mathbf{g}}^T(\tau_1)\mathbf{U}_2^t \\ \tilde{\mathbf{g}}^T(\tau_2)\mathbf{U}_2^t \end{bmatrix} + \mathbf{N} \cdot \mathbf{U}_2^t$$

and

$$\mathbf{X}_2 = \mathbf{X}_t \cdot \mathbf{U}_1^t \tag{7.25}$$

$$\approx \beta_3 \mathbf{a}(\theta_3) \tilde{\mathbf{g}}^T(\tau_3) \mathbf{U}_1^t + \mathbf{N} \cdot \mathbf{U}_1^t$$

where $\approx$ in (7.24) and (7.25) indicates that the residue signals from ray 3, and from rays 1 and 2, respectively, are neglected. Discussions about the magnitude of the neglected residue signal will be given later in this section. It will be shown in Appendix A that the transformed noise matrices in (7.24) and (7.25) are still temporally and spatially white within the projected subspace.

Note that the two dominant rays contained in (7.24) have their DOAs $\theta_2 > \theta_1$. The DOA estimates $\hat{\theta}_1$ and $\hat{\theta}_2$ thus can be accurately obtained by applying the S-ESPRIT to $\mathbf{X}_1$. Similarly, $\hat{\theta}_3$ is obtained by applying the S-ESPRIT to $\mathbf{X}_2$. Note that $\hat{\theta}_1 \approx \hat{\theta}_3$, but the signals of rays 1 and 3 are separated into two different signal groups before the S-ESPRIT is applied. Therefore, θ_1 and θ_3 also can be accurately estimated with the help of the temporal filtering following the first T-ESPRIT. Note that, right after estimating $\{\theta_k\}$ in the S-ESPRIT, the estimated array vectors $\{\mathbf{a}(\hat{\theta}_k)\}$ are determined, and will be used in the spatial beamforming described below.

To further divide each group matrix into several single-ray matrices, the spatial beamforming matrices $\mathbf{U}_i^s$ can be defined as

$$\mathbf{U}_1^s = \mathbf{I} - \mathbf{a}(\hat{\theta}_1) \cdot \mathbf{a}^H(\hat{\theta}_1), \ \mathbf{U}_2^s = \mathbf{I} - \mathbf{a}(\hat{\theta}_2) \cdot \mathbf{a}^H(\hat{\theta}_2)$$

for $\mathbf{X}_1$, and

$$\mathbf{U}_3^s = \mathbf{I} - \mathbf{a}(\hat{\theta}_3) \cdot \mathbf{a}^H(\hat{\theta}_3)$$

for $\mathbf{X}_2$, respectively. Note that $\mathbf{U}_i^s$ nulls the signal from ray i as

$$\mathbf{U}_1^s \cdot \mathbf{a}(\hat{\theta}_1) = 0, \ \mathbf{U}_2^s \cdot \mathbf{a}(\hat{\theta}_2) = 0 \text{ and } \mathbf{U}_3^s \cdot \mathbf{a}(\hat{\theta}_3) = 0$$

Similar to the temporal filtering process, the TST-ESPRIT algorithm premultiplies $\mathbf{X}_1$ by $\mathbf{U}_1^s$ and $\mathbf{U}_2^s$, to null the corresponding ray, which is referred to as the spatial beamforming process. It follows that two single-ray matrices, $\mathbf{X}_{1,1}$ and $\mathbf{X}_{1,2}$, are formed, respectively, as

$$\mathbf{X}_{1,1} = \mathbf{U}_2^s \cdot \mathbf{X}_1 \qquad (7.26)$$
$$\approx \beta_1 \cdot \mathbf{U}_2^s \mathbf{a}(\theta_1) \cdot \tilde{\mathbf{g}}^T(\tau_1) \mathbf{U}_2^t + \mathbf{U}_2^s \cdot \mathbf{N} \cdot \mathbf{U}_2^t$$

and

$$\mathbf{X}_{1,2} = \mathbf{U}_1^s \cdot \mathbf{X}_1 \qquad (7.27)$$
$$\approx \beta_2 \cdot \mathbf{U}_1^s \mathbf{a}(\theta_2) \cdot \tilde{\mathbf{g}}^T(\tau_2) \mathbf{U}_2^t + \mathbf{U}_1^s \cdot \mathbf{N} \cdot \mathbf{U}_2^t$$

Again, the residues of rays 2 and 1 are neglected in (7.26) and (7.27), respectively. The noise matrices in (7.26) and (7.27) are again temporally and spatially white within the projected subspace. In (7.26) and (7.27), the single-ray structure of $\mathbf{X}_{1,1}$ and $\mathbf{X}_{1,2}$ implies that the two rays with close ray delays, rays 1 and 2, are separated into different subgroups by the spatial beamforming process. As a result, by applying the T-ESPRIT algorithm to $\mathbf{X}_{1,1}$ and $\mathbf{X}_{1,2}$, ray delay estimates $\hat{\tau}_1$ and

$\hat{\tau}_2$ then can be accurately estimated, respectively. It also follows that the pairing of $(\hat{\tau}_1, \hat{\theta}_1)$ and $(\hat{\tau}_2, \hat{\theta}_2)$ is automatically achieved. For the other branch of the signal with respect to $\mathbf{X}_2$, only one single ray is found. Therefore, no spatial beamforming is needed; that is, $\mathbf{X}_{2,1} = \mathbf{X}_2$ and $\hat{\tau}_3 = \hat{t}_2$.

Next, we investigate the magnitude of the residue signals neglected in (7.24) to (7.27). It is obvious that these residue signals are the results of the leakage in the filtering and beamforming processes. For example, in (7.24), the neglected residue signal for $\mathbf{X}_1$ is equal to $\beta_3 \mathbf{a}(\theta_3) \tilde{\mathbf{g}}^H(\tau_3) \cdot \mathbf{U}_2^t$. Similar terms also exist in $\mathbf{X}_2$, $\mathbf{X}_{1,1}$, and $\mathbf{X}_{1,2}$. By including the residue signals, the covariance matrices of $\mathbf{X}_k$ can be expressed as

$$\mathbf{R}_{\mathbf{X}_k} = [\mathbf{a}(\theta_1), \mathbf{a}(\theta_2), \mathbf{a}(\theta_3)] \cdot \begin{bmatrix} \tilde{\sigma}_{1k}^2 & 0 & 0 \\ 0 & \tilde{\sigma}_{2k}^2 & 0 \\ 0 & 0 & \tilde{\sigma}_{3k}^2 \end{bmatrix} \cdot \begin{bmatrix} \mathbf{a}^H(\theta_1) \\ \mathbf{a}^H(\theta_2) \\ \mathbf{a}^H(\theta_3) \end{bmatrix} + \sigma_n^2 \frac{N_t - 1}{N_t} \mathbf{I}$$

(7.28)

where

$$\tilde{\sigma}_{mk}^2 \triangleq \sigma_m^2 \cdot \left\| \tilde{\mathbf{g}}^H(\tau_m) \cdot \mathbf{U}_{3-k}^t \right\|^2$$

(7.29)

is the average signal power of the mth ray in the kth group after filtering for $m = 1, 2, 3$, and $k = 1, 2$. Assume the variances of these three fading amplitudes are equal before filtering. We then have $\tilde{\sigma}_{11} \approx \tilde{\sigma}_{21} \gg \tilde{\sigma}_{31}$ and $\tilde{\sigma}_{32} \gg \tilde{\sigma}_{12} \approx \tilde{\sigma}_{22}$. This explains why ray 3 can be neglected in $\mathbf{X}_1$, and why rays 1 and 2 can be neglected in $\mathbf{X}_2$. Similarly, the covariance matrices of the rows of $\mathbf{X}_{1,1}$ and $\mathbf{X}_{1,2}$ are

$$\mathbf{R}_{\mathbf{X}_{1,k}^T} = [\overline{\mathbf{g}}(\tau_1), \overline{\mathbf{g}}(\tau_2), \overline{\mathbf{g}}(\tau_3)] \cdot \begin{bmatrix} \overline{\sigma}_{1k}^2 & 0 & 0 \\ 0 & \overline{\sigma}_{2k}^2 & 0 \\ 0 & 0 & \overline{\sigma}_{3k}^2 \end{bmatrix} \cdot \begin{bmatrix} \overline{\mathbf{g}}^H(\tau_1) \\ \overline{\mathbf{g}}^H(\tau_2) \\ \overline{\mathbf{g}}^H(\tau_3) \end{bmatrix} + \sigma_n^2 \frac{M-1}{M} \mathbf{U}_2^t$$

(7.30)

for $k = 1, 2$. In (7.30), the modified temporal array vector is defined as

$$\overline{\mathbf{g}}(\tau_m) \triangleq \frac{\left(\mathbf{U}_2^t\right)^T \cdot \tilde{\mathbf{g}}(\tau_m)}{\left\| \left(\mathbf{U}_2^t\right)^T \cdot \tilde{\mathbf{g}}(\tau_m) \right\|}$$

and the average signal power of the mth ray in the kth subgroup after beamforming is

$$\overline{\sigma}_{mk}^2 \triangleq \tilde{\sigma}_{m1}^2 \cdot \left\| \mathbf{U}_{3-k}^s \cdot \mathbf{a}(\theta_m) \right\|^2$$

(7.31)

for $m = 1, 2, 3$, and $k = 1, 2$. Thus, we have $\overline{\sigma}_{11} \gg \overline{\sigma}_{21}$, $\overline{\sigma}_{11} \gg \overline{\sigma}_{31}$, $\overline{\sigma}_{22} \gg \overline{\sigma}_{12}$, and $\overline{\sigma}_{22} \gg \overline{\sigma}_{32}$ for the 3-ray scenario. Note that the signals from rays 1 and 2 are isolated. Finally, the covariance matrix of the signal from ray 3 is given by

$$\mathbf{R}_{\mathbf{X}_{2,1}^T} = [\check{\mathbf{g}}(\tau_1), \check{\mathbf{g}}(\tau_2), \check{\mathbf{g}}(\tau_3)] \cdot \begin{bmatrix} \tilde{\sigma}_{12}^2 & 0 & 0 \\ 0 & \tilde{\sigma}_{22}^2 & 0 \\ 0 & 0 & \tilde{\sigma}_{32}^2 \end{bmatrix} \cdot \begin{bmatrix} \check{\mathbf{g}}^H(\tau_1) \\ \check{\mathbf{g}}^H(\tau_2) \\ \check{\mathbf{g}}^H(\tau_3) \end{bmatrix} + \sigma_n^2 \mathbf{U}_1^t \quad (7.32)$$

where

$$\check{\mathbf{g}}(\tau_m) \triangleq \frac{\mathbf{U}_1^t \cdot \tilde{\mathbf{g}}(\tau_m)}{\|\mathbf{U}_1^t \cdot \tilde{\mathbf{g}}(\tau_m)\|}$$

for $m = 1, 2, 3$. Because of $\tilde{\sigma}_{32} \gg \tilde{\sigma}_{12} \approx \tilde{\sigma}_{22}$, as shown earlier, signals from ray 3 remain isolated due to the application of the temporal filtering.

The above discussion can be readily extended to more general cases. Suppose that in total there are Q rays distributed in q temporal groups, and that the number of rays contained in the kth group is denoted by $r(k)$, where Q is assumed to be known a priori. Assume that the rays are not close in both DOA and delay. The TST-ESPRIT described above can be generalized as follows.

TST-ESPRIT Algorithm

Step 1: Grouping
Applying the T-ESPRIT algorithm to the received $\mathbf{X}_t$, we obtain the group delays $\{\hat{t}_1, \ldots, \hat{t}_q\}$.

Step 2: Temporal Filtering
The temporal excluding matrices $\mathbf{U}_n^t$, $n = 1, \ldots, q$, are generated by

$$\mathbf{U}_n^t = \left\{ \left[\tilde{\mathbf{g}}^T(\hat{t}_k) \right]_{k \neq n; k=1,\ldots,q} \right\} \quad (7.33)$$

while the output of the kth temporal filter is given by

$$\mathbf{X}_k = \mathbf{X}_t \cdot \left(\mathbf{U}_n^t \right)^\perp \quad (7.34)$$

for $k = 1, 2, \ldots, q$, where the superscript $(\cdot)^\perp$ denotes the orthogonal complementary projection of the embraced matrix.

Step 3: DOA Estimation
Applying the S-ESPRIT algorithm to each $\mathbf{X}_k$, we estimate the DOAs of the rays in group k. The results are $\hat{\boldsymbol{\theta}}_k = [\hat{\theta}_{k,1}, \ldots, \hat{\theta}_{k,r(k)}]$ for $k = 1, 2, \ldots, q$, where $r(k)$ is the number of rays in the kth group, and

$$\sum_{k=1}^{q} r(k) = Q$$

Step 4: Spatial Beamforming
The spatial beamforming matrix, $\mathbf{U}^s_{k,n}$, can be written as

$$\mathbf{U}^s_{k,n} = \left\{ \left[\mathbf{a}(\hat{\theta}_{k,m}) \right]_{m \neq n; m = 1, \ldots, r(k)} \right\} \quad (7.35)$$

while the signal at the output of the mth spatial beamformer is given by

$$\mathbf{X}_{k,m} = \left(\mathbf{U}^s_{k,n} \right)^\perp \cdot \mathbf{X}_k \quad (7.36)$$

for $k = 1, 2, \ldots, q$, and $m = 1, 2, \ldots, r(k)$.

Step 5: Delay Estimation
Apply the T-ESPRIT algorithm again but with different temporal array manifolds,

$$\left(\prod_{n=1; n \neq k}^{q} \mathbf{U}^t_n \right)^T \tilde{\mathbf{g}}(\tau)$$

to each $\mathbf{X}_{k,m}$. If a single ray is in the mth subgroup, and $\{\hat{\tau}_{k,m}\}$ is obtained, then pair it with $\{\hat{\theta}_{k,m}\}$; otherwise, pair $\hat{\tau}_k$ with $\hat{\theta}_{k,m}$.

Remarks
Since the DOA of each ray is estimated in the group data matrices, $\mathbf{X}_k$, the number of antennas required in the TST-ESPRIT is smaller than the number of rays. More specifically, in an uncorrelated propagation environment, $r(k) + 1$ antennas are required in the S-ESPRIT algorithm to identify $r(k)$ rays [21]. As a result, the minimum number of antennas required in the TST-ESPRIT algorithm for estimating Q rays is

$$M_{\min} = \max_k r(k) + 1 \leq Q \text{ provided } N_t > Q \quad (7.37)$$

If all the Q rays have distinct delays, we have $r(k) = 1$ for all ks. In such a scenario, only two antennas are needed for the TST-ESPRIT algorithm. On the other hand, since $(N_t - 1)$ groups can be resolved by the T-ESPRIT algorithm, and if M antennas are employed, then the TST-ESPRIT algorithm can identify a maximum number of $Q_{\max} = (N_t - 1)(M - 1)$ rays.

The computational complexity of the TST-ESPRIT includes (a) the eigendecomposition of $\mathbf{R}^s$ and $\mathbf{R}^t$, respectively, of order $O(M3)$ and order $O(N_t^3)$; and (b) the formulation of the temporal filtering matrices and the spatial beamforming matrices, respectively, of order $O(M^2)$ and order $O(N_t^2)$. In general, the length of the temporal array is greater than the number of the employed antennas, (i.e., $N_t > M$). We thus conclude that the computational complexity of the TST-ESPRIT is order $O(N_t^3)$. On the other hand, the JADE-ESPRIT requires eigendecomposition on covariance matrices, the complexity of which is order $O[(MN_t)^3]$. It is obvious that the computational burden of the TST-ESPRIT is substantially less than that of the JADE-ESPRIT.

In general, wireless communication channels might be much more complicated than the 3-ray example described above. Depending on the antenna array geometry and the signal bandwidth, respectively, the one-dimensional S-ESPRIT and the one-dimensional T-ESPRIT only can provide the DOA estimates and the delay estimates up to a certain resolution. The TST-ESPRIT can resolve rays that are extremely close in DOA or in delay, as long as these rays are not close in both parameters. As for the scenario in which rays are close in both DOAs and delays, the proposed algorithm will obtain the estimate of only one path. It also has been shown that distinguishing rays that are close in both DOA and delay does not improve the data demodulation accuracy at the receivers in a wireless communication system [22]. In other words, to maintain the diversity of a wireless signal, we have to distinguish (a) rays with close DOAs but with far-apart delays, or (b) rays with very close delays but with far-apart DOAs. The TST-ESPRIT is developed with these goals in mind to maintain the channel diversity.

7.4 Simulations and Discussions

In this section, we conduct several simulations to evaluate the TST-ESPRIT algorithm. We assume narrowband signals that are transmitted through four rays ($Q = 4$), and received by a three-element uniform linear array ($M = 3$). The antennas are of equal gains, and are spaced one-half wavelength apart, corresponding to the carrier frequency. Assuming the GSM system model [16], the GMSK modulated signals are tested. The received signal $\mathbf{x}(t)$ is sampled over 20 data bursts. In the basic setup, we set the angles of arrival to be [−43°, 27°, −40°, 30°], and the propagation delays to be $[0.03, 0.1, 0.86, 0.94]T_s$, where $T_s = 3.68\ \mu s$ is the symbol period of the GSM system. The oversampling factor $P = 2$, and the training sequence of each burst was truncated from the 6th training bit to the 21st training bit to keep the samples of the training sequence from being corrupted by the data bits. The average fading amplitudes of the four rays are equal, and normalized to 0 dB, with randomly selected but constant fading phases. The average power of the additive Gaussian noise is adjusted to achieve the required SNR.

Figure 7.5 compares the root-mean-square-error (RMSE) of the DOA and the delay estimates of the TST-ESPRIT, the JADE-ESPRIT, and the JADE-MUSIC algorithms, with respect to SNRs from 0 to 27 dB. For each specific SNR, 200 Monte Carlo trials are conducted. As shown in Figure 7.5, compared to the JADE-ESPRIT algorithm (two-dimensional ESPRIT), the TST-ESPRIT has better accuracy in the DOA estimation at low SNR, and has similar performance in the delay estimation. This is because the JADE-ESPRIT stacks each $M \times N_t$ (3×32) data matrix into an $MN_t \times 1$ snapshot vector, and 20 observation data bursts can provide only 20 snapshot vectors for them. Thus, the sample covariance matrices

7.4 Simulations and Discussions

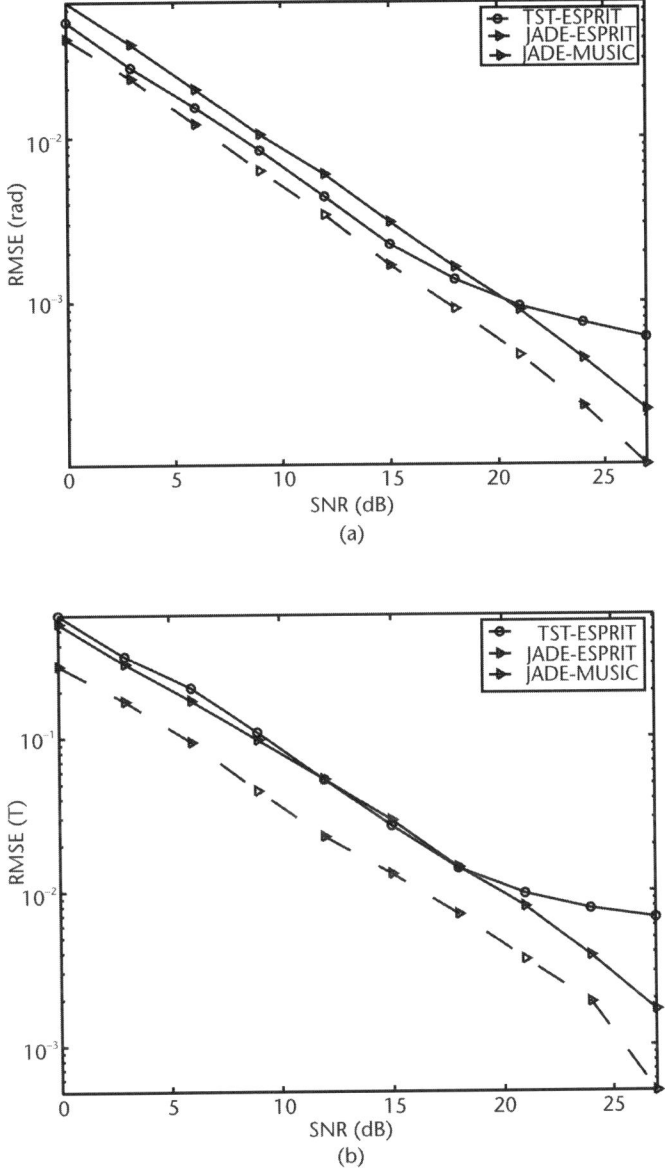

Figure 7.5 Comparison of (a) the root mean square error of the estimated DOAs between TST-ESPRIT, JADE-ESPRIT, and JADE-MUSIC; and (b) the root mean square error of the estimated delays between TST-ESPRIT, JADE-ESPRIT, and JADE-MUSIC.

of the JADE-ESPRIT are quite noisy. On the other hand, 20 bursts can offer $20 \cdot N_t = 640$ spatial snapshot vectors for the S-ESPRIT, and $20 \cdot M = 60$ temporal snapshot vectors for the T-ESPRIT to estimate the associated sample covariance matrix. However, the RMSE curves of the TST-ESPRIT become flat at high SNRs, where the residue signals caused by the temporal filtering and spatial beamforming processes dominate the RMSE of the TST-ESPRIT performance.

Figure 7.6 illustrates the parameter estimates obtained by the JADE-ESPRIT, JADE-MUSIC, and TST-ESPRIT algorithms. Each point on the DOA-Delay plane

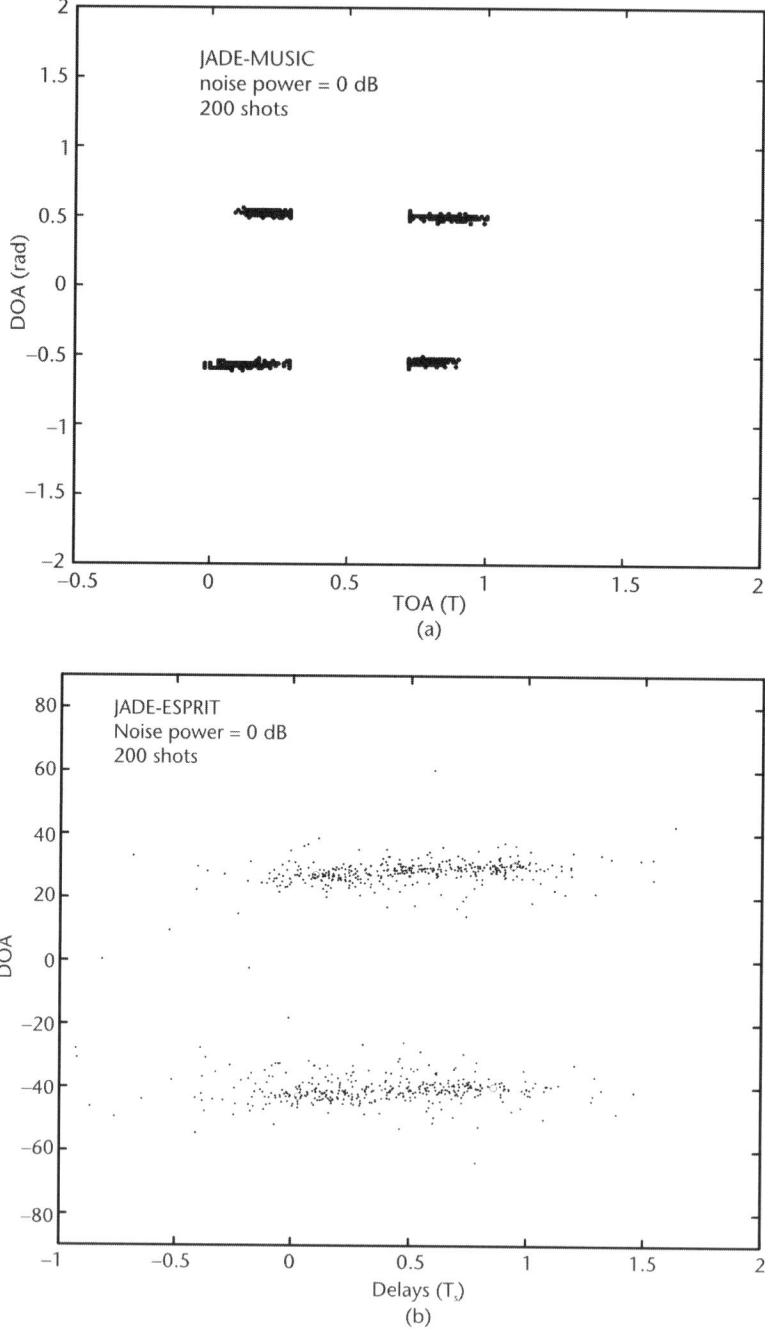

Figure 7.6 (a) Distribution of the JADE-MUSIC estimates; (b) distribution of the JADE-ESPRIT estimates; and (c) distribution of the TST-ESPRIT estimates.

represents an independent trial with $\sigma_n^2 = 0$ dB. Note that in all the simulation scenarios above, we assume time-varying channels, which makes the temporal array vectors inaccurate and results in higher RMSE than expected. However, the simulation results prove the resolvability of the proposed TST-ESPRIT.

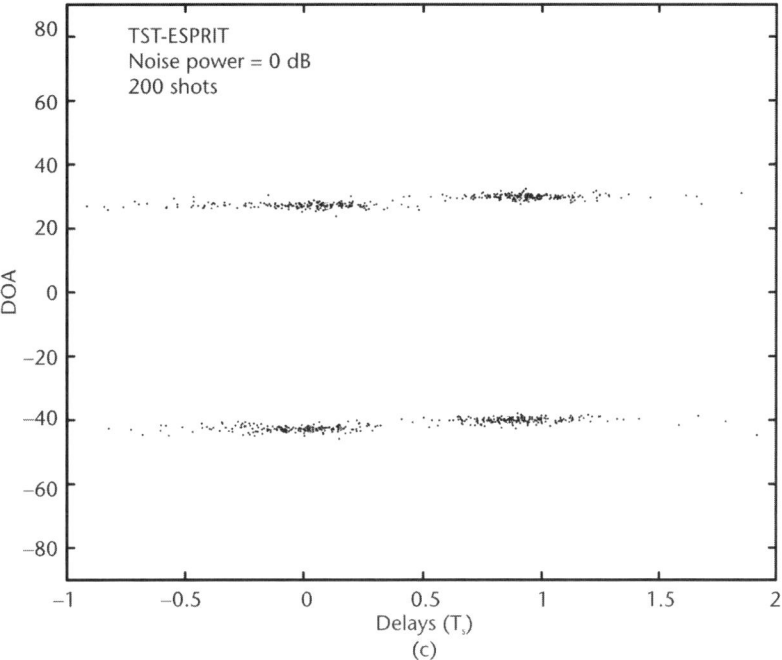

Figure 7.6 (Continued.)

7.5 Conclusions

This chapter proposes a novel algorithm—the TST-ESPRIT—which combines three one-dimensional ESPRITs with the temporal filtering and spatial beamforming techniques to jointly estimate the DOAs and delays of the multiple rays in a wireless channel. The S-ESPRIT and the T-ESPRIT algorithms in the TST-ESPRIT are used to estimate the DOAs and the propagation delays, respectively. The TST-ESPRIT is biased due to the propagation of the residue signals, which come from the leakage of the spatial beamforming process and of the temporal filtering process. The amount of bias depends on how accurately the group delays and the DOAs are estimated. Compared to the JADE-MUSIC, the TST-ESPRIT algorithm is relatively poor in parameter estimation accuracy, but requires considerably lower computational complexity.

References

[1] Rappaport, T. S., J. H. Reed, and B. D. Woerner, "Position Location Using Wireless Communications on Highways of the Future," *IEEE Commun. Mag.*, October 1996.

[2] Chen, J.-T., A. Paulraj, and U. Reddy, "Multichannel MLSE Equalizer for GSM Using a Parametric Channel Model," *IEEE Trans. on Communications*, January 1999, pp. 53–63.

[3] Chen, J.-T., J. Kim, and J. Liang, "Multichannel MLSE Equalizer with Parametric FIR Channel Identification," *IEEE Trans. on Vehicular Technology*, November 1999.

[4] Swindlehurst, A. L., "Time Delay and Spatial Signature Estimation Using Known Asynchronous Signals," *IEEE Trans. on Signal Processing*, Vol. 46, February 1998, pp. 449–461.

[5] Jakobsson, A., A. L. Swindlehurst, and P. Stoica, "Subspace-Based Estimation of Time Delays and Doppler Shifts," *IEEE Trans. on Signal Processing*, Vol. 46, September 1998, pp. 2472–2483.

[6] Swindlehurst, A. L., and J. H. Gunther, "Methods for Blind Equalization and Resolution of Overlapping Echos of Unknown Shape," *IEEE Trans. on Signal Processing*, Vol. 47, May 1999, pp. 1245–1254.

[7] Clark, M. P., and L. L. Scharf, "Two-Dimensional Modal Analysis Based on Maximum Likelihood," *IEEE Trans. on Signal Processing*, Vol. 42, June 1994, pp. 1443–1452.

[8] Bertaux, N., et al., "A Parameterized Maximum Likelihood Method for Multipaths Channels Estimation," *Proc. Signal Processing Advances in Wireless Comm.*, May 1999, pp. 391–394.

[9] Ogawa, Y., et al., "High-Resolution Analysis of Indoor Multipath Propagation Structure," *IEICE Trans. Communications*, E78B, November 1995, pp. 1450–1457.

[10] Vanderveen, M. C., C. B. Papadias, and A. Paulraj, "Joint Angle and Delay Estimation (JADE) for Multipath Signals Arriving at an Antenna Array," *IEEE Comm. Letters*, Vol. 1, January 1997, pp. 12–14.

[11] van der Veen, A. J., M. C. Vanderveen, and A. Paulraj, "Joint Angle and Delay Estimation (JADE) Using Shift-Invariance Techniques," *IEEE Signal Processing Letters*, Vol. 4, May 1997, pp. 142–145.

[12] Vanderveen, M. C., A. J. van der Veen, and A. Paulraj, "Estimation of Multipath Parameters in Wireless Communications," *IEEE Trans. on Signal Processing*, Vol. 46, March 1998, pp. 682–690.

[13] van der Veen, A. J., M. C. Vanderveen, and A. Paulraj, "Joint Angle and Delay Estimation (JADE) Using Shift-Invariance Properties," *IEEE Trans. on Signal Processing*, Vol. 46, February 1998, pp. 405–418.

[14] Roy, R. H., and T. Kailath, "ESPRIT-Estimation of Signal Parameters Via Rotational Invariance Techniques," *IEEE Trans. on Acoustics, Speech, and Signal Processing*, Vol. 37, July 1989, pp. 984–995.

[15] *EIA/TIA Interim Standard IS-136*, Telecommunication Association, December 1994.

[16] *European Telecommunication Standard Institute (ETSI), rec. ETSI/GSM 05.02*, European Telecommunication Standard Institute, 1990.

[17] Raleigh, G., et al., "Characterization of Fast Fading Vector Channels for Multiantenna Communication Systems," *28th Asilomar Conference on Signals, Systems and Computers*, 1994.

[18] Feher, K., *Wireless Digital Communications*, Upper Saddle River, NJ: Prentice Hall, 1995.

[19] Wax, M., and T. Kailath, "Detection of Signals by Information Theoretic Criteria," *IEEE Trans. on Acoustics, Speech, and Signal Processing*, Vol. 33, April 1985, pp. 387–392.

[20] Stark, H., and J. W. Woods, *Probability, Random Process and Estimation Theory for Engineers*, Prentice Hall, 1986.

[21] Bresler, Y., and A. Macovski, "On the Number of Signals Resolvable by a Uniform Linear Array," *IEEE Trans. on Acoustics, Speech, and Signal Processing*, Vol. 34, December 1986, pp. 1361–1375.

[22] Chen, J.-T., "Robustness of the Parametric MLSE Algorithm Against Nonparametric Channels," *Proc. GLOBECOM*, Sydney, Australia, 1998.

Appendix 7.A Noise Property in (7.24)

In this appendix, we prove that $\mathbf{N} \cdot \mathbf{U}_2^t$ in (7.24) is spatially and temporally white with unchanged noise power within the row space of $\mathbf{U}_2^t$ (i.e., within the projected subspace).

Appendix 7.A Noise Property in (7.24)

Proof: Let $\mathbf{N} = [\mathbf{n}_1, \ldots, \mathbf{n}_N]^T$, where $\mathbf{n}_k^T$ is a $1 \times N_t$ noise vector, and $\mathbf{U}_2^t = [\mathbf{u}_1, \ldots, \mathbf{u}_{N_t}] = [u_{qm}]$. We then have $\mathbf{N} \cdot \mathbf{U}_1^t = [\tilde{n}_{lm}]_{M \times N_t}$ with

$$\tilde{n}_{lm} = \mathbf{n}_l^T \cdot \mathbf{u}_m = \sum_{k=1}^{N} n_{lk} \cdot u_{km}$$

The correlation between n_{lm} and n_{pq} is given by

$$\begin{aligned} E\{\tilde{n}_{lm} \cdot \tilde{n}_{pq}^H\} &= E\left\{\left(\sum_{k=1}^{N} n_{lk} \cdot u_{km}\right) \cdot \left(\sum_{i=1}^{N} n_{pi} \cdot u_{iq}\right)^H\right\} \\ &= \sum_{k=1}^{N} \sum_{i=1}^{N} \sigma_n^2 \delta(l-p) \delta(k-i) u_{km} \cdot u_{iq}^H \\ &= \sigma_n^2 \cdot \delta(l-p) \mathbf{u}_q^H \cdot \mathbf{u}_m \\ &= \sigma_n^2 \cdot u_{qm} \cdot \delta(l-p) \end{aligned}$$

where $\delta(\cdot)$ is the Kronecker delta function. It thus proves that the transformed noise matrix is spatially white. Furthermore, since u_{qm} is the (q, m)th element of the projection matrix $\mathbf{U}_2^t$, the noise is temporally white with the same power within the projected subspace.

CHAPTER 8
Direction-of-Arrival Estimation Approach Using Switched Parasitic Arrays

Pantelis K. Varlamos, Stelios A. Mitilineos, Stylianos C. Panagiotou, Apostolos I. Sotiriou, and Christos N. Capsalis

In this chapter, two methods for DOA estimation are presented. The first method resolves the two strongest signals in a wideband environment with their relative strengths, using the tool of genetic algorithms (GA) and a switched parasitic array (SPA) as a receiving antenna. The method concentrates on the two dominant paths, while incorporating the rest in the model as noise. It is based on the concept of electronic beam steering, and is mainly aimed at low-mobility applications [e.g., wireless local area networks (WLAN)]. The second method applies the MUSIC algorithm to SPAs, with appropriate modifications. The major change consists in the steering matrix, which corresponds to the different voltage patterns of the SPA. The specific method combines the high-resolution capability of MUSIC with the possibility of sampling the received signals with several radiation patterns generated by a SPA. Numerical results are presented to demonstrate both methods' performances. Various SPAs and DOA distributions are selected, under different signal-to-noise ratios.

8.1 Introduction

Recent research toward 3G and 4G wireless communications systems requires thorough knowledge of the basic propagation mechanisms in the mobile radio channel. One of the most important aspects into investigation is the estimation of the location of sources that transmit signals of interest. Hence, numerous direction-finding methods have been developed, which are useful in a variety of radar or sonar applications and location-based services (LBS) [1]. Moreover, direction-finding algorithms are deployed to locate the positions of mobile users. Such real-time algorithms are applicable in most space division multiple access (SDMA) schemes. Furthermore, smart antennas capable of separating signals from multiple sources can improve the performance of global positioning systems (GPS) [2]. In addition, DOA estimation algorithms provide the number, the DOAs, and relative strengths of the incident signals [3, 4]. On the other hand, knowledge of the DOAs of incoming signals in the uplink can be used for effective transmit diversity in the downlink.

SPAs are suitable for electronic beam steering [5–9]. An appropriate selection of the switch settings renders each element active (driven with voltage) or parasitic (short-circuited). A different feed configuration changes the currents flowing in the elements, and produces a completely different radiation pattern. The speed of the RF switches sets a lower limit on the time required to perform the operation of beam steering.

Direction-finding methods using SPAs have been developed in the past [1, 9, 10]. Some of them are based on the detection of the total received power for each combination of the switch positions [9, 10]. The angular location of a single source may be determined by the order and relative magnitudes of the total received power levels. However, when multiple sources are present, these methods are inadequate.

The first method that is proposed in this chapter is able to reliably find the DOAs and relative powers of two wideband signals (e.g., the desired plus the strongest multipath component, signals transmitted by two different users) [11]. It is mostly suitable for low-mobility applications (e.g., WLANs). This method utilizes the concept of beamforming, and does not exploit either the model of the received signal vector or the statistical properties of signals and noise [3, 11]. The direction-finding problem is solved by employing the method of GAs, which uses the mechanisms of genetics (e.g., selection, crossover, and mutation) to conduct a global search of nonlinear and discontinuous solution spaces [12]. In this chapter, a GA estimates the angular positions and relative strengths of two signals.

The second method applies, with appropriate modifications, the MUSIC algorithm to SPAs, according to the idea discussed in [1]. MUSIC is a signal parameter estimation subspace-based algorithm that provides information about the number of incident signals, DOA of each signal, and strengths and cross correlations between incident signals [3, 13]. The major change consists in the steering matrix, which corresponds to the different voltage patterns of the SPA. On the contrary, in the classical MUSIC, the steering matrix is associated with the steering vectors related to each incoming signal [3, 13]. The new approach exploits the high-resolution capability of MUSIC, combined with SPAs' electronic control of the radiation pattern, through the insertion of a "digital word" (specific feed configuration) in the antenna feeding circuit. This concept may be extended to other DOA estimation schemes derived for a general antenna array, by inserting a new steering matrix.

In this chapter, both approaches to DOA estimation are described, and numerical results are presented to analyze their performance. The SPAs deployed offer either nonsymmetric or circularly symmetric radiation patterns. Two incoming signals are assumed, following uniform and nonuniform DOA distributions.

8.2 Switched Parasitic Arrays

Switched parasitic arrays have the advantage of an electronically controllable radiation pattern through the insertion of a "digital word" in the antenna feeding circuit. The term "digital word" is equivalent to a predefined feed configuration. The

8.2 Switched Parasitic Arrays

length of the digital word equals the number of array elements. The ones and zeros in the digital word represent the active and parasitic elements of the array, respectively.

The seven-element SPA used in this paper has been designed with the aid of a GA and analyzed with the induced EMF method [8, 14]. Each array element is a dipole having length $l = \lambda/2$ and radius $ra = 0.001\lambda$, where λ is the wavelength. Each dipole axis is parallel to the z-axis, while its feeding point lies in the x-y plane. Figure 8.1 depicts the SPA configuration in the horizontal plane. Table 8.1 shows the coordinates (x_m, y_m) in the azimuth plane, and voltage phase values δ_m of each element. One phase shifter is connected to each array element, and when the RF switch is "on," the input voltage is equal to $V_m = \exp(j \cdot \delta_m)$; otherwise, it is equal to 0.

The normalized radiation patterns in the azimuth plane and the combinations of active and parasitic elements that produce them are shown in Figure 8.2.

A GA also was used to design a six-element switched parasitic circular array (SPCA), with one element driven and five short-circuited, all having length $l = \lambda/2$ and radius $ra = 0.001\lambda$. The dipoles are parallel to the z-axis. Their feeding points are located on a circle of radius $\lambda/2$ in the x-y plane, while each voltage phase value is equal to 0V (unitary voltage of the driven element). The SPCA covers the azimuth plane with six identical radiation patterns, due to circular symmetry (see Figure 8.3). Each radiation pattern has 3-dB beamwidth slightly higher than 60° and maximum relative sidelobe level approximately equal to −9 dB. Similar GAs

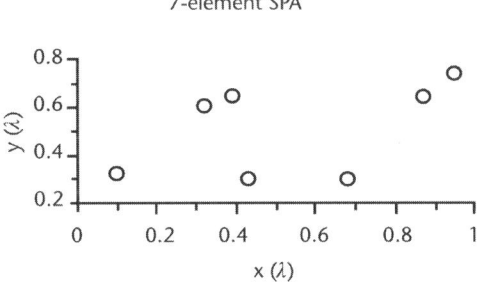

Figure 8.1 Configuration in the x-y plane of an SPA of seven identical dipoles ($l = \lambda/2$, $ra = 0.001\lambda$), parallel to the z-axis. (*From:* [11]. © 2004 Springer Science and Business Media. Reprinted with permission.)

Table 8.1 Element Positions in the Horizontal Plane and Constant Voltage Phase Values of a Seven-Element SPA

$x_m(\lambda)$	$y_m(\lambda)$	δ_m (°)
0.683	0.291	24.26
0.875	0.640	149.80
0.950	0.743	20.47
0.433	0.299	272.15
0.101	0.317	154.17
0.395	0.646	253.12
0.323	0.603	327.94

Source: [11].

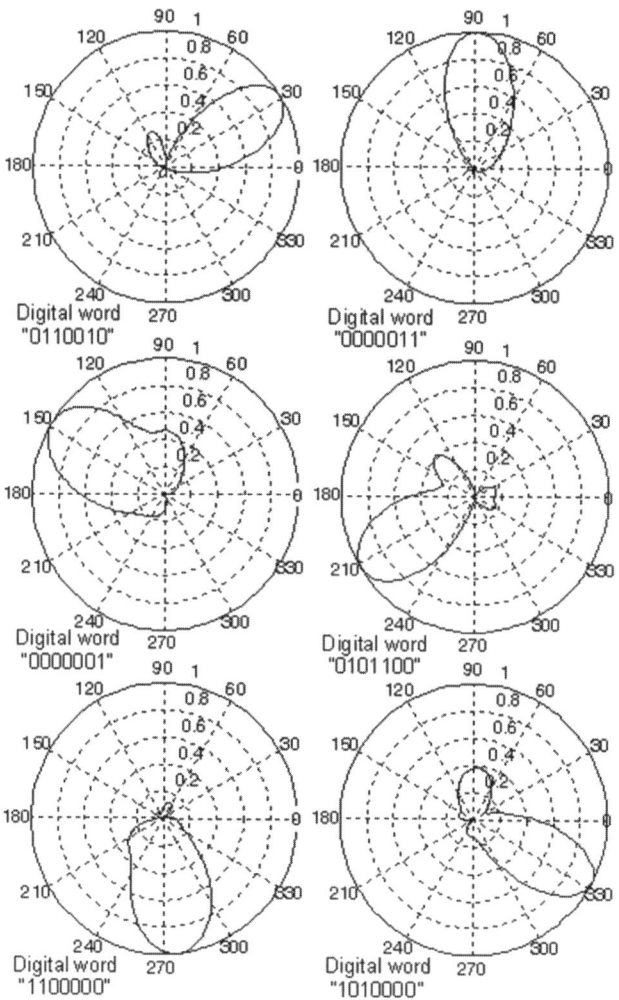

Figure 8.2 Normalized radiation patterns in the azimuth plane of a seven-element SPA. (*From:* [11]. © 2004 Springer Science and Business Media. Reprinted with permission.)

were deployed to design 4-, 8-, and 10-element SPCAs, offering 4, 8 and 10 circularly symmetric radiation patterns, respectively.

8.3 Wideband Direction-Finding Method for a Two-Path Model

The present method provides information about the DOAs and relative power levels of the two strongest signals in a wideband (multipath or multiuser) environment, using an SPA and the method of GAs. Signals may arrive at the receiving antenna from more than two paths, but the method concentrates on the two dominant ones. The remaining weaker signals are incorporated in the model as noise.

8.3 Wideband Direction-Finding Method for a Two-Path Model

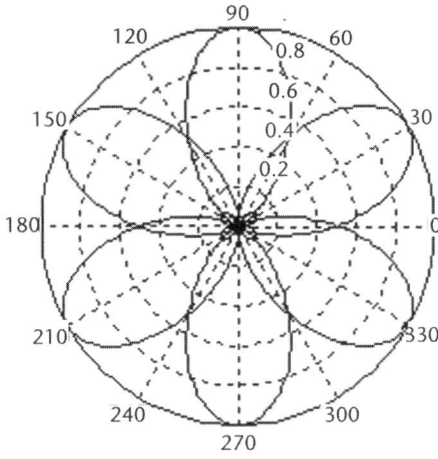

Figure 8.3 Normalized radiation patterns in the azimuth plane of a six-element SPCA. (*From:* [11]. © 2004 Springer Science and Business Media. Reprinted with permission.)

8.3.1 Wideband Signaling in Multipath Environments

In wideband signaling, the ideal transmitted signal with infinite bandwidth is the impulse function. In a multipath environment, a transmitted impulse $\delta(t)$ arrives at the receiver as the sum of a number of impulses with different amplitudes and phases. The composite impulse response is given by [15]

$$h(\tau, t) = \sum_{i=1}^{L} \beta_i e^{j\theta_i} \delta(t - \tau_i) \tag{8.1}$$

where β_i, θ_i, and τ_i represent the amplitude, phase, and delay of the ith path, respectively. For ideal wideband communication, the paths are assumed isolated and independent of one another. Therefore, the received power from L multipath directions can be expressed as the sum of squares of all path amplitudes [15]:

$$P_r = \sum_{i=1}^{L} |\beta_i|^2 \tag{8.2}$$

The result of (8.2) is valid, assuming pulses of very short duration. Moreover, it can be extended to the case of L users in a multiuser environment.

8.3.2 Method Description

This section describes the direction-finding method using an M-element SPA, which offers N_{RP} radiation patterns in the azimuth plane. Exploiting the electronically controllable radiation pattern of the SPA, the total received signal power for each digital word is detected sequentially. The method aims at determining s_1, s_2, the powers of the two dominant incoming signals; and ϕ_1, ϕ_2, the respective DOAs in the horizontal plane. The total measured signal power P_k for the kth digital word, according to (8.2), is

$$P_1 = s_1 G_1(\phi_1) + s_2 G_1(\phi_2) \tag{8.3a}$$

$$P_2 = s_1 G_2(\phi_1) + s_2 G_2(\phi_2) \tag{8.3b}$$

$$P_k = s_1 G_k(\phi_1) + s_2 G_k(\phi_2), \quad k = 3, \ldots, N_{RP} \tag{8.3c}$$

where $G_k(\phi_1), G_k(\phi_2)$ are the normalized kth radiation pattern values at ϕ_1, ϕ_2, respectively. Signals arriving at the receiver from other than the two main paths are incorporated in the model as noise with total average power s_r, compared to the power of the weakest signal. This is achieved by introducing an additional noise factor in (8.3a–c).

The estimation of the DOAs and relative strengths of the two strongest signals are based on the optimization technique of GAs. In this case, each individual in the GA population represents DOA estimates, $\hat{\phi}_1, \hat{\phi}_2$. Replacing ϕ_1, ϕ_2, with their estimates and solving (8.4) for s_1, s_2, two estimates $\hat{s}_1, \hat{s}_2$, of the incoming signal powers are obtained:

$$\left\{ \begin{aligned} P_1 &= s_1 G_1(\hat{\phi}_1) + s_2 G_1(\hat{\phi}_2) \\ P_2 &= s_1 G_2(\hat{\phi}_1) + s_2 G_2(\hat{\phi}_2) \end{aligned} \right\} \tag{8.4}$$

This is feasible, since P_1, P_2 are the total measured signal powers of (8.3a) and (8.3b), and $G_k(\hat{\phi}_1), G_k(\hat{\phi}_2)$ are the normalized kth radiation pattern values at $\hat{\phi}_1, \hat{\phi}_2$, respectively ($k = 1, 2$). Consequently, estimates of the remaining ($N_{RP} - 2$) total measured power levels may be produced, using $\hat{s}_1, \hat{s}_2$ and $\hat{\phi}_1, \hat{\phi}_2$:

$$\hat{P}_k = \hat{s}_1 G_k(\hat{\phi}_1) + \hat{s}_2 G_k(\hat{\phi}_2), \quad k = 3, \ldots, N_{RP} \tag{8.5}$$

Hence, when each $\hat{P}_k$ estimate given by (8.5) equals the respective measured P_k of (8.3c) ($k = 3, \ldots, N_{RP}$), the solutions $\hat{s}_1, \hat{s}_2$ and $\hat{\phi}_1, \hat{\phi}_2$ terminate the algorithm. Therefore, an appropriate objective function, normalized between 0 and 1, would be:

$$of = \frac{1}{1 + \sqrt{rerr}} \tag{8.6}$$

where

$$rerr = \left[\frac{1}{(N_{RP} - 2)} \right] \sum_{k=3}^{N_{RP}} \left[\frac{(P_k - \hat{P}_k)}{P_k} \right]^2 \tag{8.7}$$

Knowledge of the DOAs of the two strongest signals, which stems from the application of the above method, can be exploited in various possible ways, such as:

- In a multipath environment, if the two DOA estimates belong to the same beam of the SPA, then selection of the appropriate digital word activates the radiation pattern that receives power from both paths. If DOAs are found to lie in different beams, then an adaptive two-beam antenna, with each lobe pointing in the direction of the relative path, should be deployed to maximize received power. In a multiuser environment, a narrow beam could be employed to receive the desired user, and to suppress the undesired one.

- The DOA estimation accomplished by the present method also can be used for selective transmission in the downlink, especially in low-mobility applications, because the angle parameters are relatively stationary. Therefore, the angle of arrival in the uplink should be the direction of downlink transmission. In the two-path model, the downlink beamformer should cover the two dominant DOAs [11].

8.3.3 Method Performance Analysis

The numerical results presented here show the direction finding method reliability for different types of SPAs. More specifically, 500 Monte Carlo (MC) trials were conducted for each type of receiving antenna to analyze the method's performance. Each MC trial aimed at resolving two dominant incoming signals, with power ratio uniformly distributed between 0 and 7 dB, and with DOAs following various distributions. Binary GAs consisting of 400 generations of 50 individuals, each with $p_{crossover} = 0.8$ and $p_{mutation} = 0.1$, were used. Normalized geometric ranking, one-point crossover, and binary mutation were deployed [12]. The total average power s_r of the remaining signals was assumed either 20 or 50 dB below the power of the weakest signal.

Table 8.2 shows the method resolution capability using the seven-element SPA and the 6-element SPCA, both presented in Section 8.2 ($N_{RP} = 6$). The DOAs were assumed uniformly distributed in $(0, 2\pi)$. The probability p_{corr} ($p_{corr} = N_{corr}/500$) implies the number of trials, N_{corr}, when both DOAs and signal powers were estimated with a maximum deviation of $\pm 10°$ and $\pm 0.2°$, respectively, provided that $s_2 = 1$ and s_1 is uniformly distributed between 1 and 5. In addition, p_{beam} ($p_{beam} = N_{beam}/500$) implies the number of trials, N_{beam}, when the beams in which the incoming signals arrived were estimated correctly. Moreover, σ_{ang} and σ_{str}, in all performance analysis results presented hereinafter, are given by

$$\sigma_{ang} = \sqrt{\left(\frac{1}{N_{corr}}\right) \sum_{i_S=1}^{N_{corr}} 0.5\left[\left(\phi_{i_S,1} - \hat{\phi}_{i_S,1}\right)^2 + \left(\phi_{i_S,2} - \hat{\phi}_{i_S,2}\right)^2\right]} \quad (8.8a)$$

$$\sigma_{str} = \sum \sqrt{\left(\frac{1}{N_{corr}}\right) \sum_{i_S=1}^{N_{corr}} 0.5\left[\left(s_{i_S,1} - \hat{s}_{i_S,1}\right)^2 + \left(s_{i_S,2} - \hat{s}_{i_S,2}\right)^2\right]} \quad (8.8b)$$

The seven-element SPA is far superior for $s_r = -20$ dB, in terms of p_{corr}, p_{beam}, and σ_{str}. As for $s_r = -50$ dB, the six-element SPCA offers better p_{corr}, σ_{ang}, and σ_{str} values. The superiority of the seven-element SPA for higher noise levels may be explained from the SPCA's circularly symmetric radiation patterns. More specifi-

Table 8.2 Performance Analysis of the GA-Based Method, for Two Types of SPAs and Uniformly Distributed DOAs

Antenna Type	s_r (dB)	p_{corr}	σ_{ang} (°)	σ_{str}	p_{beam}
7-el SPA ($N_{RP} = 6$)	−50	0.886	0.621	0.027	0.952
	−20	0.686	1.890	0.057	0.838
6-el SPCA ($N_{RP} = 6$)	−50	0.918	0.364	0.026	0.940
	−20	0.534	1.650	0.073	0.740

Source: [11].

cally, these symmetries lead the GA to good fitness values, but are related to incorrect signal powers and DOA values.

Table 8.3 shows the resolution capability with the seven-element SPA and three DOA distributions. The two dominant incoming signals arrive in 3-dB beamwidths of different radiation patterns, or their DOAs follow the geometrically-based single bounce elliptical model (GBSBEM) pdf [3]:

$$f_\phi(\phi) = \frac{1}{2\pi\beta} \frac{(r_m^2 - 1)^2}{(r_m - \cos\phi)^2} \quad (8.9)$$

where $\beta = r_m\sqrt{r_m^2 - 1}$ and $r_m = 10^{(T-L_r)/(10n)}$ is the maximum normalized delay. The power margin of the incoming signals is $T = 7$ dB, the reflection loss is $L_r = 6$ dB, and the path loss exponent n is either 2 or 4.

Comparing Tables 8.2 and 8.3, it is obvious that p_{corr} and p_{beam} for the seven-element SPA are higher in the case of two signals arriving in different beams compared to the uniform case, especially for $s_r = -20$ dB, while σ_{ang} and σ_{str} maintain similar values. As for the GBSBEM cases, the method reliability slightly deteriorates as the angular discrimination of the incoming signals becomes smaller, but the results remain satisfying. The resolution capability is better for $n = 2$ than for $n = 4$, because the ensemble average angle spread is higher in the former case (29.70√ and 20.78√, respectively) [3, 11]. Hence, it is worth mentioning that even for nonuniformly distributed DOAs, the method operates with good accuracy.

Furthermore, the effect of using additional radiation patterns for uniformly distributed DOAs was examined. Table 8.4 shows the improvement achieved on method reliability when the SPCA radiation patterns or the total measured signal powers are sequentially increased. Taking into account (8.7), it is clear that by adding terms in *rerr*, the probability that good fitness values occur for incorrect signal powers and angular locations diminishes. No further improvement is

Table 8.3 Performance Analysis of the GA-Based Method, for the Seven-Element SPA and Three DOA Distributions

DOA Distribution	s_r (dB)	p_{corr}	σ_{ang} (√)	σ_{str}	p_{beam}
Different beams	−50	0.918	0.797	0.019	0.958
	−20	0.822	1.866	0.057	0.950
GBSBEM ($n = 2$)	−50	0.864	0.593	0.040	0.878
GBSBEM ($n = 4$)	−50	0.852	0.623	0.040	0.842

Source: [11].

Table 8.4 Performance Analysis of the GA-Based Method, for NRP-Element SPCAs, Uniformly Distributed DOAs, and $s_r = -50$ dB

Antenna Type	p_{corr}	σ_{ang} (°)	σ_{str}
4-el SPCA ($N_{RP} = 4$)	0.470	1.021	0.036
6-el SPCA ($N_{RP} = 6$)	0.918	0.364	0.026
8-el SPCA ($N_{RP} = 8$)	0.934	0.681	0.033
10-el SPCA ($N_{RP} = 10$)	0.884	0.679	0.034

Source: [11].

achieved, as far as p_{corr} is concerned, when $N_{RP} > 8$, while the best σ_{ang} and σ_{str} values are noticed for $N_{RP} = 6$.

8.4 A Modified MUSIC Algorithm for Switched Parasitic Arrays

The present method applies the MUSIC algorithm to SPAs, with appropriate modifications. The concept is introduced in [1], and combines the high-resolution capability of MUSIC with SPAs' advantage of sampling the received signals with different radiation patterns. The latter seems realistic, provided that the speed of the RF switches that control the SPA is sufficiently high, usually of the order of a few nanoseconds. In this section, the idea of the modified MUSIC algorithm proposed in [1] is presented in a comprehensive and analytical way, and its performance is demonstrated for different types of SPAs and DOA distributions.

Classical MUSIC is an array-based signal parameter estimation algorithm that exploits the eigenstructure of the input data matrix [3, 13]. Assuming that there are D signals impinging on an M-element antenna array ($D < M$), the received input data vector $\mathbf{u}(t)$ can be expressed as a linear combination of the D incident waveforms and noise. More specifically:

$$\mathbf{u}(t) = \mathbf{A}\mathbf{b}(t) + \mathbf{n}_0(t) \tag{8.10}$$

where $\mathbf{A}$ is an $M \times D$ matrix containing the array steering vectors corresponding to the DOAs of the incident signals ($d = 1, \ldots, D$). In addition, $\mathbf{b}$ is a $D \times 1$ vector of the baseband complex envelopes of the incident signals, while $\mathbf{n}_0$ is a noise vector.

In the SPA-based MUSIC algorithm, the major alteration consists in the formulation of the steering matrix $\mathbf{A}$. Given a SPA with N_{RP} radiation patterns, (8.10) remains unchanged, with the exception of matrix $\mathbf{A}$. The latter contains the SPA voltage patterns corresponding to the DOAs of D incident signals ($D < N_{RP}$). That is,

$$\mathbf{A} = \begin{bmatrix} g_1(\phi_1) & g_1(\phi_2) & \cdots & g_1(\phi_D) \\ \cdot & \cdot & \cdot & \cdot \\ \cdot & \cdot & \cdot & \cdot \\ g_{N_{RP}}(\phi_1) & g_{N_{RP}}(\phi_2) & \cdots & g_{N_{RP}}(\phi_D) \end{bmatrix} \tag{8.11}$$

where $g_k(\phi)$ ($k = 1, \ldots, N_{RP}$) are the SPA-normalized azimuth voltage patterns.

Taking into account the new steering matrix introduced in (8.11), the modified MUSIC algorithm is summarized as follows [3].

1. The input covariance matrix $\mathbf{R}_{uu}$ may be estimated through N_S input samples. The input data vector $\mathbf{u}(t)$ represents the total measured signal and noise level for the N_{RP} digital words inserted in the antenna feeding circuit. Assuming uncorrelated signals and noise, $\mathbf{R}_{uu}$ can be expressed as

$$\mathbf{R}_{uu} = E\left[\mathbf{u}\mathbf{u}^H\right] = \mathbf{A}E\left[\mathbf{b}\mathbf{b}^H\right]\mathbf{A}^H + E\left[\mathbf{n}_0\mathbf{n}_0^H\right] \quad (8.12)$$

where $[\ ]^H$ denotes matrix complex conjugate transpose.

2. The eigendecomposition of the estimated covariance matrix $\hat{\mathbf{R}}_{uu}$ is

$$\hat{\mathbf{R}}_{uu}\mathbf{V} = \mathbf{V}\mathbf{\Lambda} \quad (8.13)$$

where $\mathbf{\Lambda} = diag\{\lambda_1, \lambda_2, \ldots, \lambda_{N_{RP}}\}$, $\lambda_1 \geq \lambda_2 \geq \ldots \geq \lambda_{N_{RP}}$, are the eigenvalues, and $\mathbf{V}$ is the matrix containing the corresponding eigenvectors of $\hat{\mathbf{R}}_{uu}$. The accuracy of the estimated covariance matrix depends on the number of input samples.

3. Estimation of the number of incident signals, $\hat{D}$, from the multiplicity, K, of the smallest eigenvalue, $\lambda_{\min}$, as

$$\hat{D} = N_{RP} - K \quad (8.14)$$

Since in practice $\hat{\mathbf{R}}_{uu}$ is formed using a finite number of samples, the K smallest eigenvalues are not exactly equal. In order to test for the closeness of eigenvalues, the Rissanen minimum descriptive length (MDL) criterion is employed. In the MDL-based approach, the number of signals is determined as the argument that minimizes the following criterion [3]:

$$\text{MDL}(\hat{D}) = -\log\left\{\frac{\prod_{k=\hat{D}+1}^{N_{RP}} \lambda_k^{1/(N_{RP}-\hat{D})}}{\frac{1}{N_{RP}-\hat{D}}\sum_{k=\hat{D}+1}^{N_{RP}} \lambda_k}\right\}^{(N_{RP}-\hat{D})N_S} + \frac{1}{2}\hat{D}(2N_{RP} - \hat{D})\log N_S \quad (8.15)$$

where $\hat{D} \in \{0, 1, \ldots, N_{RP} - 1\}$.

4. The computation of the MUSIC spectrum is

$$P_{\text{MUSIC}}(\phi) = \frac{\boldsymbol{\alpha}^H(\phi)\boldsymbol{\alpha}(\phi)}{\boldsymbol{\alpha}^H(\phi)\mathbf{V}_n\mathbf{V}_n^H\boldsymbol{\alpha}(\phi)} \quad (8.16)$$

where $\boldsymbol{\alpha}(\phi) = [g_1(\phi)\ g_2(\phi)\ \ldots\ g_{N_{RP}}(\phi)]^T$ where $[\ ]^T$ denotes matrix transpose, and $\mathbf{V}_n$ is the matrix containing the K noise eigenvectors.

5. The DOA estimates are the $\hat{D}$ largest peaks of P_{MUSIC}.

8.4.1 Method Performance Analysis

The numerical results presented in this section demonstrate the performance and resolution capability of the modified MUSIC algorithm applied to different types of SPAs. More specifically, 500 MC trials were conducted for each type of receiving antenna in the same manner as in the GA-based method ($D = 2$). The vector $\mathbf{b}$ of the baseband complex envelopes was assumed to be temporally white and circularly Gaussian distributed. This is equivalent to $\mathbf{b}(t) \in N(0, \mathbf{R}_{bb})$, where $\mathbf{R}_{bb}$ is the signal correlation matrix. Spatially, temporally white and circularly Gaussian distributed noise was assumed, with average noise power s_n equal to -10 or -20 dB, compared to the average power of the weakest signal. Each MC trial was conducted four times, sequentially increasing the number of input samples:

$$N_S = 500 \cdot i_{N_S}, \; i_{N_S} = 1, \ldots, 4 \qquad (8.17)$$

Figure 8.4 shows the method resolution capability using the seven-element SPA and the six-element SPCA presented in Section 8.2 ($N_{RP} = 6$). The DOAs were assumed to be uniformly distributed in $(0, 2\pi)$. The probability p_{corr} implies the number of trials, N_{corr}, when both DOAs were estimated with a maximum deviation of $\pm 10°$, and the number of incident signals was accurately detected ($\hat{D} = D$). Similar p_{corr} values are obtained for both SPAs, especially for larger samples. As s_n decreases and N_S increases, p_{corr} asymptotically tends to unity. The seven-element SPA is slightly superior in terms of the σ_{ang} results contained in Table 8.5. Hereinafter, σ_{ang} is given by (8.8a), with N_{corr} as previously described.

Moreover, the effect of increasing the number of radiation patterns N_{RP} of the SPCA on method reliability was examined for uniformly distributed DOAs. Higher

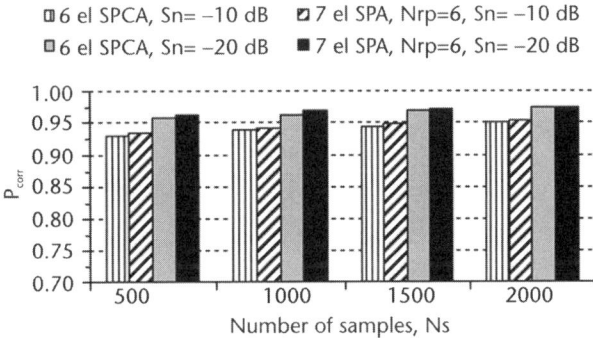

Figure 8.4 Probability of successful MC trials with the MUSIC-based method as a function of N_S, for two types of SPAs and uniformly distributed DoAs.

Table 8.5 Performance Analysis of the MUSIC-Based Method, for Two Types of SPAs, Uniformly Distributed DOAs, and $N_S = 2{,}000$ Input Samples

Antenna Type	$s_n = -10$ dB		$s_n = -20$ dB	
	p_{corr}	σ_{ang} (°)	p_{corr}	σ_{ang} (°)
7-el SPA ($N_{RP} = 6$)	0.954	0.289	0.974	0.111
6-el SPCA ($N_{RP} = 6$)	0.950	0.473	0.974	0.211

p_{corr} values as N_{RP} increases, for the same number of samples N_S and noise powers s_n, are depicted in Figure 8.5. Table 8.6 shows similar improvement as far as σ_{ang} values are concerned.

Furthermore, the eight-element SPCA was deployed to estimate DOAs following two GBSBEM pdfs, as presented in (8.9). Selecting the same T, L_r, and n values as in Section 8.3.3, the results shown in Figure 8.6 and Table 8.7 are obtained, proving that the resolution capability deteriorates as the ensemble average angle spread decreases. This is due to the fact that when the two signals arrive from DOAs that are too closely spaced (e.g., even closer than 5°), which is the case in many MC trials, the MUSIC spectrum gives rise to a one peak instead of two. The difficulty in distinguishing between two signals with small angular discrimination is more obvious for higher noise levels.

In addition, the effect of angular separation $\Delta \phi$ on method reliability was studied, by performing 100 successful MC trials, in which both signals were correctly estimated, ($p_{\text{corr}} = 1$) for each one of the following cases.

- Three pairs of incoming signals, with different $\Delta \phi$ separations and uniformly distributed average power ratios between 0 and 7 dB, were resolved using the four-, six-, and eight-element SPCAs, with $s_n = -10$ dB and varying input sample size. It is obvious from Figure 8.7 that σ_{ang} diminishes as $\Delta \phi$ increases, for the same number of samples N_S and number of patterns N_{RP}. As expected, σ_{ang} decreases with larger samples and more radiation patterns.

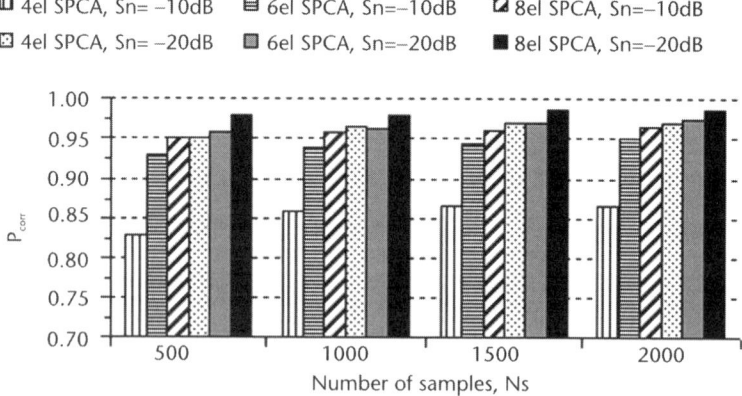

Figure 8.5 Probability of successful MC trials with the MUSIC-based method as a function of N_S, for three types of SPCAs and uniformly distributed DoAs.

Table 8.6 Performance Analysis of the MUSIC-Based Method, for Three Types of SPCAs, Uniformly Distributed DOAs, and $N_S = 2,000$ Input Samples

Antenna Type	$s_n = -10$ dB		$s_n = -20$ dB	
	p_{corr}	σ_{ang} (°)	p_{corr}	σ_{ang} (°)
4-el SPCA ($N_{RP} = 4$)	0.866	0.960	0.970	0.422
6-el SPCA ($N_{RP} = 6$)	0.950	0.473	0.974	0.211
8-el SPCA ($N_{RP} = 8$)	0.966	0.393	0.986	0.096

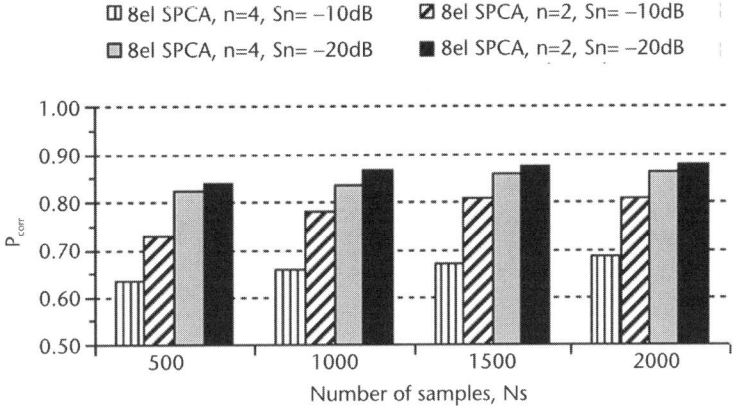

Figure 8.6 Probability of successful MC trials with the MUSIC-based method as a function of N_S, for the eight-element SPCA and two GBSBEM pdfs.

Table 8.7 Performance Analysis of the MUSIC-Based Method, for the Eight-Element SPCA, Three DOA Distributions, and $N_S = 2,000$ Input Samples

DOA Distribution	$s_n = -10$ dB		$s_n = -20$ dB	
	p_{corr}	σ_{ang} (°)	p_{corr}	σ_{ang} (°)
Uniform	0.966	0.393	0.986	0.096
GBSBEM ($n = 2$)	0.808	0.647	0.880	0.327
GBSBEM ($n = 4$)	0.686	0.848	0.862	0.448

- Two equally powered incoming signals, assuming $\phi_1 = 30°$ and $\phi_2 = \phi_1 + \Delta\phi$ ($\Delta\phi = 6°$ ... 30° with a 2° step), were resolved using the six- and eight-element SPCAs ($s_n = -20$ dB and $N_S = 1,000$). In Figure 8.8, RMSE results are depicted for the first DOA ϕ_1:

$$\text{RMSE} = \sqrt{\frac{1}{100} \sum_{i_S=1}^{100} \left(\phi_{i_S,1} - \hat{\phi}_{i_S,1}\right)^2} \qquad (8.18)$$

The six-element SPCA achieves $p_{corr} = 1$ (i.e., resolves both signals in all MC trials), provided that $\Delta\phi \geq 8\Sigma$ (RMSE = 0.36Σ for $\Delta\phi = 8\Sigma$). On the other hand, the eight-element SPCA improves resolution capability, offering $p_{corr} = 1$, even for $\Delta\phi = 6\Sigma$ (where RMSE = 0.48Σ). The RMSE values gradually decrease for higher angular separations, reaching a lower limit of approximately 0.08Σ.

8.5 Conclusions

Two direction-finding algorithms are presented, based on electronic beam steering, which is achieved with the aid of switched parasitic arrays. The radiation pattern of a switched parasitic array is controlled by a digital word or specific feed configuration.

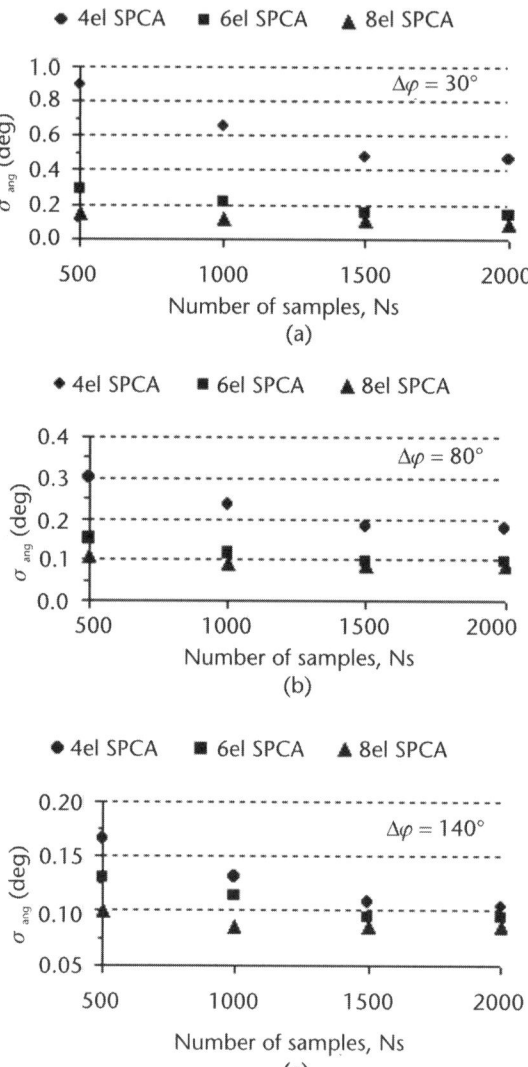

Figure 8.7 Values of σ_{ang} as a function of N_S for 100 MC trials ($p_{corr} = 1$) with the MUSIC-based method, three types of SPCAs, $s_n = -10$ dB, and: (a) $\Delta\phi = 30°$, $\phi_1 = 10°$, $\phi_2 = 40°$; (b) $\Delta\phi = 80°$, $\phi_1 = 40°$, $\phi_2 = 120°$; and (c) $\Delta\phi = 140°$, $\phi_1 = 20°$, $\phi_2 = 160°$.

The first method deploys switched parasitic arrays to determine the azimuth angular positions and relative powers of the two strongest signals in a wideband environment. The method concentrates on the two dominant paths, while incorporating the remainder in the model as noise. The tool of GAs is used, and the method is based on the sequential detection of the total received signal power for each digital word inserted in the antenna feeding circuit. Comparisons are made between two kinds of switched parasitic arrays, either with nonsymmetric or circularly symmetric radiation patterns. The former type demonstrates superior performance for higher noise levels, whereas the latter type behaves better for lower noise levels.

8.5 Conclusions

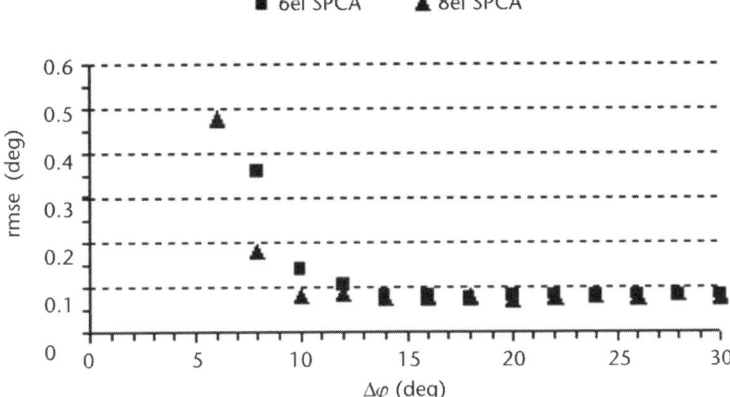

Figure 8.8 Values of RMSE as a function of $\Delta\phi$ for 100 MC trials ($p_{corr} = 1$) with the MUSIC-based method, two types of SPCAs, $s_n = -20$ dB, $N_S = 1{,}000$, $\phi_1 = 30°$, and $\phi_2 = \phi_1 + \Delta\phi$.

Increasing the number of radiation patterns up to eight improves method reliability, which slightly deteriorates as the angular discrimination of the incoming signals becomes smaller.

The second method deploys switched parasitic arrays in conjunction with the MUSIC algorithm to estimate the DOAs of two incoming signals. The major modification, compared to the array-based classical MUSIC, consists in the steering matrix, which corresponds to the different voltage patterns of the switched parasitic array. This method maintains the high-resolution capability and basic characteristics of MUSIC; that is, improved reliability for larger samples, more radiation patterns, and higher signal-to-noise ratios. Closely separated signals are less distinguishable from each other. The MUSIC algorithm is combined with the possibility of sampling the received signals with different radiation patterns, provided by a switched parasitic array. This concept may be extended to other DOA estimation schemes derived for a general antenna array by inserting a new steering matrix.

The GA-based method offers comparable performance results with the MUSIC-based method for lower noise levels. It combines GAs and SPAs in a flexible manner, and the proposed resolution procedure is simpler than with the MUSIC-based method. However, the modified MUSIC algorithm is superior for higher noise levels.

References

[1] Svantesson, T., and M. Wennström, "High-Resolution Direction Finding Using a Switched Parasitic Antenna," *Proc. 11th IEEE Signal Processing Workshop on Statistical Signal Processing*, Singapore, August 2001, pp. 508–511.

[2] El Zooghby, A. H., C. G. Christodoulou, and M. Georgiopoulos, "A Neural Network-Based Smart Antenna for Multiple Source Tracking," *IEEE Trans. on Antennas and Propagation*, Vol. 48, No. 5, May 2000, pp. 768–775.

[3] Liberti, J. C., and T. S. Rappaport, *Smart Antennas for Wireless Communications*, Upper Saddle River, NJ: Prentice Hall, 1999, Chs. 7 and 9.

[4] Godara, L. C., "Application of Antenna Arrays to Mobile Communications, Part II: Beam-Forming and Direction-of-Arrival Considerations," *IEEE Proc.*, Vol. 85, No. 8, August 1997, pp. 1195–1245.

[5] Thiel, D. V., S. G. O'Keefe, and J. W. Lu, "Electronic Beam Steering in Wire and Patch Antenna Systems Using Switched Parasitic Elements," *Proc. IEEE Antennas Propagation Society URSI Radio Science Meeting*, Baltimore, MD, July 1996, pp. 534–537.

[6] Schlub, R., et al., "Dual-Band Six-Element Switched Parasitic Array for Smart Antenna Cellular Communications Systems," *Electronics Letters*, Vol. 36, No. 16, August 2000, pp. 1342–1343.

[7] Varlamos, P. K., and C. N. Capsalis, "Electronic Beam Steering Using Switched Parasitic Smart Antenna Arrays," *Progress in Electromagnetics Research*, PIER 36, 2002, pp. 101–119.

[8] Varlamos, P. K., and C. N. Capsalis, "Design of a Six-Sector Switched Parasitic Planar Array Using the Method of Genetic Algorithms," *Wireless Personal Communications Journal*, Vol. 26, No. 1, August 2003, pp. 77–88.

[9] Preston, S. L., et al., "Base-Station Tracking in Mobile Communications Using a Switched Parasitic Antenna Array," *IEEE Trans. on Antennas and Propagation*, Vol. 46, No. 6, June 1998, pp. 841–844.

[10] Preston, S. L., and D. V. Thiel, "Direction Finding Using a Switched Parasitic Antenna Array," *Proc. IEEE Antennas Propagation Society URSI Radio Science Meeting*, Montreal, Canada, July 1997, pp. 1024–1027.

[11] Varlamos, P. K., and C. N. Capsalis, "Direction-of-Arrival Estimation (DOA) Using Switched Parasitic Planar Arrays and the Method of Genetic Algorithms," *Wireless Personal Communications Journal*, Vol. 28, No. 1, January 2004, pp. 59–75.

[12] Goldberg, D. E., *Genetic Algorithms in Search, Optimization, and Machine Learning*, Reading, MA: Addison-Wesley, 1989.

[13] Schmidt, R. O., "Multiple Emitter Location and Signal Parameter Estimation," *IEEE Trans. on Antennas and Propagation*, Vol. AP-34, No. 3, March 1986, pp. 276–280.

[14] Balanis, C. A., *Antenna Theory, Analysis and Design*, 2nd ed., New York: John Wiley & Sons, 1997, Ch. 8.

[15] Pahlavan, K., and A. Levesque, *Wireless Information Networks*, New York: John Wiley & Sons, 1995, pp. 50–56.

CHAPTER 9
DOA Estimation for Noncircular Signals: Performance Bounds and Algorithms

Jean-Pierre Delmas

9.1 Introduction

There is considerable literature about second-order statistics-based algorithms for estimating DOAs of narrowband sources impinging on an array of sensors. These algorithms, however, have been examined only under the complex circular Gaussian assumption. The interest in these algorithms stems from a large number of applications, including mobile communications systems [1]. In this application, after frequency downshifting the sensor signals to baseband, the in-phase and quadrature components are paired to obtain complex signals. Complex noncircular signals [2], such as binary phase shift keying (BPSK) and offset quadrature phase shift keying (OQPSK) modulated signals, are often used. Naturally, the second-order algorithms devoted to complex circular signals relying on the positive definite Hermitian covariance matrix $E(\mathbf{y}_t \mathbf{y}_t^H)$ can be used in this context. Because the second-order statistical characteristics are also contained in the complex symmetric unconjugated spatial covariance matrix $E(\mathbf{y}_t \mathbf{y}_t^T)$ for noncircular signals, a potential performance improvement ought to be obtained if these two covariance matrices are used. However, only a few contributions [3–5] have been devoted to noncircular signals in DOA estimation in the last decade.

The aim of this chapter is to present an overview of algorithms and performance bounds of DOA estimates of noncircular signals. This chapter is organized as follows. The array signal model with some notations and the statement of the problem are given in Section 9.2. The potential benefit due to the noncircularity property is underscored by the help of subspace-based algorithms built only from the unconjugated spatial covariance matrix. Three attractive MUSIC-like algorithms and an optimally weighted MUSIC algorithm built from the two covariance matrices are then presented with their asymptotic performances in Section 9.3. To assess the performance and efficiency of these algorithms, asymptotically (in the number of measurements) minimum variance (AMV) algorithms, in the class of consistent algorithms based on the two covariance matrices or on their associated orthogonal projectors and AMV bounds, are described in Section 9.4. Because the Gaussian stochastic CRB matrix is, under rather general conditions, the largest of all CRB matrices among the class of arbitrary distributions with given mean and covariance [6, p. 293], the noncircular Gaussian stochastic CRB is derived in Section 9.5 as a tight upper bound of the stochastic CRB under discrete distributions. The case of BPSK sources is specifically treated in Section 9.6. Finally, this chapter ends by illustrative examples and conclusion.

9.2 Array Signal Model

Let an array of M sensors receive the signals emitted by K narrowband sources. The observation vectors are modeled as

$$\mathbf{y}_t = \mathbf{A}\mathbf{x}_t + \mathbf{n}_t \quad t = 1, \ldots, T$$

where $(\mathbf{y}_t)_{t=1,\ldots,T}$ are independent and identically distributed. $\mathbf{A}(\theta) = [\mathbf{a}_1, \ldots, \mathbf{a}_K]$ is the steering matrix, where each vector $\mathbf{a}_k = \mathbf{a}(\theta_k)$ is parameterized by the real scalar parameter θ_k to avoid unnecessary notational complexity, but the results presented here apply to a general parameterization. $\mathbf{x}_t = (x_{t,1}, \ldots, x_{t,K})^T$ and $\mathbf{n}_t$ model the source signals and additive measurement noise, respectively. $\mathbf{x}_t$ and $\mathbf{n}_t$ are multivariate independent and zero-mean. $\mathbf{n}_t$ is assumed Gaussian complex circular, spatially uniformly uncorrelated with $\mathrm{E}(\mathbf{n}_t\mathbf{n}_t^H) = \sigma_n^2 \mathbf{I}_M$, or spatially correlated with unknown covariance matrix $\mathrm{E}(\mathbf{n}_t\mathbf{n}_t^H) = \mathbf{Q}_n(\boldsymbol{\sigma})$ parameterized by the vector of real unknown coefficients $\boldsymbol{\sigma} \stackrel{\text{def}}{=} (\sigma_1, \ldots, \sigma_N)^T$. This general noise model was introduced in [7] and used in [8, 9]. $\mathbf{x}_t$ is complex noncircular, not necessarily Gaussian, and possibly spatially correlated or even coherent with spatial covariance matrices $\mathbf{R}_x \stackrel{\text{def}}{=} \mathrm{E}(\mathbf{x}_t\mathbf{x}_t^H)$ and $\mathbf{R}'_x \stackrel{\text{def}}{=} \mathrm{E}(\mathbf{x}_t\mathbf{x}_t^T)$. Consequently, this leads to two covariance matrices of $\mathbf{y}_t$ that contain information about $\boldsymbol{\theta} \stackrel{\text{def}}{=} (\theta_1, \ldots, \theta_K)^T$:

$$\mathbf{R}_y = \mathbf{A}\mathbf{R}_x\mathbf{A}^H + \mathbf{Q}_n \quad \text{and} \quad \mathbf{R}'_y = \mathbf{A}\mathbf{R}'_x\mathbf{A}^T \neq \mathbf{O} \quad (9.1)$$

If no a priori information is available concerning the spatial covariance of the sources, then $\mathbf{R}_x$ and $\mathbf{R}'_x$ are generically parameterized by the real parameters $\boldsymbol{\rho} = ((\Re([\mathbf{R}_x]_{i,j}), \Im([\mathbf{R}_x]_{i,j}), \Re([\mathbf{R}'_x]_{i,j}), \Im([\mathbf{R}'_x]_{i,j}))_{1 \leq j < i \leq K}, ([\mathbf{R}_x]_{i,i}, \Re([\mathbf{R}'_x]_{i,i}), \Im([\mathbf{R}'_x]_{i,i}))_{i=1,\ldots,K})^T$. Thus, the couple $(\mathbf{R}_y, \mathbf{R}'_y)$ is parameterized by the real parameter $\boldsymbol{\alpha} \stackrel{\text{def}}{=} (\boldsymbol{\theta}^T, \boldsymbol{\rho}^T, \boldsymbol{\sigma}^T)^T \in \mathbb{R}^L$. This parameter is assumed identifiable from $[\mathbf{R}_y(\boldsymbol{\alpha}), \mathbf{R}'_y(\boldsymbol{\alpha})]$, in the following sense:

$$\mathbf{R}_y(\boldsymbol{\alpha}) = \mathbf{R}_y(\boldsymbol{\alpha}') \quad \text{and} \quad \mathbf{R}'_y(\boldsymbol{\alpha}) = \mathbf{R}'_y(\boldsymbol{\alpha}') \Rightarrow \boldsymbol{\alpha} = \boldsymbol{\alpha}'$$

These covariance matrices are traditionally estimated by

$$\mathbf{R}_{y,T} = \frac{1}{T}\sum_{t=1}^{T} \mathbf{y}_t\mathbf{y}_t^H$$

and

$$\mathbf{R}'_{y,T} = \frac{1}{T}\sum_{t=1}^{T} \mathbf{y}_t\mathbf{y}_t^T$$

respectively.

For a performance analysis, some extra hypotheses are needed. We suppose that the signal waveforms stem from $\overline{K}$ independent signals $\overline{x}_{t,k}$, with $\overline{K} \leq K$, with strict inequality implying linear dependence among the signal waveforms emanating from, for example, specular multipath or smart jamming in communication applications. We suppose that the signal waveforms have finite fourth-order moments. The noncircularity rate ρ_k of the kth source is defined by $E(x_{t,k}^2) = \rho_k e^{i\phi_k} \sigma_{s_k}^2$, where ϕ_k is its circularity phase that satisfies $0 \leq \rho_k \leq 1$ and $\sigma_{s_k}^2 \stackrel{\text{def}}{=} E|s_{t,k}^2|$.

The problem of interest in this work concerns estimating $\boldsymbol{\theta}$ from the two sample covariance matrices $\mathbf{R}_{y,T}$ and $\mathbf{R}'_{y,T}$. The number K of sources is assumed to be known.

9.3 MUSIC-Like Algorithms

We assume in this section that $\mathbf{Q}_n = \sigma_n^2 \mathbf{I}_M$, $\mathbf{A}$ is full column rank, and $\mathbf{R}_x$ and $\mathbf{R}'_x$ are nonsingular. To prove the potential benefit due to the noncircularity of the sources, we first propose MUSIC-like algorithms built only from the unconjugated spatial covariance matrix.

9.3.1 MUSIC-Like Algorithms Built Only from $\mathbf{R}'_{y,T}$

Because $\mathbf{R}'_y$ and $\mathbf{R}_y$ have a common noise subspace [see (9.1)] with associated orthogonal projection matrices $\boldsymbol{\Pi}' = \boldsymbol{\Pi}$, the first idea for estimating $\boldsymbol{\theta}$ from $\mathbf{R}'_{y,T}$ alone is to apply the following steps. Estimate the projection matrix $\boldsymbol{\Pi}'_T$ associated with the noise subspace of $\mathbf{R}'_{y,T}$ by the singular value decomposition of the symmetric complex-valued matrix $\mathbf{R}'_{y,T}$, and then use the standard MUSIC algorithm based on $\boldsymbol{\Pi}'_T$, where the DOA $(\theta_{k,T})_{k=1,\ldots,K}$ are estimated as the locations of the K smallest minima of the function:

$$\theta_{k,T}^{\text{Alg}_0} = \arg\min_\theta \mathbf{a}^H(\theta) \boldsymbol{\Pi}'_T \mathbf{a}(\theta) \tag{9.2}$$

Compared to the standard MUSIC algorithm based on $\boldsymbol{\Pi}_T$ associated with the noise subspace of $\mathbf{R}_{y,T}$, we prove in [10] the following

Theorem 9.1
The sequences $\sqrt{T}(\boldsymbol{\theta}_T - \boldsymbol{\theta})$, where $\boldsymbol{\theta}_T$ is the DOA estimate given by these two standard MUSIC algorithms, converge in distribution to the zero-mean Gaussian distribution of covariance matrix of the similar structure

$$(\mathbf{C}_{\boldsymbol{\theta}})_{k,l} = \frac{2}{\alpha_k \alpha_l} \Re\left[\left(\mathbf{a}_l^H \mathbf{U} \mathbf{a}_k\right)\left(\mathbf{a}_k'^H \boldsymbol{\Pi} \mathbf{a}_l'\right)\right] \tag{9.3}$$

with $\mathbf{a}'_k \stackrel{\text{def}}{=} d\mathbf{a}_k/d\theta_k$ and $\alpha_k \stackrel{\text{def}}{=} 2\mathbf{a}_k'^H \boldsymbol{\Pi} \mathbf{a}'_k$, where $\mathbf{U} \stackrel{\text{def}}{=} \sigma_n^2 \mathbf{S}^\# \mathbf{R}_y \mathbf{S}^\#$ with $\mathbf{S} \stackrel{\text{def}}{=} \mathbf{A}\mathbf{R}_x\mathbf{A}^H$, and $\mathbf{U} \stackrel{\text{def}}{=} \sigma_n^2 \mathbf{S}'^\# \mathbf{R}_y^T \mathbf{S}'^{*\#}$ with $\mathbf{S}' \stackrel{\text{def}}{=} \mathbf{A}\mathbf{R}'_x\mathbf{A}^T$ for the MUSIC algorithm built on $\mathbf{R}_{y,T}$ and $\mathbf{R}'_{y,T}$, respectively.

As a result, the asymptotic performance of the estimates given by these two standard MUSIC algorithms can be very similar. In particular, for only one source, it is proved in [10] that these asymptotic variances are, respectively, given by:

$$C_{\theta_1} = \frac{1}{\alpha_1 r_1}\left(1 + \frac{1}{\|\mathbf{a}_1\|^2 r_1}\right) \quad \text{and} \quad C_{\theta_1} = \frac{1}{\alpha_1 r_1 \rho_1^2}\left(1 + \frac{1}{\|\mathbf{a}_1\|^2 r_1}\right)$$

with

$$r_1 \stackrel{\text{def}}{=} \frac{\sigma_{s_1}^2}{\sigma_n^2}$$

We note that for $\rho_1 = 1$ (e.g., for an unfiltered BPSK modulated source), these two variances are equal. Naturally, when ρ_1 approaches zero, C_{θ_1} is unbounded, and the unconjugated spatial covariance matrix $\mathbf{R}'_y$ conveys no information about θ_1. This result raises the following query: How does one combine the statistics $\mathbf{\Pi}_T$ and $\mathbf{\Pi}'_T$ to improve the estimate of $\boldsymbol{\theta}$? A possible solution is proposed in the following section.

9.3.2 MUSIC-Like Algorithms Built from Both $\mathbf{R}_{y,T}$ and $\mathbf{R}'_{y,T}$

To devise simple subspace-based algorithms built from both $\mathbf{R}_{y,T}$ and $\mathbf{R}'_{y,T}$, we consider the extended covariance matrix $\mathbf{R}_{\tilde{y}} \stackrel{\text{def}}{=} E(\tilde{\mathbf{y}}_t \tilde{\mathbf{y}}_t^H)$, where $\tilde{\mathbf{y}}_t \stackrel{\text{def}}{=} (\mathbf{y}_t^T, \mathbf{y}_t^H)^T$, for which

$$\mathbf{R}_{\tilde{y}} = \tilde{\mathbf{A}} \mathbf{R}_{\tilde{x}} \tilde{\mathbf{A}}^H + \sigma_n^2 \mathbf{I}_{2M} \tag{9.4}$$

with

$$\tilde{\mathbf{A}} \stackrel{\text{def}}{=} \begin{pmatrix} \mathbf{A} & \mathbf{O} \\ \mathbf{O} & \mathbf{A}^* \end{pmatrix}$$

and

$$\mathbf{R}_{\tilde{x}} \stackrel{\text{def}}{=} \begin{pmatrix} \mathbf{R}_x & \mathbf{R}'_x \\ \mathbf{R}'^*_x & \mathbf{R}^*_x \end{pmatrix}$$

From the assumptions of Section 9.3, $K \leq \text{rank}(\mathbf{R}_{\tilde{x}}) \leq 2K$, and, depending on this rank, many situations may be considered. We concentrate first on a particular case [Case (1)] for which the sources are uncorrelated and with noncircularity rate ρ_k equal to 1, because very attractive algorithms have been devised for this case [3, 4]. This corresponds, for example, to unfiltered BPSK or OQPSK uncorrelated modulated signals. In this case, $\mathbf{R}_x = \boldsymbol{\Delta}_\sigma$ and $\mathbf{R}'_x = \boldsymbol{\Delta}_\sigma \boldsymbol{\Delta}_\phi$, with $\boldsymbol{\Delta}_\sigma \stackrel{\text{def}}{=} \text{diag}(\sigma_{s_1}^2, \ldots, \sigma_{s_K}^2)$ and $\boldsymbol{\Delta}_\phi \stackrel{\text{def}}{=} \text{diag}(e^{i\phi_1}, \ldots, e^{i\phi_K})$.

9.3 MUSIC-Like Algorithms

Consequently,

$$\mathbf{R}_{\tilde{x}} = \begin{pmatrix} \boldsymbol{\Delta}_\sigma & \boldsymbol{\Delta}_\sigma \boldsymbol{\Delta}_\phi \\ \boldsymbol{\Delta}_\sigma \boldsymbol{\Delta}_\phi^* & \boldsymbol{\Delta}_\sigma \end{pmatrix} = \begin{pmatrix} \mathbf{I}_K \\ \boldsymbol{\Delta}_\phi^* \end{pmatrix} \boldsymbol{\Delta}_\sigma \begin{pmatrix} \mathbf{I}_K \\ \boldsymbol{\Delta}_\phi^* \end{pmatrix}^H$$

and rank$(\mathbf{R}_{\tilde{x}}) = K$. Subsequently, we consider the general case for which rank$(\mathbf{R}_{\tilde{x}}) = 2K$ [Case (2)]. This case corresponds, for example, to filtered BPSK or OQPSK modulated signals. In these two cases, the orthogonal projector matrix onto the noise subspace is structured as $\tilde{\mathbf{A}}$ and $\mathbf{R}_{\tilde{x}}$. More precisely, the following lemma is proved in [10].

Lemma 9.1
In Cases (1) and (2), the orthogonal projector matrix $\tilde{\boldsymbol{\Pi}}$ onto the noise subspace is structured 5 as

$$\tilde{\boldsymbol{\Pi}} = \begin{pmatrix} \boldsymbol{\Pi}_1 & \boldsymbol{\Pi}_2 \\ \boldsymbol{\Pi}_2^* & \boldsymbol{\Pi}_1^* \end{pmatrix}$$

where $\boldsymbol{\Pi}_1$ and $\boldsymbol{\Pi}_2$ are Hermitian and complex symmetric, respectively; and where $\boldsymbol{\Pi}_1$ is the orthogonal projector onto the column space of $\mathbf{A}$ and $\boldsymbol{\Pi}_2 = \mathbf{O}$ in Case (2).[1] Furthermore, the orthogonal projector onto the noise subspace $\tilde{\boldsymbol{\pi}}_T$ associated with the sample estimate $\mathbf{R}_{\tilde{y},T}$ of $\mathbf{R}_{\tilde{y}}$ has the same structure

$$\tilde{\boldsymbol{\Pi}}_T = \begin{pmatrix} \boldsymbol{\Pi}_{1,T} & \boldsymbol{\Pi}_{2,T} \\ \boldsymbol{\Pi}_{2,T}^* & \boldsymbol{\Pi}_{1,T}^* \end{pmatrix} \quad (9.5)$$

where $\boldsymbol{\Pi}_{1,T}$ and $\boldsymbol{\Pi}_{2,T}$ are Hermitian and complex symmetric, respectively.

Case (1): Uncorrelated Sources with $\rho_k = 1$

Consider now three MUSIC-like algorithms, for which (9.4) becomes

$$\mathbf{R}_{\tilde{y}} = \begin{pmatrix} \mathbf{A} \\ \mathbf{A}^* \boldsymbol{\Delta}_\phi^* \end{pmatrix} \boldsymbol{\Delta}_\sigma \begin{pmatrix} \mathbf{A} \\ \mathbf{A}^* \boldsymbol{\Delta}_\phi^* \end{pmatrix}^H + \sigma_n^2 \mathbf{I}_{2M} \quad (9.6)$$

An algorithm (denoted Alg$_1$), devised in [3], has been derived from the standard MUSIC algorithm. Specifically, the estimated DOA $(\theta_{k,T})_{k=1,\ldots,K}$ are obtained as the locations of the K smallest minima of the following function:

$$\theta_{k,T}^{\text{Alg}_1} = \arg\min_\theta \left[\min_\phi \tilde{\mathbf{a}}^H(\theta,\phi) \tilde{\boldsymbol{\Pi}}_T \tilde{\mathbf{a}}(\theta,\phi) \right] \quad (9.7)$$

$$= \arg\min_\theta \left[\mathbf{a}^H(\theta) \boldsymbol{\Pi}_{1,T} \mathbf{a}(\theta) - \left| \mathbf{a}^T(\theta) \boldsymbol{\Pi}_{2,T}^* \mathbf{a}(\theta) \right| \right]$$

---

1. We note that $\boldsymbol{\Pi}_1$ is not a projection matrix in Case (1).

with the extended steering vector

$$\tilde{\mathbf{a}}(\theta, \phi) \stackrel{\text{def}}{=} \begin{pmatrix} \mathbf{a}(\theta) \\ \mathbf{a}^*(\theta) e^{-i\phi} \end{pmatrix}$$

Because

$$\begin{bmatrix} \mathbf{a}^H(\theta) & e^{i\phi} \mathbf{a}^T(\theta) \end{bmatrix} \tilde{\mathbf{\Pi}} \begin{pmatrix} \mathbf{a}(\theta) \\ \mathbf{a}^*(\theta) e^{-i\phi} \end{pmatrix} = \begin{pmatrix} 1 & e^{i\phi} \end{pmatrix} \mathbf{M} \begin{pmatrix} 1 \\ e^{-i\phi} \end{pmatrix} = 0$$

with

$$\mathbf{M} \stackrel{\text{def}}{=} \begin{pmatrix} \mathbf{a}^H(\theta) & \mathbf{0}^T \\ \mathbf{0}^T & \mathbf{a}^T(\theta) \end{pmatrix} \tilde{\mathbf{\Pi}} \begin{pmatrix} \mathbf{a}(\theta) & 0 \\ 0 & \mathbf{a}^*(\theta) \end{pmatrix}$$

the matrix

$$\mathbf{M}_T \stackrel{\text{def}}{=} \begin{pmatrix} \mathbf{a}^H(\theta) & \mathbf{0}^T \\ \mathbf{0}^T & \mathbf{a}^T(\theta) \end{pmatrix} \tilde{\mathbf{\Pi}}_T \begin{pmatrix} \mathbf{a}(\theta) & 0 \\ 0 & \mathbf{a}^*(\theta) \end{pmatrix}$$

is positive definite and a consistent estimate of the rank deficient 2×2 matrix $\mathbf{M}$. Using this property, a subspace-based algorithm (denoted Alg_2) is proposed in [10] defined by

$$\theta_{k,T}^{\text{Alg}_2} = \arg \min_{\theta} g_{2,T}(\theta)$$

with

$$g_{2,T}(\theta) \stackrel{\text{def}}{=} \det(\mathbf{M}_T) \tag{9.8}$$

$$= \left[\mathbf{a}^H(\theta) \mathbf{\Pi}_{1,T} \mathbf{a}(\theta) \right]^2 - \left[\mathbf{a}^T(\theta) \mathbf{\Pi}_{2,T}^* \mathbf{a}(\theta) \right] \left[\mathbf{a}^H(\theta) \mathbf{\Pi}_{2,T} \mathbf{a}^*(\theta) \right]$$

In the particular case of a uniform linear array, replacing the generic steering vector $\mathbf{a}(\theta) = (1, e^{i\theta}, \ldots, e^{i(M-1)\theta})^T$ by $\mathbf{a}(z) \stackrel{\text{def}}{=} (1, z, \ldots, z^{M-1})^T$ in (9.8), [4] proposed a root-MUSIC-like algorithm (denoted Alg_3) defined by

$$\theta_{k,T}^{\text{Alg}_3} = \arg(z_k) \text{ with } z_k \text{ } K \text{ roots}_{|z|<1} \text{ of } g_{3,T}(z) \text{ closest to the unit circle} \tag{9.9}$$

where $g_{3,T}(z)$ is the following polynomial[2] of degree $4(M-1)$ whose roots appear in reciprocal conjugate pairs z_k and $(z_k^*)^{-1}$:

---

2. We note that this procedure allows one to estimate up to $2(M-1)$ possible DOA.

$$g_{3,T}(z) \stackrel{\text{def}}{=} \left[\mathbf{a}^T(z^{-1})\mathbf{\Pi}_{1,T}\mathbf{a}(z)\right]^2 - \left[\mathbf{a}^T(z)\mathbf{\Pi}_{2,T}^*\mathbf{a}(z)\right]\left[\mathbf{a}^T(z^{-1})\mathbf{\Pi}_{2,T}\mathbf{a}(z^{-1})\right]$$

Considering the performance of these three algorithms, the following is proved in [10].

Theorem 9.2
The sequences $\sqrt{T}(\boldsymbol{\theta}_T - \boldsymbol{\theta})$, where $\boldsymbol{\theta}_T$ are the DOA estimates given by these three MUSIC-like algorithms [respectively, first two MUSIC-like algorithms] for a uniform linear array [respectively, arbitrary array], converge in distribution to the same zero-mean Gaussian distribution[3] of covariance matrix

$$(\mathbf{C}_{\boldsymbol{\theta}})_{k,l} = \frac{1}{\gamma_k \gamma_l} \begin{pmatrix} \alpha_{\phi,\phi}^{(k)} & -\alpha_{\theta,\phi}^{(k)} \end{pmatrix} \mathbf{B}^{(k,l)} \begin{pmatrix} \alpha_{\phi,\phi}^{(l)} \\ -\alpha_{\theta,\phi}^{(l)} \end{pmatrix} \tag{9.10}$$

with $(\mathbf{B}^{(k,l)})_{i,j} \stackrel{\text{def}}{=} 4\Re\left[\left(\tilde{\mathbf{a}}_k^T \tilde{\mathbf{U}}^* \tilde{\mathbf{a}}_l^*\right)\left(\tilde{\mathbf{a}}_{i,k}^{'H} \tilde{\mathbf{\Pi}} \tilde{\mathbf{a}}_{j,l}'\right)\right]$, $i, j = \theta, \phi$, where

$$\tilde{\mathbf{a}}_k \stackrel{\text{def}}{=} \begin{pmatrix} \mathbf{a}_k \\ \mathbf{a}_k^* e^{-i\phi_k} \end{pmatrix}$$

$$\tilde{\mathbf{a}}_{\theta,k}' \stackrel{\text{def}}{=} \frac{d\tilde{\mathbf{a}}_k}{d\theta_k}$$

$$\tilde{\mathbf{a}}_{\phi,k}' \stackrel{\text{def}}{=} \frac{d\tilde{\mathbf{a}}_k}{d\phi_k}$$

$$\tilde{\mathbf{U}} \stackrel{\text{def}}{=} \sigma_n^2 \tilde{\mathbf{S}}^{\#} \mathbf{R}_{\tilde{y}} \tilde{\mathbf{S}}^{\#}$$

with $\tilde{\mathbf{S}} \stackrel{\text{def}}{=} \tilde{\mathbf{A}} \mathbf{R}_{\tilde{x}} \tilde{\mathbf{A}}^H$, and with $\left(\alpha_{i,j}^{(k)}\right)_{i,j=\theta,\phi}$ and γ_k are the purely geometric factors $\alpha_{i,j}^{(k)} \stackrel{\text{def}}{=} \Re\left(\tilde{\mathbf{a}}_{i,k}^{'H} \tilde{\mathbf{\Pi}} \tilde{\mathbf{a}}_{j,k}'\right)$ and $\gamma_k \stackrel{\text{def}}{=} \alpha_{\theta,\theta}^{(k)} \alpha_{\phi,\phi}^{(k)} - \left(\alpha_{\theta,\phi}^{(k)}\right)^2$. In particular

$$(\mathbf{C}_{\boldsymbol{\theta}})_{k,k} = \frac{2\alpha_{\phi,\phi}^{(k)}}{\gamma_k} \left(\tilde{\mathbf{a}}_k^H \tilde{\mathbf{U}} \tilde{\mathbf{a}}_k\right), \quad k = 1, \ldots, K \tag{9.11}$$

which gives, in the case of a single source,

$$C_{\theta_1} = \frac{1}{\alpha_1 r_1} \left(1 + \frac{1}{2\|\mathbf{a}_1\|^2 r_1}\right) \tag{9.12}$$

Remark 9.1
If the case of a single noncircular complex Gaussian distributed source of maximum circularity rate ($\rho_1 = 1$), asymptotic variance (9.12) attains the noncircular Gaussian Cramer-Rao bound (9.23). Consequently, these three MUSIC-like algorithms previously described are efficient for a single source.

---

3. These three algorithms have different behavior outside the asymptotic regime, as will be stressed in Section 9.7.

Case (2): Arbitrary Full Rank Spatial Extended Covariance Matrix

In this case, based on

$$\tilde{\Pi}\tilde{A} = \begin{pmatrix} \Pi_1 & \Pi_2 \\ \Pi_2^* & \Pi_1^* \end{pmatrix} \begin{pmatrix} A & O \\ O & A^* \end{pmatrix} = O$$

different MUSIC-like algorithms can be proposed. Using $\Pi_2 = O$, the simplest one[4] (denoted Alg$_4$) is the following

$$\theta_{k,T}^{\text{Alg}_4} = \arg\min_{\theta} \mathbf{a}^H(\theta)\Pi_{1,T}\mathbf{a}(\theta) \tag{9.13}$$

Because this algorithm is always outperformed by the standard MUSIC algorithm based only on $\mathbf{R}_{y,T}$ [10], the following column weighting[5] MUSIC (denoted Alg$_5$) was proposed in [10], by using the ideas of the weighted MUSIC algorithm introduced for DOA estimation [11], then applied for frequency estimation in [12, 13]:

$$\theta_{k,T}^{\text{Alg}_5} = \arg\min_{\theta} g_{5,T}(\theta) \quad \text{with} \quad g_{5,T}(\theta) \stackrel{\text{def}}{=} \text{Tr}\left[\mathbf{W}\overline{\mathbf{A}}^H(\theta)\tilde{\Pi}_T\overline{\mathbf{A}}(\theta)\right]$$

where $\mathbf{W}$ is a 2×2 nonnegative definite weighting matrix, and $\overline{\mathbf{A}}(\theta)$ is the steering matrix

$$\begin{pmatrix} \mathbf{a}(\theta) & 0 \\ 0 & \mathbf{a}^*(\theta) \end{pmatrix}$$

To derive the optimal weighting matrix

$$\mathbf{W} = \begin{pmatrix} w_{1,1} & w_{1,2} \\ w_{1,2}^* & w_{2,2} \end{pmatrix}$$

in the next section, the weighted MUSIC cost function can be written as

$$g_{5,T}(\theta) = (w_{1,1} + w_{2,2})\left\{\mathbf{a}^H(\theta)\Pi_{1,T}\mathbf{a}(\theta) + \Re\left[z\mathbf{a}^T(\theta)\Pi_{2,T}\mathbf{a}(\theta)\right]\right\} \tag{9.14}$$

with

$$z \stackrel{\text{def}}{=} \frac{2w_{1,2}^*}{w_{1,1} + w_{2,2}}$$

4. We note that unlike Π_1, $\Pi_{1,T}$ is not a projection matrix.
5. Because $\tilde{\Pi}_T$ is an orthogonal projector, the cost function $g_{5,T}(\theta)$ reduces to $\|\tilde{\Pi}_T\tilde{\mathbf{A}}(\theta)\mathbf{W}^{1/2}\|_{\text{Fro}}^2$.

9.3 MUSIC-Like Algorithms

Consequently, the performance of this algorithm depends only on z. By choosing $\mathbf{W}$ diagonal, we have $z = 0$, and this algorithm reduces to Alg4. Considering the performance of this algorithm, the following is proved in [10].

Theorem 9.3
The sequence $\sqrt{T}(\boldsymbol{\theta}_T - \boldsymbol{\theta})$, where $\boldsymbol{\theta}_T$ is the DOA estimate given by this weighted MUSIC algorithm, converges in distribution to the zero-mean Gaussian distribution of covariance matrix

$$(\mathbf{C}_{\boldsymbol{\theta}})_{k,l} = \frac{1}{2\alpha_k \alpha_l}(1 \quad z^* \quad z \quad 1)$$
$$\left[(\overline{\mathbf{A}}_k^T \tilde{\mathbf{U}}^* \overline{\mathbf{A}}_l^*) \otimes (\overline{\mathbf{A}}_k'^H \tilde{\mathbf{\Pi}} \overline{\mathbf{A}}_l') + (\overline{\mathbf{A}}_k'^T \tilde{\mathbf{\Pi}}^* \overline{\mathbf{A}}_l'^*) \otimes (\overline{\mathbf{A}}_k^H \tilde{\mathbf{U}} \overline{\mathbf{A}}_l)\right] \quad (9.15)$$
$$(1 \quad z^* \quad z \quad 1)^H$$

with

$$\overline{\mathbf{A}}_k \stackrel{\text{def}}{=} \overline{\mathbf{A}}(\theta_k)$$

and

$$\overline{\mathbf{A}}'_k \stackrel{\text{def}}{=} \frac{d\overline{\mathbf{A}}_k}{d\theta_k}$$

Furthermore, the value z_k^{opt} that minimizes $(\mathbf{C}_{\boldsymbol{\theta}})_{k,k}$ is given by

$$z_k^{\text{opt}} = -\frac{\mathbf{a}_k^T \mathbf{U}_2^* \mathbf{a}_k}{\mathbf{a}_k^H \mathbf{U}_1 \mathbf{a}_k} \quad (9.16)$$

with

$$\tilde{\mathbf{U}} = \begin{pmatrix} \mathbf{U}_1 & \mathbf{U}_2 \\ \mathbf{U}_2^* & \mathbf{U}_1^* \end{pmatrix}$$

for which the minimum value of $(\mathbf{C}_{\boldsymbol{\theta}})_{k,k}$ is

$$\min_z (\mathbf{C}_{\boldsymbol{\theta}})_{k,k} = \frac{\det(\overline{\mathbf{A}}_k^H \tilde{\mathbf{U}} \overline{\mathbf{A}}_k)}{2(\mathbf{a}_k^H \mathbf{U}_1 \mathbf{a}_k)(\mathbf{a}_k'^H \mathbf{\Pi}_1 \mathbf{a}_k')} \quad (9.17)$$

For a single source, we have the following [10].

Corollary 9.1
The asymptotic variance of the DOA estimate given by this optimal weighting MUSIC algorithm attains the noncircular Gaussian Cramer Rao bound (9.22) for all values of the noncircularity rate in the single source case.

Remark 9.2

The optimal value of the weight previously derived depends on the specific DOA whose variance is to be minimized, which means that the optimal weight is not the same for all DOAs. This, however, might have been expected, since MUSIC estimates the DOAs one by one. In addition, it should be noted that z_k^{opt} is sample-dependent. Consequently, this value ought to be replaced by a consistent estimate in the implementation of the optimal weighting MUSIC algorithm. We note that this replacement of z_k^{opt} by a consistent estimate has no effect on the asymptotic variance of the weighting MUSIC algorithm, as proved in [10].

Remark 9.3

For circular sources, $\mathbf{R}_{\tilde{y}}$ is block diagonal. This successively implies that $\tilde{\mathbf{S}}, \tilde{\mathbf{S}}^{\#}$, and $\tilde{\mathbf{U}}$ are block diagonal. Consequently, $\mathbf{U}_2 = \mathbf{O}$, $z_k^{\mathrm{opt}} = 0$, $\mathbf{W}_{\mathrm{opt}}$ is diagonal, and the optimal weighting MUSIC algorithm reduces to the standard MUSIC algorithm. Then, (9.17) becomes

$$\min_z (\mathbf{C}_\theta)_{k,k} = \frac{\mathbf{a}_k^H \mathbf{U} \mathbf{a}_k}{2 \mathbf{a}_k'^H \mathbf{\Pi}_1 \mathbf{a}_k'}$$

which is the asymptotic variance given by (9.3).

Remark 9.4

All the performance results presented until now depend on the distribution of the sources only through their second-order moments. These results can be extended. Following a functional analysis [14] and assuming some regularity conditions, the following is proved in [10, 15].

Theorem 9.4

The asymptotic performance given by an arbitrary subspace-based algorithm built from $\tilde{\mathbf{\Pi}}_T$ [respectively, $(\mathbf{\Pi}_T, \mathbf{\Pi}_T')$] associated with the noise subspace of $\mathbf{R}_{\tilde{y},T}$ [respectively, $(\mathbf{R}_{y,T}, \mathbf{R}_{y,T}')$] depends only on the distribution of the sources through their second-order moments. Furthermore, using the same approach, it is proved in [10] that the noncircularity of the sources does not change the asymptotic performance of the standard second-order algorithms. More precise details are given in Theorem 9.5.

Theorem 9.5

All DOA-consistent estimates given by second-order algorithms based only on $\mathbf{R}_{y,T}$, that do not suppose explicitly the sources to be spatially uncorrelated, are robust to the distribution and to the noncircularity of the sources; that is, the asymptotic performances are those of the standard complex circular Gaussian case.

9.4 Asymptotically Minimum Variance Estimation

To assess the performance and the efficiency of the previous MUSIC-like algorithms, their performances are compared in this section to those of the AMV estimators

9.4 Asymptotically Minimum Variance Estimation

in the class of consistent estimators based on $\mathbf{R}_{\tilde{y},T}$—on $(\mathbf{R}_{y,T}, \mathbf{R}'_{y,T})$ [16], on $(\mathbf{\Pi}_T, \mathbf{\Pi}'_T)$, and on $\tilde{\mathbf{\Pi}}_T$ [15]. Considering the statistic $(\mathbf{R}_{y,T}, \mathbf{R}'_{y,T})$, the following theorem is proved in [15].

Theorem 9.6
The covariance matrix $\mathbf{C}_{\boldsymbol{\alpha}}$ of the asymptotic distribution of an estimator of $\boldsymbol{\alpha}$, given by an arbitrary consistent algorithm based on $(\mathbf{R}_{y,T}, \mathbf{R}'_{y,T})$, is bounded below by the real symmetric matrix $\mathbf{C}_{\boldsymbol{\alpha}}^{\text{AMV(R,R')}} = \left[S^H \mathbf{C}_s^{-1}(\boldsymbol{\alpha}) S\right]^{-1}$

$$\mathbf{C}_{\boldsymbol{\alpha}} \geq \left[S^H \mathbf{C}_s^{-1}(\boldsymbol{\alpha}) S\right]^{-1} \tag{9.18}$$

where[6]

$$S \stackrel{\text{def}}{=} \frac{d\mathbf{s}(\boldsymbol{\alpha})}{d\boldsymbol{\alpha}}$$

$$\mathbf{s}(\boldsymbol{\alpha}) \stackrel{\text{def}}{=} \left\{\text{vec}^T\left[\mathbf{R}_y(\boldsymbol{\alpha})\right], v^T\left[\mathbf{R}'_y(\boldsymbol{\alpha})\right], v^H\left[\mathbf{R}'_y(\boldsymbol{\alpha})\right]\right\}^T$$

and where $\mathbf{C}_s(\boldsymbol{\alpha})$ is the first covariance matrix of the asymptotic distribution of $\mathbf{s}_T \stackrel{\text{def}}{=} \left[\text{vec}^T(\mathbf{R}_{y,T}), v^T(\mathbf{R}'_{y,T}), v^H(\mathbf{R}'_{y,T})\right]^T$.

Remark 9.5
A priori knowledge on the spatial correlation of the sources may be be introduced in the bound (9.18) by using adapted parameterization of $\mathbf{R}_x$ and $\mathbf{R}'_x$. For example, if the sources are supposed to be spatially uncorrelated, then $\mathbf{R}_x$ will be parameterized by $[(\mathbf{R}_x)_{k,k}]_{k=1,\ldots,K}$ and if, moreover, they are independent, $\mathbf{R}_x$ and $\mathbf{R}'_x$ will be parameterized only by $\{(\mathbf{R}_x)_{k,k}, \Re[(\mathbf{R}'_x)_{k,k}], \Im[(\mathbf{R}'_x)_{k,k}]\}_{k=1,\ldots,K}$.

Furthermore, it is proved in [15] that this lowest bound is asymptotically tight; that is, there exists an algorithm whose covariance of the asymptotic distribution of $\boldsymbol{\alpha}_T$ satisfies (9.18) with equality.

Theorem 9.7
The following nonlinear least-square algorithm is an AMV second-order algorithm:

$$\boldsymbol{\alpha}_T = \arg \min_{\boldsymbol{\alpha} \in \mathbb{R}^L} [\mathbf{s}_T - \mathbf{s}(\boldsymbol{\alpha})]^H \mathbf{C}_s^{-1}(\boldsymbol{\alpha}) [\mathbf{s}_T - \mathbf{s}(\boldsymbol{\alpha})] \tag{9.19}$$

and this lowest bound (9.18) is also obtained if an arbitrary consistent estimate $\mathbf{C}_{s,T}$ of $\mathbf{C}_s(\boldsymbol{\alpha})$ is used in (9.19).

To reduce the computational complexity due to the nonlinear minimization required by this matching approach, the covariance matching estimation techniques (COMET) can be included to simplify this algorithm if the parameterization of $\mathbf{Q}_n$ is linear in $\boldsymbol{\sigma}$, because in this case there exists a matrix $\boldsymbol{\Psi}(\boldsymbol{\theta})$, such that

---

6. vec(·) is the "vectorization" operator that turns a matrix into a vector by stacking the columns of the matrix one below another. v(·) denotes the operator obtained from vec(·) by eliminating all supradiagonal elements of the matrix.

$\mathbf{s}(\boldsymbol{\alpha}) = \boldsymbol{\Psi}(\boldsymbol{\theta})(\boldsymbol{\rho}^T, \boldsymbol{\sigma}^T)^T$. Using the approach introduced in [17] and then in [16], with the following consistent estimate

$$\mathbf{W} \stackrel{\text{def}}{=} \frac{1}{T} \sum_{t=1}^{T} \left[\left(\mathbf{s}(t) - \frac{1}{T} \sum_{t=1}^{T} \mathbf{s}(t) \right) \left(\mathbf{s}(t) - \frac{1}{T} \sum_{t=1}^{T} \mathbf{s}(t) \right)^H \right]$$

of $\mathbf{C}_s(\boldsymbol{\alpha})$, where

$$\mathbf{s}(t) \stackrel{\text{def}}{=} \begin{pmatrix} \mathbf{y}_t^* \otimes \mathbf{y}_t \\ \mathbf{y}_t \otimes \mathbf{y}_t \end{pmatrix}$$

the proof given in [16] can be extended, and $\boldsymbol{\theta}_T$ is obtained by

$$\boldsymbol{\theta}_T = \arg \min_{\boldsymbol{\theta} \in \mathbb{R}^K} \mathbf{s}_T^H \mathbf{W} \boldsymbol{\Psi}(\boldsymbol{\theta}) [\boldsymbol{\Psi}^H(\boldsymbol{\theta}) \mathbf{W} \boldsymbol{\Psi}(\boldsymbol{\theta})]^{-1} \boldsymbol{\Psi}^H(\boldsymbol{\theta}) \mathbf{W} \mathbf{s}_T \qquad (9.20)$$

Furthermore, the top-left $K \times K$ "DOA corner" of $\mathbf{C}_{\boldsymbol{\alpha}}^{\text{AMV}(R,R')}$ is given by:

$$\mathbf{C}_{\boldsymbol{\theta}}^{\text{AMV}(R,R')} = \left[S_1^H \mathbf{C}_s^{-1/2}(\boldsymbol{\alpha}) \boldsymbol{\Pi}_{\mathbf{C}_s^{-1/2}(\boldsymbol{\alpha}) \boldsymbol{\Psi}}^{\perp} \mathbf{C}_s^{-1/2}(\boldsymbol{\alpha}) S_1 \right]^{-1} \qquad (9.21)$$

with $S = [S_1, \boldsymbol{\Psi}]$, and where $\boldsymbol{\Pi}_{\mathbf{C}_s^{-1/2}(\boldsymbol{\alpha}) \boldsymbol{\Psi}}^{\perp}$ denotes the projector onto the orthogonal complement of the columns of $\mathbf{C}_s^{-1/2}(\boldsymbol{\alpha}) \boldsymbol{\Psi}$.

In the particular case where $\mathbf{Q}_n = \sigma_n^2 \mathbf{I}_M$, Theorems 9.6 and 9.7 are extended in [15] to the statistics $\mathbf{s}_T = (\boldsymbol{\Pi}_T, \boldsymbol{\Pi}_T')$ and $\mathbf{s}_T = \tilde{\boldsymbol{\Pi}}_T$, where $\boldsymbol{\theta}$ alone is identifiable from $(\boldsymbol{\Pi}, \boldsymbol{\Pi}')$ and $\tilde{\boldsymbol{\Pi}}$, and where the inverse of the matrix $\mathbf{C}_s$ (which is here singular) is replaced by its Moore Penrose inverse. Furthermore, the following is proved in [18].

Theorem 9.8
For Gaussian distributed signals, if no a priori information is available on $\mathbf{R}_x$ and $\mathbf{R}_x'$, the previous different AMV bounds coincide with the normalized[7] Gaussian CRB given by Theorem 9.9 [case (UW)]:

$$\mathbf{C}_{\boldsymbol{\theta}}^{\text{AMV}(\Pi,\Pi')} = \mathbf{C}_{\boldsymbol{\theta}}^{\text{AMV}(\tilde{\Pi})} = \mathbf{C}_{\boldsymbol{\theta}}^{\text{AMV}(R,R')} = \text{CRB}_{\text{AU}}^{\text{NCG}}(\boldsymbol{\theta})$$

This proves the efficiency of the AMV estimator based on orthogonal projectors, with respect to CRB. Naturally this result extends to the AMV estimator based on $\mathbf{R}_{y,T}$ only for circular Gaussian signals and explains the good performances of the standard DOA subspace-based algorithms.

---

7. That is, for $T = 1$ throughout this chapter.

9.5 Stochastic Cramér-Rao Bound for Noncircular Gaussian Signals

In this section, the sources are assumed Gaussian distributed. In such a case, (9.21) gives the stochastic CRB for the parameter $\boldsymbol{\theta}$ alone. This expression lacks engineering insight and applies only for a linearly parameterized noise covariance matrix $\mathbf{Q}_n$. Deriving such an interpretable expression for circular Gaussian signals has been an intensive research field. Among them, Stoica and Nehorai [19], Ottersten et al. [20], and Weiss and Friedlander [21] derived this bound indirectly as the asymptotic covariance matrix of the ML estimator for uniform white noise. Ten years later, Stoica et al. [22], Pesavento and Gershman [23], and Gershman et al. [8] derived directly this bound from the Slepian-Bangs formula [24, 25] for uniform white, nonuniform white and arbitrary unknown noise field, respectively. Using these two approaches, these results are extended in [9, 26] for noncircular Gaussian sources. The following result is proved in [9, 26].

Theorem 9.9
The normalized DOA-related block of CRB for noncircular complex Gaussian (NCG) sources in the presence of an arbitrary unknown (AU), nonuniform white (NU), or uniform white (UW) noise field is given by the following explicit expression:

$$\mathrm{CRB}_X^{\mathrm{NCG}}(\boldsymbol{\theta}) = \frac{1}{2}\left\{ \Re\left[(\check{\mathbf{D}}^H \mathbf{\Pi}_{\check{\mathbf{A}}}^\perp \check{\mathbf{D}}) \odot \left(\left[\mathbf{R}_s \check{\mathbf{A}}^H, \mathbf{R}'_s \check{\mathbf{A}}^T \right] \overline{\mathbf{R}}_{\tilde{y}}^{-1} \begin{bmatrix} \check{\mathbf{A}} \mathbf{R}_s \\ \check{\mathbf{A}}^* \mathbf{R}_s'^* \end{bmatrix} \right)^T \right] - \mathbf{M}_X \mathbf{T}_X^{-1} \mathbf{M}_X^T \right\}^{-1}$$

where

$$\check{\mathbf{A}} \stackrel{\mathrm{def}}{=} \mathbf{Q}_n^{-1/2} \mathbf{A}$$

$$\check{\mathbf{D}} \stackrel{\mathrm{def}}{=} \frac{d\check{\mathbf{A}}}{d\boldsymbol{\theta}}$$

$$\overline{\mathbf{R}}_{\tilde{y}} \stackrel{\mathrm{def}}{=} \mathbf{Q}_{\tilde{n}}^{-1/2} \mathbf{R}_{\tilde{y}} \mathbf{Q}_{\tilde{n}}^{-1/2}$$

with

$$\mathbf{Q}_{\tilde{n}} \stackrel{\mathrm{def}}{=} \begin{pmatrix} \mathbf{Q}_n & \mathbf{O} \\ \mathbf{O} & \mathbf{Q}_n^* \end{pmatrix}$$

and where the expressions of $\mathbf{M}_X$ and $\mathbf{T}_X$ are given for the (X = AU) and (X = NU) noise field cases in [9] and $\mathbf{M}_{\mathrm{UW}} = \mathbf{O}$ and $\mathbf{T}_{\mathrm{UW}} = \mathbf{O}$ [26].

In the particular case of one source, the following is proved in [9].

Theorem 9.10
The normalized CRB of θ_1 for a noncircular complex Gaussian source corrupted by nonuniform or uniform white noise field decreases monotonically as the noncircularity rate ρ_1 increases, and is given by

$$\text{CRB}_X^{\text{NCG}}(\theta_1) = \frac{1}{\alpha_1}\left[\frac{2r_1^{-1} + \|\mathbf{a}_1\|^{-2}r_1^{-2} + \|\mathbf{a}_1\|^2 - \|\mathbf{a}_1\|^2\rho_1^2}{\|\mathbf{a}_1\|^2 r_1 + 1 + (1 - \|\mathbf{a}_1\|^2 r_1)\rho_1^2}\right] \quad (9.22)$$

where the SNR is defined here by

$$r_1 \stackrel{\text{def}}{=} \frac{\sigma_{s_1}^2}{M}\sum_{m=1}^{M}\frac{1}{\sigma_m^2}$$

where $\sigma_m^2 \stackrel{\text{def}}{=} E|n_{t,m}^2|$, $m = 1, \ldots, M$ and α_1 is the noise-dependent factor

$$2M\left(\sum_{m=1}^{M}\frac{1}{\sigma_m^2}\right)^{-1}\mathbf{a}_1'^H \mathbf{\Pi}_{\check{\mathbf{a}}_1}^{\perp}\check{\mathbf{a}}_1'$$

($2\mathbf{a}_1'^H \mathbf{\Pi}_{\mathbf{a}_1}^{\perp}\mathbf{a}_1'$ for uniform white noise with $\check{\mathbf{a}}_1 \stackrel{\text{def}}{=} \mathbf{Q}_n^{-1/2}\mathbf{a}_1$ and $\check{\mathbf{a}}_1' \stackrel{\text{def}}{=} (d\check{\mathbf{a}}_1/d\theta_1)$).
Consequently, for one source, the CRB decreases from

$$\text{CRB}_{\text{NU}}^{\text{CG}}(\theta_1) = \frac{1}{\alpha_1 r_1}\left(1 + \frac{1}{\|\mathbf{a}_1\|^2 r_1}\right)$$

($\rho_1 = 0$, circular case) to

$$\text{CRB}_{\text{NU}}^{\text{NCG}}(\theta_1) = \frac{1}{\alpha_1 r_1}\left(1 + \frac{1}{2\|\mathbf{a}_1\|^2 r_1}\right) \quad (\rho = 1) \quad (9.23)$$

Furthermore, this bound has also been compared to the CRB for circular Gaussian sources and to the deterministic CRB in [9], where the following results are proved.

Theorem 9.11
The DOA-related block of CRB for noncircular complex Gaussian sources is upper bounded by the associated CRB for circular complex Gaussian sources, corresponding to the same first covariance matrix $\mathbf{R}_s$ and the same arbitrary noise covariance matrix $\mathbf{Q}_n$.

$$\text{CRB}_{\mathbf{Q}_n}^{\text{NCG}}(\boldsymbol{\theta}) \leq \text{CRB}_{\mathbf{Q}_n}^{\text{CG}}(\boldsymbol{\theta})$$

Compared to the asymptotic deterministic CRB

$$\text{CRB}_{\mathbf{Q}_n}^{\text{DET}}(\boldsymbol{\theta}) = \frac{1}{2}\{\Re[(\check{\mathbf{D}}^H\mathbf{\Pi}_{\check{\mathbf{A}}}^{\perp}\check{\mathbf{D}}) \odot \mathbf{R}_s^T]\}^{-1}$$

which is unchanged with respect to the circular case,

$$\mathrm{CRB}_{Q_n}^{\mathrm{DET}}(\boldsymbol{\theta}) \leq \mathrm{CRB}_{Q_n}^{\mathrm{NCG}}(\boldsymbol{\theta})$$

This result proves that for noncircular sources, a potential performance improvement of any second-order algorithm can be expected with respect to a second-order algorithm based only on the standard covariance. Furthermore, compared to the circular case, the deterministic CRB approaches the stochastic noncircular Gaussian CRB.

9.6 Stochastic Cramér-Rao Bound for BPSK Signals

In the case of noncircular discrete distributed sources, the stochastic CRB appears to be prohibitive to compute. However, for independent BPSK modulated sources in uniform or nonuniform white noise field with $\|\mathbf{a}(\theta)\|^2 = M$, interpretable closed-form expressions of the stochastic CRB can be derived. Compared to those associated with QPSK modulated sources, it is proved in [27] for uniform white noise, then extended to nonuniform white noise in [28] by simple whitening of the noise covariance matrix with $\boldsymbol{\alpha} = (\theta_1, \phi_1, \sigma_{s_1}, \boldsymbol{\sigma}^T)^T$.

Theorem 9.12
For a single BPSK or QPSK modulated source, the normalized stochastic CRB of the DOA alone are given by the closed-form expressions

$$\mathrm{CRB}_{\mathrm{NU}}^{\mathrm{BPSK}}(\theta_1) = \frac{1}{\alpha_1 r_1}\left(\frac{1}{1 - f(Mr_1)}\right) \quad \mathrm{CRB}_{\mathrm{NU}}^{\mathrm{QPSK}}(\theta_1) = \frac{1}{\alpha_1 r_1}\left(\frac{1}{1 - f\left(\frac{Mr_1}{2}\right)}\right)$$

where α_1 and r_1 are given in Theorem 9.10, and where $f(\rho)$ is the following decreasing function of ρ

$$f(\rho) \stackrel{\mathrm{def}}{=} \frac{e^{-\rho}}{\sqrt{2\pi}} \int_{-\infty}^{+\infty} \frac{e^{-(u^2/2)}}{\cosh(u\sqrt{2\rho})} \, du$$

We note that $\mathrm{CRB}_{\mathrm{NU}}^{\mathrm{BPSK}}(\theta_1) < \mathrm{CRB}_{\mathrm{NU}}^{\mathrm{QPSK}}(\theta_1)$, and that compared to the stochastic complex Gaussian CRB (see Theorem 9.10) associated with the same noncircularity rate (1 and 0 for a BPSK and QPSK modulated source, respectively):

$$\frac{\mathrm{CRB}_{\mathrm{NU}}^{\mathrm{BPSK}}(\theta_1)}{\mathrm{CRB}_{\mathrm{NU}}^{\mathrm{NCG}}(\theta_1)} = \frac{1}{[1 - f(Mr_1)]\left(1 + \frac{1}{2Mr_1}\right)}$$

and

$$\frac{\mathrm{CRB}_{\mathrm{NU}}^{\mathrm{QPSK}}(\theta_1)}{\mathrm{CRB}_{\mathrm{NU}}^{\mathrm{CG}}(\theta_1)} = \frac{1}{\left[1 - f\left(\frac{Mr_1}{2}\right)\right]\left(1 + \frac{1}{Mr_1}\right)}$$

We note that these ratios only depend on Mr_1, and tend to 1 when Mr_1 tends to ∞.

For two independent BPSK or QPSK modulated sources, we prove [27, 28] that for large SNRs, that is,

$$\sum_{m=1}^{M} \frac{\sigma_{s_1}^2}{\sigma_m^2} \gg 1 \text{ and } \sum_{m=1}^{M} \frac{\sigma_{s_2}^2}{\sigma_m^2} \gg 1$$

the CRB for the DOA of one source is independent of the parameters of the other source, and

$$\mathrm{CRB}_{\mathrm{NU}}^{\mathrm{BPSK}}(\theta_1, \theta_2) \approx \mathrm{CRB}_{\mathrm{NU}}^{\mathrm{QPSK}}(\theta_1, \theta_2) \approx \begin{bmatrix} \frac{1}{\alpha_1 r_1} & 0 \\ 0 & \frac{1}{\alpha_2 r_2} \end{bmatrix} \qquad (9.24)$$

We note that this property is quite different from the behavior of the CRB under the circular Gaussian distribution and the deterministic CRB, for which the normalized CRB for the DOA of one source depends on the DOA separation. More precisely, it is proved [19, result R9] that these latter two CRBs tend to the same limit as all SNRs increase. For independent sources, they are given by [19, (2.13)]

$$\mathrm{CRB}_{\mathrm{UW}}^{\mathrm{DET}}(\theta_1, \theta_2) = \mathrm{CRB}_{\mathrm{UW}}^{\mathrm{CG}}(\theta_1, \theta_2) = \begin{bmatrix} \frac{1}{\beta_1 r_1} & 0 \\ 0 & \frac{1}{\beta_2 r_2} \end{bmatrix}$$

with

$$\beta_k \stackrel{\text{def}}{=} 2\left[\|\mathbf{a}'_k\|^2 - \gamma_k(\theta_1, \theta_2)\right] \quad k = 1, 2$$

where $\gamma_k(\theta_1, \theta_2)$, $k = 1, 2$ depends on the source separation. This strange property of the stochastic CRB for independent BPSK sources is illustrated in the next section with the performance of the EM algorithm.

As a consequence, the behavior of the resolution threshold for two equipowered $\left(r \stackrel{\text{def}}{=} r_1 = r_2 \text{ and } \alpha \stackrel{\text{def}}{=} \alpha_1 = \alpha_2\right)$ closely spaced independent sources is also quite different. Despite the fact that the CRB does not directly indicate the best resolution achievable by an unbiased estimator, it can be used to define an absolute limit of resolution. Following the criterion described in [29], two sources are meaningfully resolved if the root mean square of the CRB of the estimated DOA separation $(\theta_{1,T} - \theta_{2,T})$ is less than the DOA separation; that is,

$$\sqrt{\mathrm{CRB}_{\mathrm{PSK}}(\theta_1 - \theta_2)} = \sqrt{\frac{2}{T} \frac{1}{\alpha r}} < \Delta\theta$$

because θ_1 and θ_2 are decoupled in (9.24). This resolution bounds the resolution of all unbiased DOA estimates in the regimes SNR $\gg 1$, where the CRB holds. For a ULA,

$$\alpha r = 2\sigma^2 \left[\sum_{m=1}^{M-1} \frac{m^2}{\sigma_m^2} - \left(\sum_{m=1}^{M-1} \frac{m}{\sigma_m^2} \right)^2 \left(\sum_{m=1}^{M} \frac{1}{\sigma_m^2} \right)^{-1} \right]$$

which is an extended SNR

$$r_e = \frac{M(M^2 - 1)}{6} \frac{\sigma^2}{\sigma_n^2}$$

for the specific case of uniform white noise), and we get:

$$r_e > \frac{2}{T(\Delta\theta)^2} \text{ and } \frac{\sigma^2}{\sigma_n^2} > \frac{12}{TM(M^2-1)(\Delta\theta)^2} \text{ for uniform white noise}$$

which is quite different compared to Gaussian distributed sources for which the SNR threshold varies as $(\Delta\theta)^{-4}$ or $(\Delta\theta)^{-3}$ according to the domain of T. See [30, (35)] for the MUSIC algorithm, which is asymptotically efficient as the SNR tends to infinity.

9.7 Illustrative Examples

In this section, we provide numerical illustrations and Monte Carlo simulations of the performance of the different algorithms presented in Section 9.3; numerical comparisons of the variances of these DOA estimates to the asymptotic variance of AMV estimators, based on $\mathbf{R}_{\tilde{y},T}$ (i.e., $\mathbf{R}_{y,T}$ and $\mathbf{R}'_{y,T}$) and on $\mathbf{R}_{y,T}$ alone [16]; and numerical illustrations of the AMV bounds and of the CRBs given in Sections 9.5 and 9.6, respectively. Finally, Monte Carlo simulations of the EM algorithm illustrate some strange properties presented in Section 9.6.

We consider throughout this section two equipowered $\left(\sigma^2 \stackrel{\text{def}}{=} \sigma_{s_1}^2 = \sigma_{s_2}^2\right)$ filtered or unfiltered BPSK modulated signals, with identical noncircularity rate $\left(\rho_{nc} \stackrel{\text{def}}{=} \rho_1 = \rho_2\right)$, with phases of circularity ϕ_1 and ϕ_2. These signals consist of two equipowered multipaths issued from the DOAs θ_1 and θ_2. Referenced on the first sensor and from the DOA θ_1, we have equivalently: $x_{t,1} = \overline{x}_{t,1}$ and $x_{t,2} = \cos(\alpha)\overline{x}_{t,1} + \sin(\alpha)\overline{x}_{t,2}$, with $\mathbf{R}_{\tilde{x}} = \sigma^2 \mathbf{I}_2$ and

$$\mathbf{R}'_{\tilde{x}} = \sigma^2 \rho_{nc} \begin{pmatrix} e^{i\phi_1} & 0 \\ 0 & e^{i\phi_2} \end{pmatrix}$$

Consequently,

$$\mathbf{R}_x = \sigma^2 \begin{pmatrix} 1 & \cos(\alpha) \\ \cos(\alpha) & 1 \end{pmatrix}$$

and

$$\mathbf{R}'_x = \sigma^2 \rho_{nc} \begin{pmatrix} e^{i\phi_1} & \cos(\alpha) e^{i\phi_1} \\ \cos(\alpha) e^{2i\phi_1} & \cos^2(\alpha) e^{i\phi_1} + \sin^2(\alpha) e^{i\phi_2} \end{pmatrix}$$

These signals impinge on a uniform linear array with $M = 6$ sensors separated by one-half wavelength for which $\mathbf{a}_k = (1, e^{i\theta_k}, \ldots, e^{i(M-1)\theta_k})^T$, where $\theta_k = \pi \sin(\psi_k)$, where ψ_k are the DOAs relative to the normal of the array. One thousand independent simulation runs have been performed to obtain the estimated variances, and the number of independent snapshots is $T = 500$ (unless explicitly stated otherwise).

The first experiment illustrates Theorem 9.2 for which $\rho_{nc} = 1$ and $\alpha = \pi/2$. Figures 9.1, 9.2, and 9.3 exhibit the dependence of var$(\theta_{1,T})$ given by algorithms 1, 2, and 3, and by the AMV algorithm based on $\mathbf{R}_{\tilde{y},T}$ (i.e., on $\mathbf{R}_{y,T}$ and $\mathbf{R}'_{y,T}$), with the SNR, the DOA separation $\Delta\theta = \theta_2 - \theta_1$, and the circularity phase separation $\Delta\phi = \phi_2 - \phi_1$.[8] Figures 9.1 and 9.2 show that the domain of validity of our

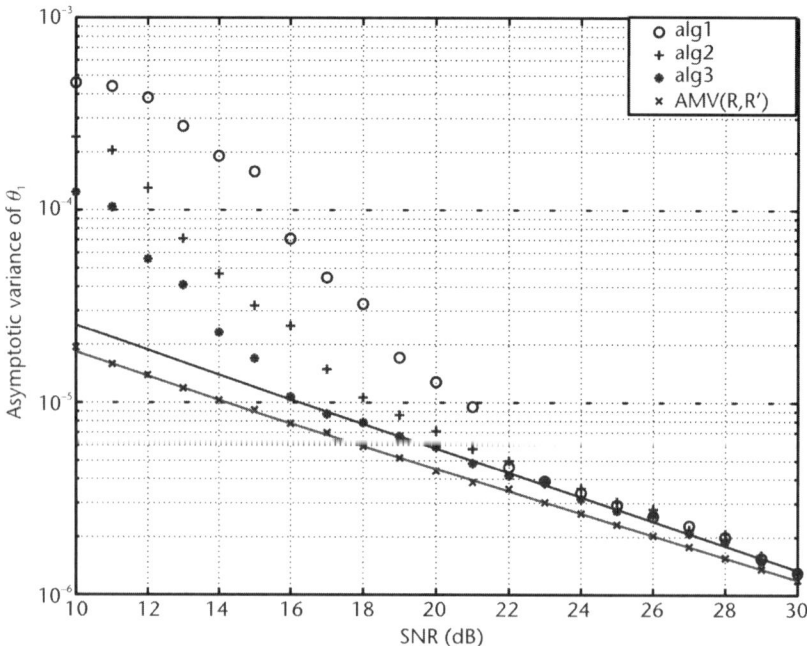

Figure 9.1 Theoretical (solid line) and empirical asymptotic variances given by algorithms 1, 2, 3, and AMV algorithms based on ($\mathbf{R}_{y,T}$, $\mathbf{R}'_{y,T}$) as a function of the SNR for $\Delta\theta = 0.05$rd, $\Delta\phi = \pi/6$rd.

8. Note that one finds, by numerical examples, for two equipowered sources with identical noncircularity rates, that the different theoretical variances depend on θ_1, θ_2, ϕ_1, ϕ_2 by only $\Delta\theta = \theta_2 - \theta_1$ and $\Delta\phi = \phi_2 - \phi_1$ in Case (1) [only $\Delta\theta = \theta_2 - \theta_1$ in Case (2)].

9.7 Illustrative Examples

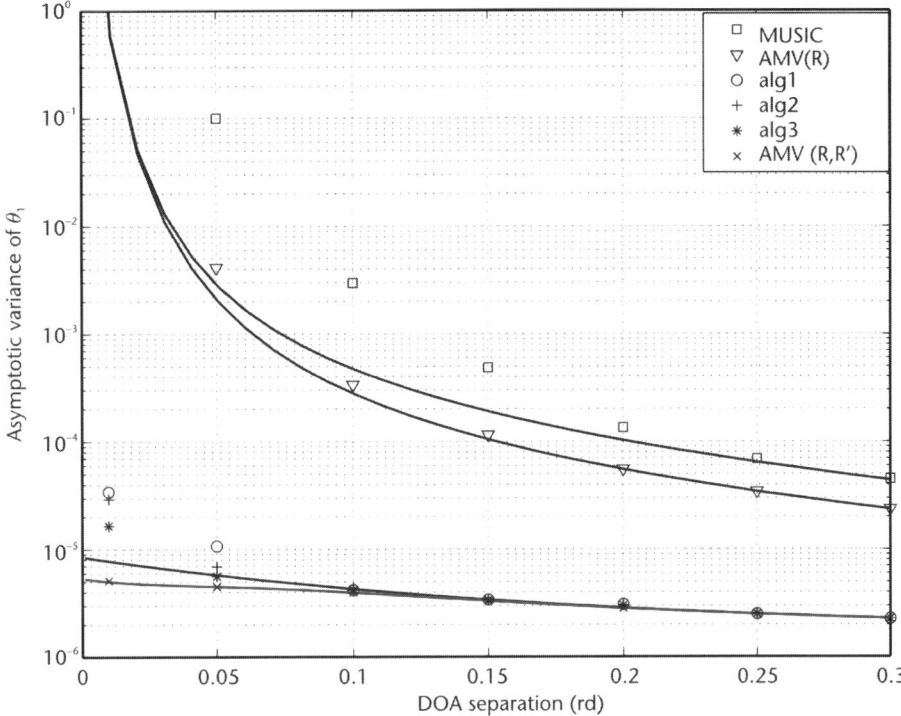

Figure 9.2 Theoretical (solid line) and empirical asymptotic variances given by algorithms 1, 2, 3, standard MUSIC and AMV algorithm based on $\mathbf{R}_{y,T}$ only and on $(\mathbf{R}_{y,T}, \mathbf{R}'_{y,T})$ as a function of the DOA separation for SNR = 20 dB, $\Delta\phi = \pi/6$ rd.

asymptotic analysis depends on the algorithm. Below an SNR threshold that is algorithm-dependent, algorithm 3 (root-MUSIC-like algorithm) outperforms algorithm 2, which outperforms algorithm 1, and naturally all three algorithms clearly outperform the standard MUSIC and the AMV algorithm based on $\mathbf{R}_{y,T}$ alone. In Figure 9.2, we note that the asymptotic variances given by algorithms 1, 2, and 3, and the AMV algorithm tend to a finite limit when the DOA separation decreases to zero. For algorithms 1, 2, and 3, this strange behavior is explained by the two nonzero eigenvalues $(\lambda_k)_{k=1,2}$ of $\tilde{\mathbf{S}}$, which interact in $\tilde{\mathbf{U}} \stackrel{\text{def}}{=} \sigma_n^2 \tilde{\mathbf{S}}^\# \mathbf{R}_{\tilde{y}} \tilde{\mathbf{S}}^\#$ given in (9.11) of Theorem 9.2. With

$$\lambda_k = 2M\sigma_{s_1}^2 \left\{ 1 + (-1)^k \cos\left[(M-1)\frac{\Delta\theta}{2} - \Delta\phi\right] \frac{\sin\left(M\frac{\Delta\theta}{2}\right)}{M \sin\left(\frac{\Delta\theta}{2}\right)} \right\} \quad k = 1, 2$$

we see that one of these eigenvalues approaches zero, and consequently the asymptotic variances increase without limit only if both $\Delta\theta$ and $\Delta\phi$ tend to zero. For the AMV algorithm, $\mathbf{C}_\theta = \left[\left(S^H \mathbf{C}_s^\# S \right)^{-1} \right]_{(1:K,1:K)}$, and S is column rank deficient only if both $\Delta\theta$ and $\Delta\phi$ tend to zero as well. Figure 9.3 illustrates the sensitivity of the performance to the circularity phase separation $\Delta\phi$, which is particularly prominent for low DOA separations. Figures 9.1 and 9.2 show the favorable efficiency of

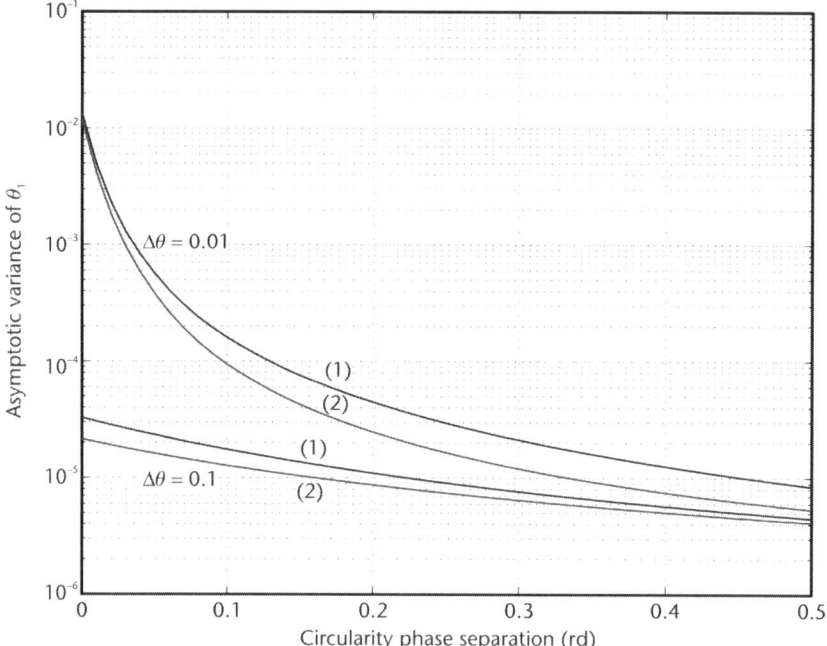

Figure 9.3 Theoretical asymptotic variances given by algorithms 1, 2, 3 (1) and by the AMV algorithm based on ($\mathbf{R}_{y,T}$, $\mathbf{R}'_{y,T}$), (2) as a function of the circularity phase separation for two DOA separations and SNR = 20 dB.

these three algorithms compared to the AMV estimator based on $\mathbf{R}_{\tilde{y},T}$, particularly for large DOA separations. To specify this point, Figure 9.4 exhibits the ratio

$$r_1 \stackrel{\text{def}}{=} \frac{\text{Var}_{\theta_1}^{\text{AMV}(R,R')}}{\text{Var}_{\theta_1}^{\text{Alg}_{1,2,3}}}$$

as a function of the SNR for different DOA separations. It shows that algorithms 1, 2, and 3 are very efficient, except for low DOA separations and low SNRs.

The second experiment illustrates Theorem 9.3. The noncircularity rate ρ_{nc} is arbitrary, and $\alpha = \pi/2$. Compared to the standard MUSIC algorithm based on $\mathbf{R}_{y,T}$, Figures 9.5 and 9.6 show that algorithm 5 outperforms this MUSIC algorithm, particularly for low SNRs and DOA separations, when the circularity rate ρ increases.

To implement the optimal weighted MUSIC algorithm, the following multistep procedure described in [13] has been proposed in [10]:

1. Determine standard MUSIC estimates of $(\theta_k)_{k=1,\ldots,K}$ from $\mathbf{R}_{y,T}$.
2. For $k = 1, \ldots, K$, perform the following: Let $\theta_{k,T}^0$ denote the estimates obtained in step 1. Use $(\theta_{k,T}^0)_{k=1,\ldots,K}$, and the estimate $\mathbf{U}_{1,T}$ and $\mathbf{U}_{2,T}$ of $\mathbf{U}_1$ and $\mathbf{U}_2$ derived from $\tilde{\mathbf{R}}_{y,T}$, to obtain consistent estimates $z_{k,T}$ of z_k^{opt}. Then determine improved estimates $\theta_{k,T}^1$ by locally minimizing the weighted MUSIC cost function (9.13) associated with $z_{k,T}$ around $\theta_{k,T}^0$.

9.7 Illustrative Examples

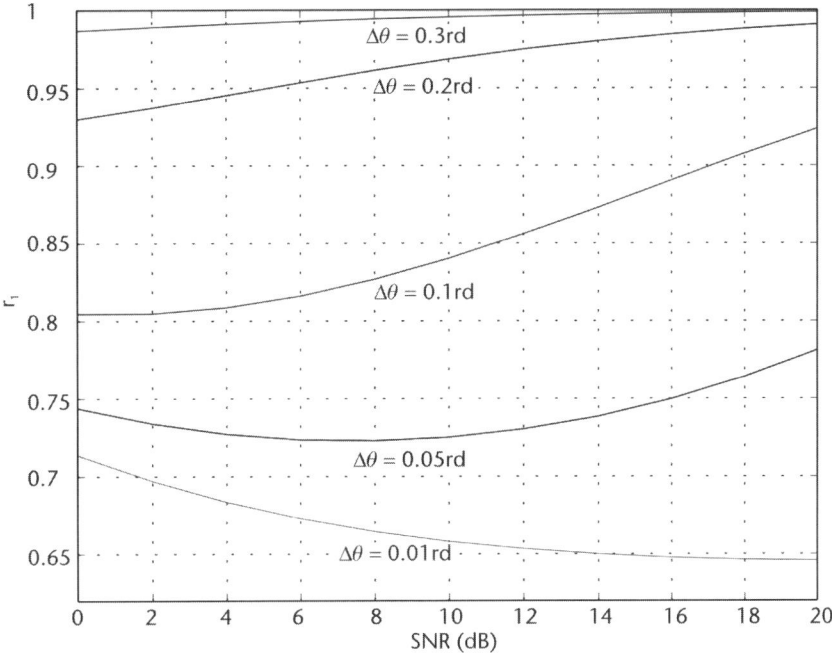

Figure 9.4 Ratio $r_1 \stackrel{\text{def}}{=} \text{Var}_{\theta_1}^{\text{AMV(R,R')}}/\text{Var}_{\theta_1}^{\text{Alg}_{1,2,3}}$ as a function of the SNR for different DOA separations, $\Delta\phi = \pi/6\text{rd}$.

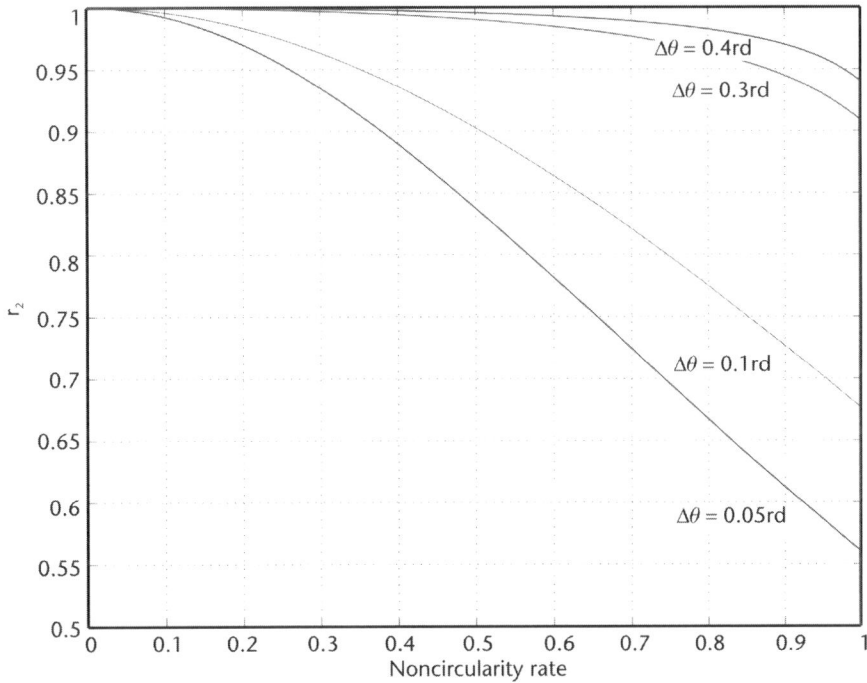

Figure 9.5 Ratio $r_2 \stackrel{\text{def}}{=} \text{Var}_{\theta_1}^{\text{Alg}_5}/\text{Var}_{\theta_1}^{\text{MUSIC(R)}}$ as a function of the noncircularity rate for different DOA separations for SNR = 5 dB.

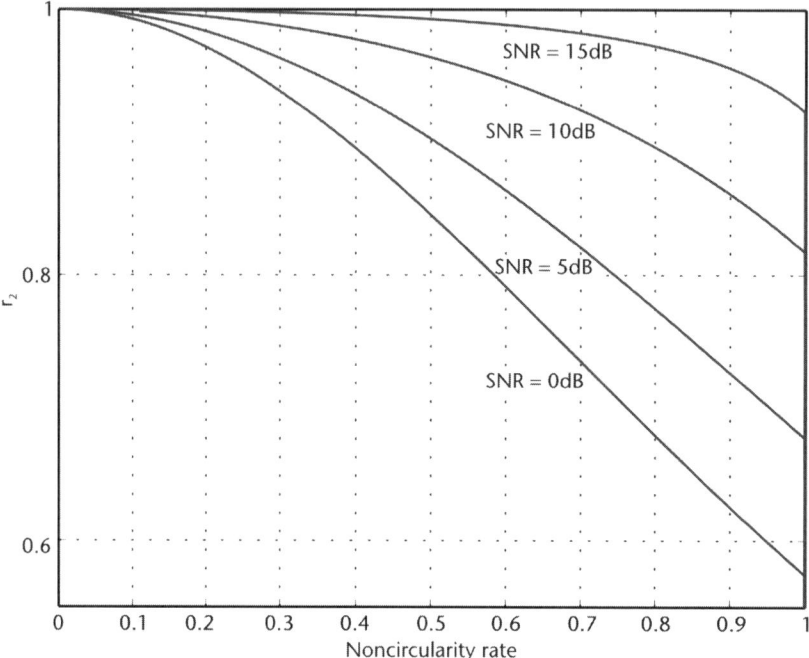

Figure 9.6 Ratio $r_2 \stackrel{\text{def}}{=} \text{Var}_{\theta_1}^{\text{Alg}_s}/\text{Var}_{\theta_1}^{\text{MUSIC(R)}}$ as a function of the noncircularity rate for different SNRs for $\Delta\theta = 0.1\text{rd}$.

Table 9.1 compares our theoretical asymptotic variance expressions with empirical mean square errors (MSE) obtained from Monte Carlo simulations for the standard MUSIC and the optimal weighted MUSIC algorithms for $\rho = 0.9$, $\Delta\theta = 0.2\text{rd}$. We see that there is an agreement between the theoretical and empirical results beyond an SNR threshold. Below this threshold, the optimal weighted MUSIC algorithm largely outperforms the standard MUSIC algorithm.

The third experiment exhibits the benefits due to the second covariance matrix $\mathbf{R}'_{y,T}$ through the comparisons between the AMV bounds based on $(\mathbf{R}_{y,T}, \mathbf{R}'_{y,T})$ and those based only on $\mathbf{R}_{y,T}$. Figures 9.7 and 9.8 show the ratio

$$\frac{\text{Var}_{\theta_1}^{\text{AMV(R,R')}}}{\text{Var}_{\theta_1}^{\text{AMV(R)}}}$$

as a function of the noncircularity rate for $\alpha = \pi/2\text{rd}$ [respectively, $\Delta\theta = 0.1\text{rd}$] fixed for different values of $\Delta\theta$ [respectively, α] when no a priori information is

Table 9.1 Empirical and Theoretical Variances Given by MUSIC Algorithms

	Standard MUSIC		Optimal Weighted MUSIC	
SNR (dB)	Empirical MSE	Theoretical Variance	Empirical MSE	Theoretical Variance
6	4.452×10^{-3}	4.589×10^{-4}	3.154×10^{-4}	4.151×10^{-4}
8	1.600×10^{-3}	2.604×10^{-4}	2.344×10^{-4}	2.449×10^{-4}
10	2.899×10^{-4}	1.527×10^{-4}	1.561×10^{-4}	1.474×10^{-4}
20	1.338×10^{-5}	1.348×10^{-5}	1.337×10^{-5}	1.347×10^{-5}

9.7 Illustrative Examples

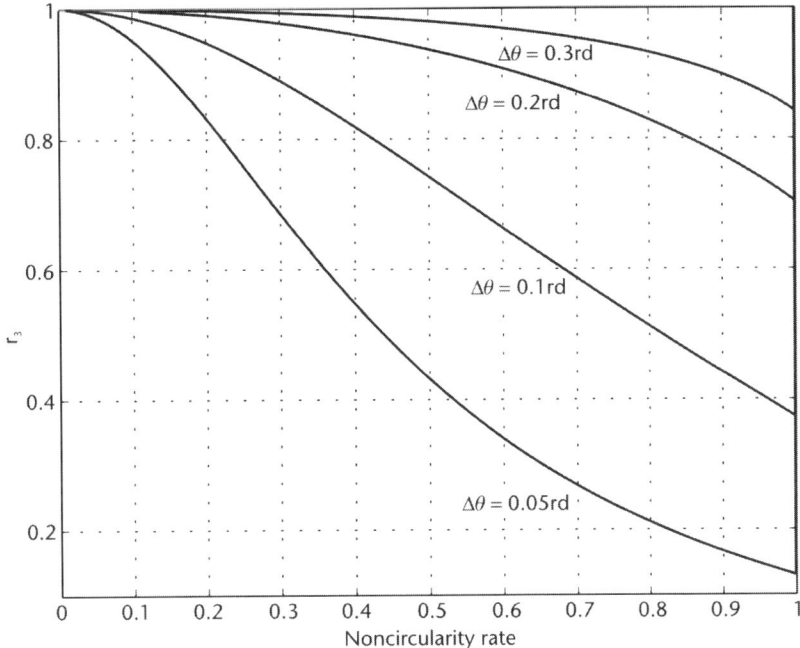

Figure 9.7 Ratio $r_3 \stackrel{\text{def}}{=} \text{Var}_{\theta_1}^{\text{AMV}(R,R')}/\text{Var}_{\theta_1}^{\text{AMV}(R)}$ as a function of the noncircularity rate for different DOA separations for SNR = 5 dB, $\alpha = \pi/2$rd, and $\Delta\phi = \pi/6$rd.

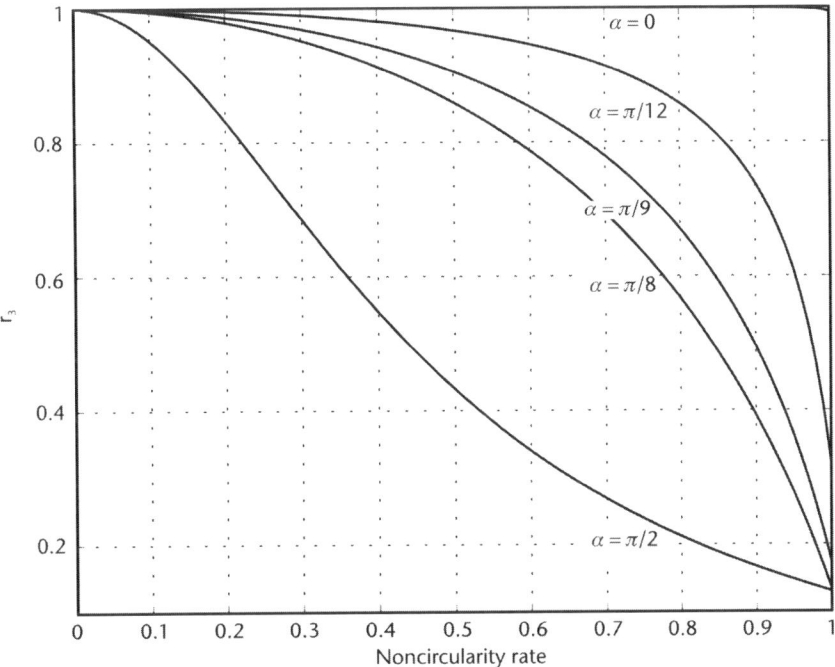

Figure 9.8 Ratio $r_3 \stackrel{\text{def}}{=} \text{Var}_{\theta_1}^{\text{AMV}(R,R')}/\text{Var}_{\theta_1}^{\text{AMV}(R)}$ as a function of the noncircularity rate for different spatial correlation α for SNR = 5 dB, $\Delta\theta = 0.1$rd, and $\Delta\phi = \pi/6$rd.

taken into account. We see that for large DOA separations and/or large spatial correlation between the sources, the second covariance matrix contributes almost no additional information beyond the information in the first covariance matrix. If this spatial uncorrelation a priori information ($\alpha = \pi/2$rd) is taken into account, Figure 9.9 shows the anticipated benefits due to the noncircularity, particularly for low DOA separations. Consequently, the subspace-based algorithms lose their efficiency in these circumstances.

The fourth experiment considers a nonwhite noise field modeled by the two following covariance matrices $Q_n^{(1)}(k, l) = \sigma_n^2 \exp[-(k - l)^2 \zeta]$ and $Q_n^{(2)}(k, l) = \sigma_n^2 \exp(-|k - l|\zeta)$ introduced in [8]. In these noise field models, $\boldsymbol{\sigma}^T = (\sigma_n^2, \zeta)$, where ζ is the "color" parameter, and the SNR is defined by $\sigma_{s_1}^2/\sigma_n^2$. Figure 9.10 shows the bounds $\text{CRB}_{\text{AU}}^{\text{NCG}}(\theta_1)$, $\text{CRB}_{\text{AU}}^{\text{CG}}(\theta_1)$, and $\text{CRB}_{\text{AU}}^{\text{DET}}(\theta_1)$[9] plotted against ζ for $\Delta\theta = \theta_2 - \theta_1 = 0.1$rd, and SNR = 0 dB. We note that when ζ decreases, all the CRBs approach zero, because Q_n becomes singular. When $\zeta \gg 1$, the two noise models tend to a uniform white noise model, and the two CRBs associated with the models merge. We see that the stochastic CRB under noncircular complex Gaussian distributed sources is visibly larger than the deterministic CRB.

The last experiment illustrates the stochastic CRB for BPSK sources for uniform white noise. Figure 9.11 shows the ratios

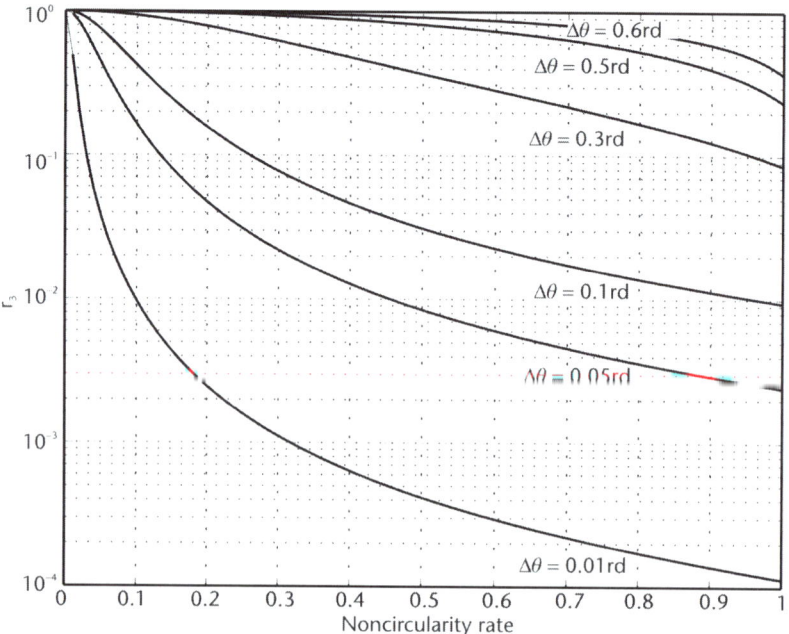

Figure 9.9 Ratio $r_3 \stackrel{\text{def}}{=} \text{Var}_{\theta_1}^{\text{AMV(R, R')}}/\text{Var}_{\theta_1}^{\text{AMV(R)}}$ as a function of the noncircularity rate for different DOA separations for SNR = 5 dB. $\Delta\phi = \pi/6$rd.

9. Note that you find by simulation that the different CRBs depend on $\theta_1, \theta_2, \phi_1, \phi_2$ by only $\Delta\theta = \theta_2 - \theta_1$ and $\Delta\phi = \phi_2 - \phi_1$ for two equipowered sources with identical noncircularity rates.

9.7 Illustrative Examples

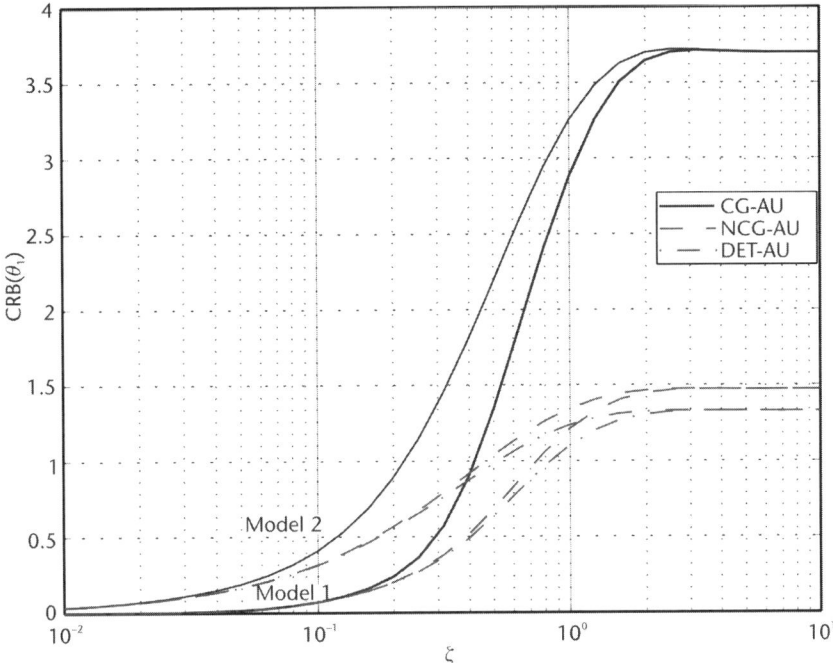

Figure 9.10 $\text{CRB}_{\text{AU}}^{\text{NCG}}(\theta_1)$, $\text{CRB}_{\text{AU}}^{\text{CG}}(\theta_1)$, and $\text{CRB}_{\text{AU}}^{\text{DET}}(\theta_1)$ as a function of ζ for the first and second models with $\Delta\theta = 0.2$rd, SNR = 0 dB, and $\Delta\phi = \pi/6$rd.

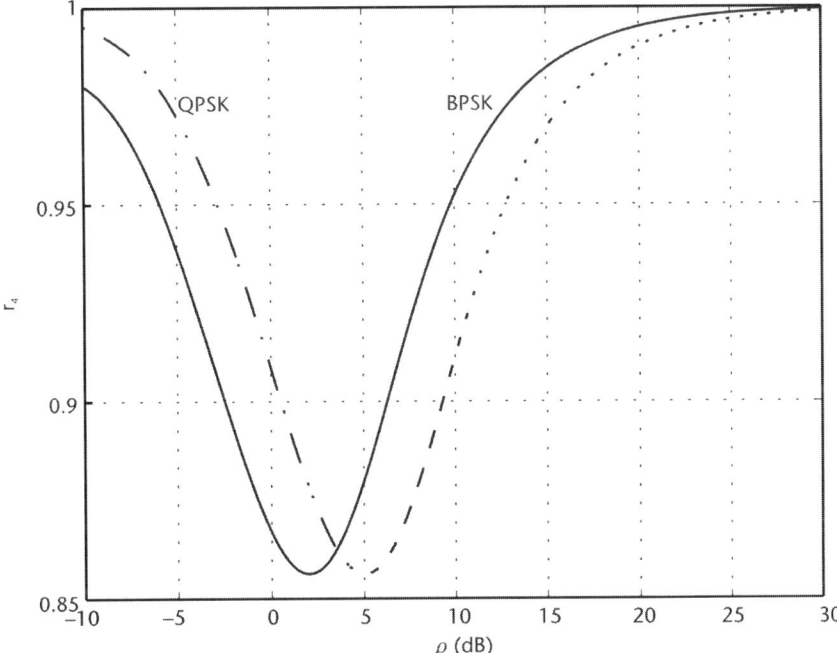

Figure 9.11 Ratios $r_4 \stackrel{\text{def}}{=} \text{CRB}_{\text{BPSK}}(\theta_1)/\text{CRB}_{\text{NCG}}(\theta_1)$ and $r_1(\theta_1) \stackrel{\text{def}}{=} \text{CRB}_{\text{QPSK}}(\theta_1)/\text{CRB}_{\text{CG}}(\theta_1)$ as a function of $\rho \stackrel{\text{def}}{=} Mr_1$.

$$\frac{\text{CRB}_{\text{BPSK}}(\theta_1)}{\text{CRB}_{\text{NCG}}(\theta_1)} \text{ and } \frac{\text{CRB}_{\text{QPSK}}(\theta_1)}{\text{CRB}_{\text{CG}}(\theta_1)}$$

as a function of $\rho \stackrel{\text{def}}{=} Mr_1$ for a single source. We see that the CRBs under the noncircular [respectively, circular] complex Gaussian distribution are tight upper bounds on the CRBs under the BPSK [respectively, QPSK] distribution, only at very low and very high SNRs. The last three figures are devoted to two BPSK independent sources and uniform white noise. Figure 9.12 exhibits the domain of validity of the high SNR approximation.[10] We see that this domain depends not only on M, SNR, and DOA separation, but also on the distributed sources. It is shown that this domain reduces for QPSK sources compared to BPSK sources. If the DOA separation or the M is larger, then the domain of validity of the approximation is larger.

Since the CRB under the noncircular [respectively, circular] Gaussian distribution is a very loose upper bound on the CRB under the discrete BPSK [respectively, QPSK] distribution, specifically for small DOA or phase separation, the ML estimators that take these discrete distributions into account outperform the stochastic ML estimator under the circular Gaussian distribution [19] and the weighted subspace fitting estimator [20], which both reach $\text{CRB}_{\text{CG}}(\theta_1)$. Consequently, the EM approaches [31] that are iterative procedures capable of implementing the

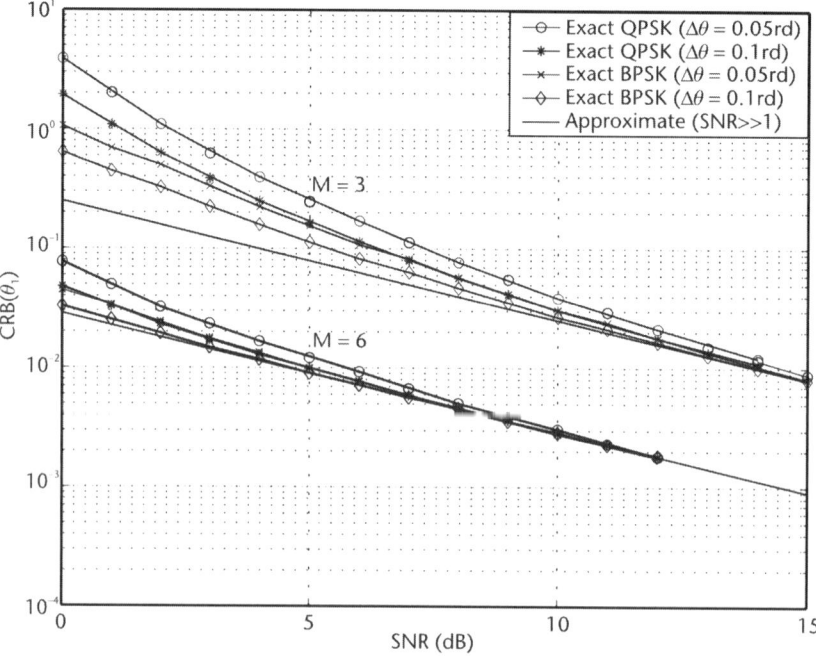

Figure 9.12 Approximate and exact value of $\text{CRB}_{\text{BPSK}}(\theta_1)$ and $\text{CRB}_{\text{QPSK}}(\theta_1)$ as a function of the SNR for different values of the DOA separation.

10. Because closed-form expressions of this stochastic CRB appears to be prohibitive to compute, a numerical approximation of the Fisher information matrix derived from the strong law of large numbers is used as a reference.

stochastic ML estimator under these discrete distributions outperform the ML estimator under noncircular or circular Gaussian distribution.

Figure 9.13 exhibits $\text{CRB}_{\text{BPSK}}(\theta_1)$ and the estimated MSE $E(\theta_{1,T} - \theta_1)^2$, which is given by the deterministic EM algorithm initialized by the estimate given by MUSIC-like algorithm 1 described in Section 9.3, as a function of the DOA separation for two SNRs. We see that contrary to $\text{CRB}_{\text{NCG}}(\theta_1)$ (see Figures 9.7 and 9.9), $\text{CRB}_{\text{BPSK}}(\theta_1)$ does not increase significantly when decreasing the DOA separation. Figure 9.14 compares the MSE $E(\theta_{1,T} - \theta_1)^2$ given by the deterministic EM algorithm (initialized as in Figure 9.13) to $\text{CRB}_{\text{BPSK}}(\theta_1)$ and $\text{CRB}_{\text{NCG}}(\theta_1)$, as a function of the SNR. We see from this figure that the EM estimate reaches $\text{CRB}_{\text{BPSK}}(\theta_1)$, which largely outperforms $\text{CRB}_{\text{NCG}}(\theta_1)$.

9.8 Conclusions

This chapter has presented an overview of DOA estimation for noncircular signals by considering algorithms and performance bounds. From the viewpoint of algorithms, we have proved that three specific MUSIC-like algorithms, built under uncorrelated sources with maximum noncircularity rate, largely outperform the standard MUSIC algorithm. In the general case of nonsingular extended spatial covariance of the sources, the optimal weighted MUSIC that we have introduced outperforms the standard MUSIC algorithm as well, but the performance gain is prominent only for low SNRs and DOA separations. Furthermore, this optimal weighted MUSIC is computationally more demanding than the standard MUSIC

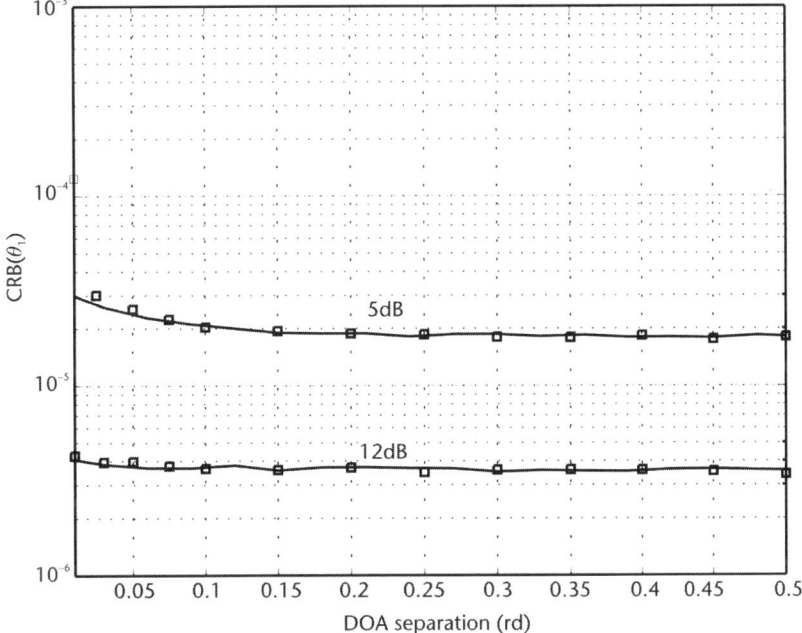

Figure 9.13 $\text{CRB}_{\text{BPSK}}(\theta_1)$ and estimated MSE $E(\theta_{1,T} - \theta_1)^2$ given by the deterministic EM algorithm (10 iterations) as a function of the DOA separation for $\Delta\phi = 0.1\text{rd}$.

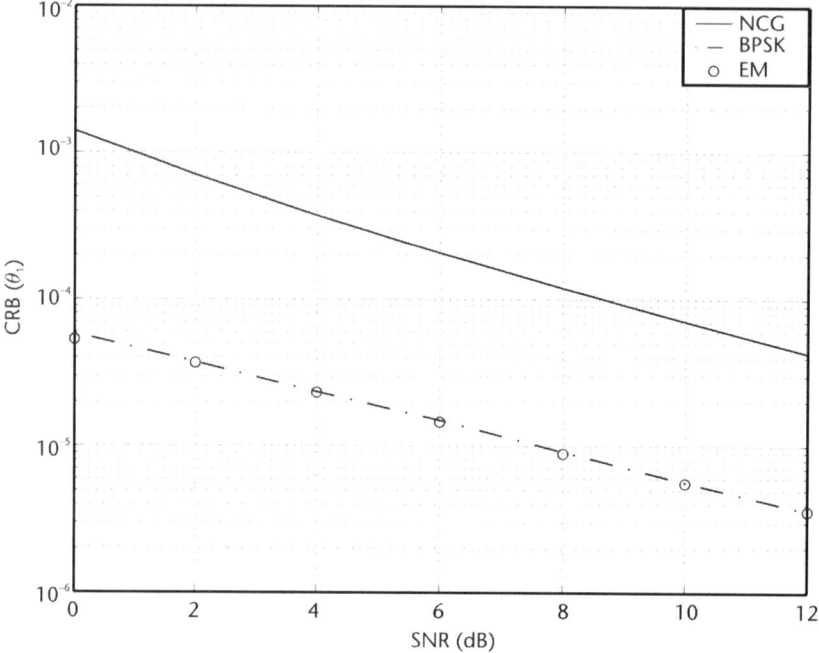

Figure 9.14 $CRB_{BPSK}(\theta_1)$, $CRB_{NCG}(\theta_1)$, and estimated MSE $E(\theta_{1,T} - \theta_1)^2$ given by the deterministic EM algorithm (5 iterations), for $\Delta\theta = 0.3$rd and $\Delta\phi = 0.1$rd, versus SNR.

algorithm. Consequently, from an application viewpoint, this performance gain might not justify the extra computational load. In this general case, only a multi-dimensional nonlinear optimization algorithm, such as the AMV estimator, is able to totally benefit from the noncircular property. From the viewpoint of performance bounds, through the stochastic Gaussian CRBs and the AMV bounds, it is proved that the benefits due to the second covariance matrix occurs primarily for low SNRs and DOA separations, particularly when the sources are uncorrelated with maximum noncircularity rates if this a priori knowledge is taken into account. In the specific case of BPSK modulated sources, it is proved that the stochastic CRB for the DOA of one source is independent of the parameters of the other source over wide SNR ranges. Consequently, ML implementations, such as the EM approaches, outperform the ML estimator under the circular Gaussian distribution, especially for small DOA or phase separations.

References

[1] Godara, L. C., "Application of Antenna Arrays to Mobile Communications, Part II: Beamforming and Direction of Arrival Considerations," *Proc. of IEEE*, Vol. 85, August 1997, pp. 1193–1245.

[2] Picinbono, B., "On Circularity," *IEEE Trans. on Signal Processing*, Vol. 42, No. 12, December 1994, pp. 3473–3482.

[3] Gounon, P., C. Adnet and J. Galy, "Angular Location for Noncircular Signals," *Traitement du Signal*, Vol. 15, No. 1, 1998, pp. 17–23.

[4] Chargé, P., Y. Wang, and J. Saillard, "A Noncircular Sources Direction Finding Method Using Polynomial Rooting," *Signal Processing*, Vol. 81, 2001, pp. 1765–1770.

[5] Grellier, O., et al., "Analytical Blind Channel Identification," *IEEE Trans. on Signal Processing*, Vol. 50, No. 9, September 2002, pp. 2196–2207.

[6] Stoica, P., and R. Moses, *Introduction to Spectral Analysis*, Upper Saddle River, NJ: Prentice Hall, 1997.

[7] Ye, H., and R. D. Degroat, "Maximum Likelihood DOA Estimation and Asymptotic Cramer-Rao Bounds for Additive Unknown Colored Noise," *IEEE Trans. on Signal Processing*, Vol. 43, No. 4, April 1995, pp. 938–949.

[8] Gershman, A. B., et al., "Stochastic Cramer-Rao Bound for Direction Estimation in Unknown Noise Fields," *IEE Proc.-Radar Sonar Navig.*, Vol. 149, No. 1, February 2002, pp. 2–8.

[9] Abeida, H., and J. P. Delmas, "Gaussian Cramer-Rao Bound for Direction Estimation of Noncircular Signals in Unknown Noise Fields," accepted to *IEEE Trans. on Signal Processing*, 2005.

[10] Abeida, H., and J. P. Delmas, "DOA MUSIC-Like Estimation for Noncircular Sources," submitted to *IEEE Trans. on Signal Processing*, 2005.

[11] Stoica, P., and A. Nehorai, "MUSIC, Maximum Likelihood, and Cramer-Rao Bound: Further Results and Comparisons," *IEEE Trans. on Acoustics, Speech, and Signal Processing*, Vol. 38, No. 12, December 1990, pp. 2140–2150.

[12] Stoica, P., and A. Eriksson, "MUSIC Estimation of Real-Valued Sine-Wave Frequencies," *Signal Processing*, Vol. 42, 1995, pp. 139–146.

[13] Stoica, P., A. Eriksson, and T. Söderström, "Optimally Weighted MUSIC for Frequency Estimation," *SIAM J. Matrix Anal. Appl.*, Vol. 16, No. 3, July 1995, pp. 811–827.

[14] Delmas, J. P., "Asymptotic Performance of Second-Order Algorithms," *IEEE Trans. on Signal Processing*, Vol. 50, No. 1, January 2002, pp. 49–57.

[15] Abeida, H., and J. P. Delmas, "Asymptotically Minimum Variance Estimator In The Singular Case," accepted to *EUSIPCO, Antalya*, September 2005.

[16] Delmas, J. P., "Asymptotically Minimum Variance Second-Order Estimation For Noncircular Signals with Application to DOA Estimation," *IEEE Trans. on Signal Processing*, Vol. 52, No. 5, May 2004, pp. 1235–1241.

[17] Ottersten, B., P. Stoica, and R. Roy, "Covariance Matching Estimation Techniques for Array Signal Processing," *Digital Signal Processing*, Vol. 8, July 1998, pp. 185–210.

[18] Abeida, H., and J. P. Delmas, "Efficiency of Subspaced-Based DOA Estimation," submitted to the *IEEE Signal Processing Letters*, 2005.

[19] Stoica, P., and A. Nehorai, "Performance Study of Conditional and Unconditional Direction of Arrival Estimation," *IEEE Trans. on Acoustics, Speech, and Signal Processing*, Vol. 38, No. 10, October 1990, pp. 1783–1795.

[20] Ottersten, B., M. Viberg, and T. Kailath, "Analysis of Subspace Fitting and ML Techniques for Parameter Estimation from Sensor Array Data," *IEEE Trans. on Signal Processing*, Vol. 40, No. 3, March 1992, pp. 590–600.

[21] Weiss, A. J., and B. Friedlander, "On the Cramer-Rao Bound for Direction Finding of Correlated Sources," *IEEE Trans. on Signal Processing*, Vol. 41, No. 1, January 1993, pp. 495–499.

[22] Stoica, P., A. G. Larsson, and A. B. Gershman, "The Stochastic CRB for Array Processing: A Textbook Derivation," *IEEE Signal Processing Letters*, Vol. 8, No. 5, May 2001, pp. 148–150.

[23] Pesavento, M., and A. B. Gershman, "Maximum-Likelihood Direction of Arrival Estimation in the Presence of Unknown Nonuniform Noise," *IEEE Trans. on Signal Processing*, Vol. 49, No. 7, July 2001, pp. 1310–1324.

[24] Slepian, D., "Estimation of Signal Parameters in the Presence of Noise," *Trans. IRE Prof. Grop Inform. Theory PG IT-3*, 1954, pp. 68–89.

[25] Bangs, W. J., "Array Processing with Generalized Beamformers," Ph.D. thesis, Yale University, New Haven, CT, 1971.

[26] Delmas, J. P., and H. Abeida, "Stochastic Cramer-Rao Bound for Noncircular Signals with Application to DOA Estimation," *IEEE Trans. on Signal Processing*, Vol. 52, No. 11, November 2004, pp. 3192–3199.

[27] Delmas, J. P., and H. Abeida, "Cramer-Rao Bounds of DOA Estimates for BPSK and QPSK Modulated Signals," accepted to *IEEE Trans. on Signal Processing*, 2005.

[28] Abeida, H., and J. P. Delmas, "Bornes de Cramer Rao de DOA pour Signaux BPSK et QPSK en Présence de Bruit Non Uniforme," accepted to *GRETSI*, Louvain, 2005.

[29] Smith, S. T., "Statistical Resolution Limits and the Complexified Cramer-Rao Bound," *IEEE Trans. on Signal Processing*, Vol. 53, No. 5, May 2005, pp. 1597–1609.

[30] Kaveh, M., and A. J. Barabell, "The Statistical Performance of MUSIC and the Minimum-Norm Algorithms in Resolving Plane Waves in Noise," *IEEE Trans. on Acoustics, Speech, and Signal Processing*, Vol. 34, No. 2, April 1986, pp. 331–341.

[31] Lavielle, M., E. Moulines, and J. F. Cardoso, "A Maximum Likelihood Solution to DOA Estimation for Discrete Sources," *Proc. 7th IEEE Workshop on Signal Processing*, 1994, pp. 349–353.

CHAPTER 10
Localization of Scattered Sources

Ahmed Zoubir, Yide Wang, and Pascal Chargé

10.1 Introduction

The problem of estimating the DOAs of sources by an array of sensors is very important. This information is used by different functions, such as sorting and routing of communications towards remote mobiles, handover management, or geographic localization of terminals.

A good model of the propagation between a source and a base station is crucial in the design of communication system employing antenna arrays. The point source model often used in the sensor array processing literature is useful for environments with open areas, where a direct transmission between the source and the base station exists. However, in many real applications, signal scattering phenomena may cause angular spreading of the source energy [1]. Indeed, for modern radio communications, the transmitted signal is often obstructed by objects such as buildings, vehicles, and trees, or reflected by, for example, rough sea surfaces. In such complex propagation environments, the distributed source model is more appropriate than the point source model [1–3].

Depending on the mobile environment, the base station–mobile distance, and the base station height, angular spreads up to $10°$ are observed in practice [4, 5]. On the other hand, depending on the relationship between the channel coherency time and observation period, the scattered sources are viewed either as coherently distributed (CD) or incoherently distributed (ID). More precisely, for a CD source, the signal components arriving from different directions are modeled as the delayed and attenuated replicas of the same signal. For an ID source, the channel coherency time is much smaller than the observation period. In this case, all signals coming from different directions are assumed to be uncorrelated [6].

The signals emitted by a scattered source are often modeled as if they were emitted from a cluster of closely spaced point sources with random complex gains [7]. The time delay difference within each cluster of point sources is small, compared with the inverse source signal bandwidth. That is, every individual point source will introduce a random attenuation and phase shift of the transmitted source signal. See [8] for an overview of spatial channel models. Two other approaches for modeling the distributed source have been proposed in [9, 10]. In the CD source case, the author of [9] proposes to model the distributed source by a linear combination of the conventional steering vector and its derivative. The second approach [10] models the distributed source by two point sources. The nominal DOA is obtained by averaging the nominal DOAs associated with the two point sources, and the angular spread is obtained by using a lookup table.

Several methods, essentially based on the spatial properties of incoming signals, have been proposed for estimating the distributed sources. Moreover, many studies about complex random variables and signals have been completed, in which a great interest for the noncircular signals has been shown [11, 12]. Many modern systems deal with noncircular incoming signals, such as telecommunication or satellite systems where amplitude modulated (AM) or BPSK modulated signals are often used. Recently, the noncircularity property has been introduced into the conventional DOA estimation methods, in order to improve the estimation performances in the case of point sources [13, 14].

The aim of this chapter is to recall the data model for both CD and ID sources, and then to present several techniques for estimating these two types of sources. In CD sources case, we show that the performance of the conventional techniques is improved by exploiting the noncircularity property of incoming signals. In the ID sources case, we clarify some problems encountered in the estimation of this type of source, and we present the solutions for these problems. Finally, some simulations are provided to compare the performances of these techniques.

10.2 Array Model

Let us consider a ULA of m sensors. The distance between two adjacent sensors is denoted by d. We assume the delay spread caused by the multipath propagation is small, compared with the inverse bandwidth of the signal. Therefore, the narrowband assumption continues to be valid, even in the presence of scattering. For simplicity, we assume that the sensors and the source are on the same plane. The baseband signals received at the antenna array are collected in the observation vector $\mathbf{x}(t) = [x_1(t), \ldots, x_m(t)]^T$. This vector is modeled as [10]

$$\mathbf{x}(t) = s(t) \sum_{n=1}^{N} \alpha_n(t) \mathbf{a}[\theta_o + \tilde{\theta}_n(t)] + \mathbf{n}(t) \tag{10.1}$$

where $s(t)$ is the signal transmitted from the source. Each contributing ray has a complex random gain factor α_n and a random angular deviation $\tilde{\theta}_n$ from the nominal DOA θ_o. The probability density function of $\tilde{\theta}_n$ is denoted by $p(\tilde{\theta}, \sigma_o)$, where σ_o is the standard deviation of the angular deviations $\tilde{\theta}_n$. $\mathbf{n}(t)$ is the additive noise vector, and $\mathbf{a}(\theta)$ is the steering vector for a point source at DOA θ,

$$\mathbf{a}(\theta) = \left[1, e^{j\frac{2\pi d}{\lambda} \sin \theta}, \ldots, e^{j\frac{2\pi d}{\lambda}(m-1) \sin \theta} \right]^T \tag{10.2}$$

where $(\cdot)^T$ denotes the transpose operator, and λ is the wavelength of the impinging signal.

Alternatively, for large N and tight cluster, the output signal (10.1) is rewritten as [6]

$$\mathbf{x}(t) = \int_{\theta \in \Theta} \mathbf{a}(\theta) s(\theta, t; \boldsymbol{\eta}) \, d\theta + \mathbf{n}(t) \tag{10.3}$$

where Θ is the angular field of view, and $s(\theta, t; \boldsymbol{\eta})$ is the angular signal density of the source. $\boldsymbol{\eta} = [\theta_o \quad \sigma_o]^T$ is the unknown parameter vector of the source, with θ_o its nominal DOA, and σ_o the corresponding angular spread. It is also equal to the standard deviation of $\tilde{\theta}_n$ in (10.1). In the presence of q sources, the continuous model (10.3) becomes

$$\mathbf{x}(t) = \sum_{i=1}^{q} \int_{\theta \in \Theta} \mathbf{a}(\theta) s_i(\theta, t; \boldsymbol{\eta}_i) \, d\theta + \mathbf{n}(t) \tag{10.4}$$

Assuming that the sources and noise are uncorrelated, the covariance matrix then can be written as

$$\mathbf{R} = E[\mathbf{x}(t)\mathbf{x}^H(t)] = \sum_{i,j=1}^{q} \int\int_{\theta, \theta' \in \Theta} p_{ij}(\theta, \theta'; \boldsymbol{\eta}_i, \boldsymbol{\eta}_j) \mathbf{a}(\theta)\mathbf{a}^H(\theta') \, d\theta \, d\theta' + \sigma_n^2 \mathbf{I} \tag{10.5}$$

where $\sigma_n^2 \mathbf{I}$ is the noise covariance matrix, σ_n^2 denotes the noise power, $E[\cdot]$ denotes the statistical expectation, and $(\cdot)^H$ denotes the complex conjugate transpose operator. The function

$$p_{ij}(\theta, \theta'; \boldsymbol{\eta}_i, \boldsymbol{\eta}_j) = E[s_i(\theta, t; \boldsymbol{\eta}_i) s_j^*(\theta', t; \boldsymbol{\eta}_j)] \tag{10.6}$$

is termed the *angular cross-correlation kernel*.

10.2.1 Coherently Distributed Source Model

For a coherently distributed source, the shape of the angular distribution does not change temporally, and the received signal components from that source at different directions are fully correlated. Thus, the angular signal density, associated with the ith CD source, is expressed as [6]

$$s_i(\theta, t; \boldsymbol{\eta}_i) = s_i(t) g_i(\theta; \boldsymbol{\eta}_i) \tag{10.7}$$

where $s_i(t)$ is the ith complex, random signal source, and $g_i(\theta; \boldsymbol{\eta}_i)$ is the corresponding deterministic angular weighting function. The sources are assumed to be uncorrelated with each other, and the deterministic angular weighting function $g_i(\theta; \boldsymbol{\eta}_i)$ is assumed to be nonvanishing only around the actual DOA θ_i. Thus, the distributed source model (10.4) is rewritten as [15, 16]

$$\mathbf{x}(t) = \sum_{i=1}^{q} s_i(t) \int_{-\infty}^{+\infty} \mathbf{a}(\theta) g_i(\theta; \boldsymbol{\eta}_i) \, d\theta + \mathbf{n}(t) \tag{10.8}$$

$$= \sum_{i=1}^{q} s_i(t) \mathbf{c}(\boldsymbol{\eta}_i) + \mathbf{n}(t) \tag{10.9}$$

where

$$\mathbf{c}(\boldsymbol{\eta}_i) = \int_{-\infty}^{+\infty} \mathbf{a}(\theta) g_i(\theta; \boldsymbol{\eta}_i) \, d\theta \qquad (10.10)$$

is the generalized steering vector of the ith CD source. In terms of the steering matrix, the coherently distributed source model (10.9) is expressed as

$$\mathbf{x}(t) = \mathbf{C}\mathbf{s}(t) + \mathbf{n}(t) \qquad (10.11)$$

where the vector $\mathbf{s}(t) = [s_1(t), \ldots, s_q(t)]^T$ contains temporal signals transmitted by the distributed sources. The matrix $\mathbf{C} = [\mathbf{c}(\boldsymbol{\eta}_i), \ldots, \mathbf{c}(\boldsymbol{\eta}_q)]$ contains the generalized steering vectors of the impinging signals.

As a common example of the coherently distributed source, assume that the deterministic angular weighting function $g_i(\theta; \boldsymbol{\eta}_i)$ has the Gaussian shape

$$g_i(\theta; \boldsymbol{\eta}_i) = \frac{1}{\sigma_i \sqrt{2\pi}} e^{-\frac{1}{2}\left(\frac{\theta - \theta_i}{\sigma_i}\right)^2} \qquad (10.12)$$

Let us introduce the variable $\tilde{\theta} = \theta - \theta_i$, the $(k+1)_{0 \le k \le m-1}$th element of the generalized steering vector $\mathbf{c}(\boldsymbol{\eta}_i)$ is written as

$$[\mathbf{c}(\boldsymbol{\eta}_i)]_{k+1} = \frac{1}{\sigma_i \sqrt{2\pi}} \int_{-\infty}^{+\infty} e^{j\frac{2\pi d}{\lambda} k \sin\theta} e^{-\frac{1}{2}\left(\frac{\theta - \theta_i}{\sigma_i}\right)^2} d\theta \qquad (10.13)$$

$$= \frac{1}{\sigma_i \sqrt{2\pi}} \int_{-\infty}^{+\infty} e^{j\frac{2\pi d}{\lambda} k \sin(\tilde{\theta} + \theta_i)} e^{-\frac{1}{2}\left(\frac{\tilde{\theta}}{\sigma_i}\right)^2} d\tilde{\theta}$$

For small values of $\tilde{\theta}$, we can use the first order approximation $\sin(\tilde{\theta} + \theta_i) \approx \sin\theta_i + \tilde{\theta}\cos\theta_i$. Thus, (10.13) is approximated by [15, 16]

$$[\mathbf{c}(\boldsymbol{\eta}_i)]_{k+1} \approx e^{j\frac{2\pi d}{\lambda} k \sin\theta_i} e^{-\frac{1}{2}\left(\frac{2\pi d}{\lambda} k \sigma_i \cos\theta_i\right)^2} = [\mathbf{a}(\theta_i)]_{k+1}[\boldsymbol{\beta}(\boldsymbol{\eta}_i)]_{k+1}$$

$$(10.14)$$

with

$$[\boldsymbol{\beta}(\boldsymbol{\eta}_i)]_{k+1} = e^{-\frac{1}{2}\left(\frac{2\pi d}{\lambda} k \sigma_i \cos\theta_i\right)^2}$$

The generalized steering vector (10.10) can then be written in the following matrix form

$$\mathbf{c}(\boldsymbol{\eta}_i) \approx \boldsymbol{\Phi}(\theta_i)\boldsymbol{\beta}(\boldsymbol{\eta}_i) \qquad (10.15)$$

where $\boldsymbol{\Phi}(\theta_i) = diag[\mathbf{a}(\theta_i)]$.

We can observe that the elements of the real-valued vector $\boldsymbol{\beta}(\boldsymbol{\eta}_i)$ depend on the shape of the angular weighting function $g_i(\theta; \boldsymbol{\eta}_i)$.

Note that, in the case of uniform shape, we have:

$$g_i(\theta; \boldsymbol{\eta}_i) = \frac{1}{2\sqrt{3}\sigma_i} \operatorname{Rect}\left(\theta_i - \sqrt{3}\sigma_i, \theta_i + \sqrt{3}\sigma_i\right) \qquad (10.16)$$

where the $(k+1)_{0 \leq k \leq m-1}$th element of the vector $\boldsymbol{\beta}(\boldsymbol{\eta}_i)$ is expressed as

$$[\boldsymbol{\beta}(\boldsymbol{\eta}_i)]_{k+1} = \operatorname{sinc}\left(\sqrt{3}\,\frac{2\pi d}{\lambda}\,k\sigma_i \cos\theta_i\right)$$

Assuming that the noise and the signals are uncorrelated, and that the noise is spatially and temporally white, the data model (10.11) allows us to write the covariance matrix of the array measurements as:

$$\mathbf{R}_{coh} = E[\mathbf{x}(t)\mathbf{x}^H(t)] = \mathbf{C}\boldsymbol{\Gamma}_s\mathbf{C}^H + \sigma_n^2\mathbf{I} \qquad (10.17)$$

where

$$\boldsymbol{\Gamma}_s = E[\mathbf{s}(t)\mathbf{s}^H(t)] \qquad (10.18)$$

is the emitted signal covariance matrix.

10.2.2 Incoherently Distributed Source Model

A distributed source is said to be ID if its components arriving from different directions are uncorrelated. That is, for the ith source, (10.6) becomes

$$p_{ii}(\theta, \theta'; \boldsymbol{\eta}_i) = \sigma_{s_i}^2 \rho_i(\theta; \boldsymbol{\eta}_i)\delta(\theta - \theta') \qquad (10.19)$$

where $\delta(\theta - \theta')$ is the Dirac delta-function, $\sigma_{s_i}^2$ is the power of the ith source, and $\rho_i(\theta; \boldsymbol{\eta}_i)$ is the corresponding normalized angular power density. For $i = 1, \ldots, q$

$$\int_{\theta \in \Theta} \rho_i(\theta; \boldsymbol{\eta}_i)\,d\theta = 1 \qquad (10.20)$$

Let us assume that all distributed sources are mutually uncorrelated. Then we can rewrite (10.6) as

$$p_{ij}(\theta, \theta'; \boldsymbol{\eta}_i, \boldsymbol{\eta}_j) = \sigma_{s_i}^2 \rho_i(\theta; \boldsymbol{\eta}_i)\delta(\theta - \theta')\delta_{ij} \qquad (10.21)$$

where δ_{ij} is the Kronecker delta. Inserting (10.21) in (10.5), one obtains

$$\mathbf{R}_{inc} = \sum_{i=1}^{q} \int_{\theta \in \Theta} \sigma_{s_i}^2 \rho_i(\theta; \boldsymbol{\eta}_i) \mathbf{a}(\theta) \mathbf{a}^H(\theta) \, d\theta + \sigma_n^2 \mathbf{I} \qquad (10.22)$$

As for a CD case, the angular power density $\rho_i(\theta; \boldsymbol{\eta}_i)$ is assumed to be nonvanishing only around the nominal DOA θ_i. Thus, the covariance matrix is written as

$$\mathbf{R}_{inc} = \sum_{i=1}^{q} \int_{-\infty}^{+\infty} \sigma_{s_i}^2 \rho_i(\theta; \boldsymbol{\eta}_i) \mathbf{a}(\theta) \mathbf{a}^H(\theta) \, d\theta + \sigma_n^2 \mathbf{I} = \sum_{i=1}^{q} \sigma_{s_i}^2 \boldsymbol{\Lambda}_i + \sigma_n^2 \mathbf{I} \qquad (10.23)$$

where

$$\boldsymbol{\Lambda}_i = \int_{-\infty}^{+\infty} \rho_i(\theta; \boldsymbol{\eta}_i) \mathbf{a}(\theta) \mathbf{a}^H(\theta) \, d\theta$$

is the normalized ($m \times m$) noise-free covariance matrix of ith source. The elements $[\boldsymbol{\Lambda}_i]_{kl}$ of this matrix are given by

$$[\boldsymbol{\Lambda}_i]_{kl} = \int_{-\infty}^{+\infty} \rho_i(\theta; \boldsymbol{\eta}_i) a_k(\theta) a_l^H(\theta) \, d\theta = \int_{-\infty}^{+\infty} \rho_i(\theta; \boldsymbol{\eta}_i) e^{j\frac{2\pi d}{\lambda}(k-l)\sin\theta} \, d\theta \qquad (10.24)$$

In the previous section, two types of distribution have been studied for the CD case, namely, the Gaussian and the uniform distributions. We use these two distributions in order to find the analytical expression of the matrix $\boldsymbol{\Lambda}_i$. Using the assumption that the angular spread is small, the elements of $\boldsymbol{\Lambda}_i$ in (10.24) is written as [10]

$$[\boldsymbol{\Lambda}_i]_{kl} \approx \begin{cases} e^{j\frac{2\pi d}{\lambda}(k-l)\sin\theta_i} \, e^{-\frac{1}{2}\left[\frac{2\pi d}{\lambda}(k-l)\sigma_i\cos\theta_i\right]^2} & \text{Gaussian distribution} \\ e^{j\frac{2\pi d}{\lambda}(k-l)\sin\theta_i} \, \text{sinc}\left[\sqrt{3}\frac{2\pi d}{\lambda}(k-l)\sigma_i\cos\theta_i\right] & \text{Uniform distribution} \end{cases} \qquad (10.25)$$

For the sake of convenience, let us introduce the matrix $\mathbf{B}_i$, such that:

$$[\mathbf{B}_i]_{kl} \approx \begin{cases} e^{-\frac{1}{2}\left[\frac{2\pi d}{\lambda}(k-l)\sigma_i \cos\theta_i\right]^2} & \text{Gaussian distribution} \\ \text{sinc}\left[\sqrt{3}\frac{2\pi d}{\lambda}(k-l)\sigma_i \cos\theta_i\right] & \text{Uniform distribution} \end{cases} \quad (10.26)$$

Using this notation, and assuming small angular spreads, the normalized noise-free covariance matrix of the ith ID source is rewritten as

$$\mathbf{\Lambda}_i = \mathbf{\Phi}_i \mathbf{B}_i \mathbf{\Phi}_i^H \quad (10.27)$$

where $\mathbf{\Phi}_i = diag[\mathbf{a}(\theta_i)]$. It is easy to show that $\mathbf{B}_i$ is a real-valued symmetric Toeplitz matrix, which is uniquely determined by the elements of its first column vector, denoted by $\boldsymbol{\beta}_i = [\beta_{i,0}, \ldots, \beta_{i,m-1}]^T$ with $(\beta_{i,k})_{0 \leq k \leq m-1} = [\mathbf{B}_i]_{k+1,1}$.

10.3 Parameter Estimation Techniques

10.3.1 Estimation Techniques for CD Sources

In this section, we present two subspace-based techniques for estimating the parameters of CD sources. The first method, called the distributed source parameter estimator DPSE) method [6], only relies on the spatial properties of impinging signals. The second technique, called the noncircular distributed source parameter estimator (NC-DSPE) method [16], simultaneously exploits the spatial and the statistical properties of signals.

According to the subspace-based methods principle, and by using the generalized steering vector $\mathbf{c}(\boldsymbol{\eta})$ in (10.15), the parameter vectors $\boldsymbol{\eta}_1, \boldsymbol{\eta}_2, \ldots, \boldsymbol{\eta}_q$ of q CD sources are found by using the DSPE algorithm [6]. They are given by the q minima of the following cost function

$$P(\boldsymbol{\eta}) = \|\hat{\mathbf{E}}_n^H \mathbf{c}(\boldsymbol{\eta})\|^2 \quad (10.28)$$

where $\hat{\mathbf{E}}_n$ is the noise subspace spanned by the $(m-q)$ orthonormal eigenvectors associated with the $(m-q)$ smallest eigenvalues of the sample covariance matrix given by

$$\hat{\mathbf{R}} = \frac{1}{T}\sum_{t=1}^{T} \mathbf{x}(t)\mathbf{x}^H(t) \quad (10.29)$$

In order to use both the spatial and the noncircular properties of the incoming signals, the observation vector and its conjugate vector are concatenated into a single extended vector as in [13, 14]. This extended observation vector is expressed as:

$$\mathbf{x}_{nc}(t) = \begin{bmatrix} \mathbf{x}(t) \\ \mathbf{x}^*(t) \end{bmatrix} = \begin{bmatrix} \mathbf{C} & 0 \\ 0 & \mathbf{C}^* \end{bmatrix} \begin{bmatrix} \mathbf{s}(t) \\ \mathbf{s}^*(t) \end{bmatrix} + \begin{bmatrix} \mathbf{n}(t) \\ \mathbf{n}^*(t) \end{bmatrix} \quad (10.30)$$

When the sources emit AM or BPSK modulated signals, $\mathbf{s}(t)$ is a noncircular complex vector, and its elliptic covariance matrix is written as $E[\mathbf{s}(t)\mathbf{s}^T(t)] = \mathbf{\Gamma}_s \mathbf{\Psi}$, where $\mathbf{\Psi}$ is a diagonal matrix, and each diagonal element is a natural phase $e^{j\phi_k}$ relative to an impinging source. The diagonal matrix $\mathbf{\Gamma}_s$ is defined by (10.18). Then, the extended covariance matrix of the new observation vector is formed as [16]:

$$\mathbf{R}_{nc} = E\left[\mathbf{x}_{nc}(t)\mathbf{x}_{nc}^H(t)\right] = \begin{bmatrix} \mathbf{C} \\ \mathbf{C}^* \mathbf{\Psi}^* \end{bmatrix} \mathbf{\Gamma}_s \begin{bmatrix} \mathbf{C} \\ \mathbf{C}^* \mathbf{\Psi}^* \end{bmatrix}^H + \sigma_n^2 \mathbf{I} \quad (10.31)$$

Using eigendecomposition of $\mathbf{R}_{nc}$, a q-dimensional signal subspace and an orthogonal $(2m - q)$-dimensional subspace is determined. This extended model allows an increase in the dimension of the observation space, while keeping that of the signal subspace unchanged.

Let the matrix $\mathbf{C}_{nc}$ formed by the q extended steering vectors corresponding to the q incoming sources,

$$\mathbf{C}_{nc} = \begin{bmatrix} \mathbf{C} \\ \mathbf{C}^* \mathbf{\Psi}^* \end{bmatrix} \quad (10.32)$$

According to the subspace processing principle, and by exploiting the extended steering matrix $\mathbf{C}_{nc}$, the DOAs and angular spreads of the sources are given by the minima of the following cost function

$$P_{nc}(\boldsymbol{\eta}, \phi) = \mathbf{c}_{nc}^H(\boldsymbol{\eta}, \phi) \hat{\mathbf{E}}_{nc} \hat{\mathbf{E}}_{nc}^H \mathbf{c}_{nc}(\boldsymbol{\eta}, \phi) \quad (10.33)$$

where

$$\mathbf{c}_{nc}(\boldsymbol{\eta}, \phi) = \begin{bmatrix} \mathbf{c}(\boldsymbol{\eta}) \\ \mathbf{c}^*(\boldsymbol{\eta})e^{-j\phi} \end{bmatrix} \quad (10.34)$$

$\hat{\mathbf{E}}_{nc}$ denotes the sample noise subspace obtained by eigendecomposition of the sample extended covariance matrix $\hat{\mathbf{R}}_{nc}$. This multidimensional minimization problem is reduced to a two-dimensional minimization problem. It is shown that the DOAs and angular spreads of the sources are estimated as [16]:

$$\hat{\boldsymbol{\eta}} = \arg\min_{\boldsymbol{\eta}} \left[\mathbf{c}^H(\boldsymbol{\eta}) \hat{\mathbf{E}}_{nU} \hat{\mathbf{E}}_{nU}^H \mathbf{c}(\boldsymbol{\eta}) - |\mathbf{c}^T(\boldsymbol{\eta}) \hat{\mathbf{E}}_{nL} \hat{\mathbf{E}}_{nU}^H \mathbf{c}(\boldsymbol{\eta})| \right] \quad (10.35)$$

where $\hat{\mathbf{E}}_{nU}$ and $\hat{\mathbf{E}}_{nL}$ are the $m \times (m - q)$ submatrices of $\hat{\mathbf{E}}_{nc}$, such as $\hat{\mathbf{E}}_{nc} = \begin{bmatrix} \hat{\mathbf{E}}_{nU}^T & \hat{\mathbf{E}}_{nL}^T \end{bmatrix}^T$.

10.3.1.1 Data Examples

Here we evaluate the performances of the DSPE (10.28) and the NC-DSPE (10.35) methods, when the sources emit BPSK modulated signals with the raised-cosine spectrum of a roll-off factor equal to 0.22, and with the bit rate is 3.84 Mbps. Incoming signals are sampled with frequency 38.4 MHz, during approximately 8 μs, so that the number of data samples taken at each sensor output is 300. The shape of the angular weighting function $g_i(\theta; \boldsymbol{\eta}_i)$ is assumed to be Gaussian (10.12).

Figure 10.1 represents the results obtained by the classical DSPE and the NC-DSPE methods on 100 independent trials. In this example, three sources, located at $-12°$, $0°$, and $+12°$, with the same angular spread $\sigma = 1°$ and the same SNR = 5 dB, are simulated. As we can observe, the variance of the NC-DSPE estimator is much smaller than that of the DSPE estimator.

The performances of the estimators in the next simulation are obtained from 500 Monte Carlo simulations, by calculating the RMSE of the DOA estimates. Figure 10.2 illustrates how the number of sources affects the performance of the NC-DSPE method, compared with that of the DSPE. The SNR of sources is 15 dB, and the angular separation between the sources is $15°$. All sources have the same angular spread $1°$. With a six-sensor uniform linear array, the NC-DSPE algorithm is able to handle 10 sources with an error lower than $0.52°$. The error remains lower than $0.12°$ in a nine-source case, while the classical DSPE method is limited to five sources.

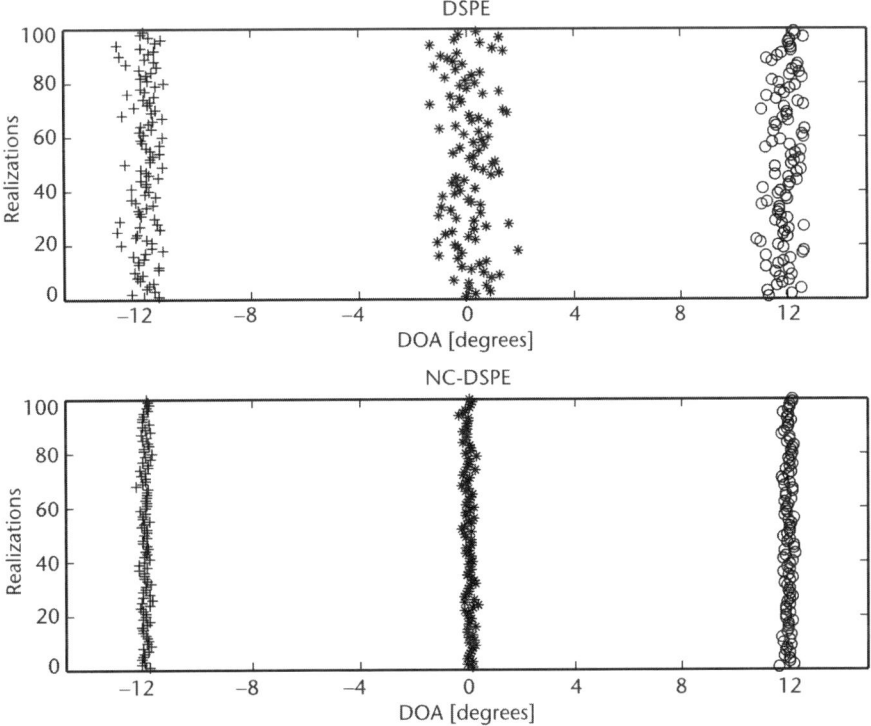

Figure 10.1 DOA estimation, six sensors, $1°$ angular spread, SNR = 5 dB, and $T = 300$ snapshots.

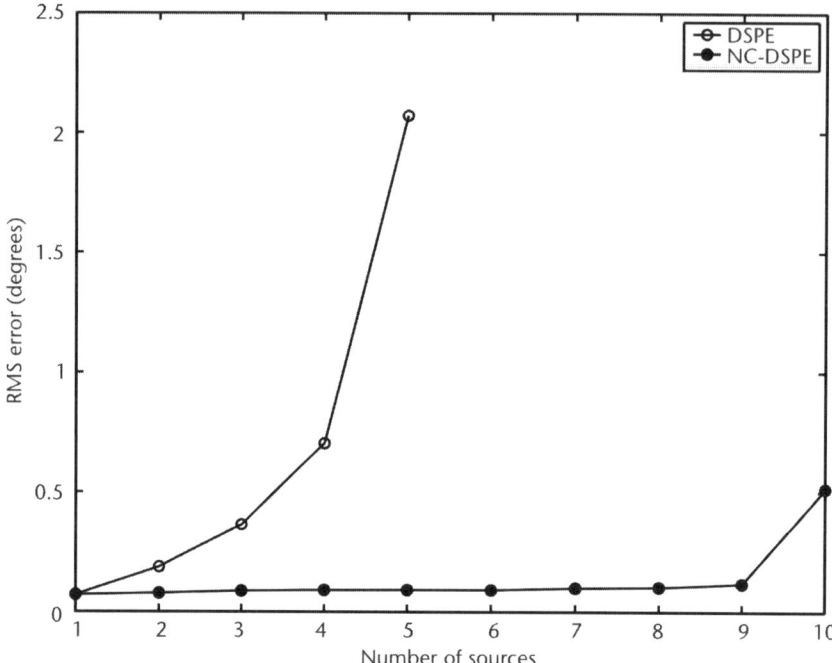

Figure 10.2 RMS error of DOA estimates versus number of sources, six sensors, 1° angular spread, SNR = 15 dB, and $T = 300$ snapshots.

10.3.2 Estimation Techniques for ID Sources

10.3.2.1 Subspace-Based Methods

In the point source case, the basic idea of subspace-based methods is that the signal contribution should be confined to an identifiable and lower-dimensional subspace. For example, the MUSIC method [17] relies on the fact that a vector that is orthogonal to the steering matrix or signal subspace belongs to the noise subspace. Consequently, the problem of finding the q true steering vectors with corresponding DOAs is formulated as finding the q minima of the squared norm $\|\hat{\mathbf{E}}_n^H \mathbf{a}(\theta)\|_2^2$, where $\hat{\mathbf{E}}_n$ is the sample noise subspace, and $\mathbf{a}(\theta)$ denotes the steering vector (10.2). However, a great deterioration in performance of this method in the case of ID sources has been observed [18]. In fact, it is not possible to directly apply the usual subspace-based methods to ID sources. Indeed, in a single ID case source, the rank of the noise-free covariance matrix increases as the angular spread increases. However, most of the signal energy is concentrated within the first few eigenvalues of the noise-free covariance matrix (see Figure 10.3). The number of these eigenvalues is referred to as the *effective dimension* ϵ_d of the signal subspace. It is generally smaller than the number of sensors.

This pseudosubspace decomposition has been exploited in some algorithms, including DSPE [6] and dispersed signal parametric estimator (DISPARE) [19], which are seen as extensions of the MUSIC algorithm. The parameter vector is estimated as

$$\hat{\boldsymbol{\eta}} = \arg \min_{\boldsymbol{\eta}} F_k(\boldsymbol{\eta}) = \arg \min_{\boldsymbol{\eta}} \mathrm{Tr}\left[\hat{\mathbf{E}}_{pn}^H \boldsymbol{\Lambda}^k(\boldsymbol{\eta}) \hat{\mathbf{E}}_{pn}\right] \qquad (10.36)$$

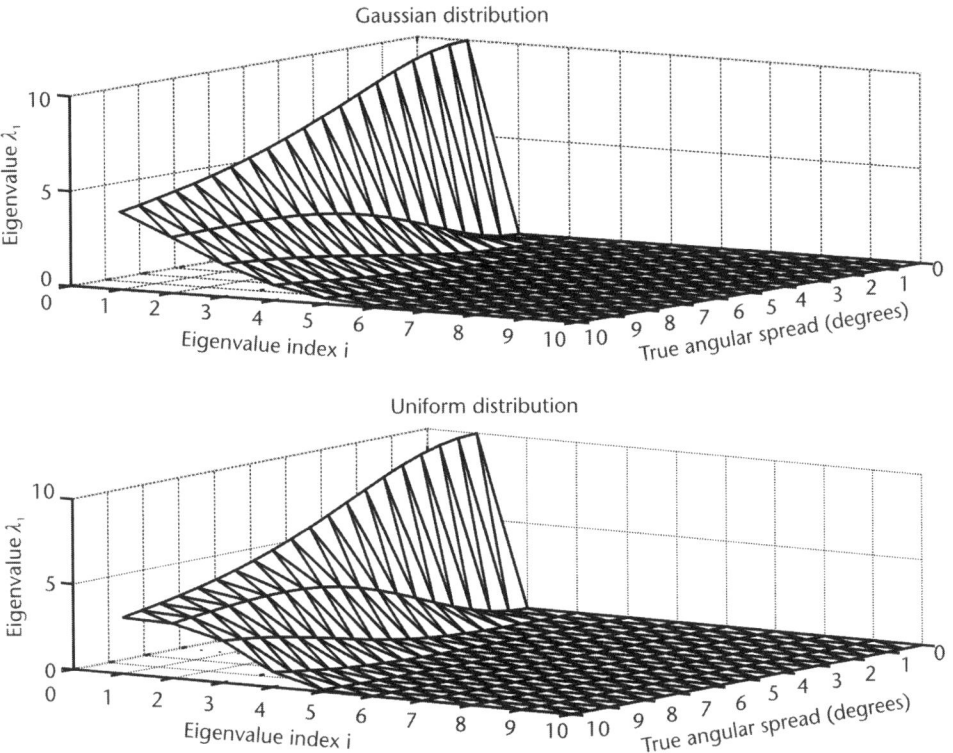

Figure 10.3 Eigenvalues of the true normalized noise-free covariance matrix of a (top) Gaussian and (bottom) uniform distributed source, $m = 6$.

where $k = 1$ gives the DSPE algorithm, and $k = 2$ gives the DISPARE algorithm. $\hat{\mathbf{E}}_{pn}$ denotes the estimated pseudonoise subspace of dimension $m \times (m - \epsilon_d)$, and $\mathbf{\Lambda} = \mathbf{\Phi B \Phi}^H$ is the normalized noise-free covariance matrix (10.27). Performing a two-dimensional search, the DOAs and angular spreads of q ID sources are found by locating the q minima of (10.36). A problem with the DSPE and DISPARE methods is that they fail to give a consistent estimate when the number of snapshots T goes to infinity [10].

To improve the performances of these two techniques, [10] proposes a generalization of the WSF method [20] to handle full-rank data models. The resulting methods are called weighted pseudosubspace fitting (WPSF). We describe one weighted version that attempts to preserve the orthogonality between the estimated pseudosignal subspace and the parameterized pseudonoise subspace. The parameter vector is estimated as [10]

$$\hat{\boldsymbol{\eta}} = \arg\min_{\boldsymbol{\eta}} \|\mathbf{E}_{pn}^H(\boldsymbol{\eta})\hat{\mathbf{E}}_{ps}\|_{\mathbf{W}} \qquad (10.37)$$

$$\arg\min_{\boldsymbol{\eta}} \text{Vec}^H\left[\mathbf{E}_{pn}^H(\boldsymbol{\eta})\hat{\mathbf{E}}_{ps}\right] \mathbf{W} \text{ Vec}\left[\mathbf{E}_{pn}^H(\boldsymbol{\eta})\hat{\mathbf{E}}_{ps}\right]$$

where $\hat{\mathbf{E}}_{ps}$ and $\mathbf{E}_{pn}$ are the estimated pseudosignal subspace and the theoretical pseudonoise subspace, respectively. It is shown that an optimal $\mathbf{W}$ exists, and the resulting criterion is formulated as [10]

$$\hat{\boldsymbol{\eta}} = \arg\min_{\boldsymbol{\eta}} \left\| \hat{\boldsymbol{\Sigma}}_{pn}^{-1/2} \left[\hat{\boldsymbol{\Sigma}}_{pn} \mathbf{E}_{pn}^H(\boldsymbol{\eta}) \hat{\mathbf{E}}_{ps} - \mathbf{E}_{pn}^H(\boldsymbol{\eta}) \hat{\mathbf{E}}_{ps} \hat{\boldsymbol{\Sigma}}_{ps} \right] \hat{\boldsymbol{\Sigma}}_{ps}^{-1/2} \right\|_F^2 \quad (10.38)$$

where $\|\cdot\|_F$ denotes the Frobenius norm, and $\hat{\boldsymbol{\Sigma}}_{pn}$ and $\hat{\boldsymbol{\Sigma}}_{ps}$ are the diagonal eigenvalue matrices for the estimated pseudonoise and pseudosignal subspaces, respectively. Note that this algorithm gives consistent estimates when the number of samples T approaches infinity.

However, due to its very high computational complexity, this technique is not supposed to be practical. It is intended to work as a benchmark for any subspace-based technique. It is interesting to note that a major problem with these subspace methods (i.e., DSPE, DISPARE, and WPSF) is the choice of the effective dimension ϵ_d of the pseudosignal subspace, since the optimal choice will depend on the unknown parameters. Due to the complexity of these subspace-based methods in the ID sources case, an interesting alternative is to use the beamforming methods, in which the estimations do not require ϵ_d. Furthermore, it is shown in [21] that some adaptive beamforming methods can give better results than with those of the subspace-based methods.

10.3.2.2 Beamforming Methods

Generalized Capon-1: The minimum variance distortionless response (MVDR) spectrum estimator was originally developed for frequency-wavenumber analysis by Capon [22]. In the point source DOA estimation field, this estimator is considered as a spatial filter whose output power is minimized, with the constraint that its response is equal to unity in the direction of arrival under consideration.

To estimate the parameters of ID sources, the MVDR principle is generalized as [23]

$$\min_{\mathbf{w}} \mathbf{w}^H \hat{\mathbf{R}} \mathbf{w} \quad \text{subject to} \quad \mathbf{w}^H \boldsymbol{\Lambda}(\boldsymbol{\eta}) \mathbf{w} = 1 \quad (10.39)$$

The solution to (10.39) is given by the following two-dimensional search problem [23]

$$\hat{\boldsymbol{\eta}} = \arg\max_{\boldsymbol{\eta}} \lambda_{\max}^{-1}[\hat{\mathbf{R}}^{-1} \boldsymbol{\Lambda}(\boldsymbol{\eta})] \quad (10.40)$$

where $\lambda_{\max}(\mathbf{A})$ denotes the maximal eigenvalue of the matrix $\mathbf{A}$. According to (10.39), the generalized Capon spatial filter has a gain of power equal to unity for a hypothetical source with parameter vector $\boldsymbol{\eta}$. In the point source case, the matrix $\boldsymbol{\Lambda}(\boldsymbol{\eta})$ transforms to $\mathbf{a}(\theta)\mathbf{a}^H(\theta)$, and the generalized Capon (GC-1) technique (10.40) coincides with the classical Capon method:

$$\hat{\theta} = \arg\min_{\theta} \lambda_{\max}^{-1}[\hat{\mathbf{R}}^{-1} \mathbf{a}(\theta) \mathbf{a}^H(\theta)] = \arg\min_{\theta} \mathbf{a}^H(\theta) \hat{\mathbf{R}}^{-1} \mathbf{a}(\theta)$$

Generalized Capon-2: In the point source case, the rank of the normalized noise-free covariance matrix $\mathbf{a}(\theta)\mathbf{a}^H(\theta)$ is equal to one. A single steering vector $\mathbf{a}(\theta)$

represents the source. On the other hand, in an ID source case, the rank of this matrix increases as the angular spread of the source increases. Consequently, the use of several pseudosteering vectors is necessary to better represent the source. The approach proposed in [24] consists first to make the estimated covariance matrix (10.29) Toeplitz, so that the first column vector of the sample covariance matrix contains all sources information; then to use adaptive processing, by exploiting only the first column of the normalized noise-free covariance matrix as pseudosteering vector of the distributed source, denoted by $\mathbf{b}(\boldsymbol{\eta}) = \boldsymbol{\Phi}(\theta)\boldsymbol{\beta}(\boldsymbol{\eta})$. This generalization of the MVDR beamformer is considered as a spatial filter whose output power is minimized, with the constraint that the information contained in the pseudosteering vector $\mathbf{b}(\boldsymbol{\eta})$ remains distortionless to a hypothetical distributed source with parameter vector $\boldsymbol{\eta}$:

$$\min_{\mathbf{w}} \mathbf{w}^H \hat{\mathbf{R}} \mathbf{w} \quad \text{subject to} \quad \mathbf{w}^H \mathbf{b}(\boldsymbol{\eta}) = \text{cst} \quad (10.41)$$

The estimates of parameter vector $\hat{\boldsymbol{\eta}}$ are obtained by the following two-dimensional search [24]

$$\hat{\boldsymbol{\eta}} = \arg\min_{\boldsymbol{\eta}} \mathbf{b}^H(\boldsymbol{\eta}) \hat{\mathbf{R}}^{-1} \mathbf{b}(\boldsymbol{\eta}) \quad (10.42)$$

In the point source case, the vector $\boldsymbol{\beta} = [1, 1, \ldots, 1]^T$, the pseudosteering vector $\mathbf{b}(\boldsymbol{\eta}) = \mathbf{a}(\theta)$, and the generalized Capon (GC-2) method (10.42) simplifies to the classical Capon's method. Note that this method (10.42) has a lower computational load than the GC-1 method (10.40).

It is interesting to note that the main weakness of all parametric approaches previously presented is that they assume the angular distribution to be known, which clearly is not the case, since the propagation environment is unknown.

10.3.2.3 Covariance Fitting Methods

We briefly present two nonparametric covariance fitting methods proposed in [25, 26]. The first method is called the covariance matching estimation technique-extended invariance principle (COMET-EXIP) [25], in which the consideration is to estimate the parameter vector $\boldsymbol{\zeta} = \begin{pmatrix} \theta & \sigma & \sigma_s^2 & \sigma_n^2 \end{pmatrix}^T$ for a single distributed source. However, we focus our attention here on the DOA estimation. For convenience, this parameter vector is reparametrized as $\boldsymbol{\zeta} = [\theta \quad \boldsymbol{\beta}^T]^T$. The COMET [27] estimates are obtained by minimizing the following multidimensional cost function

$$C(\boldsymbol{\zeta}) = \|\mathbf{W}^{H/2}[\hat{\mathbf{R}} - \mathbf{R}(\boldsymbol{\zeta})]\mathbf{W}^{1/2}\|_F^2 \quad (10.43)$$

where $\mathbf{W}$ is a positive definite weighting matrix, and $\hat{\mathbf{R}}$ and $\mathbf{R}$ are the sample and theoretical covariance matrices, respectively. Differentiating (10.43) with respect to $\boldsymbol{\beta}$, the optimal real-valued vector that minimizes the cost function $C(\boldsymbol{\zeta})$ is $\hat{\boldsymbol{\beta}} = \mathbf{Y}^{-1}\mathbf{y}$, and the DOA of the source is obtained by [25]

$$\hat{\theta} = \arg \min_{\theta} \text{Vec}^H(\hat{\mathbf{R}}) \check{\mathbf{W}} \text{Vec}(\hat{\mathbf{R}}) - \mathbf{y}^T \mathbf{Y}^{-1} \mathbf{y} \quad (10.44)$$

$$= \arg \max_{\theta} \mathbf{y}^T \mathbf{Y}^{-1} \mathbf{y} \quad (10.45)$$

where $\check{\mathbf{W}} = \mathbf{W}^T \otimes \mathbf{W}$, and $\otimes$ is the Kronecker matrix product. The vector $\mathbf{y}$ and the matrix $\mathbf{Y}$ are given by

$$\mathbf{y} = \mathbf{J}^T \check{\mathbf{\Psi}}^H \check{\mathbf{W}} \text{Vec}(\hat{\mathbf{R}}) \quad \text{and} \quad \mathbf{Y} = \mathbf{J}^T \check{\mathbf{\Psi}}^H \check{\mathbf{W}} \check{\mathbf{\Psi}} \mathbf{J} \quad (10.46)$$

where $\check{\mathbf{\Psi}} = \mathbf{\Phi}^H \otimes \mathbf{\Phi}$, and $\mathbf{J}$ is an $m^2 \times m$ logical matrix, such as $\mathbf{J}[(n - 1)m + l, k)] = 1$ for $|l - n| = k - 1$, with $1 \leq n, l, k \leq m$. This method estimates efficiently the DOA of a single source. Unfortunately, it has an ambiguity problem that limits its utilization in practice (see Figure 10.4). This ambiguity problem has been identified and solved in [28, 29]. It is shown that the ambiguity directions, as shown in Figure 10.5, are expressed as

$$\theta_{amb} = \begin{cases} \sin^{-1}[\sin \theta - \text{sgn}(\theta)] & \text{for } \theta \neq 0 \\ \pm \pi/2 & \text{for } \theta = 0 \end{cases} \quad (10.47)$$

where $\text{sgn}(\theta)$ is the Signum function defined for nonzero direction θ by $\text{sgn}(\theta) = \theta/|\theta|$. The approach proposed in [28, 29] consists of first adding a constraint to the original COMET-EXIP cost function (10.43), and then replacing the constrained problem with an unconstrained problem by using a penalty function method.

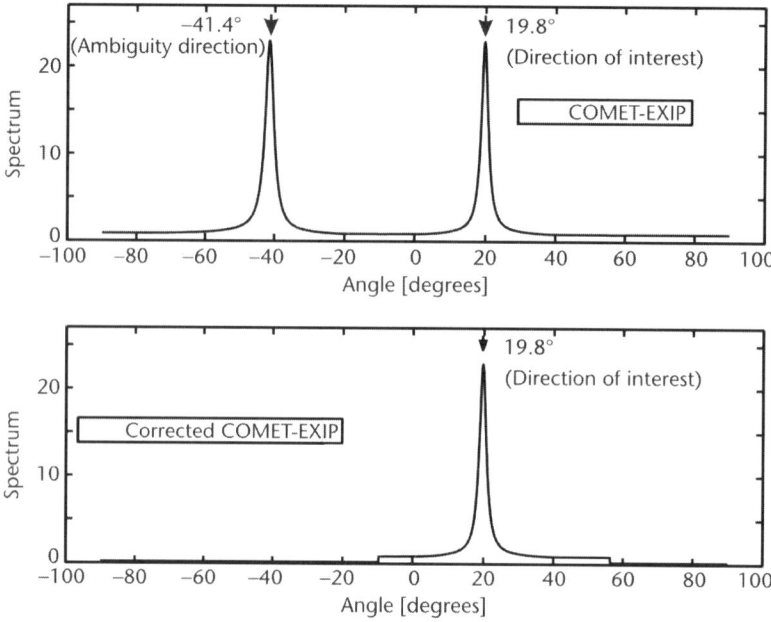

Figure 10.4 COMET-EXIP (10.44) and corrected COMET-EXIP (10.48), $\mathbf{W} = \hat{\mathbf{R}}^{-1}$, $\theta = 20°$, $\sigma = 5°$, $m = 6$, Gaussian angular distribution, $T = 1{,}000$ snapshots, and SNR = 10 dB.

10.3 Parameter Estimation Techniques

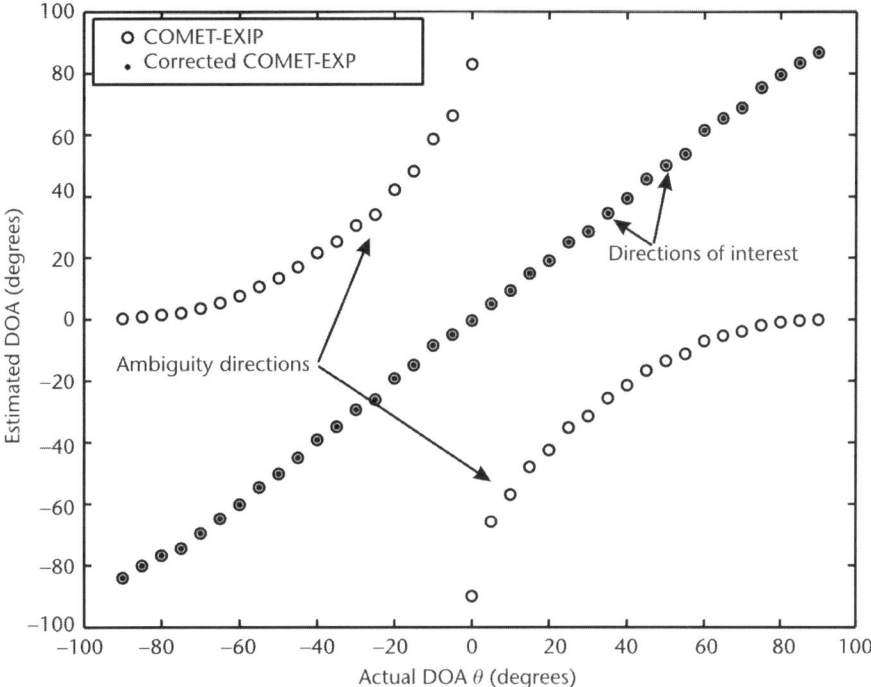

Figure 10.5 Estimated DOA versus the actual DOA, $\mathbf{W} = \hat{\mathbf{R}}^{-1}$, $\sigma = 5°$, $m = 6$, Gaussian angular distribution, $T = 1{,}000$ snapshots, and SNR = 10 dB.

Despite its efficiency, this approach requires an important computational load compared with the original method. To solve this problem, the authors have proposed a simple and fast technique to eliminate this ambiguity. This method estimates the DOA of the source as

$$\hat{\theta} = \arg \min_{\theta} \text{Vec}^H(\hat{\mathbf{R}}) \check{\mathbf{W}} \text{Vec}(\hat{\mathbf{R}}) - H([\mathbf{Y}^{-1}\mathbf{y}]_2) \mathbf{y}^T \mathbf{Y}^{-1} \mathbf{y} \qquad (10.48)$$

where $H(\cdot)$ is the Heaviside function, defined as $H(x) = 1$ if $x \geq 0$, and $H(x) = 0$ otherwise, and $[\mathbf{Y}^{-1}\mathbf{y}]_2$ is the second element of the vector $\mathbf{Y}^{-1}\mathbf{y}$.

The second method (called Subdiag) has been proposed for estimating the DOA of a single source in the case of imperfect spatial coherence; that is, when the amplitude and phase of the wavefront vary randomly along the array aperture [26]. This method also can be applied for estimating the DOA of a single ID source. This estimator is based on a reduced statistic obtained from the subdiagonals of the estimated covariance matrix. It was proposed to jointly estimate the DOA θ and vector $\overline{\boldsymbol{\beta}} = [\beta_1, \ldots, \beta_{m-1}]^T$ of the source as

$$\hat{\theta}, \hat{\overline{\boldsymbol{\beta}}} = \arg \min_{\theta, \overline{\beta}} \sum_{k=1}^{m-1} |\hat{z}_k - z_k|^2 \qquad (10.49)$$

$$= \arg \min_{\theta, \overline{\beta}} \sum_{k=1}^{m-1} \left| \hat{z}_k - \beta_k e^{j\frac{2\pi d}{\lambda} k \sin \theta} \right|^2$$

where

$$\hat{z}_k = \sum_{l=1}^{m-k} [\hat{\mathbf{R}}]_{k+l,l}$$

From (10.49), the authors show that the DOA of the source is given by [26]

$$\hat{\theta} = \arg\max_{\theta} \text{Re}\left[\sum_{k=1}^{m-1} \hat{z}_k^2 e^{-j\frac{4\pi d}{\lambda} k \sin\theta}\right] \quad (10.50)$$

This estimator also suffers from the same ambiguity problem as for the COMET-EXIP estimator. Note that to avoid this problem, a new unambiguous method has been proposed in [30]. This new method estimates the DOA of the source as

$$\hat{\theta} = \arg\max_{\theta} \text{Re}\left[\sum_{k=1}^{m-1} (m-k)\hat{z}_k e^{-j\frac{2\pi d}{\lambda} k \sin\theta}\right] \quad (10.51)$$

This method estimates unambiguously the DOA of the source. However, it gives bad results compared with the original method, especially for a large source spreading.

10.3.2.4 Data Examples

In this section, we will compare the simulation results provided by the GC-1 (10.40), GC-2 (10.42), COMET-EXIP (10.45), Subdiag-1 (10.50), and Subdiag-2 (10.51) methods. These results are obtained by using eight-element ULA, the same BPSK signals previously used for estimating the CD sources, and the toeplitzed sample covariance matrix. The single ID source is located at broadside (i.e., $\theta = 0°$). The performances of the estimators are obtained from 500 Monte Carlo simulations, by calculating the RMSE of DOA estimates.

Let us examine how these estimators behave as the angular spread increases for SNR = 0 dB, and $T = 100$ snapshots. Figure 10.6 shows that the GC-2 estimator performs better for large source spreading than the other methods. For a small value of angular spread, the results given by all methods are very similar.

In Figure 10.7, the influence of the number of snapshots T is investigated when the SNR = 0 dB with 5° angular spread. As we can observe, the GC-2 estimator performs better, especially for a small number of samples. In Figure 10.8, the performance of estimators is given in terms of SNR for $T = 100$ snapshots with 5° source spreading. The main point is that the Subdiag-1 method performs better for low SNRs. When SNR increases, both the GC-2 and COMET-EXIP estimators give the best results, compared to the other estimators.

Finally, the resolution powers of GC-1 and GC-2 estimators are compared when two ID sources are located at −8° and +8°. These sources both have SNR = 10 dB, and angular spread $\sigma = 5°$, which is assumed to be known. Figure 10.9

10.3 Parameter Estimation Techniques

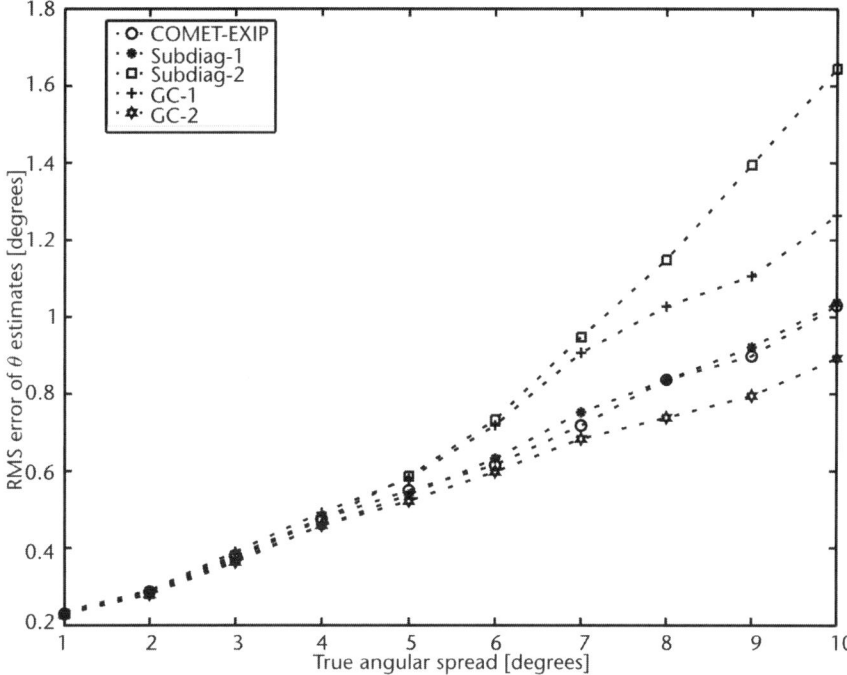

Figure 10.6 RMS error of DOA estimates versus the source angular spread, $m = 8$, Gaussian angular distribution, $T = 100$ snapshots, and SNR = 0 dB.

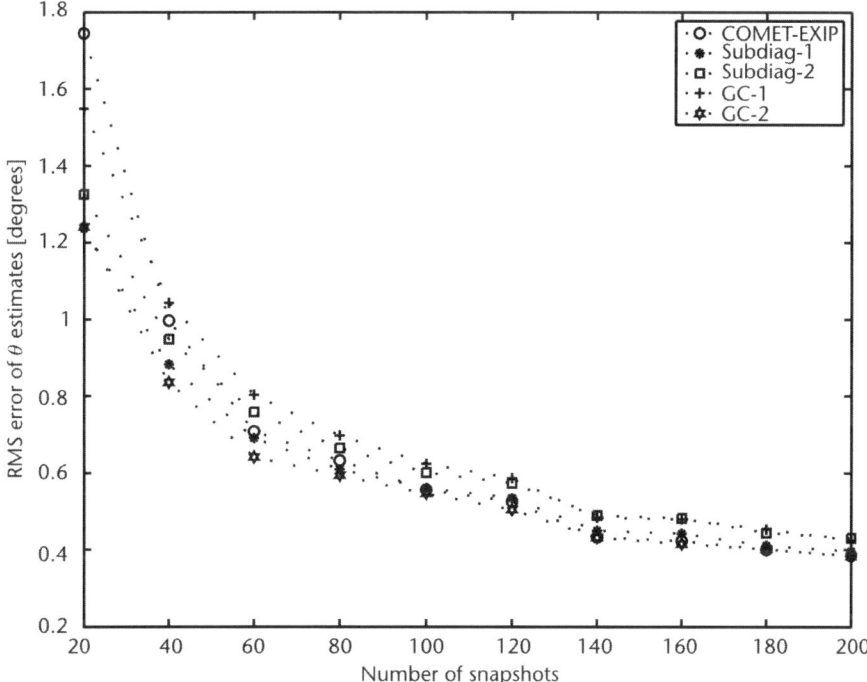

Figure 10.7 RMS error of DOA estimates versus the number of snapshots, $m = 8$, Gaussian angular distribution, $\sigma = 5°$, and SNR = 0 dB.

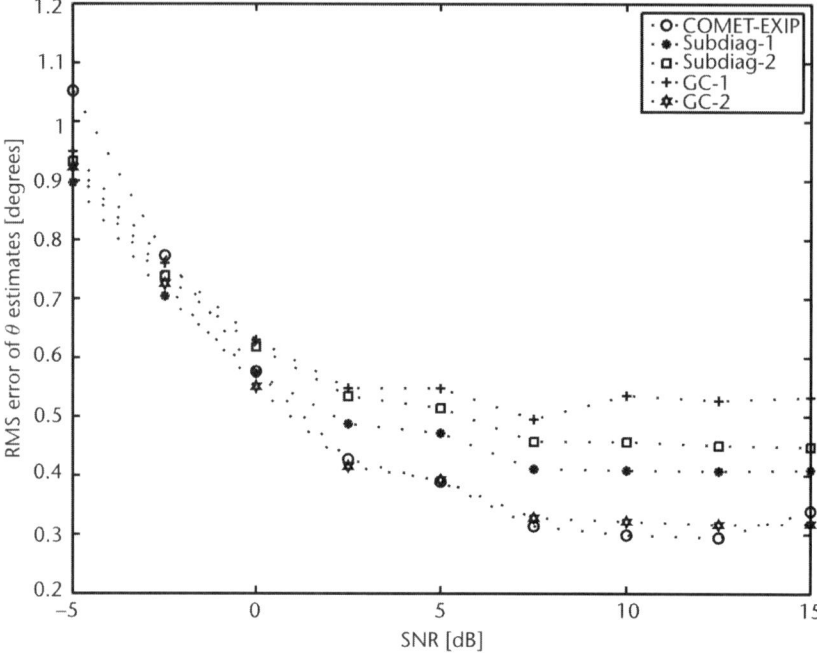

Figure 10.8 RMS error of DOA estimates versus the signal to noise ratio, $m = 8$, Gaussian angular distribution, $\sigma = 5°$, and $T = 100$ snapshots.

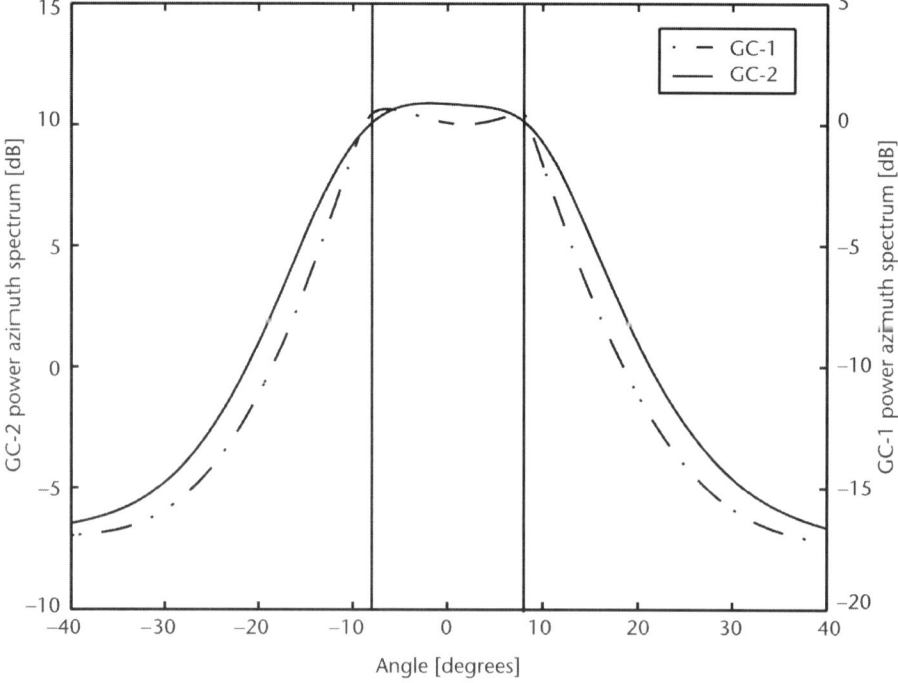

Figure 10.9 GC-1 and GC-2 power azimuth spectra, $\eta_1 = [-8° \quad 5°]^T$, $\eta_2 = [8° \quad 5°]^T$, $m = 8$, Gaussian angular distribution, and $T = 100$ snapshots.

gives the power azimuth spectra of these two techniques. It is observed that the GC-1 estimator allows the separation of the two sources, while the GC-2 estimator fails. Note that the others estimators are not presented in this figure because they are limited to the single ID source case.

10.4 Conclusions

In this chapter, the models for CD and ID sources are described, and several estimation techniques for these two types of sources have been presented. In the CD sources case, we have shown that the performance of the classical techniques is improved significantly by exploiting the noncircularity property of incoming signals. For the ID sources, we have shown that the use of the subspace-based techniques remains difficult. We also have shown that the use of beamforming techniques is a good alternative for estimating the ID sources. We have presented some covariance-fitting techniques that provide a fast and consistent estimate. However, these techniques suffer from two drawbacks—they are entirely based on the single ID source assumption, and they are ambiguous. Two methods to solve the ambiguity problem have been presented. The main point to be noted with ID source cases is that most methods have been proposed for single source cases, while multiple source localization remains an open area.

References

[1] Zetterberg, P., "Mobile Cellular Communications with Base Station Antenna Arrays: Spectrum, Algorithms and Propagation Models," Ph.D. thesis, Royal Institute of Technology, Sweden, 1997.

[2] Thull, D., and M. Fattouche, "Angle of Arrival Analysis of Indoor Radio Propagation Channel," *Proc. Int. Conf. on Univ. Pers. Comm.*, Vol. 1, October 1993, pp. 79–83.

[3] Jeong, J. S., et al., "Performance of MUSIC and ESPRIT for Joint Estimation of DOA and Angular Spread in Slow Fading Environment," *IEICE Trans. Commun.*, Vol. E85-B, No. 5, May 2002, pp. 972–977.

[4] Pedersen K. I., P. E. Mogensen, and B. H. Fleury, "Spatial Channel Characteristics in Outdoor Environments and Their Impact on BS Antenna System Performance," in *Proc. Veh. Technol. Conf.*, Vol. 2, May 1998, pp. 719–723.

[5] Pedersen K. I., P. E. Mogensen, and B. H. Fleury, "A Stochastic Model of the Temporal and Azimuthal Dispersion Seen at the Base Station in Outdoor Propagation Environments," *IEEE Trans. on Vehicular Technology*, Vol. 49, March 2000, pp. 437–447.

[6] Valaee, S., B. Champagne, and P. Kabal, "Parametric Localization of Distributed Sources," *IEEE Trans. on Signal Processing*, Vol. 43, September 1995, pp. 2144–2153.

[7] Zetterberg, P., and B. Ottersten, "The Spectrum Efficiency of a Base-Station Antenna Array System for Spatially Selective Transmission," *IEEE Trans. on Vehicular Technology*, Vol. 44, No. 3, 1995, pp. 651–660.

[8] Ertel, R. B., et al., "Overview of Spatial Channel Models for Antenna Array Communication Systems," *IEEE Personal Communication Magazine*, Vol. 5, No. 1, February 1998, pp. 10–22.

[9] Asztély, D., "Spatio and Spatio-Temporal Processing with Antenna Arrays in Wireless Systems," Ph.D. dissertation, Royal Institute of Technology, Sweden, May 1999.

[10] Bengtsson, M., "Antenna Array Signal Processing for High Rank Models," Ph.D. thesis, Royal Institute of Technology, Sweden, 1999.

[11] Lacoume, J. L., "Complex Random Variables and Signals," *Traitement du Signal*, Vol. 15, 1998, pp. 535–544.

[12] Picinbono, B., "On Circularity," *IEEE Trans. on Signal Processing*, Vol. 42, December 1994, pp. 3473–3482.

[13] Chargé, P., "Traitement d'antenne pour les Télécommunications: Localisation de Sources et Autocalibration," Ph.D. thesis, University of Nantes, France, 2001.

[14] Galy, J., "Localisation Angulaire de Signaux Noncirculaires," *Revue Traitement du Signal*, 1997.

[15] Lee, J., et al., "Low-Complexity Estimation of 2D DOA for Coherently Distributed Sources," *Elsevier Signal Processing*, Vol. 83, 2003, pp. 1789–1802.

[16] Zoubir, A., Y. Wang, and P. Chargé, "Coherently Distributed Noncircular Sources Estimation with DSPE," submitted to *Annals of Telecommunications*, November 2004.

[17] Schmidt, R. O., "Multiple Emitter Location and Signal Parameter Estimation," *IEEE Trans. on Antennas and Propagation*, Vol. AP-34, March 1986, pp. 276–280.

[18] Asztély, D., and B. Ottersten, "The Effect of Local Scattering on Direction of Arrival Estimation with MUSIC," *IEEE Trans. on Signal Processing*, Vol. 47, No. 12, December 1999, pp. 3220–3234.

[19] Meng, Y., P. Stoica, and K. Wong, "Estimation of the Directions of Arrival of Spatially Dispersed Signals in Array Processing," *IEE Proc.—Radar, Sonar and Navigation*, Vol. 143, No. 1, February 1996, pp. 1–9.

[20] Viberg, B., and B. Ottersten, "Sensors Array Processing Based on Subspace Fitting," *IEEE Trans. on Signal Processing*, Vol. 39, No. 5, May 1991, pp. 1110–1121.

[21] Tapio, M., "Channel Estimation in Wireless Communication Systems Employing Multiple Antennas," Ph.D. thesis, Chalmers University of Technology, Sweden, 2004.

[22] Capon, J., "High Resolution Frequency-Wavenumber Spectrum Analysis," *IEEE Proc.*, Vol. 57, August 1969, pp. 1408–1418.

[23] Hassanien, A., S. Shahbazpanahi, and A. B. Gershman, "A Generalized Capon Estimator for Localization of Multiple Spread Sources," *IEEE Trans. on Signal Processing*, Vol. 52, No. 1, January 2004, pp. 280–283.

[24] Zoubir, A., Y. Wang, and P. Chargé, "Robust Beamforming Method for Estimating the Incoherently Distributed Source," submitted to *ECWT*, October 2005.

[25] Besson, O. and P. Stoica, "Decoupled Estimation of DOA and Angular Spread for a Spatially Distributed Source," *IEEE Trans. on Signal Processing*, Vol. 48, No. 7, July 2000, pp. 1872–1882.

[26] Besson, O., P. Stoica, and A. B. Gershman, "Simple and Accurate Direction of Arrival Estimator in the Case of Imperfect Spatial Coherence," *IEEE Trans. on Signal Processing*, Vol. 49, No. 4, July 2001, pp. 730–737.

[27] Ottersten, B., P. Stoica, and R. Roy, "Covariance Matching Estimation Techniques for Array Signal Processing Applications," *Digital Signal Processing*, Vol. 8, July 1998, pp. 185–210.

[28] Zoubir, A., Y. Wang, and P. Chargé, "On the Ambiguity of COMET-EXIP Algorithm for Estimating a Scattered Source," *Proc. of ICASSP*, March 2005.

[29] Zoubir, A., Y. Wang, and P. Chargé, "A Modified COMET-EXIP Method for Estimating a Scattered Source," submitted to *Elsevier Signal Processing*, November 2004.

[30] Monakov, A., and O. Besson, "Direction Finding for an Extended Target with Possibly Nonsymmetric Spatial Spectrum," *IEEE Trans. on Signal Processing*, Vol. 52, No. 1, January 2004, pp. 283–287.

PART III
Source Localization Problems

CHAPTER 11
Direct Position Determination of Multiple Radio Transmitters

Anthony J. Weiss, and Alon Amar

The most common methods for position determination of radio signal emitters, such as communications or radar transmitters, are based on the measurement of specified parameters, such as angle-of-arrival (AOA) or time-of-arrival (TOA) of the signal. The measured parameters are then used to estimate the transmitter's location. The measurements are done at each base station independently, without using the constraint that the AOA/TOA estimates at different base stations should correspond to the same transmitter's location. This is a suboptimal location determination technique. Further, if the number of array elements at each base station is M, and the signal waveforms are unknown, then the number of cochannel simultaneous transmitters that can be localized by AOA is limited to $M - 1$. Most AOA algorithms fail when the sources are not angularly separated very well.

We propose a technique that uses exactly the same data as the common AOA methods, but the position determination is direct. The proposed method can handle more than $M - 1$ cochannel simultaneous signals. Although there are many stray parameters, only a two-dimensional search is required for a planar geometry. The technique provides a natural solution to the measurements-sources association problem that is encountered in AOA-based location systems. In addition to new algorithms, we provide analytical performance analysis, Cramér-Rao bounds, and Monte Carlo simulations. We demonstrate that the proposed approach frequently outperforms the traditional AOA methods for unknown, as well as known, signal waveforms.

11.1 Introduction

The problem of emitter location attracts much interest in the signal processing, communications, and underwater acoustics literature. Since World War I, there have been plenty of reports about defense-oriented location systems. Civilian systems are now in use for the localization of cellular phone callers, spectrum monitoring, and law enforcement. Perhaps the first paper on the mathematics of emitter location using AOA is by Stansfield [1]. Many other publications have followed, including a fine review paper by Torrieri [2]. The papers by Krim and Viberg [3] and Wax [4] are comprehensive review papers on antenna array processing for location by AOA. Recently, Van Trees [5] published a book that is fully devoted to array processing. Positioning by TOA and its derivatives (DTOA, EOTD) is used extensively in cellular phone localization [6], radar systems [7], and underwater acoustics [8]. In underwater acoustics, matched-field processing (MFP) is often proposed for source localization [9]. MFP is interpreted as the maximum a posteriori (MAP) estimate of location, given the observed signal at the output of an array of sensors [9, 10]. Other interpretation of MFP is the well-known beamforming extended to wide bandwidth signals, nonplanar wave fields, and unknown environmental parameters [11]. The majority of the literature on MFP focuses on single-source localization.

In this chapter, we discuss a method that solves the localization problem using the data collected at all sensors at all base stations *together*. This is in contradiction to the traditional AOA/TOA approach that is composed of two separate steps (1) AOA/TOA independent estimates, and (2) triangulation based on the results of the first step. The traditional techniques are classified as *decentralized* processing methods [12–15].

Wax and Kailath [12] discussed eigenstructure algorithms for narrowband signals observed by multiple arrays, assuming perfect spatial coherence across each array but no coherence between arrays. Stoica et al. [13] proposed variants of the method of direction estimation (MODE) algorithm for decentralized processing. Weinstein [14] discussed pair-wise processing as an alternative to centralized processing of a wideband single array. The results have indicated that pair-wise processing may be used at high SNR without significant loss of performance. Recently, Kozick and Sadler [15] presented performance analysis for localization of a single wideband source using multiple distributed arrays. They have assumed perfect spatial coherence over each array, and frequency selective coherence between arrays. The proposed method is based on bearing estimation at each array, and delay estimation using the signal observed by a single element at each array.

As indicated in [16], it is rather obvious that measuring AOA/TOA at each base station separately and independently is suboptimal, since this approach ignores the constraint that the measurements must correspond to the same source position. Moreover, the base stations are geographically separated, and often the desired signal appears weak or absent in some of the base stations. Thus, the system must somehow ensure that all AOA/TOA measurements used to locate a specific source correspond to the same source. In the case of cochannel simultaneous sources, the localization system confronts an association problem of deciding which of the multiple AOA/TOA estimates reported by the base stations correspond to which source.

The direct position determination (DPD) method that we propose takes advantage of the rather simple propagation assumptions that are usually used for RF signals. We assume line-of-sight propagation with unknown complex attenuation at each base station. We also assume that all base stations are time-synchronized to the level provided by GPS (approximately 50 ns). The proposed method may implicitly use both the array response at each station and the time of arrival at each station. We derive the maximum likelihood estimate (MLE) of the position of the sources. However, the cost function associated with MLE requires a multidimensional search in the multiple sources case. Thus, for multiple signals with *unknown* waveforms, we resort to a method based on the ideas of Schmidt [17], also known as the MUSIC algorithm. For multiple signals with *known* waveforms, we use the ideas of [18] to simplify the cost function. We show that, for a planar geometry of sources and base stations, a two-dimensional search is sufficient to localize all sources. For a general geometry, only a three-dimensional search is needed. A side benefit of the DPD, in contrast to AOA, is its ability to determine the positions of more sources than the number of sensors at each base station. The DPD technique requires the transmission of the received signals (possibly sampled) to a central processing location, in contrast to AOA and TOA, which require only the transmission of the measured parameters to the central processing location. This is the cost of employing DPD. This chapter focuses on the multiple signals case.

In Section 11.2, we define the models that we use. Section 11.3 describes potential algorithms for known and unknown waveforms, Section 11.4 includes some numerical examples that demonstrate the potential of the advocated approach, and Section 11.5 contains our conclusions. The appendices include derivations of the Cramér-Rao bounds, the performance analysis of the algorithms, and a discussion of the frequency domain model.

11.2 Mathematical Preliminaries

Consider Q transmitters and L base stations intercepting the transmitted signals. Each base station is equipped with an antenna array consisting of M elements. The bandwidth of the signal is small compared to the inverse of the propagation time over the array aperture. Denote the qth transmitter's position by the $D \times 1$ vector of coordinates, $\mathbf{p}_q$. Obviously, for planar geometry, $D = 2$, and for the general case, $D = 3$. The complex envelopes of the signals observed by the lth base station array is given by

$$\mathbf{r}_l(t) = \sum_{q=1}^{Q} b_{lq} \mathbf{a}_l(\mathbf{p}_q) s_q \left[t - \tau_l(\mathbf{p}_q) - t_q^{(0)} \right] + \mathbf{n}_l(t) \quad 0 \leq t \leq T \quad (11.1)$$

where:

$\mathbf{r}_l(t)$ is a time-dependent $M \times 1$ vector.

b_{lq} is an *unknown* complex scalar representing the channel attenuation between the qth transmitter and the lth base station.

$\mathbf{a}_l(\mathbf{p}_q)$ is the lth array response to a signal transmitted from position $\mathbf{p}_q$.

$s_q \left[t - \tau_l(\mathbf{p}_q) - t_q^{(0)} \right]$ is the qth signal waveform, transmitted at time $t_q^{(0)}$ and delayed by $\tau_l(\mathbf{p}_q)$.

$\mathbf{n}_l(t)$ represents noise and interference observed by the array.

The observed signal can be partitioned into K sections, each of length $T/K \gg \max\{\tau_l(\mathbf{p}_q)\}$. The maximum propagation time of interest is the propagation time between the base stations. For example, if the largest separation between the base stations is 10 km, then $T/K \gg 34$ μs, and T/K of approximately 7 μs will satisfy the requirement. See [5], and Appendix 11.D for further discussion of this model. When the total observation time, T, is long, the sources are assumed to be stationary. Otherwise, the location accuracy might be degraded. Each section can be Fourier transformed, and the result of this process is given by

$$\mathbf{r}_l(j, k) = \sum_{q=1}^{Q} b_{lq} \mathbf{a}_l(\mathbf{p}_q) s_q(j, k) e^{-i\omega_j \left[\tau_l(\mathbf{p}_q) + t_q^{(0)} \right]} + \mathbf{n}_l(j, k) \qquad (11.2)$$

$$j = 1, 2, \ldots, J; k = 1, 2, \ldots, K$$

where $\mathbf{r}_l(j, k)$ is the Fourier coefficient of the kth section of the observed signal corresponding to frequency ω_j, $s(j, k)$ is the jth Fourier coefficient of the kth section of the signal, and $\mathbf{n}_l(j, k)$ represents the jth Fourier coefficient of the kth section of the noise waveform.

For easy exhibition, we define the following vectors and scalars

$$\bar{s}_q(j, k) \triangleq s_q(j, k) e^{-i\omega_j t_q^{(0)}} \qquad (11.3)$$

$$\bar{\mathbf{a}}_l(j, \mathbf{p}_q, b_{lq}) \triangleq b_{lq} \mathbf{a}_l(\mathbf{p}_q) e^{-i\omega_j \tau_l(\mathbf{p}_q)}$$

We observe that all information about the transmitter's position is embedded in the vector $\bar{\mathbf{a}}_l(j, \mathbf{p}_q, b_{lq})$. This leads to the following representation of equation (11.2)

$$\mathbf{r}_l(j, k) = \sum_{q=1}^{Q} \bar{\mathbf{a}}_l(j, \mathbf{p}_q, b_{lq}) \bar{s}_q(j, k) + \mathbf{n}_l(j, k) \qquad (11.4)$$

In matrix notation, (11.4) becomes

$$\mathbf{r}_l(j, k) = \mathbf{A}_l \bar{\mathbf{s}}(j, k) + \mathbf{n}_l(j, k)$$

$$\mathbf{A}_l(j) \triangleq \left[\bar{\mathbf{a}}_l(j, \mathbf{p}_1, b_{l1}), \ldots, \bar{\mathbf{a}}_l(j, \mathbf{p}_Q, b_{lQ}) \right] \qquad (11.5)$$

$$\bar{\mathbf{s}}(j, k) \triangleq \left[\bar{s}_1(j, k), \ldots, \bar{s}_Q(j, k) \right]^T$$

Since the vector $\bar{\mathbf{s}}(j, k)$ is the same at all base stations, we can concatenate the observed vectors at all base stations and form the following equation that encompasses all the data and all the information of the location system at hand

$$\mathbf{r}(j, k) = \mathbf{A}(j) \bar{\mathbf{s}}(j, k) + \mathbf{n}(j, k)$$

$$\mathbf{r}(j, k) \triangleq \left[\mathbf{r}_1^T(j, k), \ldots, \mathbf{r}_L^T(j, k) \right]^T \qquad (11.6)$$

$$\mathbf{n}(j, k) \triangleq \left[\mathbf{n}_1^T(j, k), \ldots, \mathbf{n}_L^T(j, k) \right]^T$$

$$\mathbf{A}(j) \triangleq \left[\mathbf{A}_1^T(j), \ldots, \mathbf{A}_L^T(j) \right]^T$$

Note that (11.6) is based on the assumption that the envelopes of the signals are the same at all base stations, up to delay and amplitude caused by the propagation channel. Although this assumption is realistic in most cases of interest, we can solve the L sets of equations represented by (11.5) without relying on this assumption, and still get improved localization with respect to traditional AOA. We do not proceed along this line in this work. This approach was adopted in [12].

If we assume that the waveforms of the signals are unknown, then it is impossible to uniquely determine the complex attenuation coefficients at all base stations and the signal waveforms. We therefore assume that the attenuation coefficients at one of the arrays (e.g., the first array) are all real, and

$$\sum_{l=1}^{L} |b_{lq}|^2 = 1$$

We now address the problem of how to efficiently estimate the locations of the emitters under various assumptions.

11.3 Position Determination

In this section, we discuss potential location algorithms for the common case of signals with unknown waveforms, and for the less common case of signals with known waveforms. We make use of results presented in the literature for the AOA case by introducing modifications as necessary.

11.3.1 Unknown Waveforms Signals

In this section, we assume that the receivers do not know the waveforms a priori. This is the case in most of the applications. Under appropriate assumptions, it is straightforward to write the probability density function of the observations presented in (11.6) as a function of the unknown parameters.

The unknown parameters include:

- QJK snapshots of the complex signals, $\{\bar{s}(j, k)\}$;
- $(L-1)Q$ complex attenuation factors of the signals at the base stations, $\{b_{lq}\}$;
- The two-dimensional real valued location vector of each transmitter, $\{\mathbf{p}_q\}$.

Thus, we have an overall of $2QJK + 2(L-1)Q + 2Q$ real unknown parameters. The MLE will therefore require a complex multidimensional search over the parameter space.

For a single source, an MLE is presented in [19], where the multidimensional search is replaced by a D-dimensional search. In order to avoid the multidimensional search for the multisource case, we can follow the steps leading to the MUSIC algorithm [17]. First note that

$$\mathbf{R}(j) \triangleq E[\mathbf{r}(j,k)\mathbf{r}^H(j,k)] = \mathbf{A}(j)\mathbf{\Lambda}(j)\mathbf{A}^H(j) + \eta\mathbf{I}$$

$$\mathbf{\Lambda}(j) \triangleq E[\bar{\mathbf{s}}(j,k)\bar{\mathbf{s}}^H(j,k)] \tag{11.7}$$

$$E[\mathbf{n}(j,k)\mathbf{n}^H(j,k)] = \eta\mathbf{I}$$

We assumed that the noise is temporally and spatially white, is uncorrelated between sensors and between frequencies, is uncorrelated with the signals, and is zero mean with variance η. The column vectors of $\mathbf{A}$ are orthogonal to the noise subspace of $\mathbf{R}(j)$ and contained in the signal subspace.

Following the MUSIC algorithm, we propose the cost function,

$$F(\mathbf{p},\mathbf{b}) \triangleq \sum_j \bar{\mathbf{a}}^H(j,\mathbf{p},\mathbf{b})\mathbf{U}_s(j)\mathbf{U}_s^H(j)\bar{\mathbf{a}}(j,\mathbf{p},\mathbf{b})$$

$$\bar{\mathbf{a}}(j,\mathbf{p},\mathbf{b}) \triangleq \left[\bar{\mathbf{a}}_1^T(j,\mathbf{p},\mathbf{b}_1), \ldots, \bar{\mathbf{a}}_L^T(j,\mathbf{p},\mathbf{b}_L)\right]^T \tag{11.8}$$

$$\mathbf{b} \triangleq [b_1, \ldots, b_L]^T$$

where $\mathbf{U}_s(j)$ is an $ML \times Q$ matrix consisting of the eigenvectors of $\mathbf{R}(j)$ corresponding to the Q largest eigenvalues. Here, $\mathbf{p}$ and $\mathbf{b}$ are variable vectors representing the unknown position and unknown attenuations, respectively. Recall that the vectors $\bar{\mathbf{a}}(j,\mathbf{p},\mathbf{b})$ contain the L unknown complex attenuation coefficients in addition to the unknown location. The minimum points of $F(\mathbf{p},\mathbf{b})$ depend on all unknowns, and therefore require a $2(L-1) + D$ dimensional search.

In order to reduce this search, we propose to represent $\bar{\mathbf{a}}(j,\mathbf{p},\mathbf{b})$ as follows,

$$\bar{\mathbf{a}}(j,\mathbf{p},\mathbf{b}) = \mathbf{\Gamma}(j)\mathbf{H}\mathbf{b}$$

$$\mathbf{\Gamma}(j) \triangleq diag\left\{\mathbf{a}_1^T(\mathbf{p})e^{-i\omega_j\tau_1(\mathbf{p})}, \ldots, \mathbf{a}_L^T(\mathbf{p})e^{-i\omega_j\tau_L(\mathbf{p})}\right\} \tag{11.9}$$

$$\mathbf{H} \triangleq \mathbf{I}_L \otimes \mathbf{1}_M$$

where $\mathbf{\Gamma}(j)$ is a diagonal matrix whose elements are the elements of the response vectors of the arrays at all base stations, $\mathbf{I}_L$ stands for the $L \times L$ identity matrix, $\mathbf{1}_M$ stands for a $M \times 1$ column vector of ones, and finally $\otimes$ stands for the Kronecker product. Substituting (11.9) in (11.8), we get

$$F(\mathbf{p},\mathbf{b}) = \mathbf{b}^H\mathbf{H}^H\left[\sum_j \mathbf{\Gamma}^H(j)\mathbf{U}_s(j)\mathbf{U}_s^H(j)\mathbf{\Gamma}(j)\right]\mathbf{H}\mathbf{b} \tag{11.10}$$

Recall that we assumed that the norm of $\mathbf{b}$ is one, in order to facilitate a unique solution. Hence, for any assumed position $\mathbf{p}$, the maximum of $F(\mathbf{p},\mathbf{b})$ corresponds to the maximal eigenvalue of the matrix $\mathbf{D}(\mathbf{p})$ defined by

$$\mathbf{D}(\mathbf{p}) \triangleq \mathbf{H}^H\left[\sum_j \mathbf{\Gamma}^H\mathbf{U}_s(j)\mathbf{U}_s^H(j)\mathbf{\Gamma}\right]\mathbf{H} \tag{11.11}$$

Therefore, (11.10) reduces to

$$F(\mathbf{p}) = \lambda_{\max}[\mathbf{D}(\mathbf{p})] \tag{11.12}$$

where the right side of (11.12) denotes the largest eigenvalue of $\mathbf{D}(\mathbf{p})$, and the matrix $\mathbf{D}(\mathbf{p})$ is a function of the observed data [i.e., $\mathbf{U}_s(j)$], and the array response at each base station to an emitter is located at $\mathbf{p}$. It is clear that the maximization of (11.12) requires only a D-dimensional search. It is interesting to note that the dimensions of the matrix $\mathbf{D}(\mathbf{p})$ are $L \times L$, which are usually rather small.

Obviously, one can adjust many known algorithms to handle the problem at hand, including conventional beamforming, Capon's method, Min-Norm, and so forth.

11.3.2 Known Waveform Signals

In certain applications, the transmitted waveforms are known to the location system. For example, in cellular systems, synchronization and training sequences are transmitted periodically, and are known a priori. Moreover, it is possible to detect the data sequence of a digitally modulated signal and then restore the complex signal envelope based on the known modulation scheme. In this section, we examine the position determination problem for known waveforms. We start by following the algebraic steps taken in [18].

We assume that the noise, $\mathbf{n}(j, k)$, is a circularly symmetric complex Gaussian random vector with zero mean and second orders statistics given by

$$E\{\mathbf{n}(i, k)\mathbf{n}^H(j, l)\} = \eta \mathbf{I} \delta_{ij} \delta_{kl} \tag{11.13}$$

$$E\{\mathbf{n}(i, k)\mathbf{n}^T(j, l)\} = 0$$

Define the signal sample covariance

$$\hat{\mathbf{R}}_{ss}(j) \triangleq \frac{1}{K} \sum_{k=1}^{K} \bar{\mathbf{s}}(j, k) \bar{\mathbf{s}}^H(j, k) \tag{11.14}$$

We further assume that asymptotically as $K \to \infty$, the signal's sample covariance is diagonal.

The log-likelihood function of the array output vectors, $\mathbf{r}(j, k)$, is proportional to

$$\begin{aligned} F &= \sum_{j=1}^{J} \frac{1}{K} \sum_{k=1}^{K} \|\mathbf{r}(j, k) - \mathbf{A}(j)\bar{\mathbf{s}}(j, k)\|^2 \\ &= \sum_{j=1}^{J} \frac{1}{K} \sum_{k=1}^{K} [\mathbf{r}(j, k) - \mathbf{A}(j)\bar{\mathbf{s}}(j, k)]^H [\mathbf{r}(j, k) - \mathbf{A}_j \bar{\mathbf{s}}(j, k)] \\ &= \sum_{j=1}^{J} \frac{1}{K} \sum_{k=1}^{K} \mathbf{r}^H(j, k)\mathbf{r}(j, k) - \mathbf{r}^H(j, k)\mathbf{A}(j)\bar{\mathbf{s}}(j, k) - \bar{\mathbf{s}}^H(j, k)\mathbf{A}^H(j)\mathbf{r}(j, k) \\ &\quad + \bar{\mathbf{s}}^H(j, k)\mathbf{A}^H(j)\mathbf{A}(j)\bar{\mathbf{s}}(j, k) \end{aligned} \tag{11.15}$$

Define

$$\hat{\mathbf{R}}_{sr}(j) \triangleq \frac{1}{K} \sum_{k=1}^{K} \bar{\mathbf{s}}(j, k) \mathbf{r}^H(j, k) \quad (11.16)$$

Substituting (11.14) and (11.16) in (11.15) and ignoring the first term, which is constant, we get

$$F_1 = \operatorname{tr}\left[\sum_{j=1}^{J} \frac{1}{K} \sum_{k=1}^{K} -\bar{\mathbf{s}}(j, k)\mathbf{r}^H(j, k)\mathbf{A}(j) - \mathbf{A}^H(j)\mathbf{r}(j, k)\bar{\mathbf{s}}^H(j, k)\right.$$
$$\left. + \mathbf{A}^H(j)\mathbf{A}(j)\bar{\mathbf{s}}(j, k)\bar{\mathbf{s}}^H(j, k)\right]$$

$$= \operatorname{tr}\left[\sum_{j=1}^{J} -\hat{\mathbf{R}}_{sr}(j)\mathbf{A}(j) - \mathbf{A}^H(j)\hat{\mathbf{R}}_{sr}^H(j) + \mathbf{A}^H(j)\mathbf{A}(j)\hat{\mathbf{R}}_{sr}(j)\right]$$

$$= \operatorname{tr}\left[\sum_{j=1}^{J} -\hat{\mathbf{R}}_{ss}(j)\hat{\mathbf{R}}_{ss}^{-1}(j)\hat{\mathbf{R}}_{sr}(j)\mathbf{A}(j) - \mathbf{A}^H \hat{\mathbf{R}}_{sr}^H(j)\hat{\mathbf{R}}_{ss}^{-1}(j)\hat{\mathbf{R}}_{ss}(j) + \mathbf{A}^H(j)\mathbf{A}(j)\hat{\mathbf{R}}_{ss}(j)\right]$$

(11.17)

Note that

$$\operatorname{tr}\left[\hat{\mathbf{R}}_{ss}(j)\hat{\mathbf{R}}_{ss}^{-1}(j)\hat{\mathbf{R}}_{sr}(j)\mathbf{A}(j)\right] = \operatorname{tr}\left[\hat{\mathbf{R}}_{ss}^{-1}(j)\hat{\mathbf{R}}_{sr}(j)\mathbf{A}(j)\hat{\mathbf{R}}_{ss}(j)\right] \quad (11.18)$$

Therefore, (11.17) can be displayed as

$$F_1 = \operatorname{tr}\left\{\sum_{j=1}^{J} \left[-\hat{\mathbf{R}}_{ss}^{-1}(j)\hat{\mathbf{R}}_{sr}(j)\mathbf{A}(j) - \mathbf{A}^H(j)\hat{\mathbf{R}}_{sr}^H(j)\hat{\mathbf{R}}_{ss}^{-1}(j) + \mathbf{A}^H(j)\mathbf{A}(j)\right]\hat{\mathbf{R}}_{ss}(j)\right\}$$

(11.19)

Define

$$\hat{\mathbf{A}}(j) = \hat{\mathbf{R}}_{sr}^H(j)\hat{\mathbf{R}}_{ss}^{-1}(j) \quad (11.20)$$

Then, minimizing F_1 is equivalent to minimizing the following:

$$F_2 = \operatorname{tr}\left\{\sum_{j=1}^{J} \left[\hat{\mathbf{A}}^H(j)\hat{\mathbf{A}}(j) - \hat{\mathbf{A}}^H(j)\mathbf{A}(j) - \mathbf{A}^H(j)\hat{\mathbf{A}}(j) + \mathbf{A}^H(j)\mathbf{A}(j)\right]\hat{\mathbf{R}}_{ss}(j)\right\}$$

$$= \operatorname{tr}\left\{\sum_{j=1}^{J} \left[\mathbf{A}(j) - \hat{\mathbf{A}}(j)\right]^H \left[\mathbf{A}(j) - \hat{\mathbf{A}}(j)\right]\hat{\mathbf{R}}_{ss}(j)\right\} \quad (11.21)$$

Obviously, the term $\hat{\mathbf{A}}^H(j)\hat{\mathbf{A}}(j)$ is constant and can be added to the cost function.

11.3 Position Determination

Since we assumed that the signals are uncorrelated, $\hat{\mathbf{R}}_{ss}(j)$ is asymptotically diagonal, and the cost function can be decoupled

$$F_3 = \sum_{q=1}^{Q} F_3(q) \qquad (11.22)$$

$$F_3(q) \triangleq \sum_{j=1}^{J} \|\bar{\mathbf{a}}(j, \mathbf{p}_q, \mathbf{b}_q) - \hat{\mathbf{a}}_q(j)\|^2$$

where $\bar{\mathbf{a}}(j, \mathbf{p}_q, \mathbf{b}_q)$ and $\hat{\mathbf{a}}_q(j)$ represent the qth column of $\mathbf{A}(j)$ and $\hat{\mathbf{A}}(j)$, respectively. The minimization of $F_3(q)$ can be done as follows

$$F_3(q) = \sum_{j=1}^{J} \|\bar{\mathbf{a}}(j, \mathbf{p}_q, \mathbf{b}_q) - \hat{\mathbf{a}}_q(j)\|^2 = \sum_{j=1}^{J} \|\mathbf{\Gamma}_q(j)\mathbf{H}\mathbf{b}_q - \hat{\mathbf{a}}_q(j)\|^2 \qquad (11.23)$$

The vector $\mathbf{b}$ that minimizes the cost function is given by

$$\hat{\mathbf{b}}_q = \left(\sum_j \mathbf{H}^H \mathbf{\Gamma}_q^H(j) \mathbf{\Gamma}_q(j) \mathbf{H}\right)^{-1} \mathbf{H}^H \sum_j \mathbf{\Gamma}_q^H(j) \hat{\mathbf{a}}_q(j) \qquad (11.24)$$

Replacing $\mathbf{b}$ with $\hat{\mathbf{b}}$ in (11.23), we get a cost function that depends only on $\mathbf{p}$. The cost function can be simplified by relying on the assumption that

$$\|\mathbf{a}_l(j)\| = 1; \ \forall \ j, l, \mathbf{p} \qquad (11.25)$$

which immediately leads to

$$\sum_j \mathbf{H}^H \mathbf{\Gamma}_q^H(j) \mathbf{\Gamma}_q(j) \mathbf{H} = \mathbf{I}_L \qquad (11.26)$$

Define the vector

$$\mathbf{w}_q \triangleq \sum_j \mathbf{\Gamma}_q^H(j) \hat{\mathbf{a}}_q(j) \qquad (11.27)$$

Substituting (11.27) and (11.26) in (11.24), and then in (11.23), yields

$$F_3(q) = \sum_{j=1}^{J} \|\mathbf{\Gamma}_q(j)\mathbf{H}\mathbf{H}^H \mathbf{w}_q - \hat{\mathbf{a}}_q(j)\|^2 \qquad (11.28)$$

It is easy to verify that minimizing (11.28) is equivalent to maximizing

$$F_4(q) = \mathbf{w}_q^H \mathbf{H}\mathbf{H}^H \mathbf{w}_q = \|\mathbf{H}^H \mathbf{w}_q\|^2 \qquad (11.29)$$

$$= \sum_{l=1}^{L} \left| \sum_{j=1}^{J} e^{i\omega_j \tau_l(\mathbf{p})} \mathbf{a}_l^H(\mathbf{p}) \hat{\mathbf{a}}_q^{(l)}(j) \right|^2$$

where $\hat{\mathbf{a}}_q^{(l)}(j)$ is the *l*th subvector of $\hat{\mathbf{a}}_q(j)$ (i.e., the subvector associated with the *l*th base station.)

Note that (11.29) indicates that the cost function is in fact a sum of *L* distinct cost functions, each associated with a distinct base station. This is the reason that DPD outperforms methods that maximize the cost function at each base station independently.

Thus, we can locate each of the emitters by a simple *D*-dimensional search.

11.4 Numerical Examples

In order to examine the performance of the advocated methods and compare it with the traditional AOA approach, we performed extensive Monte Carlo simulations. Some examples are shown here.

We applied two different techniques in order to locate the transmitter:

1. AOA estimation using MUSIC or beamforming at each base station independently;
2. DPD according to the algorithms described in the previous section.

The performance evaluation is based on the RMS error defined as

$$\text{RMS} = \sqrt{\frac{1}{N} \sum_{i=1}^{N} (\hat{x}_i - x_t)^2 + (\hat{y}_i - y_t)^2}$$

where (x_t, y_t) is the emitter location, $(\hat{x}_i, \hat{y}_i)$ is the *i*th location estimate, and *N* is the number of experiments.

11.4.1 Example 1

Consider three base stations placed at coordinates (2, –2), (2, 0), and (2, 2), and a single emitter located at (0, 1.5). All coordinates are in kilometers. The transmitted signal is a carrier amplitude modulated by narrowband random Gaussian waveform. The signal is unknown to the receivers. Each base station is equipped with a uniform linear array of only three antenna elements. Each location determination is based on 200 snapshots, each of 4.5 ms at a single frequency (i.e., $K = 200$, $J = 1$). The snapshot length ensures that the errors introduced by the finite length FFT are 30 dB below the signal level. The SNR (at the base station receiving the strongest signal) is varied between 3 and 23 dB, using 2 dB steps. At each SNR value, we performed 100 experiments in order to obtain the statistical properties of the performance. The attenuation vector is selected as $\mathbf{b} = [1, 0.8, 0.4]^T$.

Figure 11.1 shows the experimental results, using (11.12), the Cramér-Rao bound, derived in Appendix B, and the performance analysis results described in Appendix 11.C. The plots indicate that DPD is superior to the traditional approach

11.4 Numerical Examples

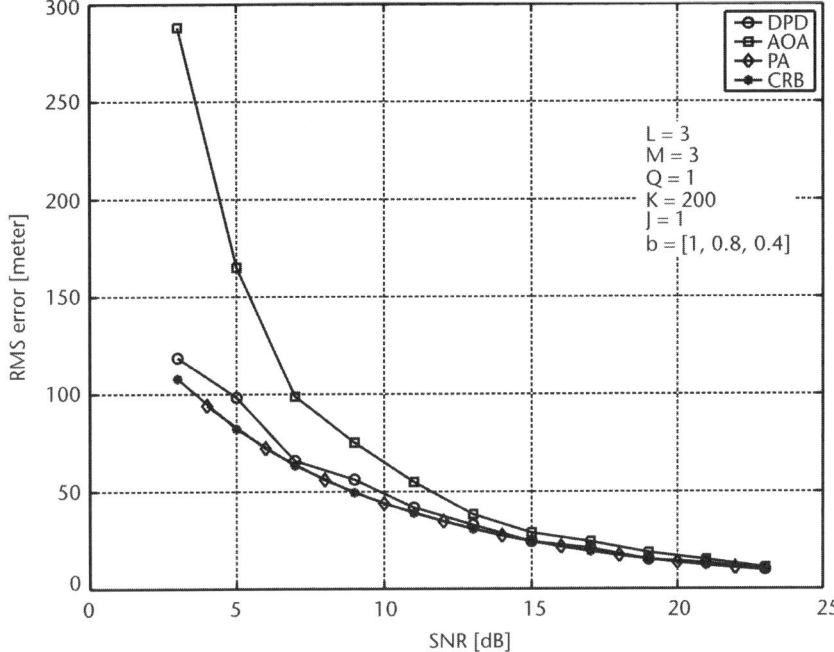

Figure 11.1 RMS error of DPD and traditional AOA, Cramér-Rao bound, and performance analysis results for three base stations, a single source, and unknown waveform.

of independent AOA estimates at each base station. The advantage of DPD is at low SNR. At high SNR, both methods give results that coincide with the theoretical bound.

11.4.2 Example 2

In a second experiment, we kept the base stations at the same locations, and we used two emitters placed at (0, +1.5) and (0, −1.5). The results for each of the sources are shown in Figure 11.2. It is again clear that the DPD outperforms AOA at low SNR, while both methods are equivalent at high SNR. At very high SNR, modeling errors will dominate the performance. Modeling errors include finite length FFT, calibration errors, synchronization errors, propagation errors, and so forth

11.4.3 Example 3

In a third experiment, we kept the base stations at the same locations, and we used two emitters placed at (0, +Y) and (0, −Y), and 100 snapshots and SNR = 20 dB. The channel attenuation to all base stations is equal. We changed Y from 200m to 1,200m and the results are plotted in Figure 11.3. As expected, it can be seen that the traditional AOA accuracy is very sensitive to sources that are not well separated, as opposed to the DPD method.

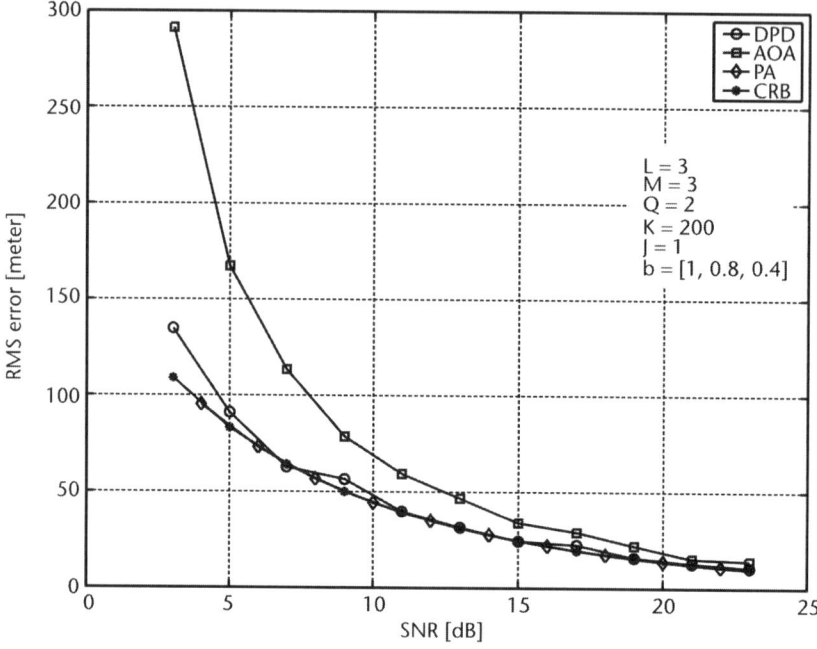

Figure 11.2 RMS error of DPD and traditional AOA, Cramér-Rao bound, and performance analysis results for three base stations, two sources, and unknown waveforms.

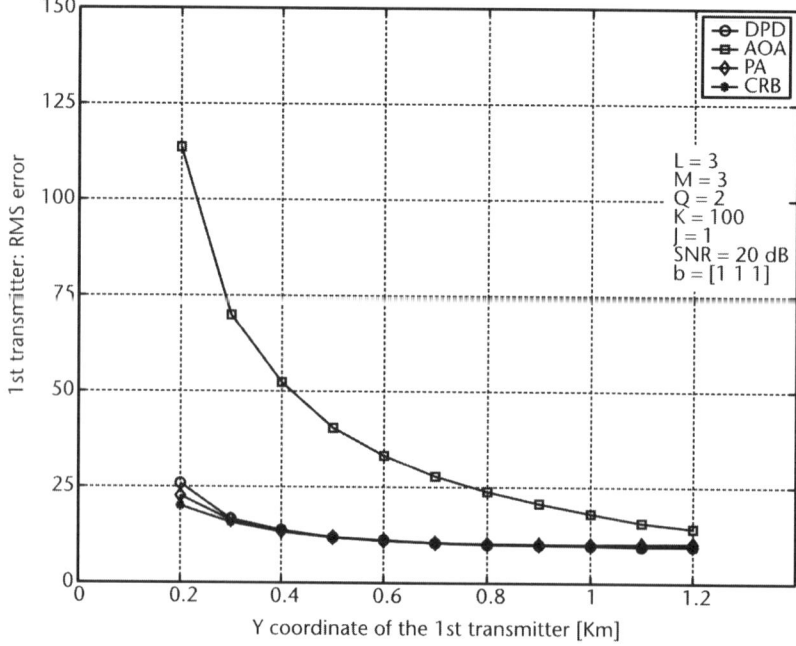

Figure 11.3 RMS error of DPD and traditional AOA, Cramér-Rao bound, and performance analysis results for three base stations, two sources, with increasing separation and unknown waveforms.

11.4.4 Example 4

In a fourth experiment, we kept the base stations at the same location, and placed three transmitters at (0, 1.5), (0, –1.5), and (–1, 0). Each base station collects 1,000 snapshots, and the attenuation is equal at all base stations. Since each base station is equipped with an array of only three elements, traditional AOA based on MUSIC fails. However, DPD works fine, as shown in Figure 11.4.

11.4.5 Example 5

In a fifth experiment, we used four base stations, located at (–2, –2), (–2, +2), (2, –2), and (2, +2), and a single source at (1, 1). Each base station is equipped with a circular array of five elements. The waveforms are known. The number of snapshots is 1,000. The attenuation to two base stations is 0 dB, and for the other two is –10 dB. The accuracy results are plotted in Figure 11.5.

In Figure 11.6, we show how the cost function looks for unknown waveforms, four base stations, each equipped with an array of only three elements, and three transmitters. Common AOA methods cannot handle three transmitters with three-element arrays, contrary to the advocated method.

11.5 Conclusions

We have proposed a direct position determination technique for localizing multiple narrowband radio frequency sources. The technique can locate more sources than

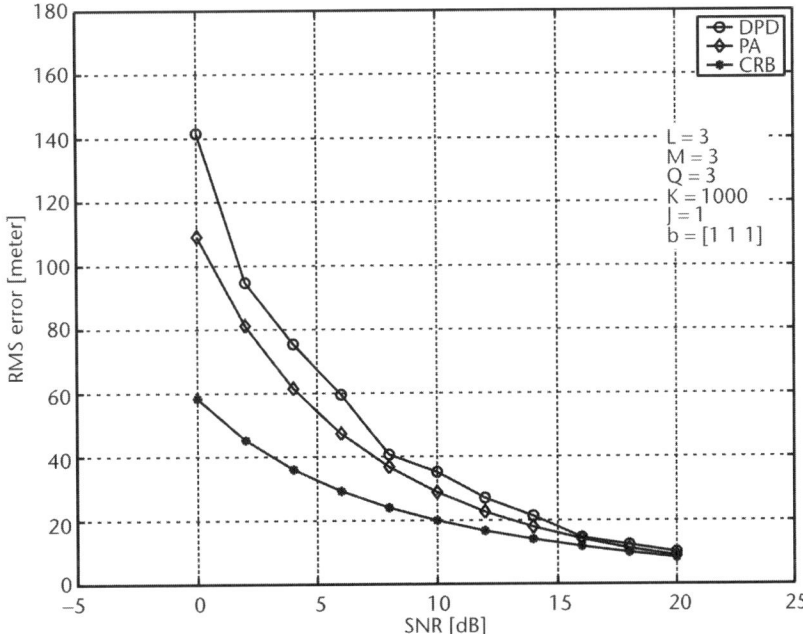

Figure 11.4 RMS error of DPD, Cramér-Rao bound, and performance analysis results for three base stations, three sources, and unknown waveforms.

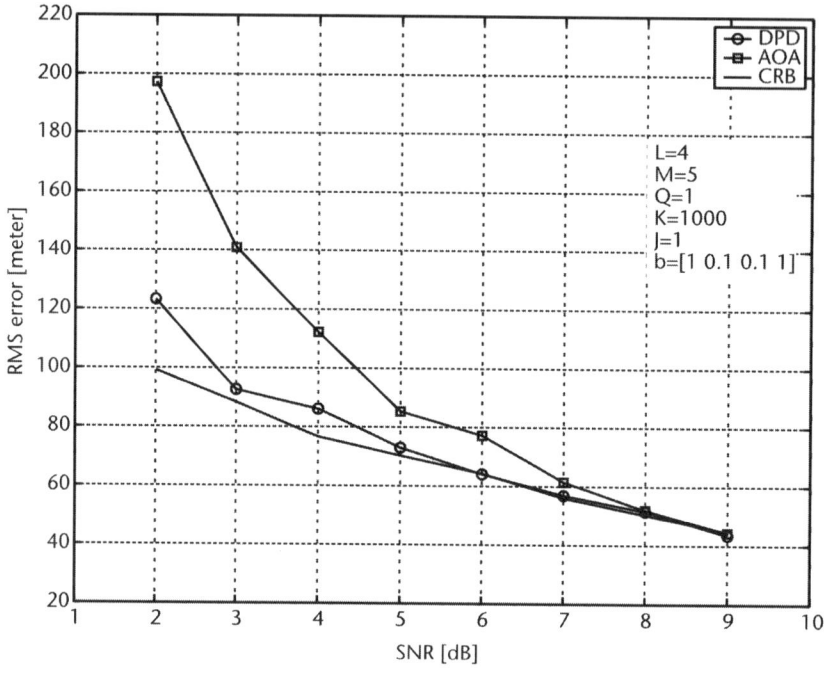

Figure 11.5 RMS error of DPD and traditional AOA for known waveforms. Four base stations, each equipped with five-element circular arrays, are used.

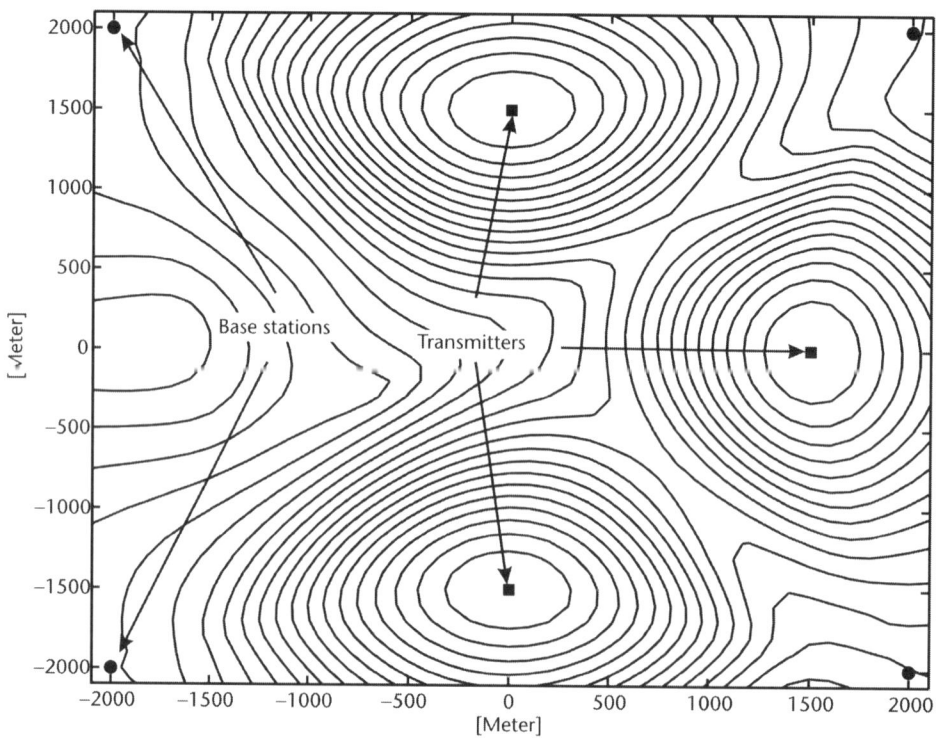

Figure 11.6 Contour description of the cost function for unknown waveforms, four base stations and three transmitters.

the traditional AOA approach. Moreover, DPD provides better accuracy than traditional AOA, and it does not encounter the association problem of independent AOA measurements at each base station. The proposed technique uses the MUSIC approach, in order to reduce the complexity of the algorithm in the case of unknown waveforms. The advantages of DPD do not come without a price. While in traditional methods, only AOA estimates must be transferred to a central processing location for triangulation, the DPD requires raw signal data to be transferred to a common processor.

References

[1] Stansfield, R. G., "Statistical Theory of DF Fixing," *Journal IEE*, Vol. 94, Part 3A, No. 15, March 1947, pp. 762–770.

[2] Torrieri, D. L., "Statistical Theory of Passive Location Systems," *IEEE Trans. on Aerospace and Electronic Systems*, Vol. AES-20, No. 2, March 1984.

[3] Krim, H., and M. Viberg, "Two Decades of Array Signal Processing Research," *IEEE Signal Processing Magazine*, Vol. 13, No. 4, July 1996.

[4] Wax, M., "Model-Based Processing in Sensor Arrays," in *Advances in Spectrum Analysis and Array Processing, Vol. III*, S. Haykin, (ed.), Englewood Cliffs, NJ: Prentice Hall, 1995.

[5] Van Trees, H. L., *Detection, Estimation, and Modulation Theory, Part IV: Optimum Array Processing*, New York: John Wiley & Sons, 2002.

[6] Liberti, J. C., and T. S. Rappaport, *Smart Antennas for Wireless Communications: IS-95 and Third Generation CDMA Applications*, Upper Saddle River, NJ: Prentice Hall, 1999.

[7] Skolnik, M. I., *Introduction to Radar Systems*, 3rd ed., New York: McGraw-Hill, 2000.

[8] Carter, G. C., (ed.), *Coherence and Time Delay Estimation*, New York: IEEE Press, 1993.

[9] Shorey, J. A., and L. W. Nolte, "Wideband Optimal a Posteriori Probability Source Localization in an Uncertain Shallow Ocean Environment," *J. Acoust. Soc. Am.*, Vol. 103, No. 1, January 1998.

[10] Harrison, B. F., "An L°-Norm Estimator for Environmentally Robust, Shallow-Water Source Localization," *J. Acoust. Soc. Am.*, Vol. 105, No. 1, January 1999.

[11] Baggeroer, A. B., W. A. Kuperman, and P. N. Mikhalevsky, "An Overview of Matched Field Methods in Ocean Acoustics," *IEEE J. of Oceanic Engineering*, Vol. 18, No. 4, October 1993, pp. 401–424.

[12] Wax, M., and T. Kailath, "Decentralized Processing in Sensor Arrays," *IEEE Trans. on Acoustics, Speech, and Signal Processing*, Vol. ASSP-33, No. 4, October 1985, pp. 1123–1129.

[13] Stoica, P., A. Nehorai, and T. Sodersrom, "Decentralized Array Processing Using the MODE Algorithm," *Circuits, Systems, and Signal Processing*, Vol. 14, No. 1, 1995, pp. 17–38.

[14] Weinstein, E., "Decentralization of the Gaussian Maximum Likelihood Estimator and its Applications to Passive Array Processing," *IEEE Trans. on Acoustics, Speech, and Signal Processing*, Vol. ASSP-29, No. 4, October 1981, pp. 945–951.

[15] Kozick, R. J., and B. M. Sadler, "Source Localization with Distributed Sensor Arrays and Partial Spatial Coherence," *IEEE Trans. Sig. Proc.*, Vol. 52, No. 3, March 2004, pp. 601–615.

[16] Wax, M., and T. Kailath, "Optimum Localization of Multiple Sources by Passive Arrays," *IEEE Trans. on Acoustics, Speech, and Signal Processing*, Vol. ASSP-31, No. 5, October 1983, pp. 1210–1217.

[17] Schmidt, R. O., "Multiple Emitter Location and Signal Parameter Estimation," *IEEE Trans. on Antennas and Propagation*, Vol. 34, No. 3, March 1986, pp. 276–280.

[18] Li, J., et al., "Computationally Efficient Angle Estimation for Signals with Known Waveforms," *IEEE Trans. on Signal Processing*, Vol. 43, No. 9, September 1995, pp. 2154–2163.

[19] Weiss, A. J., "Direct Position Determination of Narrowband Radio Frequency Transmitters," *IEEE Signal Processing Letters*, Vol. 11, No. 5, May 2004, pp. 513–516.

[20] Porat, B., and B. Friedlander, "Analysis of the Asymptotic Relative Efficiency of the MUSIC Algorithm," *IEEE Trans. on Acoustics, Speech, and Signal Processing*, Vol. 36, No. 4, April 1988, pp. 532–544.

[21] Brandwood, D. H., "A Complex Gradient Operator and Its Application in Adaptive Array Theory," *Proc. Inst. Elec. Eng.*, Vol. 130, Parts F and H, No. 1, February 1983, pp. 11–16.

[22] Weiss, A. J., and B. Friedlander, "Performance Analysis of Spatial Smoothing with Interpolated Arrays," *IEEE Trans. on Acoustics, Speech, and Signal Processing*, Vol. 41, No. 5, May 1993, pp. 1881–1892.

Appendix 11.A Cramér-Rao Bound for Known Waveforms

We start with (11.6) repeated here for easy reference

$$\mathbf{r}(j, k) = \mathbf{A}(j)\bar{\mathbf{s}}(j, k) + \mathbf{n}(j, k) \qquad (11.A.1)$$

The unknown parameters are the entries of $\mathbf{P}$ and $\mathbf{B}$ defined as

$$\mathbf{P} \triangleq [\mathbf{p}_1, \mathbf{p}_2, \ldots, \mathbf{p}_Q]$$
$$\mathbf{B} \triangleq [\mathbf{b}_1, \mathbf{b}_2, \ldots, \mathbf{b}_Q] \qquad (11.A.2)$$
$$\mathbf{b}_q \triangleq [b_{1,q}, b_{2,q}, \ldots, b_{L,q}]^T$$

The log likelihood function is given by

$$\ln L = \text{const.} - \frac{1}{\eta} \sum_{j,k} \|\mathbf{r}(j, k) - \mathbf{A}(j)\bar{\mathbf{s}}(j, k)\|^2 \qquad (11.A.3)$$

It is easy to verify the following derivatives

$$\frac{\partial \ln L}{\partial P_{nq}} = \frac{2}{\eta} \text{Re} \left\{ \sum_{j,k} \bar{s}_q^*(j, k) \frac{\partial \mathbf{a}_q^H(j)}{\partial P_{nq}} \mathbf{n}(j, k) \right\} \qquad (11.A.4)$$

where P_{nq} is the n, q element of the matrix $\mathbf{P}$, and $\mathbf{a}_q(j)$ is the qth column of $\mathbf{A}(j)$. Define the matrices

Appendix 11.A Cramér-Rao Bound for Known Waveforms

$$\mathbf{D}_n(j) \triangleq \left[\frac{\partial \mathbf{a}_1(j)}{\partial P_{n1}}, \frac{\partial \mathbf{a}_2(j)}{\partial P_{n2}}, \ldots, \frac{\partial \mathbf{a}_Q(j)}{\partial P_{nQ}} \right] \quad (11.A.5)$$

$$\mathbf{X}(j,k) \triangleq diag[\bar{\mathbf{s}}(j,k)]$$

Now (11.A.4) can be written as

$$\frac{\partial \ln L}{\partial \mathbf{P}_n} = \frac{2}{\eta} \operatorname{Re}\left\{ \sum_{j,k} \mathbf{X}^H(j,k) \mathbf{D}_n^H(j) \mathbf{n}(j,k) \right\} \quad (11.A.6)$$

where $\mathbf{P}_n$ is the nth row of $\mathbf{P}$.

Define now the following variables

$$\overline{\mathbf{C}}_n(j) \triangleq \left[\frac{\partial \mathbf{a}_1(j)}{\partial \bar{b}_{n1}}, \frac{\partial \mathbf{a}_2(j)}{\partial \bar{b}_{n2}}, \ldots, \frac{\partial \mathbf{a}_Q(j)}{\partial \bar{b}_{nQ}} \right] \quad (11.A.7)$$

$$\tilde{\mathbf{C}}_n(j) \triangleq \left[\frac{\partial \mathbf{a}_1(j)}{\partial \tilde{b}_{n1}}, \frac{\partial \mathbf{a}_2(j)}{\partial \tilde{b}_{n2}}, \ldots, \frac{\partial \mathbf{a}_Q(j)}{\partial \tilde{b}_{nQ}} \right] = i\overline{\mathbf{C}}_n(j)$$

We can now write

$$\frac{\partial \ln L}{\partial \overline{\mathbf{B}}_n} = \frac{2}{\eta} \operatorname{Re}\left[\sum_{j,k} \mathbf{X}^H(j,k) \overline{\mathbf{C}}_n^H(j) \mathbf{n}(j,k) \right] \quad (11.A.8)$$

$$\frac{\partial \ln L}{\partial \tilde{\mathbf{B}}_n} = \frac{2}{\eta} \operatorname{Im}\left[\sum_{j,k} \mathbf{X}^H(j,k) \overline{\mathbf{C}}_n^H(j) \mathbf{n}(j,k) \right]$$

where $\overline{\mathbf{B}}_n$, $\tilde{\mathbf{B}}_n$ are the real and imaginary parts of the nth row of $\mathbf{B}$, respectively.

The blocks of the Fisher information matrix (FIM) are given by

$$\operatorname{FIM}_{11}(n,m) \triangleq E\left(\frac{\partial \ln L}{\partial \overline{\mathbf{B}}_n}\right)\left(\frac{\partial \ln L}{\partial \overline{\mathbf{B}}_m}\right)^T = \frac{2}{\eta} \operatorname{Re}\left[\sum_{j,k} \mathbf{X}^H(j,k) \overline{\mathbf{C}}_n^H(j) \overline{\mathbf{C}}_m(j) \mathbf{X}(j,k) \right]$$

$$\operatorname{FIM}_{12}(n,m) \triangleq E\left(\frac{\partial \ln L}{\partial \overline{\mathbf{B}}_n}\right)\left(\frac{\partial \ln L}{\partial \tilde{\mathbf{B}}_m}\right)^T = -\frac{2}{\eta} \operatorname{Im}\left[\sum_{j,k} \mathbf{X}^H(j,k) \overline{\mathbf{C}}_n^H(j) \overline{\mathbf{C}}_m(j) \mathbf{X}(j,k) \right]$$

$$\operatorname{FIM}_{22}(n,m) \triangleq E\left(\frac{\partial \ln L}{\partial \tilde{\mathbf{B}}_n}\right)\left(\frac{\partial \ln L}{\partial \tilde{\mathbf{B}}_m}\right)^T = \frac{2}{\eta} \operatorname{Re}\left[\sum_{j,k} \mathbf{X}^H(j,k) \overline{\mathbf{C}}_n^H(j) \overline{\mathbf{C}}_m(j) \mathbf{X}(j,k) \right]$$

$$(11.A.9)$$

where $\operatorname{FIM}_{i,j}(n,m)$ stands for the n, m subblock of the $\operatorname{FIM}_{i,j}$ block.

$$\text{FIM}_{31} \triangleq E\left(\frac{\partial \ln L}{\partial \mathbf{P}_n}\right)\left(\frac{\partial \ln L}{\partial \overline{\mathbf{B}}_m}\right)^T = \frac{2}{\eta}\text{Re}\left[\sum_{j,k}\mathbf{X}^H(j,k)\mathbf{D}_n^H(j)\overline{\mathbf{C}}_m(j)\mathbf{X}(j,k)\right]$$

$$\text{FIM}_{32} \triangleq E\left(\frac{\partial \ln L}{\partial \mathbf{P}_n}\right)\left(\frac{\partial \ln L}{\partial \tilde{\mathbf{B}}_m}\right)^T = -\frac{2}{\eta}\text{Im}\left[\sum_{j,k}\mathbf{X}^H(j,k)\mathbf{D}_n^H(j)\overline{\mathbf{C}}_m(j)\mathbf{X}(j,k)\right]$$

$$\text{FIM}_{33} \triangleq E\left(\frac{\partial \ln L}{\partial \mathbf{P}_n}\right)\left(\frac{\partial \ln L}{\partial \mathbf{P}_m}\right)^T = \frac{2}{\eta}\text{Re}\left[\sum_{j,k}\mathbf{X}^H(j,k)\mathbf{D}_n^H(j)\mathbf{D}_m(j)\mathbf{X}(j,k)\right]$$

(11.A.10)

The CRB bound is obtained by inverting the Fisher information matrix.

Appendix 11.B   CRB for Unknown Gaussian Waveforms

It is well known that the Fisher information matrix for zero-mean Gaussian signals is given by

$$[\text{FIM}]_{i,j} = \text{tr}\left(\mathbf{R}^{-1}\frac{\partial \mathbf{R}}{\partial \theta_i}\mathbf{R}^{-1}\frac{\partial \mathbf{R}}{\partial \theta_j}\right) \quad (11.\text{B}.1)$$

where $\mathbf{R}$ is the observation covariance and θ_i is the ith parameter. The covariance matrix for a given frequency is

$$\mathbf{R} = \mathbf{A}\mathbf{\Lambda}\mathbf{A}^H + \eta\mathbf{I} \quad (11.\text{B}.2)$$

The unknown parameters are the entries of $\mathbf{P}$, $\mathbf{B}$, and $\mathbf{\Lambda}$ defined in (11.A.2). We will frequently use the notation $\mathbf{e}_n$ for nth column vector of the identity matrix. First note that

$$\begin{aligned}
\text{tr}\left(\mathbf{R}^{-1}\frac{\partial \mathbf{R}}{\partial \Lambda_i}\mathbf{R}^{-1}\frac{\partial \mathbf{R}}{\partial \Lambda_j}\right) &= \text{tr}\left(\mathbf{R}^{-1}\mathbf{A}\mathbf{e}_i\mathbf{e}_i^T\mathbf{A}^H\mathbf{R}^{-1}\mathbf{A}\mathbf{e}_j\mathbf{e}_j^T\mathbf{A}^H\right) \\
&= \left(\mathbf{e}_j^T\mathbf{A}^H\mathbf{R}^{-1}\mathbf{A}\mathbf{e}_i\right)\left(\mathbf{e}_i^T\mathbf{A}^H\mathbf{R}^{-1}\mathbf{A}\mathbf{e}_j\right) \quad (11.\text{B}.3)\\
&= \left(\mathbf{e}_j^T\mathbf{A}^H\mathbf{R}^{-1}\mathbf{A}\mathbf{e}_i\right)\left(\mathbf{e}_j^T\mathbf{A}^H\mathbf{R}^{-1}\mathbf{A}\mathbf{e}_i\right)^*
\end{aligned}$$

Thus, the FIM related to the diagonal elements of the signal covariance are given by

$$\text{FIM}_{\Lambda\Lambda} = (\mathbf{A}^H\mathbf{R}^{-1}\mathbf{A}) \times (\mathbf{A}^H\mathbf{R}^{-1}\mathbf{A})^* \quad (11.\text{B}.4)$$

where $\times$ denotes element-by-element multiplication.

Appendix 11.B CRB for Unknown Gaussian Waveforms

We can write

$$\text{tr}\left(\mathbf{R}^{-1}\frac{\partial \mathbf{R}}{\partial \bar{b}_{ql}}\mathbf{R}^{-1}\frac{\partial \mathbf{R}}{\partial \bar{b}_{km}}\right)$$
$$= \text{tr}\left[\mathbf{R}^{-1}(\overline{\mathbf{C}}_l\mathbf{e}_q\mathbf{e}_q^T\mathbf{\Lambda}\mathbf{A}^H + \mathbf{A}\mathbf{\Lambda}\mathbf{e}_q\mathbf{e}_q^T\overline{\mathbf{C}}_l^H)\mathbf{R}^{-1}(\overline{\mathbf{C}}_m\mathbf{e}_k\mathbf{e}_k^T\mathbf{\Lambda}\mathbf{A}^H + \mathbf{A}\mathbf{\Lambda}\mathbf{e}_k\mathbf{e}_k^T\overline{\mathbf{C}}_m^H)\right] \quad (11.\text{B}.5)$$

Rearranging terms, we get

$$\text{tr}\left(\mathbf{R}^{-1}\frac{\partial \mathbf{R}}{\partial \bar{b}_{ql}}\mathbf{R}^{-1}\frac{\partial \mathbf{R}}{\partial \bar{b}_{km}}\right) = \text{tr}\left(\mathbf{R}^{-1}\overline{\mathbf{C}}_l\mathbf{e}_q\mathbf{e}_q^T\mathbf{\Lambda}\mathbf{A}^H\mathbf{R}^{-1}\overline{\mathbf{C}}_m\mathbf{e}_k\mathbf{e}_k^T\mathbf{\Lambda}\mathbf{A}^H\right)$$
$$+ \text{tr}\left(\mathbf{R}^{-1}\overline{\mathbf{C}}_l\mathbf{e}_q\mathbf{e}_q^T\mathbf{\Lambda}\mathbf{A}^H\mathbf{R}^{-1}\mathbf{A}\mathbf{\Lambda}\mathbf{e}_k\mathbf{e}_k^T\overline{\mathbf{C}}_m^H\right) \quad (11.\text{B}.6)$$
$$+ \text{tr}\left(\mathbf{R}^{-1}\mathbf{A}\mathbf{\Lambda}\mathbf{e}_q\mathbf{e}_q^T\overline{\mathbf{C}}_l^H\mathbf{R}^{-1}\overline{\mathbf{C}}_m\mathbf{e}_k\mathbf{e}_k^T\mathbf{\Lambda}\mathbf{A}^H\right)$$
$$+ \text{tr}\left(\mathbf{R}^{-1}\mathbf{A}\mathbf{\Lambda}\mathbf{e}_q\mathbf{e}_q^T\overline{\mathbf{C}}_l^H\mathbf{R}^{-1}\mathbf{A}\mathbf{\Lambda}\mathbf{e}_k\mathbf{e}_k^T\overline{\mathbf{C}}_m^H\right)$$

Taking advantage of the trace operator, we can write

$$\text{tr}\left(\mathbf{R}^{-1}\frac{\partial \mathbf{R}}{\partial \bar{b}_{ql}}\mathbf{R}^{-1}\frac{\partial \mathbf{R}}{\partial \bar{b}_{km}}\right) = \left(\mathbf{e}_k^T\mathbf{\Lambda}\mathbf{A}^H\mathbf{R}^{-1}\overline{\mathbf{C}}_l\mathbf{e}_q\right)\left(\mathbf{e}_q^T\mathbf{\Lambda}\mathbf{A}^H\mathbf{R}^{-1}\overline{\mathbf{C}}_m\mathbf{e}_k\right)$$
$$+ \left(\mathbf{e}_k^T\overline{\mathbf{C}}_m^H\mathbf{R}^{-1}\overline{\mathbf{C}}_l\mathbf{e}_q\right)\left(\mathbf{e}_q^T\mathbf{\Lambda}\mathbf{A}^H\mathbf{R}^{-1}\mathbf{A}\mathbf{\Lambda}\mathbf{e}_k\right) \quad (11.\text{B}.7)$$
$$+ \left(\mathbf{e}_k^T\mathbf{\Lambda}\mathbf{A}^H\mathbf{R}^{-1}\mathbf{A}\mathbf{\Lambda}\mathbf{e}_q\right)\left(\mathbf{e}_q^T\overline{\mathbf{C}}_l^H\mathbf{R}^{-1}\overline{\mathbf{C}}_m\mathbf{e}_k\right)$$
$$+ \left(\mathbf{e}_k^T\overline{\mathbf{C}}_m^H\mathbf{R}^{-1}\mathbf{A}\mathbf{\Lambda}\mathbf{e}_q\right)\left(\mathbf{e}_q^T\overline{\mathbf{C}}_l^H\mathbf{R}^{-1}\mathbf{A}\mathbf{\Lambda}\mathbf{e}_k\right)$$

Rearranging terms again, we get

$$\text{tr}\left(\mathbf{R}^{-1}\frac{\partial \mathbf{R}}{\partial \bar{b}_{ql}}\mathbf{R}^{-1}\frac{\partial \mathbf{R}}{\partial \bar{b}_{km}}\right) = \left(\mathbf{e}_k^T\mathbf{\Lambda}\mathbf{A}^H\mathbf{R}^{-1}\overline{\mathbf{C}}_l\mathbf{e}_q\right)\left(\mathbf{e}_k^T\overline{\mathbf{C}}_m^H\mathbf{R}^{-1}\mathbf{A}\mathbf{\Lambda}\mathbf{e}_q\right)^*$$
$$+ \left(\mathbf{e}_k^T\overline{\mathbf{C}}_m^H\mathbf{R}^{-1}\overline{\mathbf{C}}_l\mathbf{e}_q\right)\left(\mathbf{e}_k^T\mathbf{\Lambda}\mathbf{A}^H\mathbf{R}^{-1}\mathbf{A}\mathbf{\Lambda}\mathbf{e}_q\right)^* \quad (11.\text{B}.8)$$
$$+ \left(\mathbf{e}_k^T\mathbf{\Lambda}\mathbf{A}^H\mathbf{R}^{-1}\mathbf{A}\mathbf{\Lambda}\mathbf{e}_q\right)\left(\mathbf{e}_k^T\overline{\mathbf{C}}_m^H\mathbf{R}^{-1}\overline{\mathbf{C}}_l\mathbf{e}_q\right)^*$$
$$+ \left(\mathbf{e}_k^T\overline{\mathbf{C}}_m^H\mathbf{R}^{-1}\mathbf{A}\mathbf{\Lambda}\mathbf{e}_q\right)\left(\mathbf{e}_k^T\mathbf{\Lambda}\mathbf{A}^H\mathbf{R}^{-1}\overline{\mathbf{C}}_l\mathbf{e}_q\right)^*$$

Finally

$$\text{tr}\left(\mathbf{R}^{-1}\frac{\partial \mathbf{R}}{\partial \bar{b}_{ql}}\mathbf{R}^{-1}\frac{\partial \mathbf{R}}{\partial \bar{b}_{km}}\right) = 2\,\text{Re}\left[\left(\mathbf{e}_k^T\mathbf{\Lambda}\mathbf{A}^H\mathbf{R}^{-1}\overline{\mathbf{C}}_l\mathbf{e}_q\right)\left(\mathbf{e}_k^T\overline{\mathbf{C}}_m^H\mathbf{R}^{-1}\mathbf{A}\mathbf{\Lambda}\mathbf{e}_q\right)^*\right] \quad (11.\text{B}.9)$$
$$+ 2\,\text{Re}\left[\left(\mathbf{e}_k^T\overline{\mathbf{C}}_m^H\mathbf{R}^{-1}\overline{\mathbf{C}}_l\mathbf{e}_q\right)\left(\mathbf{e}_k^T\mathbf{\Lambda}\mathbf{A}^H\mathbf{R}^{-1}\mathbf{A}\mathbf{\Lambda}\mathbf{e}_q\right)^*\right]$$

Thus, the corresponding blocks of the FIM are given by

$$\text{FIM}_{\overline{B}_m \overline{B}_l} = 2 \operatorname{Re}\left[(\boldsymbol{\Lambda}\mathbf{A}^H \mathbf{R}^{-1} \overline{\mathbf{C}}_l) \times (\overline{\mathbf{C}}_m^H \mathbf{R}^{-1} \mathbf{A}\boldsymbol{\Lambda})^* + (\overline{\mathbf{C}}_m^H \mathbf{R}^{-1} \overline{\mathbf{C}}_l) \times (\boldsymbol{\Lambda}\mathbf{A}^H \mathbf{R}^{-1} \mathbf{A}\boldsymbol{\Lambda})^*\right]$$

$$= 2 \operatorname{Re}\left[(\mathbf{A}_1^H \overline{\mathbf{C}}_l) \times (\overline{\mathbf{C}}_m^H \mathbf{A}_1)^* + (\overline{\mathbf{C}}_m^H \mathbf{R}^{-1} \overline{\mathbf{C}}_l) \times \mathbf{A}_2^*\right]$$

$$\text{FIM}_{\overline{B}_m \widetilde{B}_l} = 2 \operatorname{Re}\left[(\mathbf{A}_1^H \widetilde{\mathbf{C}}_l) \times (\overline{\mathbf{C}}_m^H \mathbf{A}_1)^* + (\overline{\mathbf{C}}_m^H \mathbf{R}^{-1} \widetilde{\mathbf{C}}_l) \times \mathbf{A}_2^*\right]$$

$$\text{FIM}_{\widetilde{B}_m \widetilde{B}_l} = 2 \operatorname{Re}\left[(\mathbf{A}_1^H \widetilde{\mathbf{C}}_l) \times (\widetilde{\mathbf{C}}_m^H \mathbf{A}_1)^* + (\widetilde{\mathbf{C}}_m^H \mathbf{R}^{-1} \widetilde{\mathbf{C}}_l) \times \mathbf{A}_2^*\right]$$

$$\text{FIM}_{P_m P_l} = 2 \operatorname{Re}\left[(\mathbf{A}_1^H \mathbf{D}_l) \times (\mathbf{D}_m^H \mathbf{A}_1)^* + (\mathbf{D}_m^H \mathbf{R}^{-1} \mathbf{D}_l) \times \mathbf{A}_2^*\right]$$

$$\text{FIM}_{P_m \overline{B}_l} = 2 \operatorname{Re}\left[(\mathbf{A}_1^H \overline{\mathbf{C}}_l) \times (\mathbf{D}_m^H \mathbf{A}_1)^* + (\mathbf{D}_m^H \mathbf{R}^{-1} \overline{\mathbf{C}}_l) \times \mathbf{A}_2^*\right]$$

$$\text{FIM}_{P_m \widetilde{B}_l} = 2 \operatorname{Re}\left[(\mathbf{A}_1^H \widetilde{\mathbf{C}}_l) \times (\mathbf{D}_m^H \mathbf{A}_1)^* + (\mathbf{D}_m^H \mathbf{R}^{-1} \widetilde{\mathbf{C}}_l) \times \mathbf{A}_2^*\right]$$

(11.B.10)

where

$$\mathbf{A}_1 \triangleq \mathbf{R}^{-1} \mathbf{A}\boldsymbol{\Lambda} \quad (11.B.11)$$

$$\mathbf{A}_2 \triangleq \boldsymbol{\Lambda}\mathbf{A}^H \mathbf{R}^{-1} \mathbf{A}\boldsymbol{\Lambda}$$

We can also obtain

$$\operatorname{tr}\left(\mathbf{R}^{-1} \frac{\partial \mathbf{R}}{\Lambda_i} \mathbf{R}^{-1} \frac{\partial \mathbf{R}}{\partial \overline{b}_{km}}\right)$$

$$= \operatorname{tr}\left[\mathbf{R}^{-1} \mathbf{A} \mathbf{e}_i \mathbf{e}_i^T \mathbf{A}^H \mathbf{R}^{-1} (\overline{\mathbf{C}}_m \mathbf{e}_k \mathbf{e}_k^T \boldsymbol{\Lambda}\mathbf{A}^H + \mathbf{A}\boldsymbol{\Lambda} \mathbf{e}_k \mathbf{e}_k^T \overline{\mathbf{C}}_m^H)\right]$$

$$= \operatorname{tr}(\mathbf{R}^{-1} \mathbf{A} \mathbf{e}_i \mathbf{e}_i^T \mathbf{A}^H \mathbf{R}^{-1} \overline{\mathbf{C}}_m \mathbf{e}_k \mathbf{e}_k^T \boldsymbol{\Lambda}\mathbf{A}^H + \mathbf{R}^{-1} \mathbf{A} \mathbf{e}_i \mathbf{e}_i^T \mathbf{A}^H \mathbf{R}^{-1} \mathbf{A}\boldsymbol{\Lambda} \mathbf{e}_k \mathbf{e}_k^T \overline{\mathbf{C}}_m^H)$$

$$= (\mathbf{e}_k^T \boldsymbol{\Lambda}\mathbf{A}^H \mathbf{R}^{-1} \mathbf{A} \mathbf{e}_i)(\mathbf{e}_i^T \mathbf{A}^H \mathbf{R}^{-1} \overline{\mathbf{C}}_m \mathbf{e}_k) + (\mathbf{e}_k^T \overline{\mathbf{C}}_m^H \mathbf{R}^{-1} \mathbf{A} \mathbf{e}_i)(\mathbf{e}_i^T \mathbf{A}^H \mathbf{R}^{-1} \mathbf{A}\boldsymbol{\Lambda} \mathbf{e}_k)$$

$$= (\mathbf{e}_i^T \mathbf{A}^H \mathbf{R}^{-1} \mathbf{A}\boldsymbol{\Lambda} \mathbf{e}_k)^*(\mathbf{e}_i^T \mathbf{A}^H \mathbf{R}^{-1} \overline{\mathbf{C}}_m \mathbf{e}_k) + (\mathbf{e}_i^T \mathbf{A}^H \mathbf{R}^{-1} \mathbf{A}\boldsymbol{\Lambda} \mathbf{e}_k)(\mathbf{e}_i^T \mathbf{A}^H \mathbf{R}^{-1} \overline{\mathbf{C}}_m \mathbf{e}_k)^*$$

$$= 2 \operatorname{Re}\left[(\mathbf{e}_i^T \mathbf{A}^H \mathbf{R}^{-1} \mathbf{A}\boldsymbol{\Lambda} \mathbf{e}_k)^*(\mathbf{e}_i^T \mathbf{A}^H \mathbf{R}^{-1} \overline{\mathbf{C}}_m \mathbf{e}_k)\right]$$

(11.B.12)

Thus, the associated blocks of the FIM are given by

$$\text{FIM}_{\Lambda \overline{B}_m} = 2 \operatorname{Re}\left[(\mathbf{A}^H \mathbf{R}^{-1} \mathbf{A}\boldsymbol{\Lambda})^* \times (\mathbf{A}^H \mathbf{R}^{-1} \overline{\mathbf{C}}_m)\right]$$

$$= 2 \operatorname{Re}\left[(\mathbf{A}^H \mathbf{A}_1)^* \times (\mathbf{A}^H \mathbf{R}^{-1} \overline{\mathbf{C}}_m)\right] \quad (11.B.13)$$

$$\text{FIM}_{\Lambda \widetilde{B}_m} = 2 \operatorname{Re}\left[(\mathbf{A}^H \mathbf{A}_1)^* \times (\mathbf{A}^H \mathbf{R}^{-1} \widetilde{\mathbf{C}}_m)\right]$$

$$\text{FIM}_{\Lambda P_m} = 2 \operatorname{Re}\left[(\mathbf{A}^H \mathbf{A}_1)^* \times (\mathbf{A}^H \mathbf{R}^{-1} \mathbf{D}_m)\right]$$

The CRB is obtained by inverting the complete Fisher information matrix.

Appendix 11.C DPD Performance Analysis for Unknown Waveforms

In this appendix, we introduce a small error analysis of the proposed algorithm for unknown signals. Although the algorithm does not impose any requirements on the signals' statistics, in order to facilitate the analysis, we assume that the transmitted signals are statistically independent, zero-mean, jointly Gaussian, and therefore satisfy

$$E\left[\bar{s}_q(j, k)\bar{s}_p^H(m, l)\right] = \Lambda_{q,q}(j)\,\delta_{q,p}\,\delta_{j,m}\,\delta_{k,l} \qquad (11.\text{C}.1)$$

For analysis convenience, we review briefly some of the definitions of the proposed algorithm.

Recall that the algorithm is based on maximizing the largest eigenvalue of $\mathbf{D}$, where

$$\mathbf{D} = \mathbf{H}^H \mathbf{T} \mathbf{H} \qquad (11.\text{C}.2)$$

$$\mathbf{T} \triangleq \sum_{j=1}^{J} \sum_{q=1}^{Q} \boldsymbol{\Gamma}^H(j)\mathbf{u}_q(j)\mathbf{u}_q^H(j)\boldsymbol{\Gamma}(j)$$

and $\mathbf{u}_q(j)$ denotes the qth eigenvector of the sample covariance matrix, $\hat{\mathbf{R}}(j)$, corresponding to the qth eigenvalue γ_q, where we assume that $\gamma_1 \geq \gamma_2 \geq \ldots \geq \gamma_{ML}$.

Using the eigendecomposition of the Hermitian matrix $\mathbf{D}$, we have

$$\mathbf{D} = \mathbf{W}\boldsymbol{\Phi}\mathbf{W}^H$$

$$\mathbf{W} \triangleq [\mathbf{w}_1, \ldots, \mathbf{w}_L] \qquad (11.\text{C}.3)$$

$$\boldsymbol{\Phi} \triangleq diag\{\lambda_1, \ldots, \lambda_L\}$$

where $\mathbf{w}_j, \lambda_j$ are the jth eigenvector/eigenvalue pair, and $\lambda_1 \geq \ldots \geq \gamma_L$. Thus, we have $\lambda_{max} \equiv \lambda_1$.

The computation of the covariance of $\mathbf{p}_q$ depends on the estimated eigenvectors $\{\hat{\mathbf{u}}_q, 1 \leq q \leq Q\}$. In the ideal case $\hat{\mathbf{u}}_q = \mathbf{u}_q$; $1 \leq q \leq Q$, and the local maxima of the cost function occur at the true source locations $\{\mathbf{p}_q\}_{q=1}^{Q}$. In reality, the eigenvectors $\hat{\mathbf{u}}_q$ are perturbed, and therefore the local maxima occur at locations $\hat{\mathbf{p}}_q$ that are not identical to the true locations. The covariance of the perturbations of $\hat{\mathbf{p}}_q$ is related to the covariance perturbations of the $\hat{\mathbf{u}}_q$ by the following equation taken from [20, p. 534, (15)]:

$$\text{cov}(\hat{\mathbf{p}}_q) = \left(\frac{\partial^2 \lambda_1}{\partial \mathbf{p}^2}\right)^{-1} \left(\frac{\partial^2 \lambda_1}{\partial \mathbf{p}\partial \boldsymbol{\xi}}\right) \text{cov}(\hat{\boldsymbol{\xi}}) \left(\frac{\partial^2 \lambda_1}{\partial \mathbf{p}\partial \boldsymbol{\xi}}\right)^T \left(\frac{\partial^2 \lambda_1}{\partial \mathbf{p}^2}\right)^{-T}$$

$$\boldsymbol{\xi} \triangleq \left(\boldsymbol{\xi}_1^T, \ldots, \boldsymbol{\xi}_Q^T\right)^T \qquad (11.\text{C}.4)$$

$$\boldsymbol{\xi}_q \triangleq \left[\boldsymbol{\xi}_q^T(0), \ldots, \boldsymbol{\xi}_q^T(J)\right]^T$$

$$\boldsymbol{\xi}_q(j) \triangleq \left[\bar{\mathbf{u}}_q^T(j), \tilde{\mathbf{u}}_q^T(j)\right]^T$$

where

$$\overline{\mathbf{u}}_k(j) \triangleq \text{Re}[\mathbf{u}_k(j)]; \quad \tilde{\mathbf{u}}_k(j) \triangleq \text{Im}[\mathbf{u}_k(j)]$$

We note that the right-hand side of (11.C.4) has to be evaluated at the true values of $\mathbf{p}_q$, $\mathbf{u}_q(j)$.

The (k, l) submatrix $\text{cov}(\hat{\boldsymbol{\xi}}_k, \hat{\boldsymbol{\xi}}_l)$ of the covariance matrix $\text{cov}(\boldsymbol{\xi})$ is given by [20, 22]:

$$\text{cov}(\hat{\boldsymbol{\xi}}_k, \hat{\boldsymbol{\xi}}_l) = \frac{1}{2} \begin{Bmatrix} \text{Re}[\text{cov}(\hat{\mathbf{u}}_k, \hat{\mathbf{u}}_l) + \text{cov}(\hat{\mathbf{u}}_k, \hat{\mathbf{u}}_l^*)] & -\text{Im}[\text{cov}(\hat{\mathbf{u}}_k, \hat{\mathbf{u}}_l) - \text{cov}(\hat{\mathbf{u}}_k, \hat{\mathbf{u}}_l^*)] \\ \text{Im}[\text{cov}(\hat{\mathbf{u}}_k, \hat{\mathbf{u}}_l) + \text{cov}(\hat{\mathbf{u}}_k, \hat{\mathbf{u}}_l^*)] & \text{Re}[\text{cov}(\hat{\mathbf{u}}_k, \hat{\mathbf{u}}_l) - \text{cov}(\hat{\mathbf{u}}_k, \hat{\mathbf{u}}_l^*)] \end{Bmatrix}$$

(11.C.5)

where $\text{cov}(\hat{\mathbf{u}}_k, \hat{\mathbf{u}}_l)$ and $\text{cov}(\hat{\mathbf{u}}_k, \hat{\mathbf{u}}_l^*)$ are block diagonal matrices where the jth block is given by [20]:

$$\text{cov}[\hat{\mathbf{u}}_k(j), \hat{\mathbf{u}}_l(j)] = \frac{\gamma_k(j)}{K} \delta_{kl} \sum_{\substack{i=1 \\ i \neq k}}^{ML} \frac{\gamma_i(j)}{[\gamma_k(j) - \gamma_i(j)]^2} \mathbf{u}_i(j) \mathbf{u}_i^H(j) \quad (11.C.6)$$

$$\text{cov}[\hat{\mathbf{u}}_k(j), \hat{\mathbf{u}}_l^*(j)] = -(1 - \delta_{kl}) \frac{\gamma_k(j) \gamma_l(j)}{K[\gamma_k(j) - \gamma_l(j)]^2} \mathbf{u}_l(j) \mathbf{u}_k^T(j)$$

For convenience, we use the definitions

$$\boldsymbol{\Omega} \triangleq \left(\frac{\partial^2 \lambda_1}{\partial \mathbf{p}^2} \right) \quad (11.C.7)$$

$$\boldsymbol{\Psi} \triangleq \left(\frac{\partial^2 \lambda_1}{\partial \mathbf{p} \partial \boldsymbol{\xi}} \right) \text{cov}(\hat{\boldsymbol{\xi}}) \left(\frac{\partial^2 \lambda_1}{\partial \mathbf{p} \partial \boldsymbol{\xi}} \right)^T$$

11.C.1 Expressions for Ω

We first note that

$$\frac{\partial \lambda_1}{\partial p_k} = \sum_{m,n=1}^{L} \frac{\overline{\partial \lambda_1}}{\partial \mathbf{D}_{m,n}} \frac{\partial \mathbf{D}_{m,n}}{\partial p_k} = \sum_{m,n=1}^{L} \mathbf{W}_{n,1} \mathbf{W}_{m,1}^* \frac{\partial \mathbf{D}_{m,n}}{\partial p_k} \quad (11.C.8)$$

where p_k is the kth element of the vector $\mathbf{p}$, and $\overline{\partial f / \partial \alpha}$ is the Brandwood complex derivative [21]; that is, the derivative of a real valued function of a complex variable and its conjugate $f(\alpha, \alpha^*)$ is taken with respect to α regarding α^* as a constant. We used the results obtained in [22, (A.12)] to express $\partial \lambda_1 / \partial \mathbf{D}_{m,n}$.

Using (11.C.8), we can express the (k, l) element of $\boldsymbol{\Omega}$ as

Appendix 11.C DPD Performance Analysis for Unknown Waveforms

$$\frac{\partial^2 \lambda_1}{\partial p_l \partial p_k} = \sum_{m,n=1}^{L} \frac{\partial \mathbf{W}_{n,1} \mathbf{W}_{m,1}^*}{\partial p_l} \frac{\partial \mathbf{D}_{m,n}}{\partial p_k} + \mathbf{W}_{n,1} \mathbf{W}_{m,1}^* \frac{\partial^2 \mathbf{D}_{m,n}}{\partial p_l \partial p_k} \quad (11.\text{C}.9)$$

Using (11.C.2) yields the following expressions for the derivatives

$$\frac{\partial \mathbf{D}_{m,n}}{\partial p_k} = 2 \operatorname{Re}\left[(\mathbf{e}_m \otimes \mathbf{1}_M)^T \frac{\partial \mathbf{T}}{\partial p_k} (\mathbf{e}_n \otimes \mathbf{1}_M) \right] \quad (11.\text{C}.10)$$

$$\frac{\partial^2 \mathbf{D}_{m,n}}{\partial p_l \partial p_k} = 2 \operatorname{Re}\left[(\mathbf{e}_m \otimes \mathbf{1}_M)^T \frac{\partial \mathbf{T}}{\partial p_l \partial p_k} (\mathbf{e}_n \otimes \mathbf{1}_M) \right]$$

and

$$\frac{\partial \mathbf{W}_{l,1}}{\partial p_k} = \sum_{m,n} \frac{\overline{\partial} \mathbf{W}_{l,1}}{\overline{\partial} \mathbf{D}_{m,n}} \frac{\partial \mathbf{D}_{m,n}}{\partial p_k} = \sum_{m,n} \left(\sum_{j=2}^{L} \frac{\mathbf{W}_{n,1} \mathbf{W}_{m,j}^* \mathbf{W}_{l,j}}{\lambda_1 - \lambda_j} \right) \frac{\partial \mathbf{D}_{m,n}}{\partial p_k}$$
$$(11.\text{C}.11)$$

where $\overline{\partial} \mathbf{W}_{l,1} / \overline{\partial} \mathbf{D}_{m,n}$ was expressed using the results of [22, (A.18)].

Substituting (11.C.10), (11.C.11) in (11.C.9) yields an explicit expression for

$$\frac{\partial^2 \lambda_1}{\partial p_l \partial p_k}$$

11.C.2 Expressions for Ψ

We first note that

$$\frac{\partial^2 \lambda_1}{\partial \boldsymbol{\xi}_l \partial p_k} = \left[\left(\frac{\partial^2 \lambda_1}{\partial \overline{\mathbf{u}}_l \partial p_k} \right)^T, \left(\frac{\partial^2 \lambda_1}{\partial \tilde{\mathbf{u}}_l \partial p_k} \right)^T \right]^T \quad (11.\text{C}.12)$$

using (11.C.8), we get

$$\frac{\partial^2 \lambda_1}{\partial \overline{\mathbf{u}}_l \partial p_k} = \sum_{m,n} \frac{\partial \mathbf{W}_{n,1} \mathbf{W}_{m,1}^*}{\partial \overline{\mathbf{u}}_l} \frac{\partial \mathbf{D}_{mn}}{\partial p_k} + \mathbf{W}_{n,1} \mathbf{W}_{m,1}^* \frac{\partial^2 \mathbf{D}_{mn}}{\partial \overline{\mathbf{u}}_l \partial p_k} \quad (11.\text{C}.13)$$

$$\frac{\partial^2 \lambda_1}{\partial \tilde{\mathbf{u}}_l \partial p_k} = \sum_{m,n} \frac{\partial \mathbf{W}_{n,1} \mathbf{W}_{m,1}^*}{\partial \tilde{\mathbf{u}}_l} \frac{\partial \mathbf{D}_{mn}}{\partial p_k} + \mathbf{W}_{n,1} \mathbf{W}_{m,1}^* \frac{\partial^2 \mathbf{D}_{mn}}{\partial \tilde{\mathbf{u}}_l \partial p_k}$$

where the derivatives in (11.C.13) are given by

$$\frac{\partial \mathbf{W}_{l,1}}{\partial \overline{\mathbf{u}}_k} = \sum_{m,n} \frac{\overline{\partial} \mathbf{W}_{l,1}}{\partial \mathbf{D}_{m,n}} \frac{\partial \mathbf{D}_{m,n}}{\partial \overline{\mathbf{u}}_k}$$

$$\frac{\partial \mathbf{W}_{l,1}}{\partial \tilde{\mathbf{u}}_k} = \sum_{m,n} \frac{\overline{\partial} \mathbf{W}_{l,1}}{\partial \mathbf{D}_{m,n}} \frac{\partial \mathbf{D}_{m,n}}{\partial \tilde{\mathbf{u}}_k} \qquad (11.C.14)$$

$$\frac{\partial \mathbf{D}_{m,n}}{\partial \overline{\mathbf{u}}_l} = \left[\frac{\partial \mathbf{D}_{m,n}}{\partial \overline{\mathbf{u}}_l(0)}, \frac{\partial \mathbf{D}_{m,n}}{\partial \overline{\mathbf{u}}_l(1)}, \ldots, \frac{\partial \mathbf{D}_{m,n}}{\partial \overline{\mathbf{u}}_l(J)} \right]$$

$$\frac{\partial \mathbf{D}_{m,n}}{\partial \tilde{\mathbf{u}}_l} = \left[\frac{\partial \mathbf{D}_{m,n}}{\partial \tilde{\mathbf{u}}_l(0)}, \frac{\partial \mathbf{D}_{m,n}}{\partial \tilde{\mathbf{u}}_l(1)}, \ldots, \frac{\partial \mathbf{D}_{m,n}}{\partial \tilde{\mathbf{u}}_l(J)} \right]$$

and $\overline{\partial} \mathbf{W}_{l,1}/\partial \mathbf{D}_{m,n}$ can be found in (11.C.11).

We also have

$$\frac{\partial \mathbf{D}_{mn}}{\partial \overline{\mathbf{u}}_l(j)} = \mathbf{u}_l^H(j) \mathbf{\Gamma}(j) \mathbf{E}_{n,m} \mathbf{\Gamma}^H(j) + \mathbf{u}_l^T(j) \mathbf{\Gamma}^*(j) \mathbf{E}_{m,n} \mathbf{\Gamma}^T(j)$$

$$\frac{\partial \mathbf{D}_{mn}}{\partial \tilde{\mathbf{u}}_l(j)} = i \left[\mathbf{u}_l^H(j) \mathbf{\Gamma}(j) \mathbf{E}_{n,m} \mathbf{\Gamma}^H(j) - \mathbf{u}_l^T(j) \mathbf{\Gamma}^*(j) \mathbf{E}_{m,n} \mathbf{\Gamma}^T(j) \right] \quad (11.C.15)$$

$$\mathbf{E}_{nm} \triangleq (\mathbf{e}_n \otimes \mathbf{1}_M)(\mathbf{e}_m \otimes \mathbf{1}_M)^T$$

and

$$\frac{\partial^2 \mathbf{D}_{m,n}}{\partial \overline{\mathbf{u}}_l \partial p_k} = \left[\frac{\partial^2 \mathbf{D}_{mn}}{\partial \overline{\mathbf{u}}_l(0) \partial p_k}, \ldots, \frac{\partial^2 \mathbf{D}_{mn}}{\partial \overline{\mathbf{u}}_l(J) \partial p_k} \right] \qquad (11.C.16)$$

$$\frac{\partial^2 \mathbf{D}_{mn}}{\partial \tilde{\mathbf{u}}_l \partial p_k} = \left[\frac{\partial^2 \mathbf{D}_{mn}}{\partial \tilde{\mathbf{u}}_l(0) \partial p_k}, \ldots, \frac{\partial^2 \mathbf{D}_{mn}}{\partial \tilde{\mathbf{u}}_l(J) \partial p_k} \right]$$

using (11.C.10), we can express each of the terms in the above equation as

$$\frac{\partial^2 \mathbf{D}_{m,n}}{\partial \overline{\mathbf{u}}_l(j) \partial p_k} = \mathbf{u}_l^H(j) \left[\mathbf{\Gamma}(j) \mathbf{E}_{nm} \frac{\partial \mathbf{\Gamma}^H(j)}{\partial p_k} + \frac{\partial \mathbf{\Gamma}(j)}{\partial p_k} \mathbf{E}_{nm} \mathbf{\Gamma}^H(j) \right]$$

$$+ \mathbf{u}_l^T(j) \left[\mathbf{\Gamma}^*(j) \mathbf{E}_{nm} \frac{\partial \mathbf{\Gamma}^T(j)}{\partial p_k} + \frac{\partial \mathbf{\Gamma}^*(j)}{\partial p_k} \mathbf{E}_{mn} \mathbf{\Gamma}^T(j) \right] \quad (11.C.17)$$

$$\frac{\partial^2 \mathbf{D}_{mn}}{\partial \tilde{\mathbf{u}}_l(j) \partial p_k} = i \mathbf{u}_l^H(j) \left[\mathbf{\Gamma}(j) \mathbf{E}_{nm} \frac{\partial \mathbf{\Gamma}^H(j)}{\partial p_k} + \frac{\partial \mathbf{\Gamma}(j)}{\partial p_k} \mathbf{E}_{nm} \mathbf{\Gamma}^H(j) \right]$$

$$- \mathbf{u}_l^T(j) \left[\mathbf{\Gamma}^*(j) \mathbf{E}_{nm} \frac{\partial \mathbf{\Gamma}^T(j)}{\partial p_k} + \frac{\partial \mathbf{\Gamma}^*(j)}{\partial p_k} \mathbf{E}_{mn} \mathbf{\Gamma}^T(j) \right]$$

Substituting (11.C.14) and (11.C.17) into (11.C.13), and using straightforward algebraic manipulations yields

Appendix 11.C DPD Performance Analysis for Unknown Waveforms

$$\frac{\partial^2 \lambda_1}{\partial \overline{\mathbf{u}}_l(j) \partial p_k} = \mathbf{u}_l^H(j)\mathbf{Z}_k + \mathbf{u}_l^T(j)\mathbf{Z}_k^* = 2\,\text{Re}\left[\mathbf{u}_l^H(j)\mathbf{Z}_k\right]$$

$$\frac{\partial^2 \lambda_1}{\partial \tilde{\mathbf{u}}_l(j) \partial p_k} = i\left[\mathbf{u}_l^H(j)\mathbf{Z}_k + \mathbf{u}_l^T(j)\mathbf{Z}_k^*\right] = -2\,\text{Im}\left[\mathbf{u}_l^H(j)\mathbf{Z}_k\right]$$

$$\mathbf{Z}_k \triangleq \sum_{m=1}^{L}\sum_{n=1}^{L} \mathbf{X}_{n,m}^k + \mathbf{W}_{n1}(n)\mathbf{W}_{m1}^* \mathbf{Y}_{n,m}^k$$

$$\mathbf{X}_{nm}^k \triangleq \sum_{r=1}^{L}\sum_{l=1}^{L}\left[\sum_{j=2}^{L}\frac{\mathbf{W}_{l1}\mathbf{W}_{rj}^*(\mathbf{W}_{nj}\mathbf{W}_{m1}^* + \mathbf{W}_{mj}\mathbf{W}_{n1})}{\lambda_1 - \lambda_j}\right]\mathbf{\Gamma E}_{lr}\mathbf{\Gamma}^H(\mathbf{e}_m \otimes \mathbf{J}_M)\frac{\partial \mathbf{T}}{\partial p_k}(\mathbf{e}_n \otimes \mathbf{J}_M)^T$$

$$\mathbf{Y}_{nm}^k \triangleq \mathbf{\Gamma E}_{nm}\frac{\partial \mathbf{\Gamma}^H}{\partial p_k} + \frac{\partial \mathbf{\Gamma}}{\partial p_k}\mathbf{E}_{nm}\mathbf{\Gamma}^H$$

(11.C.18)

Recalling that $\gamma_k(j) = \eta$ for all $Q + 1 \leq k \leq ML$, we can express the (m, n) element of $\mathbf{\Psi}$ using [20] as:

$$\mathbf{\Psi}_{m,n} = \frac{\partial^2 \lambda_1}{\partial \xi_m \partial p_k}\,\text{cov}(\hat{\xi}_k, \hat{\xi}_l)\,\frac{\partial^2 \lambda_1}{\partial \xi_n \partial p_l} \qquad (11.C.19)$$

$$= \frac{2\eta}{N}\,\text{Re}\left\{\sum_{j=1}^{J}\sum_{k=1}^{Q}\sum_{j=Q+1}^{ML}\frac{\gamma_k(j)}{[\gamma_k(j) - \eta]^2}\mathbf{u}_k^H(j)\mathbf{Z}_m\mathbf{u}_l(j)\mathbf{u}_l^H(j)\mathbf{Z}_n\mathbf{u}_k(j)\right\}$$

In summary, using the result of (11.C.19) and (11.C.9), we can express the matrix $\text{cov}(\hat{\mathbf{p}}_i)$ in (11.C.4).

11.C.3 Special Case: Uniform Linear Array

So far, we have considered a general array configuration. In this section, we obtain the required expressions for a uniform linear array with M elements. The coordinates of the mth element, at the lth base station, are $x = (m - 1)\Delta \cos \phi_l$; $y = (m - 1)\Delta \sin \phi_l$, where Δ is the elements spacing and ϕ_l is the counterclockwise rotation of the array baseline with respect to the x-axis.

The steering vector of the lth base station takes the form:

$$\mathbf{a}_l(\mathbf{p}) = \begin{pmatrix} 1 & e^{jk\Delta \cos \theta'_l} & \cdots & e^{jk(M-1)\Delta \cos \theta'_l} \end{pmatrix}^T \qquad (11.C.20)$$

where $k = 2\pi/\lambda$ is the signal wave number, λ is the wavelength, and θ'_l is direction of arrival (relative to the array baseline) of a signal emitted from (x_t, y_t) to the lth base station located at (x_l, y_l). In addition, the delay between the emitter and the base station is $\tau_l = d_l/c$, where $d_l = \sqrt{(x_t - x_l)^2 + (y_t - y_l)^2}$ is the distance between the emitter and the lth base station, and c is the speed of propagation.

Note that the angle of arrival with respect to the x-axis at the lth base station is $\theta_l = \theta'_l + \phi_l$.

The steering vector derivatives with respect to the x-axis are

$$\frac{d\mathbf{a}_l}{dx} = -jk\Delta \sin \theta_l' \frac{d\theta_l'}{dx} \mathbf{a}_l^T \overline{\mathbf{M}}, \quad \frac{d\mathbf{a}_l}{dy} = -jk\Delta \sin \theta_l' \frac{d\theta_l'}{dy} \mathbf{a}_l^T \overline{\mathbf{M}} \quad (11.C.21)$$

where $\overline{\mathbf{M}} \triangleq diag(0, 1, \ldots, M-1)$ and

$$\frac{d\theta_l'}{dx} = \frac{d\theta_l}{dx} = -\frac{\sin \theta_l}{d_l}, \quad \frac{d\theta_l'}{dy} = \frac{d\theta_l}{dy} = \frac{\cos \theta_l}{d_l} \quad (11.C.22)$$

We also have

$$\frac{d^2 \mathbf{a}_l}{dx^2} = -j\frac{k\Delta}{d_l^2} \mathbf{a}_l^T \overline{\mathbf{M}} \left[\sin(\theta_l' + \theta_l) \sin(\theta_l) \mathbf{I} - jk\Delta \sin^2 \theta_l' \sin^2 \theta_l \overline{\mathbf{M}} \right]$$

$$\frac{d^2 \mathbf{a}_l}{dy^2} = -j\frac{k\Delta}{d_l^2} \mathbf{a}_l^T \overline{\mathbf{M}} \left[\cos(\theta_l' + \theta_l) \cos(\theta_l) \mathbf{I} - jk\Delta \sin^2 \theta_l' \cos^2 \theta_l \overline{\mathbf{M}} \right] \quad (11.C.23)$$

$$\frac{d^2 \mathbf{a}_l}{dx\,dy} = j\frac{k\Delta}{d_l^2} \mathbf{a}_l^T \overline{\mathbf{M}} \left[\sin(\theta_l' + \theta_l) \cos(\theta_l) \mathbf{I} - j(1/2)k\Delta \sin^2 \theta_l' \sin 2\theta_l \overline{\mathbf{M}} \right]$$

This concludes the performance analysis.

Appendix 11.D The Frequency Domain Representation of Finite Length Observations

Consider the observation of $s(t)$, $0 \le t \le T_1$ at a given base station and the observation $s(t - D)$, $0 \le t \le T_1$ at a different base station. The Fourier transform of these signals are given by

$$S_1 = \int_0^{T_1} s(t) e^{-i\omega t} \, dt \quad (11.D.1)$$

$$S_2 = \int_0^{T_1} s(t - D) e^{-i\omega t} \, dt = e^{-i\omega D} \int_{-D}^{T_1 - D} s(\sigma) e^{-i\omega \sigma} \, d\sigma$$

The relation between S_1 and S_2 is given by

$$S_2 = e^{-i\omega D}(S_1 + \Delta) \quad (11.D.2)$$

$$\Delta \triangleq \int_{-D}^{0} s(\sigma) e^{-i\omega \sigma} \, d\sigma - \int_{T_1 - D}^{T_1} s(\sigma) e^{-i\omega \sigma} \, d\sigma$$

Appendix 11.D The Frequency Domain Representation of Finite Length Observations

In the main text, we used the approximation $S_2 \cong e^{-i\omega D} S_1$, under the assumption that $\Delta \ll S_1$. It is easy to verify that the energy relation between S_1 and Δ is given by

$$\frac{E(|S_1|^2)}{E(|\Delta|^2)} = \frac{T_1}{2D} \qquad (11.D.3)$$

for a random signal with flat spectral density. Thus, in order to get a ratio of 20 dB, the required observation length, T_1, should be $200D$. In the text, each snapshot is of length $T_1 = T/K$.

CHAPTER 12
Direction-of-Arrival Estimation Under Uncertainty
Visa Koivunen and Esa Ollila

12.1 Introduction

Smart antennas are commonly used in applications such as wireless communications, satellite navigation, radar, and electronic warfare systems. In particular, DOA estimates are needed in many key tasks, such as beamforming and interference cancellation, emitter or target localization, and tracking.

Signals are modeled either as deterministic sequences or random processes, and noises are typically modeled as random processes. In the case of stochastic models, statistical methods are needed for characterizing signal properties, performance of the estimation procedures, and uncertainties present in the system. In the case of deterministic models and methods, tools from matrix perturbation theory also are applied.

A key design goal for signal processing algorithms in smart antenna systems is to achieve optimal performance in a given application. Statistics is also required for defining the criteria for optimality; for example, the minimum mean square error or the maximum of the likelihood function. The optimization is typically performed using strict assumptions on the propagation environment, source signals, sensor array configuration, interference, and noise. A shortcoming of the optimal estimation procedures is that they are extremely sensitive, even to small deviations from the assumed model. In reality, the underlying assumptions may not be valid, and a significant degradation from the optimal performance is experienced. Procedures that perform reliably under uncertainty are commonly called robust [1–5]. Robust methods typically trade off optimality for reliable performance under mismatches in signal and noise models, as well as under less-than-ideal propagation environments.

In this chapter, we consider uncertainty in signal and noise models in direction of arrival estimation. In practice, the uncertainties may be caused by errors in array response or calibration; impairment in the signal waveforms caused by the propagation environment; lack of data; the time-varying nature of the propagation environment; interferers; or misspecified noise model.

In particular, so-called high-resolution direction finding methods are considered in this chapter. Such methods extensively use array output covariance matrices, the subspaces spanned by its eigenvectors, and the projection matrices to these subspaces. This is justified, since the subspaces are related to the subspaces spanned by the array steering vectors that contain the unknown angles of arrival. Hence, we pay special attention to the covariance matrix and its eigenvalues. Subspace methods and maximum likelihood methods are prominent examples of techniques that rely on subspaces and projection matrices.

This chapter is organized as follows. First, the basic signal model and the underlying assumptions used in estimating the angles of arrival are presented. The uncertainties in modeling emitted signals, array configuration, and propagation environment are considered. Some array configurations and direction-of-arrival estimation methods that are robust in the face of such uncertainties are briefly reviewed. The uncertainties in noise modeling are considered next. Robust procedures that are close to optimal in the presence of Gaussian noise, and highly robust in non-Gaussian noise, are presented. The robustness of these procedures is described using the concept of influence functions. Finally, some examples of robust angle-of-arrival estimation are given.

12.2 Basic DOA Estimation Signal Model

In the following section, we describe the basic signal model and underlying assumptions used in DOA estimation. ULA is assumed for the sake of clarity. We assume the standard narrowband low-rank signal model of K incoherent source signals impinging to an array of M sensors ($K < M$). The sources are assumed to be narrowband and to originate from a point source. The received signal vector $\mathbf{z}(n)$ is a $M \times 1$ complex vector given by

$$\mathbf{z}(n) = A\mathbf{s}(n) + \mathbf{w}(n) \tag{12.1}$$

where A is an $M \times K$ matrix, such that

$$A = [\mathbf{a}(\theta_1), \mathbf{a}(\theta_2), \ldots, \mathbf{a}(\theta_K)]$$

with $\mathbf{a}(\theta_k)$ being the $M \times 1$ array steering vector corresponding to the DOA θ_k of the kth signal. The K-vector

$$\mathbf{s}(n) = [s_1(n), s_2(n), \ldots, s_K(n)]^T$$

is the vector of incident signals, and $\mathbf{w}(n)$ is the $M \times 1$ complex noise vector.

It is assumed that for any collection of K distinct θ_i, the matrix A has full rank. When the array is a ULA, the array steering vector is given by

$$\mathbf{a}(\theta) = \left[1, e^{2\pi j(d/\lambda)\cos(\theta)}, \ldots, e^{2\pi j(M-1)(d/\lambda)\cos(\theta)}\right]^T$$

where d denotes the element spacing, and λ denotes the wavelength. The angle is here expressed against the line connecting the array elements. In case the angle is measured against the broadside of the array, $-\sin(\theta)$ should be used instead. The elements of the steering vector depend on the array structure. In the case of ULA

12.2 Basic DOA Estimation Signal Model

with identical array elements, the matrix A becomes a Vandermonde matrix. The system arrangement is illustrated in Figure 12.1.

In case of uniform circular array, the elements of the steering vector are of form

$$\mathbf{a}(\theta, \phi) = \begin{pmatrix} e^{j\omega \frac{r}{c} \sin \phi \cos(\theta - \gamma_0)} \\ e^{j\omega \frac{r}{c} \sin \phi \cos(\theta - \gamma_1)} \\ \vdots \\ e^{j\omega \frac{r}{c} \sin \phi \cos(\theta - \gamma_{M-1})} \end{pmatrix}$$

where $\omega = 2\pi \cdot f'$ is the angular frequency, r is the radius of the array,

$$\gamma_i = \frac{2\pi i}{M}$$

is the angular position of the element (counted in counterclockwise manner from x-axis), and c is the speed of light. The azimuth and elevation angles are denoted by θ_l and ϕ_l, respectively.

Assume that the signal is a zero-mean wide sense stationary time series (i.e., stochastic), and the noise is spatially and temporally white and uncorrelated with the signals. Then, the covariance matrix of $\mathbf{z}(n)$ (if it exists) is

$$\Sigma = E[\mathbf{z}(n)\mathbf{z}^H(n)] = A\Sigma_s A^H + \sigma^2 I$$

where $\Sigma_s = E[\mathbf{s}(n)\mathbf{s}^H(n)]$ is the $K \times K$ signal covariance matrix of full rank, with the signals assumed to be incoherent; σ^2 is the noise variance; superscript H

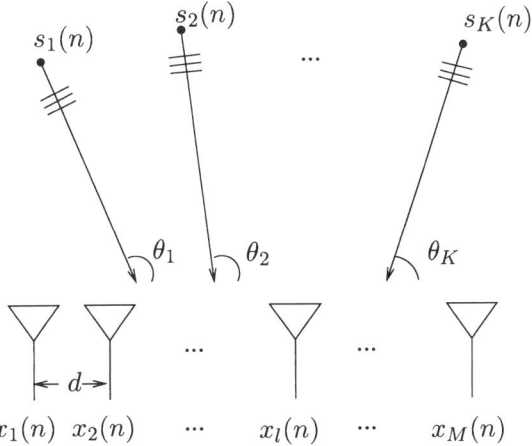

Figure 12.1 A uniform linear array of M sensors receiving plane waves from K farfield point sources.

denotes the Hermitian transpose; and I denotes the identity matrix. The eigenvalue decomposition of the array covariance matrix is given by

$$\Sigma = U \Lambda U^H$$

where Λ is a diagonal matrix of eigenvalues, and matrix U contains the corresponding eigenvectors. If $M > K$, then we have a low-rank signal model, and the $M - K$ smallest eigenvalues of Σ are equal to σ^2. The corresponding eigenvectors are orthogonal to the columns of the matrix A. These $M - K$ eigenvectors U_n span the *noise subspace*, and the K eigenvectors U_s corresponding to the K largest eigenvalues span the *signal subspace*. The same subspace also is spanned by the columns of A. This property is explicitly exploited in high-resolution methods, such as MUSIC and ESPRIT, as well as in maximum likelihood methods where projection matrices to signal or noise subspaces are employed. For example, the MUSIC method is based on the property that the signals within the signal subspace are orthogonal to the entire noise subspace spanned by U_n. In terms of matrix notation, this may be expressed by MUSIC pseudospectrum

$$V_M(\theta) = \frac{1}{\mathbf{a}^H(\theta_i) U_n U_n^H \mathbf{a}(\theta_i)}$$

at the correct DOAs θ_i, $i = 1, \ldots, K$. For other angles $\theta \neq \theta_i$, the denominator in the above expression is greater than zero. Estimates of the K angles of arrival may be found from the K highest peaks of the pseudospectrum.

The classical DOA estimation techniques based on spatial spectrum estimation, the MVDR method, as well as the high-resolution methods including different subspace and maximum likelihood methods, extensively employ the array covariance matrix and its structure. The departures from the underlying assumptions both in signal and noise models are reflected in the array covariance matrix and its eigenvalue-eigenvector decomposition, for example. Hence, the covariance matrix Σ will play a crucial role in our study. The uncertainties may cause additional variance in the DOA estimates, and they may lead to highly biased estimates and reduced resolution of the array. Moreover, some signals impinging the array may remain completely undetected. Estimating the number of signals and the dimensions of noise and signal subspaces is also an important issue. It is closely related to the problem of reliably estimating noise, variance, and eigenvalues, even though the underlying assumptions do not necessarily hold. Commonly, methods like MDL [6] are employed. Estimation of number of signals in a robust manner is not considered here due to lack of space.

12.3 Uncertainties in Signal Model and System Parameters

In this section, the impact of signal modeling errors, as well as departures from assumptions in the array configuration and propagation environment, are considered. These errors typically cause systematic error (bias) in DOA estimates, excess variance of the estimates, and loss of high-resolution property. This section addresses the problem briefly, since the issue of performance in the presence of modeling errors has been considered in recent textbooks [1, 5]. In particular, the impacts of modeling errors on beamforming, and spatial spectrum estimation, have been extensively studied.

12.3.1 Sources of Uncertainty

Some commonly occurring modeling errors are listed next.

- *Signal model errors:* Signals may not be narrowband, they may not originate from a point source, emitters may not be in the farfield, and planewave assumption is not valid;
- *Array model errors:* Uncertainty in the gain and phase responses of the sensors, uncertainty or changes in the element positions, and mutual coupling;
- *Propagation model errors:* The media may not be homogeneous, the propagation conditions may vary (e.g., due to weather), the scattering environment may be rich or change, there may not be a line-of-sight component, there may be local scatterers, and there may be correlated signals caused by multipath.

Some of the errors are angle-dependent and some angle-independent [7]. Moreover, too few observations may be available, which is a problem, since sample estimates of the array covariance matrix have to be used when the theoretical covariance matrix is not known. In particular, subspace methods may suffer from samples sizes that are too small, since both the signal covariance matrix and array covariance matrix are assumed to be of full-rank, and a sufficient number of samples are needed to build up the rank. All these errors cause perturbations in the array covariance matrix, and consequently, to the subspaces spanned by the eigenvectors. These perturbations obviously lead to errors in angle-of-arrival estimates. The impacts of these errors in terms of bias and variance are considered in detail for different algorithms [1, 7, 8], where expressions for the error also are derived. For small enough errors, first-order Taylor series formulas [8] are found, and CRB is established.

12.3.2 Improving the Robustness

Obviously, some of the array and propagation model errors may be handled by appropriate calibration of the array. However, the conditions where the calibration was handled may be significantly different from the conditions where the smart antenna system is deployed. Autocalibration procedures [9] also exist, which allow calibration of the array when it is being deployed. It is well-known that some array configurations are more sensitive to calibration errors than others.

For example, the two subarray configurations used in the ESPRIT algorithm require only partial calibration, and are more robust against calibration errors compared to the MUSIC algorithm, which assumes accurately calibrated arrays. Recently, a technique called RARE was proposed. It is more flexible than ESPRIT in terms of allowed subarray configurations, and requires only partial array calibration [10].

In case the signals are not narrowband, ambiguities in the estimates may follow. This is due to the fact that different frequencies may be differently delayed. As a result, an interval of potential angle estimates is available instead of point estimates. This may be handled by processing the data in the frequency domain, subdividing the problem into many narrowband problems, or by using space-time processing, where a tapped delay line of L coefficients is associated with each array element. Each space-time snapshot is then an $ML \times 1$ vector.

The departures from the farfield and point source assumptions may force the array response to be removed from the assumed manifold. These errors typically are due to a lack of understanding of the physical problem. Correcting these errors may require changes in the array configuration. One simple approach is to assume that these errors are random, independent, and zero-mean. Hence, they just decrease the SNR by increasing the noise variance. If it is known in advance that the sources are not point sources, or that there is a significant angular spread in the signals, then techniques used in channel sounding should be considered [11, 12].

A deterministic model considers a large number of discrete beams with associated parameter values [11], whereas a stochastic model approach uses angular distribution models in order to find maximum likelihood estimates of the mean angle and the scatter about the mean angle [12].

Assuming random errors, the perturbations may be expressed in matrix form, using the array covariance matrix as follows [8]:

$$\hat{\Sigma} = (I + \Delta)[(A + \tilde{A})\Sigma_s(A + \tilde{A})^H + \sigma^2(I + \tilde{\Sigma}_v)](I + \Delta)^H$$

where matrix Δ is associated with errors that influence both the signal and noise components of the data. Departures from the nominal array response are included in the matrix $\tilde{A}$. This matrix describes the perturbations in element positions, errors in gain and phase responses of the sensors, and mutual coupling. The term $\tilde{\Sigma}_v$ describes the deviation of the noise covariance matrix from the nominal (identity) matrix I. The effects of the perturbations to signal and noise subspace may be studied using first-order analysis introduced in [8], or by using tools from matrix perturbation theory. One such convenient measure is obtained via singular value decomposition of

$$\hat{U}_s^H U_s$$

where $\hat{U}_s$ and U_s are the perturbed and true signal subspace eigenvectors. The canonical angles of the basis vectors in $\hat{U}_s$ and U_s are obtained by $\arccos(\gamma_i)$, where γ_i is a singular value of $\hat{U}_s^H U_s$. Obviously, the true and perturbed subspaces are close to each other if the largest canonical angle $\arccos(\gamma_i)$ is small.

There are array processing techniques that are robust in the face of array and signal model errors. Beamspace processing [13] makes the subspace methods less sensitive to modeling errors. The observed data is preprocessed by a beamformer before applying the subspace method.

In case of coherent sources caused by multipath or jamming, many high-resolution methods fail. Spatial smoothing preprocessing technique [14] has to be employed in order to build up the rank of the signal covariance matrix Σ_s again. Maximum likelihood methods can deal with coherent sources as well.

The array and signal modeling errors may be taken into account by using appropriate weighting in MUSIC and subspace fitting algorithms, for example. The weighted MUSIC [8] introduces a weighting matrix in the pseudospectrum expression

$$V_M(\theta, W) = \frac{1}{\mathbf{a}^H(\theta_i) U_n W U_n^H \mathbf{a}(\theta_i)}$$

and the weights may be chosen, such that the variance of the estimates is minimized. Obviously, when $W = I$, we have the conventional MUSIC algorithm. In addition, the weighted subspace fitting (WSF) method

$$\hat{\theta} = \arg \min_{\theta} Tr\{\Pi_A^\perp \hat{U}_s W \hat{U}_s^H\}$$

where $\Pi_A^\perp = I - \Pi_A$ is a projection matrix to noise subspace, where $\Pi_A = A[A^H A]^{-1} A^H$ may be employed to reduce the effects of modeling errors [15, 16]. The eigenvectors are again weighted in order to minimize the variance.

The optimum weighting matrix, based on estimated eigenvalues and eigenvectors, is obtained as follows

$$\hat{W}_{opt} = (\hat{\Lambda}_s - \hat{\sigma}^2 I)^2 \hat{\Lambda}_s^{-1}$$

where $\hat{\Lambda}_s$ are the eigenvalues of $\hat{\Sigma}$ corresponding to $\hat{U}_s$. The weighting is impacted by the modeling errors. See [15] for a detailed derivation and further references.

12.4 Uncertainty in Noise Model

In this section, uncertainty in the noise model is considered. The influence of departures from the underlying noise model is studied both quantitatively and qualitatively. Robust estimation procedures that have desirable robustness properties also are presented. A signal processing procedure is robust if it is not sensitive to small departures from the assumed noise models [2, 3]. Typically, the departures occur in the form of outliers; that is, observations that do not follow the pattern of the majority of the data. Other causes of departures include noise model class selection errors and incorrect assumptions on noise environment. The errors in array and signal model, and possible uncertainty in propagation environment and noise model, emphasize the importance of validating all the assumptions by physical measurements. Many assumptions are made in order to simplify the algorithm design for the receiver. For example, by assuming Gaussian pdfs, the derivation of the algorithms often leads to linear structures that are easy to compute, because linear transformations of Gaussians are Gaussians.

The noise in array processing applications may be heavy-tailed for many reasons. Man-made interference typically has a heavy-tailed distribution. In outdoor radio channels, heavy-tailed noise is often encountered [4, 17]. Middleton's class A and B noises, used to describe man-made and natural noises, have a heavy-tailed component that causes severe performance loss for maximum likelihood methods derived under Gaussian noise assumption [18].

12.4.1 Qualitative and Quantitative Robustness

Next, we describe the quantitative and qualitative measures of robustness, after introducing some notations. We use $\hat{S} \in \text{PDH}(M)$ to denote a scatter matrix estimate based on the snapshot dataset $Z_N = \{\mathbf{z}_1, \ldots, \mathbf{z}_N\}$ in $\mathbb{C}^M$, and $\text{PDH}(M)$ denotes the set of all positive definite Hermitian $M \times M$ matrices. The term scatter matrix is used to account for the noise distributions, where second-order moments are not defined. For example, symmetric α-stable distributions and Cauchy-distributions belong to such distributions. We also write $\hat{S}(Z_N)$ to indicate that $\hat{S}$ is based on dataset Z_N. We denote random vectors by uppercase boldface letters. We require that $\hat{S}$ is *affine equivariant*; that is, for any nonsingular $M \times M$-matrix B, the estimate for the transformed data $BZ_N = \{B\mathbf{z}_1, \ldots, B\mathbf{z}_N\}$ is

$$\hat{S}(BZ_N) = B\hat{S}(Z_N)B^H$$

Assume that $F \in \mathscr{F}$ where $\mathscr{F}$ denotes a large subset of distributions on $\mathbb{C}^M$. By "large," we mean one that contains plausible models for the unknown population, as well as the empirical distribution F_N associated with the dataset Z_N. Then, a map $C: \mathscr{F} \to \text{PDH}(M)$ is a statistical functional corresponding to $\hat{S}$ whenever $\hat{S} = C(F_N)$. Affine equivariance implies that at centered complex elliptically symmetric (CES) distribution F,

$$C(F) = \sigma S$$

for some positive σ, the value of which depends both on the functional C and the centered CES distribution F. CES distributions are a class of distributions introduced in [19] and parameterized by a scatter matrix S, which is proportional to the covariance matrix Σ of the CES distribution when it exists. The multivariate complex Gaussian and Cauchy distributions, for example, are prominent members in this class of distributions. Here, "centered" means that the location or the symmetry center $\boldsymbol{\mu}$ of the CES distribution is assumed to be known or fixed, and without loss of generality, we assume $\boldsymbol{\mu} = 0$.

Quantitative Robustness

The concept of breakdown point is typically used for quantitative robustness. The breakdown point is defined by the smallest fraction of bad outliers that can cause the estimator to break down. In the context of covariance estimation, this could mean the smallest number k observations out of N total observations that could

12.4 Uncertainty in Noise Model

make either the largest eigenvalue over all bounds, or the smallest eigenvalue, arbitrarily close to zero. This makes the matrix ill-conditioned. The (finite sample) breakdown point for a scatter matrix estimator may be defined as follows

$$\epsilon^* = \min_{1 \leq k \leq N} \left\{ \frac{k}{N} \;\middle|\; \sup_{Z_{N,k}} D[\hat{S}(Z_N), \hat{S}(Z_{N,k})] = \infty \right\}$$

where $Z_{N,k}$ is the corrupted sample obtained by replacing k observations of Z_N with arbitrary values, and taking the supremum over all possible datasets $Z_{N,k}$. $D(,)$ above is a measure of dissimilarity of two positive definite Hermitian matrices, and defined as:

$$D(A, B) = \max[|\lambda_1(A) - \lambda_1(B)|, |\lambda_M(A)^{-1} - \lambda_M(B)^{-1}|]$$

with $\lambda_1(A) \geq \ldots \geq \lambda_M(A)$ being the ordered eigenvalues of the matrix $A \in \mathrm{PDH}(M)$. This particular measure of dissimilarity was employed in [20]. Here, the breakdown point is defined in terms of eigenvalues, but similar concepts for eigenvectors could be constructed as well. The breakdown point is always below 50%, and it often depends on the dimension. For additional definitions and detailed descriptions of breakdown points, see [2, 3, 20].

Qualitative Robustness

Qualitative robustness of an estimator may be described in terms of influence function (IF). An IF measures the effects of infinitesimal perturbations on the estimator [3]. A robust estimator should have a bounded and continuous IF. Loosely speaking, the boundedness implies that a small amount of contamination at any point does not have an arbitrarily large influence on the estimate, whereas the continuity implies that the small changes in the dataset cause only small changes in the estimate.

An influence function is essentially the first derivative of the functional version of an estimator. If we denote the point mass at z by Δ_z, and consider the contaminated distribution $F_\epsilon = (1 - \epsilon)F + \epsilon\Delta_z$, then the IF of a functional T at F is

$$\mathrm{IF}(z; T, F) = \lim_{\epsilon \downarrow 0} \frac{T(F_\epsilon) - T(F)}{\epsilon} = \frac{\partial}{\partial \epsilon} T(F_\epsilon)\Big|_{\epsilon=0} \qquad (12.2)$$

One may interpret the influence function as describing the effect of an *infinitesimal* contamination at point z on the estimator, standardized by the mass of the contamination. See [3] for a comprehensive study of the influence functions.

For any affine equivariant scatter matrix functional $C(F) \in \mathrm{PDH}(k)$, there exists functions $\alpha, \beta: \mathbb{R}^+ \to \mathbb{R}$, such that the influence function of $C(F)$ at a CES distribution F is

$$\mathrm{IF}(z; C, F) = \alpha(r) S^{1/2} [\mathbf{u}\mathbf{u}^H - (1/M)I] S^{1/2} + \beta(r) S$$

where $r^2 = \mathbf{z}^H S^{-1} \mathbf{z}$, and $\mathbf{u} = S^{-1/2}\mathbf{z}/r$. The above implies that the influence function of the scatter functional is bounded if and only if the corresponding "weight functions" α and β are bounded. See [21] for this result, and for more details and examples.

The knowledge of the influence function of the scatter matrix functional $C(F)$ allows us to obtain the influence functions of its eigenvector and eigenvalue functionals. As an example, the influence function for eigenvector functional $\mathbf{g}_i(F)$ of $C(F)$ corresponding to a simple eigenvalue λ_i is given by [21]:

$$\text{IF}(\mathbf{z}; \mathbf{g}_i, F) = \frac{\alpha(r)}{\sigma} \sum_{\substack{j=1 \\ j \neq i}}^{M} \frac{\sqrt{\lambda_j \lambda_i}}{\lambda_i - \lambda_j} u_j u_i^* \gamma_j$$

where $\gamma_1, \ldots, \gamma_M$ denotes the eigenvectors of the scatter matrix S. Eigenvectors and subspaces spanned by the eigenvectors are of crucial importance in smart antenna algorithms. The influence function and breakdown point allow the description of the qualitative and quantitative robustness of an array processing procedure in a well-defined manner.

12.4.2 Robust Procedures

Next, examples of DOA estimation methods that are both qualitatively and quantitatively robust are given. The first class of methods stems from nonparametric statistics. The semiparametric M-estimation and stochastic ML-estimation also are considered. The robust methods also can be employed in the process of reducing the impact of modeling errors. Robust estimates of covariance and subspaces can be plugged to beamspace methods, and methods using partially calibrated arrays (e.g. ESPRIT), Spatial smoothing of robustly estimate covariance matrices may be performed to build up the rank [17, 22].

Nonparametric Statistics

We begin by giving definitions for multivariate spatial sign and rank concepts. For an M-variate complex dataset, $\mathbf{z}(1), \ldots, \mathbf{z}(N)$, the *spatial rank function* is

$$\mathbf{r}(\mathbf{z}) = \frac{1}{N} \bigvee_{i=1}^{n} \mathbf{s}[\mathbf{z} - \mathbf{z}(i)]$$

where $\mathbf{s}$ is the *spatial sign function*

$$\mathbf{s}(\mathbf{z}) = \begin{cases} \dfrac{\mathbf{z}}{\|\mathbf{z}\|} & \mathbf{z} \neq 0 \\ 0 & \mathbf{z} = 0 \end{cases}$$

with $\|\mathbf{z}\| = (\mathbf{z}^H \mathbf{z})^{1/2}$. The spatial sign covariance matrix is defined as

12.4 Uncertainty in Noise Model

$$\Sigma_{\text{SCM}} = \frac{1}{N} \sum_{i=1}^{n} \mathbf{s}[\mathbf{z}(i)] \mathbf{s}^H[\mathbf{z}(i)]$$

The spatial Kendall's tau covariance matrix (TCM), and the spatial rank covariance matrix (RCM) are defined as

$$\Sigma_{\text{TCM}} = \frac{1}{N^2} \sum_{i=1}^{N} \sum_{j=1}^{N} \mathbf{s}[\mathbf{z}(i) - \mathbf{z}(j)] \mathbf{s}^H[\mathbf{z}(i) - \mathbf{z}(j)]$$

and

$$\Sigma_{\text{RCM}} = \frac{1}{N} \sum_{i=1}^{n} \mathbf{r}[\mathbf{z}(i)] \mathbf{r}^H[\mathbf{z}(i)]$$

respectively. It has been shown that these covariance matrix estimators produce convergent estimates of the eigenvectors and consequently of the subspaces [22]. Consequently, these estimators may be plugged into any subspace estimator, such as MUSIC, and by performing spatial smoothing on these covariance matrices, coherent sources also may be handled.

M-Estimation

An illustrative example derivation of a robust M-estimator of covariance is given next.

Definition

The M-estimator of scatter $\hat{S} \in \text{PDH}(M)$ based upon a sample $\mathbf{z}_1, \ldots, \mathbf{z}_N \in \mathbb{C}^M$ solve

$$S = \text{ave}\{v(d_i^2) \mathbf{z}(i) \mathbf{z}^H(i)\}$$

where ave{} denotes the average of the expression within the brackets over $i = 1, \ldots, N$, $d_i^2 = \mathbf{z}^H(i) S^{-1} \mathbf{z}(i)$, and v is a real-valued function on $[0, \infty)$.

First, we note that the estimate $\hat{S}$ is affine equivariant. The statistical functional $C(F) \in \text{PDH}(k)$ corresponding to M-estimator $\hat{S}$ is defined in an analogous manner as a solution of

$$C(F) = E\{v[\mathbf{Z}^H C(F)^{-1} \mathbf{Z}] \mathbf{Z} \mathbf{Z}^H\}$$

Note that $\hat{S} = C(F_N)$. One may easily verify that $C(F) = \sigma \Sigma$ (i.e., proportional to the covariance matrix when it exists). The scalar $\sigma > 0$ depends on the estimator and the CES distribution F.

The influence function of the scatter M-functionals $C(F)$ at a CES distribution F is as given in the previous section, with [21]:

$$\alpha(r) \propto v(r^2/\sigma) r^2 \quad \text{and} \quad \beta(r) \propto v(r^2/\sigma) r^2 - M\sigma$$

Thus, the influence function of the M-functional $C(F)$ is continuous and bounded if $v(r^2)r^2$ is continuous and bounded.

We will now consider an interesting special case—the complex $t_1 M$-estimator [21, 23, 24]. This is an M-estimator obtained using the weight function

$$v(d^2) = (2M + 1)/(1 + 2d^2)$$

This choice of weight function yields the ML-estimate of the scatter matrix S, if the underlying distribution is complex multivariate Cauchy. The actual estimates can be found conveniently using the following iterative algorithm. Given an initial estimate $\Sigma_0 \in \text{PDH}(M)$, define

$$\Sigma_{m+1} = \text{ave}\left\{v\left[\mathbf{z}^H(i) S_m^{-1} \mathbf{z}(i)\right] \mathbf{z}(i) \mathbf{z}^H(i)\right\} \tag{12.3}$$

See [23, 25] for details. In [23], the $t_1 M$-estimator is used in various sensor array signal processing applications. The IF of the $t_1 M$-estimator is smooth and bounded, since $v(r^2)r^2$ is smooth and bounded. Consequently, its eigenvector and eigenvalue functionals are bounded as well. In Figure 12.2, an example of the influence function of the eigenvector of the regular covariance matrix estimator and $t_1 M$-estimator is plotted. The regular covariance matrix clearly has an unbounded influence function, whereas the influence function of the $t_1 M$-estimator is smooth and bounded, as shown earlier.

The robust estimators, derived using the maximum likelihood principle for complex t-distribution with different degrees of freedom, have two desirable properties: they are highly robust, and have reasonable sized arrays that are very close to optimal, even in complex Gaussian cases. The efficiency of the estimator grows rapidly as a function of the array size.

Stochastic Maximum Likelihood

In the classical stochastic ML (SML) approach [6], the noise and the signal distribution are modeled as complex circular Gaussian, in which case $\mathbf{z}(i) \sim \mathbb{C}N_k(0, \Sigma)$. Recall that $\Sigma = \text{Cov}[\mathbf{z}(i)]$. The signal parameters $\theta = (\theta_1, \ldots, \theta_K)$, signal covariance matrix $\Sigma_s \in \text{PDH}(K)$, and noise variance $\sigma^2 \in \mathbb{R}^+$ are found by solving by solving

$$\{\hat{\theta}, \hat{\Sigma}_s, \hat{\sigma}^2\} = \arg \min_{\theta, \Sigma_s, \sigma^2} \{\log[\det(\Sigma)] + Tr[\Sigma^{-1}\hat{\Sigma}]\}$$

where

$$\hat{\Sigma} = \frac{1}{N} \sum_{i=1}^{n} \mathbf{z}(i)\mathbf{z}^H(i)$$

Although optimal (if the model is correct), the drawback of the SML method is that it leads to a difficult multidimensional nonlinear optimization problem, and

12.4 Uncertainty in Noise Model

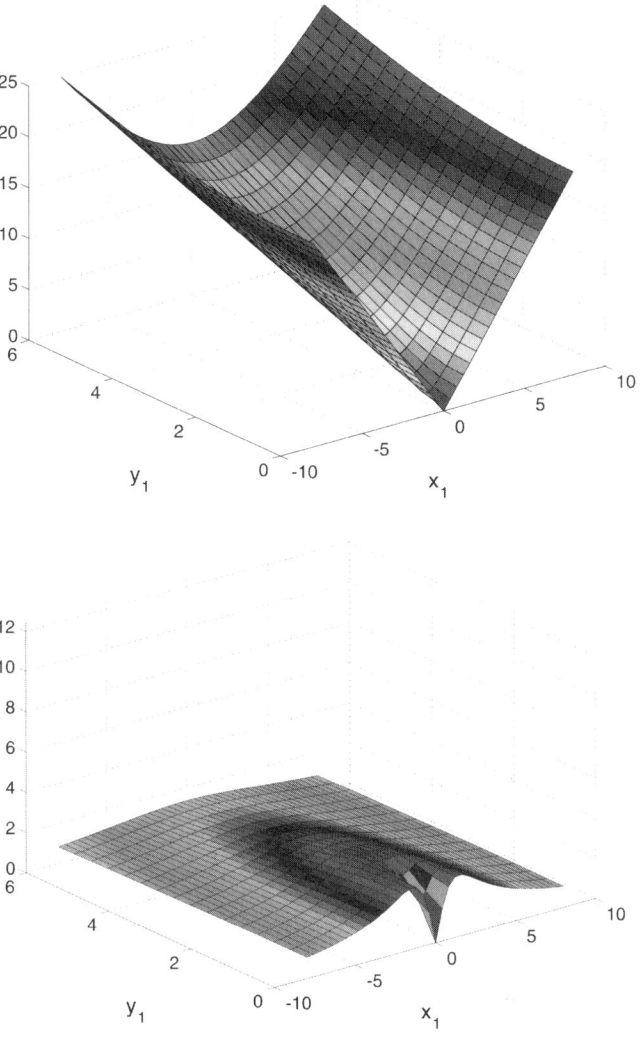

Figure 12.2 $\|IF(z; g_1, F)\|$ for the regular covariance estimator (top), and $t_1 M$-estimator (bottom) at bivariate complex normal distribution ($\lambda_1 = 1$, $\lambda_2 = 0.6$). Here, $z = [z(1), z(2)]^T$ is such that $z(2)$ is fixed and $z(1)$ varies.

it is highly sensitive to heavy-tailed noise. To obtain robust estimates of DOAs, one could use the complex multivariate t-distribution of Definition 1 as a (robust) array output model distribution. Thus, if we model $z(i) \sim \mathbb{C}t_{M,\nu}(0, S)$, then the ML-estimate of the signal parameters are found from

$$\{\hat{\theta}, \hat{\Sigma}_s, \hat{\sigma}^2\} = \arg \min_{\theta, \Sigma_s, \sigma^2} \left\{ \log[\det(S)] + [(2M + \nu)/2] \operatorname{ave}\left[\log\left(1 + 2z_i^H S^{-1} z_i/\nu\right)\right] \right\}$$

Note that scatter matrix S is used in the above expression instead of the covariance matrix Σ, only to account for the distributions where second-order moments are not defined. The stochastic ML estimation in the context of heavy-tailed models is also considered in [17, 26].

12.5 Array Processing Examples

We now demonstrate the robustness of the MUSIC DOA estimator based on robust estimate of covariance. An eight-element ($M = 8$) ULA with interelement spacing equal to $\lambda/2$ was used. Two uncorrelated signals, ($K = 2$) with SNR = 20 dB, are impinging on the array from DOAs $\theta_1 = -2°$ and $\theta_2 = +2°$ (broadside). In our study, $N = 300$ snapshots are generated from multivariate complex Gaussian and Cauchy distributions. The sample covariance matrix and $t_1 M$-estimator were used to estimate the noise subspace. Figure 12.3 depicts the MUSIC pseudospectrum associated with the two estimators of the scatter matrix for the simulated complex Gaussian and Cauchy snapshots. We notice that both the sample covariance matrix $\hat{\Sigma}$ and $t_1 M$-estimator of scatter matrix are able to resolve the two sources in the Gaussian case, but that in the Cauchy case, the sample covariance matrix is not able to resolve the sources (i.e., a loss of resolution is experienced). However, the robust $t_1 M$-estimator of scatter matrix yields reliable estimates of the DOAs with high resolution.

Another example on robustness is given, using uniform circular array (UCA). Beamspace transformation is employed, leading to a virtual array with Vandermonde structure for the steering vector matrix. The unitary root-MUSIC algorithm derived for circular arrays in [27] is employed. As a by-product of the transformation, spatial smoothing is performed, which leads to improved robustness in the face of coherent signals sources. Additional robustness in the face of heavy-tailed

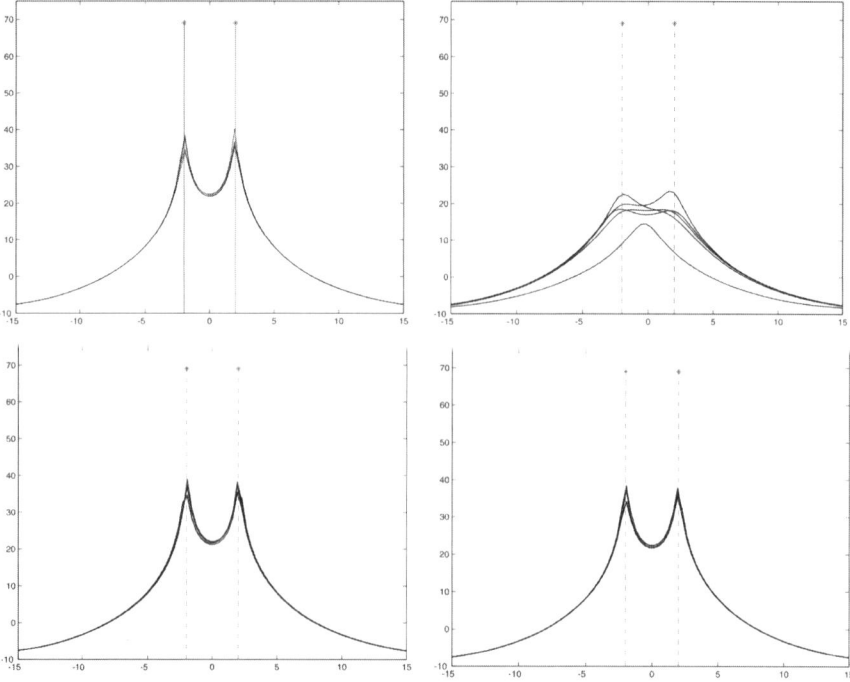

Figure 12.3 MUSIC pseudospectrum based on sample covariance matrix (upper row) and $t_1 M$-estimator of scatter matrix (lower row), for five simulated data sets generated from complex normal (first column) and Cauchy (second column) distributions.

noise is achieved by estimating the sign covariance matrix presented earlier in this chapter, instead of conventional sample covariance matrix. Next, the robustness of the method will be demonstrated. The additive noise is here assumed to be complex Cauchy distributed. The SNR in the Gaussian case would be 10 dB. This is a very demanding scenario, since the moments of the noise distribution are not even defined in Cauchy noise. The zeros of the unitary root-MUSIC for UCA are plotted in Figure 12.4, using a conventional sample covariance matrix and a robust spatial sign covariance matrix estimate. The DOA is $\phi = 280°$, and the robust method finds it very reliably, whereas the conventional estimator completely fails.

Finally, we consider an example where coherent sources and heavy-tailed (i.e., symmetric α-stable noise with characteristic exponent α) noise are present simultaneously. Figure 12.5 shows five estimation results for the cases $\alpha = 2$ and $\alpha = 1$. In the case of Gaussian noise, the performance of both algorithms is almost identical. The algorithm using robust RCM estimates of covariance for each subarray in forward-backward spatial smoothing, followed by the conventional MUSIC algorithm, estimates the angles of arrival reliably even in extremely heavy-tailed noise conditions. However, the standard spatial smoothing MUSIC algorithm typically fails when the noise is heavy-tailed (i.e., $\alpha < 2$).

12.6 Conclusions

In this chapter, the problem of direction of arrival estimation in the face of uncertainty was addressed. Uncertainties in the signal model, noise model, array model, and propagation model were considered. Sources of uncertainty were described, and methods for making the array processing procedures more robust were presented. A

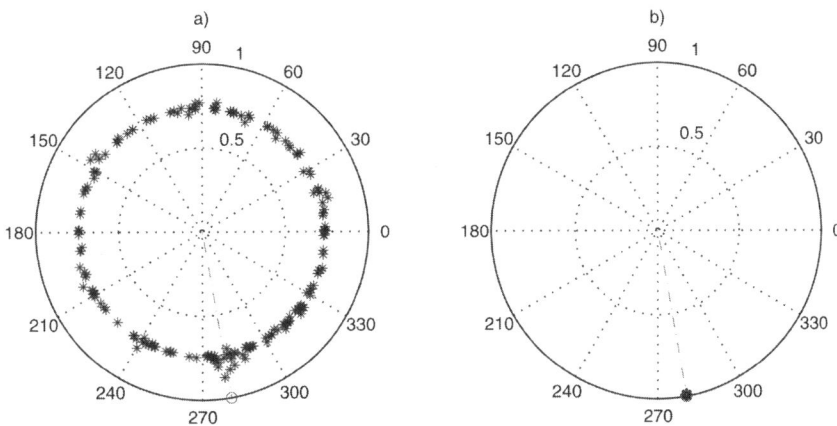

Figure 12.4 Zeros of the unitary root-MUSIC algorithm for UCA in Cauchy noise environment: (left) results using conventional sample covariance matrix estimator, and (right) result using robust spatial sign covariance matrix estimator. The robust method finds the angle of arrival reliably, and the roots are clustered about the true angle of arrival. The conventional covariance matrix–based estimator completely fails since the roots are scattered all over the unit circle. Total of 200 snapshots are used, $r = \lambda$, $N = 19$ UCA elements.

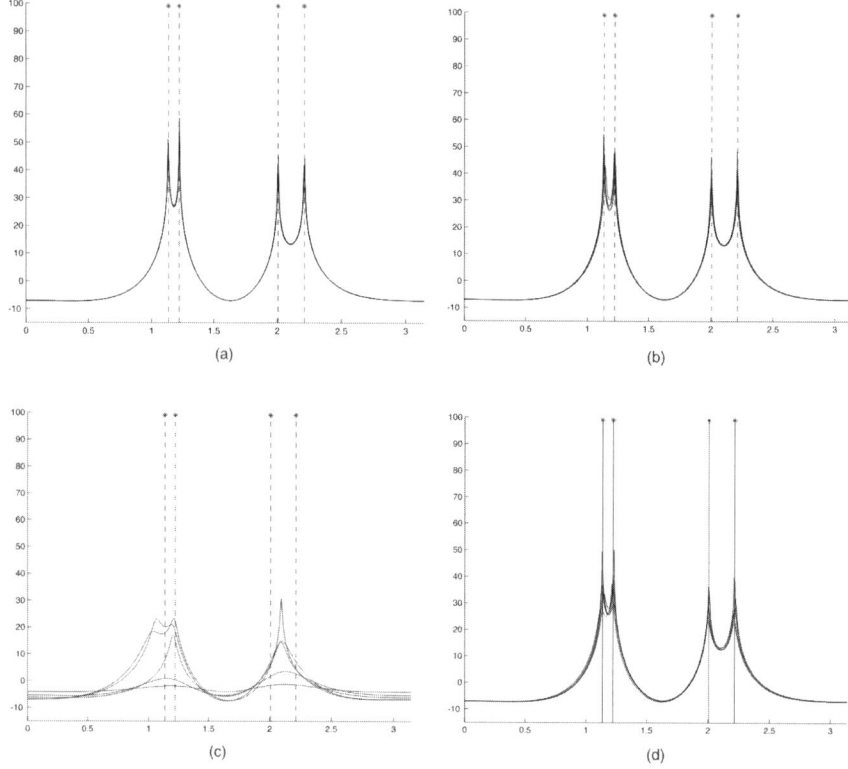

Figure 12.5 DOA estimation results for α-stable noise conditions: (a) spatial smoothing MUSIC, $\alpha = 2$; (b) RCM-based spatial smoothing MUSIC, $\alpha = 2$; (c) spatial smoothing MUSIC, $\alpha = 1$; and (d) RCM-based spatial smoothing MUSIC, $\alpha = 1$. The size of the ULA is 8, and the size of the subarray is 6. The DOAs are 65°, 70°, 115°, and 127°.

special emphasis on the robustness of array covariance matrix and the subspaces spanned by its eigenvectors was presented. In order to illustrate these ideas, robust estimator for the array covariance matrix was derived, and its qualitative robustness was demonstrated. Practical direction-finding examples were given, and the robustness of the algorithms demonstrated in a very demanding noise environment and in the presence of signal model uncertainty.

References

[1] Godara, L., *Smart Antennas*, Boca Raton, FL: CRC Press, 2004.

[2] Kassam, S. A., and H. V. Poor, "Robust Techniques for Signal Processing: A Survey," *IEEE Proc.*, Vol. 73, No. 3, 1985, pp. 433–481.

[3] Hampel, F. R., et al., *Robust Statistics: The Approach Based on Influence Functions*, New York: John Wiley & Sons, 1986.

[4] Swami, A., and B. M. Sadler, "On Some Detection and Estimation Problems in Heavy-Tailed Noise," *Signal Processing*, Vol. 82, 2002, pp. 1829–1846.

[5] Geshman, A., "Robustness Issues in Adaptive Beamforming and High-Resolution Direction Finding," in *High-Resolution and Robust Signal Processing*, Y. Hua, A. Geshman, and Q. Cheng, (eds.), New York: Marcel Dekker, 2004.

[6] Krim, H. and M. Viberg, M., "Two Decades of Array Signal Processing: The Parametric Approach," *IEEE Signal Processing Mag.*, Vol. 13, No. 4, 1996, pp. 67–94.

[7] Rao, B. D., and K. V. S. Hari, "Performance Analysis of Subspace Methods," *IEEE 4th Annual ASSP Workshop on Spectrum Estimation and Modeling*, 1988, pp. 92–97.

[8] Swindlehurst, A. L., and T. Kailath, "A Performance Analysis of Subspace-Based Methods in the Presence of Model Errors, Part I: The MUSIC Algorithm," *IEEE Trans. on Signal Processing*, Vol. 40, No. 7, 1992, pp. 1758–1774.

[9] Weiss, A. J., and B. Friedlander, "Array Shape Calibration Using Sources in Unknown Locations—A Maximum Likelihood Approach," *IEEE Trans. on Acoustics, Speech, and Signal Processing*, Vol. 37, No. 12, 1988, pp. 1958–1966.

[10] Pesavento, M., A. B. Gershman, and K. M. Wong, "Direction Finding in Partly Calibrated Sensor Arrays Composed of Multiple Subarrays," *IEEE Trans. on Signal Processing*, Vol. 50, No. 9, 2002, pp. 2103–2115.

[11] Pedersen, K. I., P. E. Mogensen, and B. H. Fleury, "A Stochastic Model of the Temporal and Azimuthal Dispersion Seen at the Base Station in Outdoor Propagation Environments," *IEEE Trans. on Vehicular Technology*, Vol. 49, No. 2, 2000, pp. 437–447.

[12] Ribeirio, C., E. Ollila, and V. Koivunen, "Propagation Parameter Estimation in MIMO Systems Using Mixture of Angular Distributions Model," *IEEE ICASSP 2005*, Philadelphia, PA, 2005, pp. 885–888.

[13] Zoltowski, M. D., G. M. Kautz, and S. D. Silverstein, "Beamspace Root-MUSIC," *IEEE Trans. on Signal Processing*, Vol. 41, No. 1, 1993, pp. 344–364.

[14] Pillai, S. U., and B. H. Kwon, "Forward/Backward Spatial Smoothing Techniques for Coherent Signal Identification," *IEEE Trans. on Acoustics, Speech, and Signal Processing*, Vol. 37, No. 1, 1989, pp. 8–15.

[15] Jansson, M., A. L. Swindlehurst, and B. Ottersten, "Weighted Subspace Fitting for General Array Error Models," *IEEE Trans. on Signal Processing*, Vol. 46, No. 9, 1998, pp. 2484–2498.

[16] Swindlehurst, A. L., and T. Kailath, "A Performance Analysis of Subspace-Based Methods in the Presence of Model Error. II. Multidimensional Algorithms," *IEEE Trans. on Signal Processing*, Vol. 41, No. 9, 1993, pp. 2882–2890.

[17] Kozick R. J., and B. M. Sadler, "Maximum-Likelihood Array Processing in Non-Gaussian Noise with Gaussian Mixtures," *IEEE Trans. on Signal Processing*, Vol. 48, No. 12, 2000.

[18] Middleton, D., *An Introduction to Statistical Communication Theory*, 1st ed., New York: Wiley-IEEE Press, 1996.

[19] Krishnaiah, P., and J. Lin, "Complex Elliptically Symmetric Distributions," *Comm. Statist.—Th. and Meth.*, Vol. 15, 1986, pp. 3693–3718.

[20] Lopuhaä, H. P., and P. J. Rousseeuw, "Breakdown Points of Affine Equivariant Estimators of Multivariate Location and Scatter," *The Annals of Statistics*, Vol. 19, 1991, pp. 229–248.

[21] Ollila, E., and V. Koivunen, "Influence Functions for Array Covariance Matrix Estimators," *Proc. IEEE Statistical Signal Processing Workshop (SSP'03)*, 2003, pp. 462–465.

[22] Visuri, S., H. Oja, and V. Koivunen, "Direction of Arrival Estimation Based on Nonparametric Statistics," *IEEE Trans. on Signal Processing*, Vol. 49, No. 9, 2001, pp. 2060–2073.

[23] Ollila, E., and V. Koivunen, "Robust Antenna Array Processing Using M-Estimators of Pseudo-Covariance," *Proc. IEEE International Symposium on Personal, Indoor and Mobile Radio Communications (PIMRC'03)*, Beijing, China, 2003.

[24] Ollila, E., and V. Koivunen, "Robust Space-Time Scatter Matrix Estimator for Broadband Antenna Arrays," in *Proc. IEEE VTC2003 Fall*, Orlando, FL, October 6–9, 2003.

[25] Kent, J. T., and D. E. Tyler, "Redescending M-Estimates of Multivariate Location and Scatter," *The Annals of Statistics*, Vol. 19, 1991, pp. 2012–2119.

[26] Williams, D.B., and D. H. Johnson, "Robust Estimation of Structured Covariance Matrices," *IEEE Trans. on Signal Processing*, Vol. 41, No. 9, 1993, pp. 2891–2906.

[27] Belloni, F., and V. Koivunen, "Unitary Root-MUSIC Technique for Uniform Circular Array," *IEEE ISSPIT 2003*, 2003, pp. 451–454.

CHAPTER 13
Coupling Model in DOA Antenna Arrays

Manuel Sierra-Pérez and Daniel Segovia-Vargas

13.1 Introduction

Direction-finding techniques based on the eigendecomposition of the covariance matrix of the received signals have been extensively discussed [1–3]. Most of these techniques require precise knowledge of the antenna behavior. In spite of the great knowledge of the eigenstructure direction-finding techniques, the application to real systems has been somewhat limited, due to the calibration associated with the data collection of the whole system. An initial calibration scheme would be applied to overcome the problems related to a lack of a precise knowledge in the signals received by the array. This could imply, for instance, a change in the position of the antennas in the array, causing a mismatch between the actual and the estimated DOA. An extensive treatment of these problems related to an incorrect estimation of the array manifold next to the presence of an imprecisely known covariance matrix has been done in [4, 5].

In addition to the problem of initial array calibration, there is the problem of the maintenance of the calibration of the array. From the electromagnetic point of view, the mutual coupling between the antennas in the array make the initial calibration no longer valid. Then, the presence of mutual coupling makes it impossible to maintain array calibration to the accuracy required for the eigenstructure DOA techniques. The reduction of mutual coupling is important for making DOA techniques useful.

Several techniques have been presented to minimize the effects of mutual coupling. A classification of these techniques can be considered as one way of overcoming the problem. There are techniques that do not deal with the mutual coupling problem itself, but with the global effect that it provokes, from the signal processing point of view [6]. In this situation, it is assumed that a mutual coupling matrix presents a banded Toeplitz structure in linear arrays or circular structure in circular arrays. The structure of the matrix can be Toeplitz or circular. In practice, the actual values of the coupling parameters differ from the assumed parameters, and a change over time will be produced. From this assumed matrix, DOA and mutual coupling with the assumed structure would be determined.

The second kind of technique deals with the electromagnetic model of the array. It estimates the mutual-coupling matrix from the actual antenna voltages or currents, and then determines the angle of arrival.

Two methods have been followed to determine the electromagnetic (EM) coupling matrix between elements in an array antenna: the infinite array, and the finite phased array. The former generally models the central elements of a large, but finite, array quite well, and implicitly includes the effect of mutual coupling between elements. This approach has been followed where an array with equally spaced elements self-calibrates using the measured mutual coupling [7]. The self-calibration is done under the assumption that all the elements have identical embedded radiation patterns (i.e., the element radiation pattern in the presence of the other elements in an infinite array) that do not vary with element position. However, this approach does not take into account the edge effects. Besides, it is not well known how large a finite array must be before it can be adequately modeled as infinite. This last topic is especially critical in adaptive array applications. Thus, the study of coupling in finite phased arrays may yield more practical information than the study of infinite arrays. However, finite arrays are considerably more difficult to analyze [8].

In this chapter, a model to compute the steering vector in a finite array of antennas used in a DOA detection system is presented. In most cases, the model used to develop the mathematical algorithms for DOA tends to simplify as much as possible the real complexity of the antenna, avoiding parameters that have no interest in the mathematical development, but are nevertheless important in a real system steering vector (SV) computation.

Here, the model takes into account all the electromagnetic antenna parameters and tries to develop a general formulation, limited only by the linear behavior in the antenna and receiver subsystems [9]. Noise signals are assumed to be random Gaussian processes, uncolored spectrum, and uncorrelated from one channel to another. The model presented is applicable to far sources planar wavefront or close sources spherical wavefronts. Besides, this model works for narrowband or wideband incoming signals, although for simplicity, the presentation will be restricted to the narrowband case.

Once the model has been introduced, the influence of several causes of steering vector errors, such as gain errors, radiating element position errors, coupling between elements, or perturbing elements close to the antennas, will be analyzed. Finally, the validity of a compensation matrix model is discussed, and the approximations in that formulation are emphasized. The matrix introduced allows a formulation to obtain further calibration methods, as presented in Chapter 19.

13.2 Antenna Steering Vector Model

In a general DOA system, we consider a set of P antennas placed in nominal positions $\vec{r}_k = \{r_{kx}, r_{ky}, r_{kz}\}$ for $k = 1$ to P, and consider a field source placed in position $\vec{r}_s = \{r_{sx}, r_{sy}, r_{sz}\}$ that produces an electric field in the array of antennas equivalent to a spherical wave, and described in the array as:

$$\vec{E}_s(\vec{r}, t) = \hat{\theta} E_{\theta s}\left(t - \frac{|\vec{r} - \vec{r}_s|}{c}\right) + \hat{\phi} E_{\phi s}\left(t - \frac{|\vec{r} - \vec{r}_s|}{c}\right) \quad (13.1)$$

where $\hat{\theta}$ and $\hat{\phi}$ are the unit vector in spherical coordinates, $E_{\theta s}(t)$ and $E_{\phi s}(t)$ are the instantaneous electric field components, and c is the wave velocity. Both functions are considered constant in relation to the position coordinates in the spherical surface of the wavefront.

A bandlimited signal can be expressed as a function of its lowpass filter response at the center frequency f_c. Both field polarization components are limited in the same frequency band B, which satisfies the following condition:

$$B \ll \frac{c}{L} = f_c \frac{\lambda}{L} \quad (13.2)$$

where L is the maximum dimension of the array, and λ is the carrier wavelength. The source field then can be written as a function of the lowpass complex field components, as follows:

$$\vec{E}_s(\vec{r}, t) = \text{Re}\left[\hat{e}_s(t) E_s(t) e^{(-jk_0|\vec{r} - \vec{r}_s|)}\right] \quad (13.3)$$

where $k_0 = 2\pi/\lambda$ is the wave number, and $\hat{e}_s(t)$ and $E_s(t)$ are the lowpass complex components of the polarization and electric field time functions.

It will also be assumed that each antenna in the receiving array has a general receiving normalized complex diagram $F_k(\theta, \phi)$ and a general receiving unit polarization vector $\hat{e}_k(\theta, \phi)$ at frequency f_c, which are constant in the source signal bandwidth. These normalization conditions can be expressed as follows:

$$\int_{4\pi} |F_k(\theta, \phi)|^2 \, d\Omega = 1 \quad |\hat{e}_k(\theta, \phi)|^2 = 1 \quad (13.4)$$

where Ω is the beam solid angle variable.

The incident field produces signals in the array element terminals that are usually denoted by a snapshot vector $\mathbf{x}(t) = \{x_1, x_2, \ldots, x_P\}^t$. These signals can be selected as a voltage, a current, a transmission line wave, or any other complex electrical parameter that could be measured at the output terminals. For a single antenna formulation, all these parameters are equivalent, and are related through a constant parameter independent of the source position, polarization, or field intensity, and are dependent only on the load impedance used in the antenna terminals. If the power wave leaving the antenna terminals is measured in a transmission line of Z_0 nominal impedance, then we can write the output signal as:

$$x_k(t) = E_s(t)[\hat{e}_s(t) \cdot \hat{e}_k(\theta_{sk}, \phi_{sk})] \frac{\lambda \rho_k}{\sqrt{2\eta_0}} F_k(\theta_{sk}, \phi_{sk}) e^{(-j\omega_c \tau_{sk})} \quad (13.5)$$

where the direction (θ_{sk}, ϕ_{sk}) corresponds to the direction of the vector $(\vec{r}_s - \vec{r}_k)$ pointing at the source from each radiator of the array, ρ_k is a complex constant antenna efficiency for passive antennas or an antenna gain for active antennas, $\eta_0 = 120\pi$ is the wave impedance in free space, $\omega_c = 2\pi f_c$ is the carrier pulsation, and τ_{sk} is the time propagation from the source to the antenna element relative to the coordinate origin at wave speed c:

$$\tau_{sk} = \frac{|\vec{r}_s - \vec{r}_k| - |\vec{r}_s|}{c} \qquad (13.6)$$

Generally, the condition of farfield plane wave is assumed, and the array of antennas satisfies that assumption for each antenna element and for the complete array. This condition can be expressed as:

$$|\vec{r}_s| \gg \frac{\lambda^2}{L} \qquad (13.7)$$

In this case, the signal at the output of each element can be written as:

$$x_k(t) = E_s(t)[\hat{e}_s(t) \cdot \hat{e}_k(\theta_s, \phi_s)] \frac{\lambda \rho_k}{\sqrt{2\eta_0}} F_k(\theta_s, \phi_s) e^{(jk_0 \hat{r}_s \vec{r}_k)} \qquad (13.8)$$

where direction (θ_s, ϕ_s) is the same for all elements, and the phase factor can be written as a function of the direction of arrival as:

$$k_0 \hat{r}_s \vec{r}_k = k_0 [r_{kx} \sin(\theta_s) \cos(\phi_s) + r_{ky} \sin(\theta_s) \sin(\phi_s) + r_{kz} \cos(\theta_s)] \qquad (13.9)$$

This phase factor represents the phase difference between each antenna and the origin of coordinates due to the physical position of the antenna.

For notation reasons, the array signal is represented as a $P \times 1$ snapshot vector, and is generally written as:

$$\mathbf{x}(t) = \begin{bmatrix} x_1(t) \\ \cdots \\ x_P(t) \end{bmatrix} = E_s(t) \frac{\lambda}{\sqrt{2\eta_0}} \begin{bmatrix} \rho_1[\hat{e}_s(t) \cdot \hat{e}_1(\theta_{s1}, \phi_{s1})] F_1(\theta_{s1}, \phi_{s1}) e^{(-j\omega_c \tau_{s1})} \\ \cdots \\ \rho_P[\hat{e}_s(t) \cdot \hat{e}_P(\theta_{sP}, \phi_{sP})] F_P(\theta_{sP}, \phi_{sP}) e^{(-j\omega_c \tau_{sP})} \end{bmatrix} \qquad (13.10)$$

From this equation, it can be seen that the snapshot generally cannot be written as an only time-dependent function, multiplied by an only direction-dependent vector, except for the case of a constant polarization wave $(\hat{e}_s)$. In general, communication systems can support both polarization waves on the same carrier, and independent modulations for each polarization result in a unit polarization vector that change in time:

$$\vec{E}(t) = \hat{\theta} E_\theta(t) + \hat{\phi} E_\phi(t) = \hat{e}(t) E(t)$$

where $\hat{\theta}$ and $\hat{\phi}$ represent two linear polarization orthogonal vectors. Any pair of orthogonal vectors could be selected to write the electric field. Using this vector decomposition, we can write the snapshot vector as:

13.2 Antenna Steering Vector Model

$$\mathbf{x}(t) = \frac{\lambda}{\sqrt{2\eta_0}} E_{s\theta}(t) \begin{bmatrix} \rho_1[\hat{\theta} \cdot \hat{e}_1(\theta_{s1}, \phi_{s1})]F_1(\theta_{s1}, \phi_{s1})e^{(-j\omega_c \tau_{s1})} \\ \cdots \\ \rho_P[\hat{\theta} \cdot \hat{e}_P(\theta_{sP}, \phi_{sP})]F_P(\theta_{sP}, \phi_{sP})e^{(-j\omega_c \tau_{sP})} \end{bmatrix} \quad (13.11)$$

$$+ \frac{\lambda}{\sqrt{2\eta_0}} E_{s\phi}(t) \begin{bmatrix} \rho_1[\hat{\phi} \cdot \hat{e}_1(\theta_{s1}, \phi_{s1})]F_1(\theta_{s1}, \phi_{s1})e^{(-j\omega_c \tau_{s1})} \\ \cdots \\ \rho_P[\hat{\phi} \cdot \hat{e}_P(\theta_{sP}, \phi_{sP})]F_P(\theta_{sP}, \phi_{sP})e^{(-j\omega_c \tau_{sP})} \end{bmatrix}$$

All these parameters can be grouped in a single function describing the behavior of the antenna elements of the array for each source impinging on the array through a direction of arrival (θ_s, ϕ_s):

$$\mathbf{x}(t) = E_{s\theta}(t) \begin{bmatrix} f_{s1\theta} \\ \cdots \\ f_{sP\theta} \end{bmatrix} + E_{s\phi}(t) \begin{bmatrix} f_{s1\phi} \\ \cdots \\ f_{sP\phi} \end{bmatrix} = E_{s\theta}(t)\mathbf{f}_{s\theta} + E_{s\phi}(t)\mathbf{f}_{s\varphi} = \vec{E}_s \vec{\mathbf{f}}_s$$

$$(13.12)$$

where $f_{sk\theta}$ and $f_{sk\phi}$ are general complex functions, independent of time, describing the behavior of the kth antenna element in the direction of source s for each polarization component. Vectors $(\mathbf{f}_{s\theta}, \mathbf{f}_{s\phi})$ represent the steering vectors for direction s at each polarization component. Finally, vector $\vec{\mathbf{f}}_s = \hat{\theta}\mathbf{f}_{s\theta} + \hat{\phi}\mathbf{f}_{s\varphi}$ is a column vector with real space vectors as components.

In the particular, but usual, situation when all the radiators in the array are the same and equally oriented, and the source is far enough away to be considered as emitting from the same pointing direction by all the array elements, the steering vector can be represented by the position phase factor and a common amplitude and phase function. This common function affects the noise and coherent incoming signal model, but can be neglected in many mathematical demonstrations. Only a complex gain constant independent of direction of arrival is considered in these cases to take into account the RF channel gain.

$$x_k(t) = f_s(t)e^{(j\omega_c \tau_{sk})} = f_s(t)e^{(jk_0 \hat{r}_s \vec{r}_k)} \quad (13.13)$$

where

$$f_s(t) = \rho E_s(t)[\hat{e}_s(t)\hat{e}(\theta_s, \phi_s)] \frac{\lambda}{\sqrt{2\eta_0}} F(\theta_s, \phi_s)$$

and $\hat{e}$ and F are the common polarization and pattern functions for all the array elements.

Function $f_s(t)$ is independent of the k index of each array element, and is proportional to the lowpass complex source signal. It is interesting to note that the function $f_s(t)$ includes not only the amplitude and phase of the receiving signal, but also the ability of the antennas to receive this signal, depending on the direction of arrival and polarization. Then, the steering vector can be written as:

$$\vec{\mathbf{f}}_s = \frac{\rho\lambda}{\sqrt{2\eta_0}} \hat{e}(\theta_s, \phi_s) F(\theta_s, \phi_s) \begin{bmatrix} e^{(-j\omega_c \tau_{s1})} \\ \cdots \\ e^{(-j\omega_c \tau_{sP})} \end{bmatrix} = \frac{\rho\lambda}{\sqrt{2\eta_0}} \hat{e}(\theta_s, \phi_s) F(\theta_s, \phi_s) \mathbf{a}_s$$

(13.14)

where $\mathbf{a}_s$ represents the commonly used formula for the steering vector phase factor due to the element position.

As far as the antenna diagram remains equal for all the antenna elements, the identical omnidirectional sensor assumption, usually taken on signal processing community, implies no limitation to their mathematical developments, except for the amplitude relation to noise or phase relation between coherent signals. The snapshot is finally simplified as the well-known formula:

$$\mathbf{x}(t) = f_s(t) \cdot \mathbf{a}_s \quad (13.15)$$

13.3 Steering Vector Errors and Their Influence in DOA Estimation

The performance of a DOA estimator severely degrades if the array differs from the ideal array. This degradation is mainly due to the presence of steering vector errors and mutual coupling between the antennas in the array. The steering errors may be caused by the look-direction-error due to the mismatch between the actual and ideal DOA when no errors were presented. Some of these errors can be alleviated by means of an initial array calibration. However, others are more difficult to overcome due to their dependence on the actual direction of arrival. These random errors may come from the gradual changes in the electronic circuitry between the antenna and the Tx/Rx module; random changes in the position of the antenna (i.e., due to vibration of the environment where the antenna is placed); and changes due to the electromagnetic environment of the array. If the changes due to the mutual coupling are excluded, then the three more common random errors in the steering vector estimation are:

- Gain (amplitude and phase) errors in the receiving system;
- Position errors in the array elements;
- Differences in the element field or polarization pattern.

Then, for the first case, the gain in each Rx chain can be written as:

$$\rho_i = \rho_0(1 + \Delta\rho_i) = \rho_0(1 + \Delta a_i) \cdot e^{j\Delta p_i} \quad (13.16)$$

where Δa_i is the gain amplitude error, and Δp_i is the zero-mean phase error in the Rx chain. If a matrix including all the gain errors in the receiving chain would be calculated, then it would result in the following diagonal matrix

$$\mathbf{G} \equiv \rho_0 \, diag[(1 + \Delta\rho_1), \ldots, (1 + \Delta\rho_P)] = \rho_0(\mathbf{I} + \Delta\mathbf{G}) \quad (13.17)$$

13.3 Steering Vector Errors and Their Influence in DOA Estimation

An active array calibration system would allow the measurement of this diagonal matrix, so that its effect could be compensated for by multiplying the estimated steering vector by **G**.

The other two error reasons are dependent on the direction of arrival. For simplicity, it will be assumed that our signal environment is composed of one desired signal impinging through θ_1, $(P-1)$ undesired signals, and white noise. Then, the snapshot in the array can be written as

$$\mathbf{x}(t) = f_s(t) \cdot \mathbf{a_s} + \mathbf{v}(t) = f_s(t) \cdot \mathbf{a_s} + \sum_{k=2}^{P} f_{sk}(t) \cdot \mathbf{a_{sk}} + \mathbf{n}(t) \qquad (13.18)$$

where $\mathbf{v}(t)$ denotes the nondesired signal plus noise vector. If there are random errors in the array, then the DOA technique will not detect the actual desired direction, but it will detect the other direction. The actual steering vector of the desired signal is no longer $\mathbf{a_s}$, and can be given by

$$\tilde{\mathbf{a}}_s = \mathbf{a_s} + \Delta \qquad (13.19)$$

where Δ denotes all the random errors, and may be considered as a Gaussian random vector with zero mean.

The Capon method [10] is one of the simplest DOA methods that can be applied to determining the direction of arrival corresponding to the steering vector $\mathbf{a_s}$. Capon's beamformer, also known as the minimum variance distortionless response filter, attempts to minimize the power contributed by noise and by any signals coming from directions other than θ_1, while maintaining a fixed gain in the desired direction θ_1. This can be expressed by:

$$P(\mathbf{w}) = \frac{1}{N} \cdot \sum_{t=1}^{N} \mathbf{w}^H \cdot \mathbf{x}(t) \cdot \mathbf{x}^H(t) \cdot \mathbf{w} = \mathbf{w}^H \cdot \hat{\mathbf{R}}_{xx} \cdot \mathbf{w}$$

$$\min[P(\mathbf{w})] \qquad (13.20)$$

$$subject\ \mathbf{w}^H \cdot \mathbf{a}(\theta_1) = 1$$

where $P(\mathbf{w})$ is the measured power at the output of the array, $\mathbf{w}$ is the weight vector that minimizes the power from noise and nondesired signals, and $\hat{\mathbf{R}}_{xx}$ is a natural estimate of the covariance matrix. The optimum weight vector can be found by using the Lagrange multipliers, resulting in:

$$\mathbf{w}_{CAP} = \frac{\hat{\mathbf{R}}_{xx}^{-1} \cdot \mathbf{a_s}(\theta_1)}{\mathbf{a}_s^H(\theta_1) \cdot \hat{\mathbf{R}}_{xx}^{-1} \cdot \mathbf{a_s}(\theta_1)} \qquad (13.21)$$

A detailed analysis of the effect of random errors in the MVDR beamformer has been done in [11]. The introduction of (13.19) in (13.21) results in

$$\mathbf{w}_{CAP} = \frac{\left[\hat{\mathbf{R}}_{xx}^{-1} \cdot \mathbf{a_s}(\theta_1) + \hat{\mathbf{R}}_{xx}^{-1} \cdot \Delta\right]}{\tilde{\mathbf{a}}_s^H(\theta_1) \cdot \hat{\mathbf{R}}_{xx}^{-1} \cdot \tilde{\mathbf{a}}_s(\theta_1)} \qquad (13.22)$$

The covariance matrix can be written as:

$$\hat{\mathbf{R}}_{xx} = \mathbf{a}_s(\theta_1) \cdot \mathbf{a}_s^H(\theta_1) \frac{1}{N} \sum_{t=1}^{N} |f_s(t)|^2 + \mathbf{a}_s(\theta_1) \cdot \frac{1}{N} \sum_{t=1}^{N} f_s^*(t) \cdot \mathbf{v}^H(t)$$

$$+ \frac{1}{N} \sum_{t=1}^{N} f_s(t) \cdot \mathbf{v}(t) \cdot \mathbf{a}_s^H(\theta_1) + \frac{1}{N} \sum_{t=1}^{N} \mathbf{v}(t) \cdot \mathbf{v}^H(t) \qquad (13.23)$$

$$= \hat{\sigma}_{s1}^2 \cdot \mathbf{a}_s(\theta_1) \cdot \mathbf{a}_s^H(\theta_1) + \mathbf{a}_s(\theta_1) \cdot \hat{\mathbf{r}}^H + \hat{\mathbf{r}} \cdot \mathbf{a}_s^H(\theta_1) + \hat{\mathbf{Q}}$$

After some algebra, the Capon weight vector results in:

$$\mathbf{w} \approx \frac{\hat{\mathbf{Q}}^{-1} \cdot \mathbf{a}_s(\theta_1)}{\mathbf{a}_s(\theta_1) \cdot \hat{\mathbf{Q}}^{-1} \cdot \mathbf{a}_s^H(\theta_1)} - \left[\mathbf{I} - \frac{\hat{\mathbf{Q}}^{-1} \cdot \mathbf{a}_s(\theta_1) \cdot \mathbf{a}_s^H(\theta_1)}{\mathbf{a}_s(\theta_1) \cdot \hat{\mathbf{Q}}^{-1} \cdot \mathbf{a}_s^H(\theta_1)} \right] \cdot \hat{\mathbf{Q}}^{-1} \qquad (13.24)$$

$$\cdot \left[(\hat{\sigma}_{s1}^2 - \hat{\mathbf{r}}^H \cdot \hat{\mathbf{Q}}^{-1} \cdot \hat{\mathbf{r}}) \cdot \Delta - \hat{\mathbf{r}} \right]$$

This expression displays clearly the decomposition of $\mathbf{w}$ into the desired value and the undesired perturbation that is caused by the random steering vectors and the finite sample size. For low SNR ratios in the desired signal, the steering vector error has no effect, and the direction of arrival can be properly estimated. However, for high SNR, the steering vector error has a dominant effect, causing an incorrect estimation of the desired DOA.

If the direction of arrival is not previously known, then these errors cannot be compensated for. In this sense, Kim [12] developed a matrix compensation algorithm to get an equivalent steering vector that compensates for angle-dependent errors. Of course, this matrix compensation is only valid for a limited number of directions or for a limited sector, depending on the error magnitude. In that paper, matrix compensation is used to process a space reference algorithm, where the desired signal direction of arrival is previously known. In DOA systems, this direction is not known, and except for sweeping direction algorithms (e.g., Capon), it cannot be used.

13.4 Coupling Model

The noncoupling model assumes that individual antennas keep their own receiving or transmitting pattern and impedance when they work isolated in space.

Array element coupling is always present in any kind of array antenna. In many array computations, such as diagram synthesis or gain computation, the effect of coupling is a second-order effect, except for some specific directions (e.g., blind spots) or some specific feeding schemes (e.g., high-input reflection). When high performance is required in some specific area (e.g., low sidelobe level, wide beamsweep, or high DOA resolution), the coupling modeling of the array becomes crucial. Coupling model characterization is very important to maintain the quality of the system.

13.4 Coupling Model

Coupling changes the receiving array behavior in two ways, as described by Kelley [13] and Mailloux [14]:

- It changes the impedance of the equivalent generator at array element terminals, depending on the load impedance of the rest of the array (active impedance).
- It changes the field diagram and polarization vector of the array element from the isolated element (active or embedded diagram).

These two changes can be modeled through an analysis of the field diagram of the transmitting or receiving array elements. Keeping the load impedance of the elements constant and equal to reference impedance Z_0, the element receiver diagram does not change, and can be evaluated for each array element. This condition is not very difficult to satisfy in active DOA systems, where a receiving chain loads the array elements.

The change of input impedance at each element terminal can be easily modeled as a passive multipole circuit, where the P antenna terminals form the P ports of the multipole circuit. Scattering parameters can be used to model the multipole circuit, and active input impedance or reflection coefficient can be computed for the transmitting or receiving array, as a function of the generators or loads connected to the ports. Figure 13.1 shows this behavior.

If the input wave vector is defined as $\mathbf{a} = (a_1, a_2, \ldots, a_P)^t$, and the received or reflected wave vector is defined as $\mathbf{b} = (b_1, b_2, \ldots, b_P)^t$, then the relation between the coupled array elements can be written as:

$$\mathbf{b} = \mathbf{S}_a \mathbf{a} \tag{13.25}$$

where $\mathbf{S}_a$ is the scattering parameters square matrix of $P \times P$ elements. This scattering matrix or any equivalent matrix [i.e., the impedance ($\mathbf{Z}_a$) or admittance ($\mathbf{Y}_a$)], can be easily measured from the antenna terminals.

For the receiving antenna case, the terminals are connected to receivers with an equivalent input reflection coefficient (Γ_k), so we can write a second relation through a diagonal matrix $\mathbf{\Gamma}_L$ as:

$$\mathbf{a} = \mathbf{\Gamma}_L \mathbf{b}, \text{ with } \mathbf{\Gamma}_L = \begin{bmatrix} \Gamma_1 & \cdots & 0 \\ \vdots & \ddots & \vdots \\ 0 & \cdots & \Gamma_P \end{bmatrix} \tag{13.26}$$

Once the load impedance has been fixed (e.g., as the normalized impedance $Z_k = Z_0$ or $\Gamma_k = 0$), the active diagram of each element is constant and independent of the direction, power, or number of signals incoming to the array. It does not mean that these diagrams are equal for all elements, or can be measured with each element apart from the array. The active diagram must take into account all the array elements loaded, the mechanical support structure, and any other perturbing objects, such as radome, limited grown planes, and so forth.

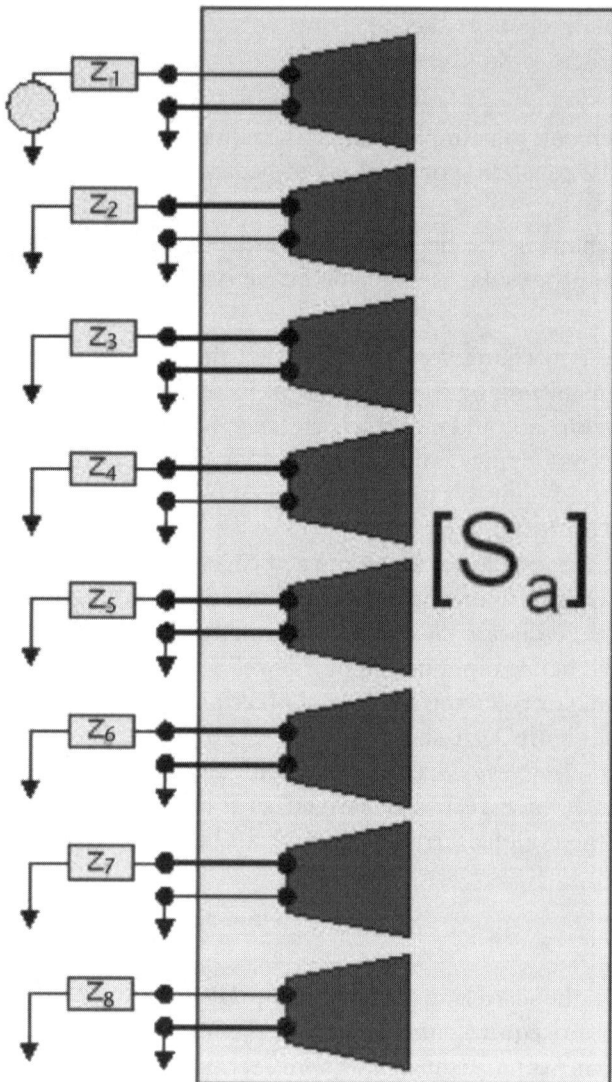

Figure 13.1 Schematic block diagram of an array as a multipole network.

If (13.12) is taken as the active diagram formulation for the equivalent output voltage wave coefficient (**b**), then we can formulate a model for a general receiving array as:

$$\mathbf{b} = \mathbf{S_a a} + E_{s\theta}\mathbf{f}_{s\theta} + E_{s\phi}\mathbf{f}_{s\varphi} \text{ or } \mathbf{b} = (\mathbf{I} - \mathbf{S_a \Gamma_L})^{-1}[E_{s\theta}\mathbf{f}_{s\theta} + E_{s\phi}\mathbf{f}_{s\varphi}] \quad (13.27)$$

where, in this case, vectors $\mathbf{f}_{s\theta}$ and $\mathbf{f}_{s\phi}$ represent the active diagrams of the array elements for each polarization component.

The next problems to be solved concern the influence of the elements of the array in the active diagram, the modeling of the active diagram from individual element measurement, and the effect of other perturbing objects on the active diagrams.

13.4 Coupling Model

Wasylkiwskij [15] showed the possibility of obtaining the coupling (Z) parameters from the element diagram for minimum scattering antennas (MSA). These particular types of antenna have no losses, are reciprocal, and do not disturb the incoming field when loaded with some specific impedance. Canonical minimum scattering antennas (CMSA) show this behavior when loaded by an open circuit. Real antennas, like dipoles, often have been characterized as MSA, obtaining a good agreement in the coupling computing parameters [16, 17]. This approximation assumes only one working mode in the antenna radiated field, and a particular reference plane for the terminals. Some authors, such as Lo and Vu [18], clearly express the approximation done: "... with the assumption that an element with its terminal open-circuited has no effect on the others in the array." Gupta [19] assumes that the induced voltage on open circuit elements is proportional to the incoming field. Steyskal [20] does the same approximation for horn antennas, using scattering parameters and matched ports. Haro [21] performs a more complete coupling model, including high-order modes in a horn cluster and finite ground plane. Most authors in signal processing assume the approximation that a coupling matrix C can always describe the coupling behavior for any situation [22, 23].

It is difficult to model the receiving antenna, but it is easy to look at the antenna as a transmitting array, and then apply the principle of reciprocity. When the generator is connected to the first element of the array, and the other elements are loaded with the reference impedance Z_0, the radiated field is created through several mechanisms.

- First, and most important, is the same current distribution that appears when the element is isolated from the rest of the antenna. That current creates a field known as the element field $[\hat{e}_A(\theta, \phi) F_A(\theta, \phi)]$.
- Another mechanism is the element field radiated by other array elements through the induced current. It can be modeled by some coupling coefficients, which are generally different from that measured in the terminals.
- Usually, other polarization or higher order modes can be induced through element-to-element coupling. These modes are usually avoided in the isolated element, but may appear when the radiators are combined.
- Currents induced in other structures, such as supporting metallic bars, covering radomes, ground plane edges, and so forth, can modify the field diagram.

If a linear behavior is assumed in the passive antenna structure, then these radiation mechanisms can be modeled by a set of coupling coefficients. Assuming the radiated field created by any radiator working in their main mode is known, or working in their higher order modes or the field radiated by any other surrounding structure modes, the field radiated by the array always can be written as a sum of contributions, resulting in:

$$\vec{E}(\theta, \phi) = \sqrt{2\eta_0} \frac{e^{(-jk_0 r)}}{r} \sum_{n=1}^{N} b_n^e \hat{e}_{An}(\theta, \phi) F_{An}(\theta, \phi) e^{[-j\omega_c \tau_n(\theta, \phi)]} \quad (13.28)$$

This effect can be seen in Figure 13.2, where the subindex An refers to the main radiating mode of the element alone in the space for $n = 1$ to P, and to the higher order modes or structure modes for $n = P + 1$ to N. Here, the parameter b_n^e represents a complex coefficient of the field radiated in some specific mode and element.

In the case of feeding only the pth element of the array and keeping the others loaded with the reference impedance, the radiation coefficients can be related to the excitation through a linear relation, such as $b_n^e = s_{p,n}^e a_p$.

For a unit input power ($a_p = 1$), the active pattern for the pth element can be written as:

$$\rho_p \hat{e}_p(\theta, \phi) F_p(\theta, \phi) e^{[-j\omega_c \tau_p(\theta, \phi)]} = \sum_{n=1}^{N} s_{p,n}^e \hat{e}_{An}(\theta, \phi) F_{An}(\theta, \phi) e^{[-j\omega_c \tau_n(\theta, \phi)]} \quad (13.29)$$

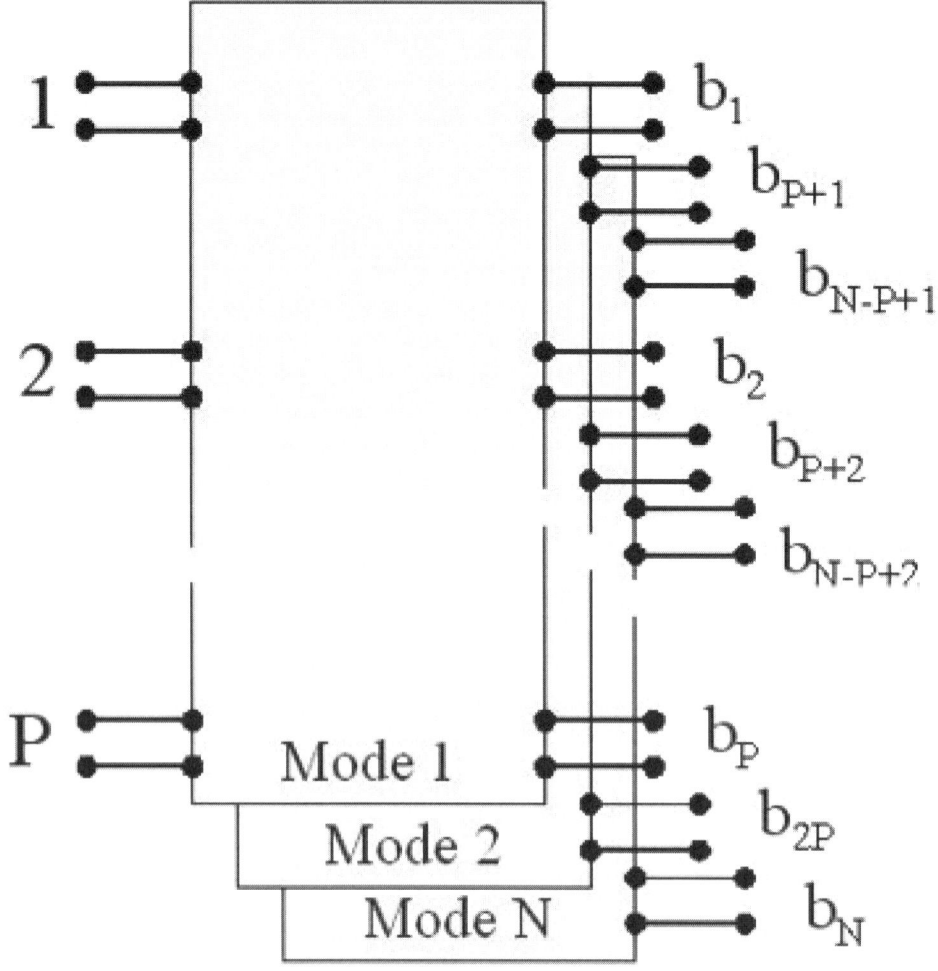

Figure 13.2 Schematic block diagram of an array as a multipole network with N radiating modes.

where ρ_p is an efficiency factor and F are the normalized diagrams. If superposition can be applied, then the overall radiated field can be computed as a sum and expressed in matrix form as:

$$\mathbf{b_e} = \mathbf{S_e a} \qquad (13.30)$$

where $\mathbf{b_e} = (b_1^e, b_2^e, \ldots, b_N^e)^t$ represents a vector of N elements, $\mathbf{S_e}$ is a rectangular matrix of N lines and P columns, and $\mathbf{a}$ is the feeding voltage wave vector at the antenna terminals.

In the receiving mode, an antenna array works exactly the same as in the transmitting mode, and a dual equation can be written. Following with the scattering parameters and voltage waves, we can write the received wave $\mathbf{b}$ vector as:

$$\mathbf{b} = \mathbf{S_r a_e} \qquad (13.31)$$

where $\mathbf{b}$ is the output vector describing the received voltage waves at terminals, $\mathbf{a_e} = (a_1^e, a_2^e, \ldots, a_N^e)^t$ represents the field incoming to the antenna array, and $\mathbf{S_r} = \mathbf{S_e^t}$ is a matrix that represents the active field pattern as a function of main-order and high-order modes radiation. From this assumption, the active pattern to compute the received signal $\mathbf{x}(t) = \mathbf{b}$, where $\mathbf{a_e}$ is an N element vector, with elements a_n^e related to the impinging field as:

$$a_n^e = \frac{\lambda}{\sqrt{2\eta_0}} \vec{E}_s(t) \hat{e}_{An}(\theta, \phi) F_{An}(\theta, \phi) e^{[-j\omega_c \tau_n(\theta, \phi)]} \qquad (13.32)$$

As a general case, it is not possible to get a coupling model that can be compensated for with a single square matrix, not taking into account the change between the single element pattern and the active pattern of the radiators.

A particular case, which is usual in the narrowband resonant elements, happens when only one radiating mode is activated in the radiators and the remaining modes are negligible. In that case, the active pattern can be written as a function of a square coupling matrix, $\mathbf{S_e}$, and the common polarization and field pattern functions, $\hat{e}_A$ and F_A:

$$\rho_p \hat{e}_p(\theta, \phi) F_p(\theta, \phi) e^{[-j\omega_c \tau_p(\theta, \phi)]} = \hat{e}_A(\theta, \phi) F_A(\theta, \phi) \sum_{n=1}^{P} s_{p,n}^e e^{[-j\omega_c \tau_n(\theta, \phi)]} \qquad (13.33)$$

In this case, the steering vector can be written as a matrix product of the position steering vector and a coupling matrix. For a source in position s:

$$\vec{\mathbf{f}}_s = \hat{e}_{As} F_{As} \mathbf{S_r a}_s \qquad (13.34)$$

The snapshot can be obtained from (13.12) as:

$$\mathbf{x}(t) = \mathbf{b} = \vec{E}_s(t)(\mathbf{I} - \mathbf{S_a \Gamma_L})^{-1} \vec{\mathbf{f}}_s = f_s(t)(\mathbf{I} - \mathbf{S_a \Gamma_L})^{-1} \mathbf{S_r a}_s \qquad (13.35)$$

The matrix $\mathbf{C} = (\mathbf{I} - \mathbf{S_a \Gamma_L})^{-1} \mathbf{S_r}$ represents the coupling, and it can be computed and compensated in the general signal process for DOA computation. It can be seen that this matrix does not depend on the incoming signal direction or the polarization, but only on the array coupling and impedance loads.

When more than a single radiating mode is considered in the array, it is not possible to build a simplified matrix equation, and the model is more complicated. One simplified option is to consider all the higher order modes as new elements, and then build a larger array matrix. The difference is that the snapshot only gets part of the array signals, and reactive impedances close the high-order modes.

13.5 Coupling Influence in DOA Estimation

The signal model used in DOA algorithms when a set of M impinging sources arrive to the antenna, and when receiver noise is considered, can be written as:

$$\mathbf{x} = \mathbf{A_s} \cdot \mathbf{s} + \mathbf{n} \qquad (13.36)$$

where $\mathbf{x}_{P \times 1}$ is the snapshot, and can represent any voltage or current; $\mathbf{A}_{sP \times M}$ is the array manifold, which is a set of steering vectors in the M arriving directions; $\mathbf{s}_{M \times 1}$ is the signal vector; and $\mathbf{n}_{P \times 1}$ is the noise vector, assumed to be internal, white, and Gaussian, and, hence, not affected by mutual coupling. When mutual coupling is present, the synthesized snapshot would be modified through a coupling matrix $\mathbf{C}$ as:

$$\mathbf{x} = \mathbf{C} \cdot \mathbf{A_s} \cdot \mathbf{s} + \mathbf{n} \qquad (13.37)$$

Resulting in a correlation matrix of:

$$\mathbf{R_{xx}} = E[\mathbf{x} \cdot \mathbf{x}^H] = \mathbf{C} \cdot \mathbf{A_s} \cdot E[\mathbf{s}(t) \cdot \mathbf{s}^H(t)] \cdot \mathbf{A_s^H} \cdot \mathbf{C}^H + \sigma^2 \mathbf{I} \qquad (13.38)$$

Two remarks must be said at this point. First, signal eigenvectors do not span the same subspace as the array manifold, but the one spanned by matrix $\mathbf{C} \cdot \mathbf{A_s}$. Then, it is necessary to estimate the coupling matrix, and introduce it into the DOA algorithm to avoid its degradation. Second, for the case in which the elements of the array are separated far enough, matrix $\mathbf{C}$ becomes diagonal. This matrix will be used in the corresponding eigenstructure algorithm to compensate mutual coupling effects. eigenspace beamformers or DOA estimation algorithms, such as MUSIC [1], only need to replace the ideal array manifold with $\mathbf{C} \cdot \mathbf{A_s}$. Matrix $\mathbf{C}$ can be obtained either via EM simulation or from direct measurements in the array, as shown in Chapter 19.

Another approach has been followed to compensate the effects of mutual coupling in DOA estimation algorithms [24]. In this approach, a corrected snapshot is introduced via the inverse of the coupling matrix in the received snapshot. This corrected snapshot ($\mathbf{x}_{corr}$) and its correlation matrix can be formulated as:

$$\mathbf{x}_{corr} = \mathbf{C}^{-1} \cdot \mathbf{x} = \mathbf{C}^{-1} \cdot \mathbf{C} \cdot \mathbf{A_s} \cdot \mathbf{s} + \mathbf{C}^{-1} \cdot \mathbf{n} = \mathbf{A_s} \cdot \mathbf{s} + \mathbf{C}^{-1} \cdot \mathbf{n} \qquad (13.39)$$

$$\mathbf{R_{xx}} = E[\mathbf{x}_{corr} \cdot \mathbf{x}_{corr}^H] = \mathbf{A_s} \cdot E[\mathbf{s} \cdot \mathbf{s}^H] \cdot \mathbf{A_s}^H + \sigma^2 \mathbf{C}^{-1} \cdot (\mathbf{C}^{-1})^H \quad (13.40)$$

It can be seen that the application of these expressions introduces a colored noise through the correlation matrix $[\mathbf{C}^{-1}(\mathbf{C}^{-1})^H]$. Its influence will depend on the coupling level. This formulation needs to solve a generalized eigenvalue, which increases the computational cost.

13.6 Error Compensation

Two experiments have been done to show the effect of the mutual coupling and the compensation techniques with the proposed model.

Experiment 1

A ULA with seven rectangular patches placed in the H-plane has been considered. The patches resonate at 3.5 GHz, and are fed through a coaxial probe. Figure 13.3 shows the schematic of the linear array used for the estimation of mutual coupling effect and the resonant patch at 3.5 GHz.

For a $\lambda/2$ separation, the module and phase of the coupling coefficients of the whole array are represented in Figure 13.4. Each C_{ij} is the (i, j) term of the coupling matrix, and represents the amount of signal coupled from element j to i in the array. It can be seen that for this array, the maximum coupling coefficient has a module lower than 0.07.

When this coupling level is introduced in a problem for estimating the direction of arrival, it can be seen that the DOA cannot be estimated properly. Two simulations are presented to show MUSIC performance in the presence of mutual coupling with and without correction techniques. In Figure 13.5, the separation between patches has been set to $\lambda/4$, while in Figure 13.6, it is set to $\lambda/2$. All the graphics have been obtained for 100 samples, with SNR = 10 dB, and a resolution of $\theta = 0.5°$. The load impedance has been set to 100Ω, and the impinging directions are assumed on the array plane ($\phi = 0°$). It can be seen that coupling effects do not allow the correct estimation of both directions of arrival (0° and 14° for a $\lambda/4$ separation, and 0° and 6° for a $\lambda/2$ separation). The introduction of the coupling matrix (estimated earlier) into the DOA estimation algorithm allows the correct discrimination of signals directions, as seen in Figures 13.5 and 13.6.

Experiment 2

To see the performance of the compensated coupling estimation, root-MUSIC and ESPRIT were implemented in a linear array composed of four double-polarized patch antennas. Using this kind of antenna has allowed us to obtain data for one linear polarization, circular polarization, or other combinations of polarizations. The scenario will be composed of two impinging sources, separated far enough apart so that the DOA could be properly estimated. The distance between patches in the array is $\lambda/2$. Figure 13.7 shows the schematic of the four double-polarized patch array antennas. The working frequency of this array will be 1.8 GHz.

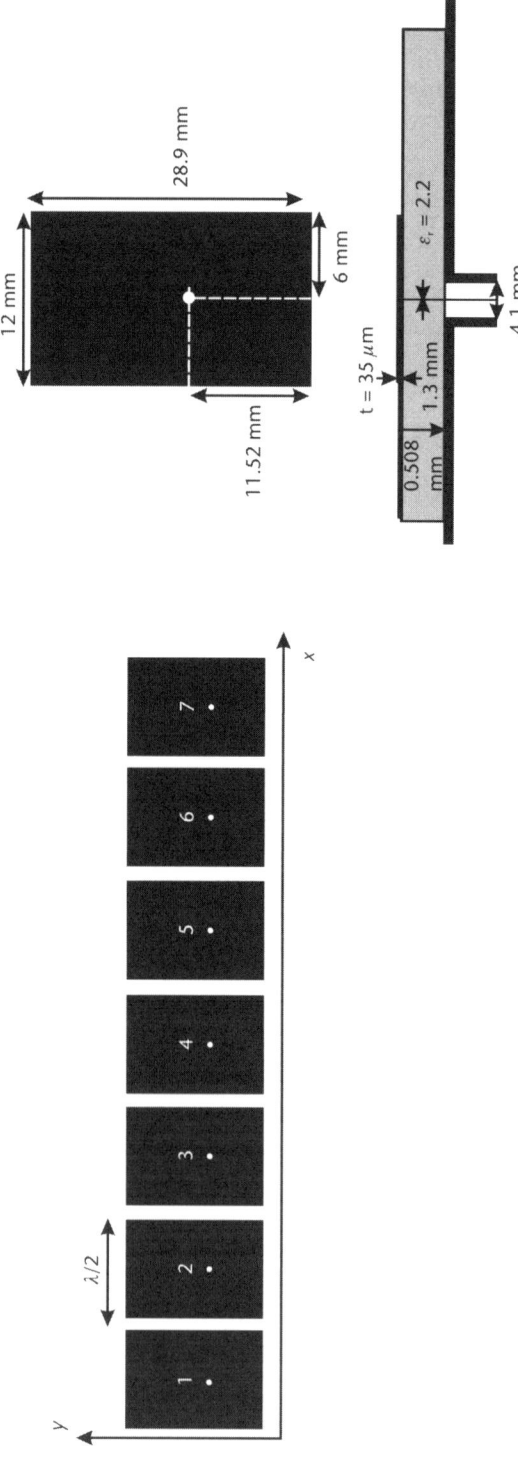

Figure 13.3 Schematic of the linear array used for estimation of mutual coupling matrix and resonant element at 3.5 GHz.

13.6 Error Compensation

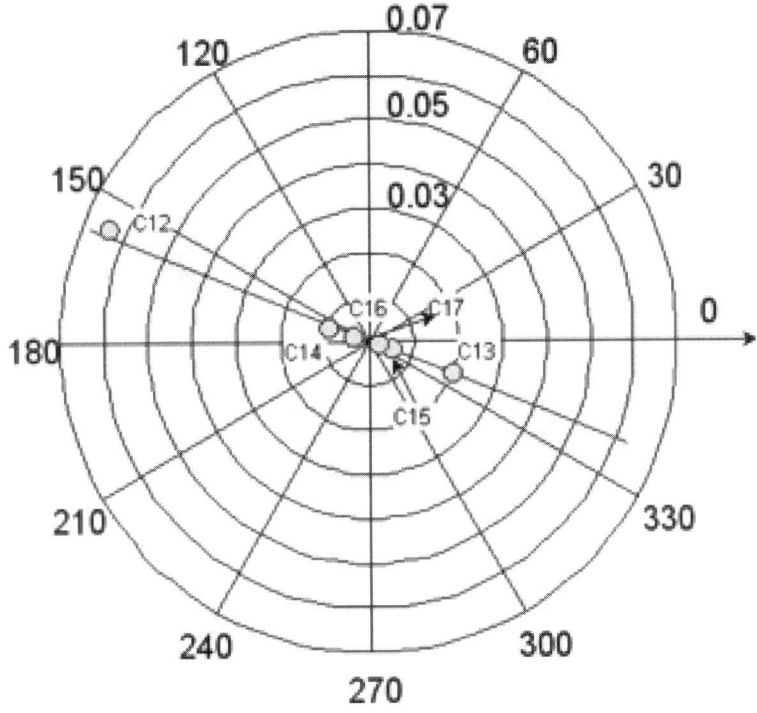

Figure 13.4 Coupling coefficients between each of the patches of the array. C_{ij} is the term (i, j) of the coupling matrix, and represents the coupling between elements i and j in polar notation. The level in radial coordinate is in natural units.

According to (13.37) and (13.39), the DOA algorithms will be compared, based on direct collected data (13.37) and on corrected collected data (13.39). The comparisons will be made based on the Cramér-Rao lower bound (CRLB), which will constitute an absolute bound. The calculation of the CRLB can be found in [25], and the theoretical value is given by

$$\sigma_{\hat{\theta}} = \frac{6}{P^3 \cdot SNR \cdot N} \qquad (13.41)$$

P is the number of antennas in the array, and N the number of samples in the DOA estimator.

The mutual couplings in the antenna have been measured for two orthogonal linear polarizations and for a circular polarization. Measurements of the RMSE in the estimation of the DOA parameter will be done in front of the SNR ratio in the incoming signal and the number of samples (N).

Figures 13.8 and 13.9 consist of plots of the CRLB and the RMSE in the estimation with root-MUSIC and ESPRIT algorithms. In the graphs, the curves denoted as *esprit* and *root-music* will represent the error in the parameter estimation with the mutual coupling effect corrected. The curves denoted as *esprit2* and *root-music2* will represent the error in the parameter estimation with no correction in the mutual coupling effect. Finally, the curves denoted as *cr* and *crcoupled* will

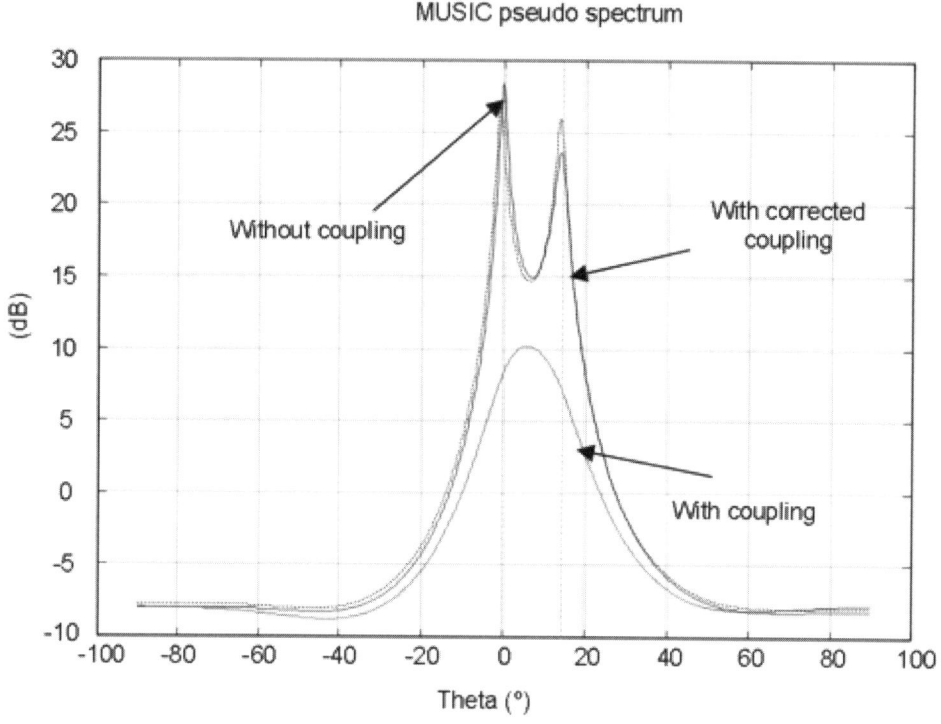

Figure 13.5 DOA estimation for a ULA with seven patches in the *H*-plane separated $\lambda/4$, including coupling effects.

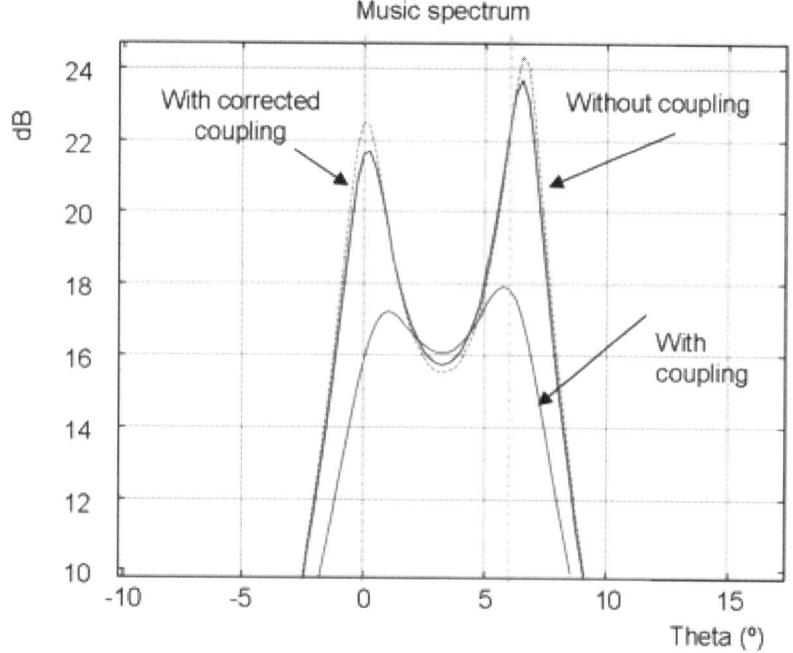

Figure 13.6 The same as Figure 13.5 for a $\lambda/2$ separation between elements.

13.6 Error Compensation

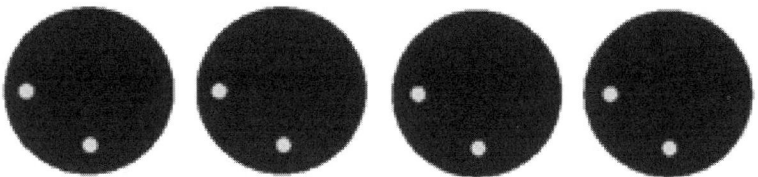

Figure 13.7 Array of four double-polarized patches.

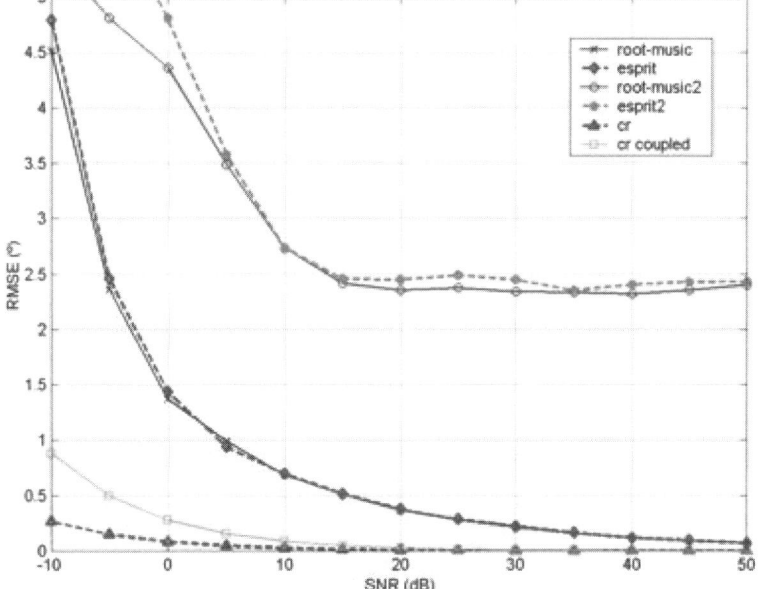

Figure 13.8 RMSE for the estimated DOA for vertical polarization.

represent the Cramér-Rao lower bound for the noncoupling and coupling cases. The number of samples for the simulation has been 128, while the number of runs for the estimation has been 1,000. The patch-antenna array, measured to determine the coupling matrix, is composed of four circular patches with double polarization. Figure 13.8 represents a linear vertical polarization, in which the horizontal case is similar, while Figure 13.9 represents a circular polarization. From these figures, it can be seen that:

- In the presence of mutual coupling, both ESPRIT and root-MUSIC reach an inaccurate estimation of DOA, resulting in a stationary error from SNR larger than 10 dB. This is more critical in the ESPRIT algorithm. This may be due to the inherent ill-conditioned problem in the four patch-antenna array. If the number of antennas were increased, then this error would be reduced.
- It has been shown that when no coupling correction scheme is done, there are no important differences in the behavior of both linear polarizations. However, when looking at the RMSE for circular polarization, a decrease

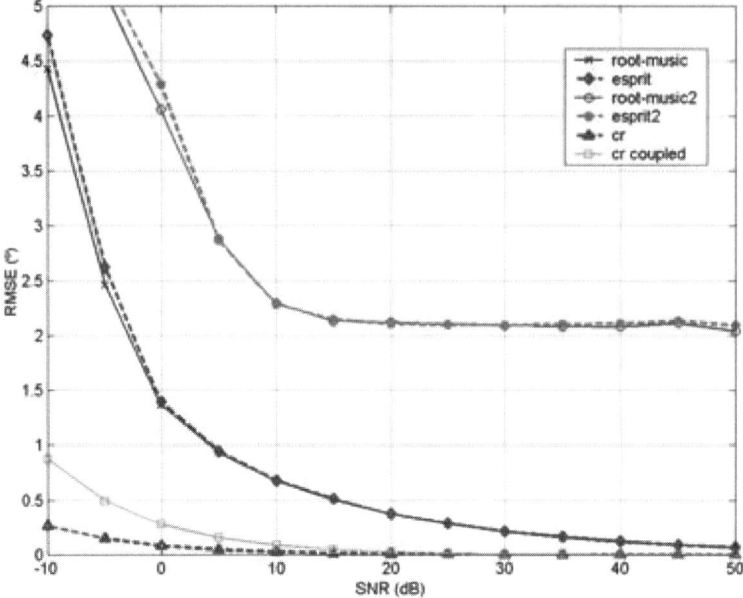

Figure 13.9 RMSE for the estimated DOA for circular polarization.

in the RMSE appears. This is due to the diversity in polarization that allows a reduction in the RMSE in the presence of mutual coupling.
- When the coupling correction technique is done, there is a clear reduction in the RMSE in the DOA estimation, and both algorithms approach the CRLB, although the behavior is somewhat better for the case of circular polarization.

Figures 13.10 and 13.11 show the RMSE dependence on the number of samples taken to do the DOA estimation. The simulations have been done on an SNR margin ranging from 0 to 15 dB, and the case for circular polarization has been chosen, since the results for the three polarizations are similar. From these figures, it can be concluded that:

- If no coupling correction technique is applied, then there is no clear dependence with the number of samples, but the error in the DOA estimation is large enough, so that the algorithm could not be used for DOA estimation.
- When the coupling correction scheme is applied, then the behavior of the corrected algorithms approaches the CRLB. This approach is closer when the number of samples in the estimation is larger. Then, a trade-off between the number of samples and the computation cost would be done. The number of 128 samples would be reasonable for the proposed array.

13.7 Conclusions

This chapter presents a steering vector model that considers possible errors and the mutual coupling among radiating elements in a real antenna. This chapter

13.7 Conclusions

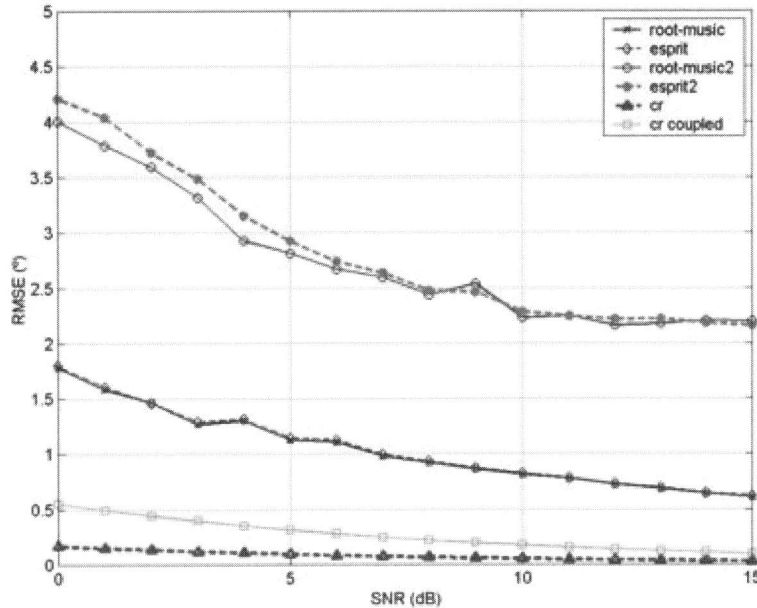

Figure 13.10 RMSE for the estimated DOA with 64 samples for circular polarization.

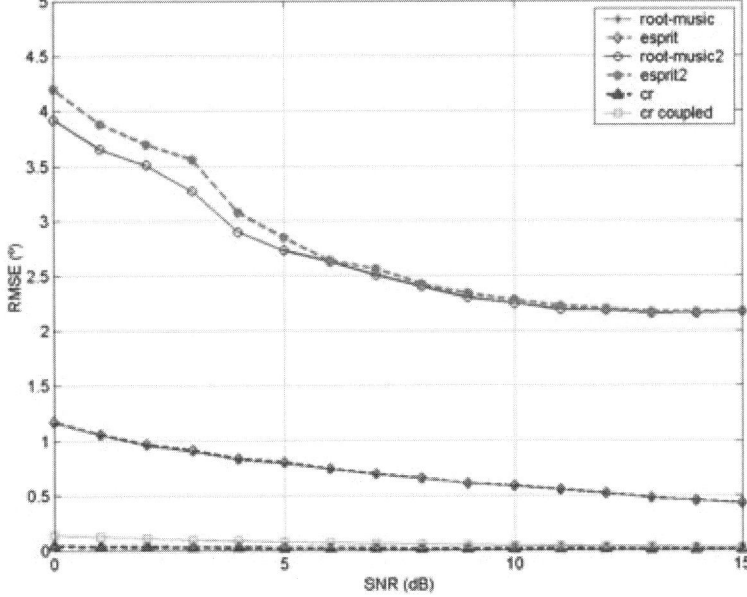

Figure 13.11 RMSE for the estimated DOA with 256 samples for circular polarization.

shows the general need of using active or embedded field patterns to obtain a good linear model of the array and to describe a dual polarization steering vector. In the particular case of single mode resonant antennas, a good approximation can be obtained with a coupling matrix model. This model is applied to estimate the DOA of an incident signal. Section 13.3 shows the influence on DOA estimation in the presence of random errors due to a nonideal antenna or RF electronic circuit behavior. Basically, errors in amplitude or phase in the receiving system, position of the array elements, and differences in the radiation element electric field or polarization pattern are studied. Section 13.4 studies the effect of the coupling among the individual receiving antennas, obtaining a matrix that characterizes this coupling. The effect of the coupling in the DOA estimation has been modeled as a colored noise on the correlation matrix. Finally, Section 13.5 provides some examples to show the effect of this coupling, and the compensation techniques used to obtain the actual DOA.

References

[1] Krim, H., and M. Viberg, "Two Decades of Array Signal Processing Research," *IEEE Signal Processing Magazine*, July 1996.

[2] Van Trees, H., *Detection, Estimation, and Modulation Theory*, New York: John Wiley & Sons, 2001.

[3] Haykin, S., *Advances in Spectrum Analysis and Array Processing, Vol. 2*, Upper Saddle River, NJ: Prentice Hall, 1991.

[4] Swindlehurst, A. L., and T. Kailath, "A Performance Analysis of Subspace-Based Methods in the Presence of Model Errors, Part I: The MUSIC Algorithm," *IEEE Trans. on Signal Processing*, Vol. SP 40, July 1992, pp. 1758–1774.

[5] Swindlehurst, A. L., and T. Kailath, "A Performance Analysis of Subspace-Based Methods in the Presence of Model Errors: Part II: Multidimensional Algorithms," *IEEE Trans. on Signal Processing*, Vol. SP 41, September 1993, pp. 2882–2890.

[6] Friedlander, B., and A. J. Weiss, "Direction Finding Techniques in the Presence of Mutual Coupling," *IEEE Trans. on Antennas and Propagation*, Vol. A-39, March 1991, pp. 273–284.

[7] Aumann, H. M., A. J. Fenn, and F. G. Willwerth, "Phased Array Antenna Calibration and Pattern Prediction Using Mutual Coupling Measurements," *IEEE Trans. on Antennas and Propagation*, Vol. AP-37, July 1989, pp. 844–851.

[8] Pozar, D. M., "Finite Phased Arrays of Rectangular Microstrip Patches," *IEEE Trans. on Antennas and Propagation*, Vol. AP-34, May 1986, pp. 658–665.

[9] Segovia, D., R. Martín, and M. Sierra, "Mutual Coupling Effects Correction in Microstrip Arrays for Direction of Arrival (DOA) Estimation," *IEE Proc. on Microwave, Antennas and Propagation*, Vol. 149-2, April 2002, pp. 113–118.

[10] Capon, J., "High Resolution Frequency Wavenumber Spectrum Analysis," *IEEE Proc.*, Vol. 57, No. 8, August 1969, pp.1408–1418.

[11] Wax, M., and Y. Anu, "Performance Analysis of the Minimum Variance Beamformer in the Presence of Steering Vector Errors," *IEEE Trans*, Vol. SP 44, No. 4, April 1996, pp. 938–947.

[12] Kim, K., T. K. Sarkar, and M. Salazar, "Adaptive Processing Using a Single Snapshot for a Nonuniformly Spaced Array in the Presence of Mutual Coupling and Near-Field Scatterers," *IEEE Trans. on Antennas and Propagation*, Vol. AP-50, No. 5, May 2002, pp. 582–590.

[13] Kelley, D. F., and W. L. Stutzman, "Array Antenna Pattern Modeling Methods That Include Mutual Coupling Effects," *IEEE Trans. on Antennas and Propagation*, Vol. AP-41, No. 12, December 1993.

[14] Mailloux, J. R., *Phased Array Antenna Handbook*, Norwood, MA: Artech House, 1994.

[15] Wasylkiwskyj, W., and W. K. Kahn, "Theory of Mutual Coupling Among Minimum-Scattering Antennas," *IEEE Trans. on Antennas and Propagation*, Vol. AP-18, No. 2, March 1970, pp. 204–216.

[16] Elliott, R. S., and G. J. Stern, "The Design of Microstrip Dipole Arrays Including Mutual Coupling, Part I: Theory," *IEEE Trans. on Antennas and Propagation*, Vol. AP-29, No. 5, September 1981, pp. 757–760.

[17] Stern, G. J., and R. S. Elliott, "The Design of Microstrip Dipole Arrays Including Mutual Coupling, Part II: Experiment," *IEEE Trans. on Antennas and Propagation*, Vol. AP-29, No. 5, September 1981, pp. 761–765.

[18] Lo, K. W., and T. B. Vu, "Simple S-Parameter Model for Receiving Antenna Array," *Electronic Letters*, Vol. 24, No. 20, September 29, 1988, pp. 1264–1266.

[19] Gupta, I. J., and A. A. Ksienski, "Effect of Mutual Coupling on the Performance of Adaptive Arrays," *IEEE Trans. on Antennas and Propagation*, Vol. AP-31, No. 5, September 1983.

[20] Steyskal, H., and J. S. Herd, "Mutual Coupling Compensation in Small Array Antennas," *IEEE Trans. on Antennas and Propagation*, Vol. 38, No. 12, December 1990.

[21] Haro, L., J. L. Besada, and B. Galocha, "On the Radiation of Horn Clusters Including Mutual Coupling and the Effects of Finite Metal Plates: Application to the Synthesis of Contoured Beam Antennas," *IEEE Trans. on Antennas and Propagation*, Vol. AP-41, No. 6, June 1993, pp. 713–722.

[22] Hui, H. T., "Compensating for the Mutual Coupling Effect in Direction Finding Base on a New Calculation Method for Mutual Impedance," *IEEE Antennas and Wireless Propagation Letters*, Vol. 2, 2003.

[23] Ludwig, A. C., "Mutual Coupling, Gain, and Directivity of an Array of Two Identical Antennas," *IEEE Trans. on Antennas and Propagation*, Vol. AP-24, No. 6, November 1976, pp. 837–841.

[24] Pasala, K. M., and E. M. Friel, "Mutual Coupling Effects and Their Reduction in Wideband Direction of Arrival Estimation," *IEEE Trans. on Aerospace Electronic Systems*, Vol. AES-30, October 1994, pp. 1116–1122.

[25] Satish, A., and R. L. Kashyap, "Maximum Likelihood Estimation and Cramer-Rao Bounds for Direction of Arrival Parameters of a Large Sensor Array," *IEEE Trans. on Antennas and Propagation*, Vol. AP-44, No. 4, April 1996, pp. 478–491.

PART IV
Specific Applications of DOA Estimation

CHAPTER 14
Maximum Likelihood Estimate of Target Angular Coordinates Under Main Beam Interference: Application to Recorded Live Data

Alfonso Farina, Giovanni Golino, and Luca Timmoneri

14.1 Introduction

Modern phased array radars are required to detect, locate, and track targets in the presence of natural and intentional electromagnetic interference. Monopulse [1] is the conventional technique to determine the target angular coordinates. This technique does not properly work in the presence of interference. The intentional interference might impinge on the sidelobes as well as on the main beam of the antenna pattern from certain directions. When there is a directional interference in the main beam, it is necessary to use auxiliary antennas that have high gain in the angular region covered by the radar main beam to form a null in the direction of the interference [2]. Thus, the shapes of the sum and difference monopulse beams will be severely distorted by adaptive processing, and the conventional monopulse technique cannot be applied [3].

This chapter illustrates the application of the MLE technique to determine the target direction-of-arrival (TDOA) in presence of main beam interference (MBI) and the related CRLB angular accuracy. Closed formula for the CRLB of the estimation of TDOA is presented for various interference conditions [4–6].

Two different sets of high-gain beams are considered. The first one includes three high-gain beams, and the sum and the two difference beams (azimuth and elevation). The second set of beams also comprises the so-called double difference beam [7].

In this chapter, the following results will be shown:

- Conception of the MLE of the target angular coordinates (Section 14.2);
- Closed form expression of the CRLB of the target angular coordinates (Section 14.3);
- MLE realization via the Newton-Raphson recursion and evaluation of the mean value, and of the standard deviation of the estimation errors via Monte Carlo simulation and comparison with the CRLB results (Section 14.4);
- Test on recorded live data (Section 14.5);
- Conclusions (Section 14.6).

14.2 MLE Algorithm for Estimation of Target Angular Coordinates

In this section, the MLE of the target angular coordinates azimuth and elevation (θ, ϕ) in the presence of a directional interference, by processing the data received by a set of high-gain beams is presented. The mathematical model of the received radar signals and the processing scheme are detailed.

Let us define the vector $\mathbf{V}(\theta, \phi)$ containing the pattern of the antenna beams in a certain direction (θ, ϕ). We considered two sets of beams. The first set is composed of the sum and difference in azimuth and difference in elevation beams. The second set is composed of the same beams, with the addition of the double difference beam. The beams are obtained by a notional two-dimensional planar array of 30-by-30 radiating elements, separated by one-half of the transmitting wavelength, and C is the carrier frequency. The pattern of such beams is shown in Figures 14.1 to 14.4. The -3-dB width of the sum beam is $3.5°$. The specification on the angular estimator is to achieve an accuracy of one-tenth of the -3-dB width of the sum beam. Thus, the standard deviation of the estimation error should not exceed $0.35°$.

The following vector can then represent the set of the received radar echoes:

$$\mathbf{z} \equiv b\mathbf{V}(\theta_T, \phi_T) + \mathbf{d} \tag{14.1}$$

where b is the complex target echo, and $\mathbf{d}$ is the disturbance, a Gaussian process with zero-mean value and covariance matrix $\mathbf{M}_d$.

The expression of the disturbance covariance matrix is:

$$\mathbf{M}_d = \sigma_n^2 \cdot [\mathbf{I} + INR \cdot \mathbf{V}(\theta_I, \phi_I) \cdot \mathbf{V}(\theta_I, \phi_I)^H] \tag{14.2}$$

where (θ_I, ϕ_I) are the angular coordinates of the interference, and

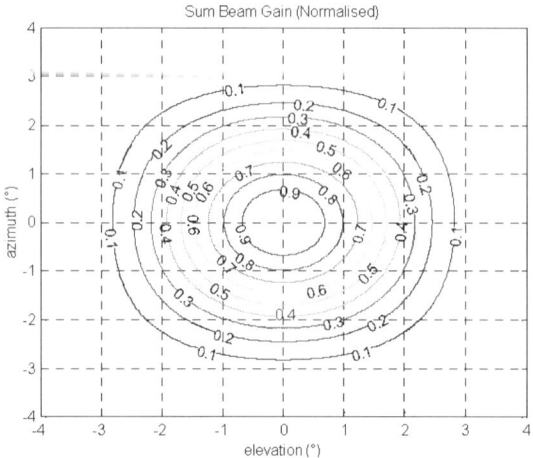

Figure 14.1 Pattern of the sum beam.

14.2 MLE Algorithm for Estimation of Target Angular Coordinates

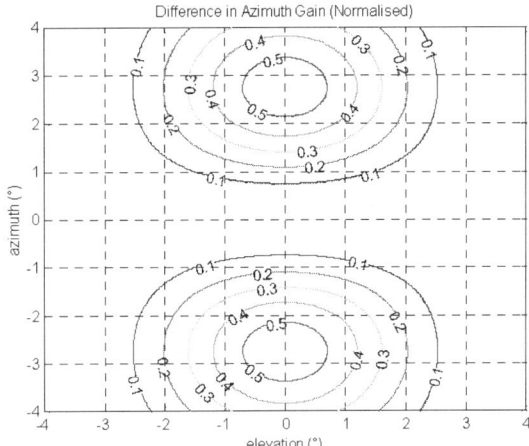

Figure 14.2 Pattern of the difference in azimuth beam.

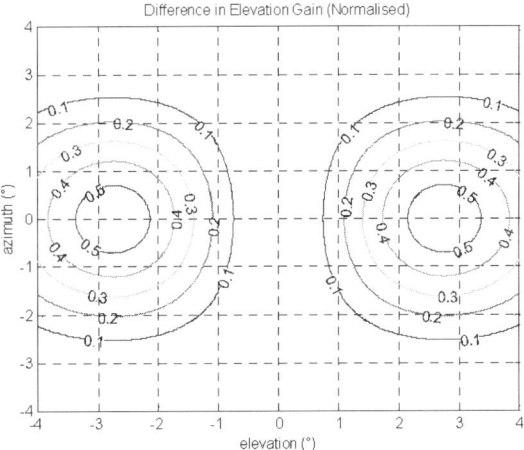

Figure 14.3 Pattern of the difference in elevation.

$$INR = \frac{P_I}{\sigma_n^2}$$

is the interference-to-noise power ratio.

Owing to the Gaussian nature of the pdf of **z**, the MLE problem brings to the following minimization problem:

$$(\hat{b}, \hat{\theta}_T, \hat{\phi}_T) = \arg\min_{b,\theta,\phi} \{[\mathbf{z} - b\mathbf{V}(\theta, \phi)]^H \mathbf{M}_d^{-1}[\mathbf{z} - b\mathbf{V}(\theta, \phi)]\} \quad (14.3)$$

$$= \arg\min_{b,\theta,\phi} F(b, \theta, \phi)$$

The minimization of the above-mentioned functional also corresponds to the maximization of the following functional:

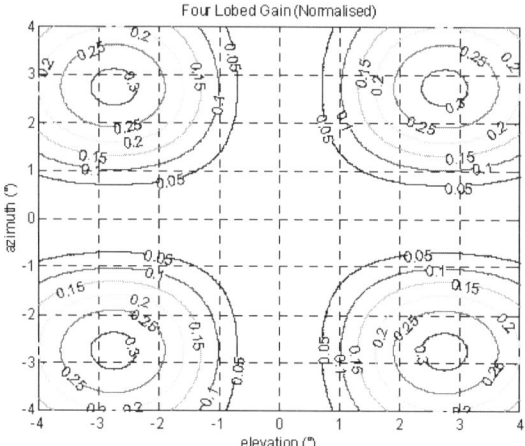

Figure 14.4 Pattern of the double difference beam.

$$Q(\theta, \phi) = \frac{|\mathbf{V}^H(\theta, \phi) \cdot \mathbf{M}_d^{-1} \cdot \mathbf{z}|^2}{\mathbf{V}^H(\theta, \phi) \cdot \mathbf{M}_d^{-1} \cdot \mathbf{V}(\theta, \phi)} \quad (14.4)$$

The algorithm needs the estimation of the disturbance covariance matrix, which is obtainable by the echoes corresponding to range cells adjacent to the cell under test. In the simulations of Section 14.4, and in the test on live data of Section 14.5, we considered 15 range cells. We also need the description of the quiescent beam shapes in the main beam region, which are available from the antenna design and measurements.

The maximum of the functional can be estimated by an exhaustive search in the range of values of interest, or by using a fast recursive algorithm, such as the Newton-Raphson algorithm:

$$\hat{\mathbf{x}}_{k+1} = \hat{\mathbf{x}}_k - \left[E\left\{ \frac{\partial^2 Q}{\partial \mathbf{x}^2} \bigg|_{\hat{\mathbf{x}}_k} \right\} \right]^{-1} \left[\frac{\partial Q}{\partial \mathbf{x}} \bigg|_{\hat{\mathbf{x}}_k} \right] \quad (14.5)$$

where $\hat{\mathbf{x}}_k = [\hat{\theta}_k \quad \hat{\phi}_k]^t$. The recursion can be initialized with the angular coordinates of the main beam pointing.

Moreover, the comparison of the Q functional with a suitable threshold λ permits the target detection to maintain the prescribed constant false alarm rate. The radar signals are taken only for the range cell in which the detection occurred, and are further processed by the MLE algorithm to produce the target DOA.

In Figure 14.5, a flowchart of the processing chain is shown.

14.3 CRLB of Estimation of Target Angular Coordinates

To calculate the CRLB, the Fisher information matrix (FIM), defined as follows, is needed [see (14.8)]:

14.3 CRLB of Estimation of Target Angular Coordinates

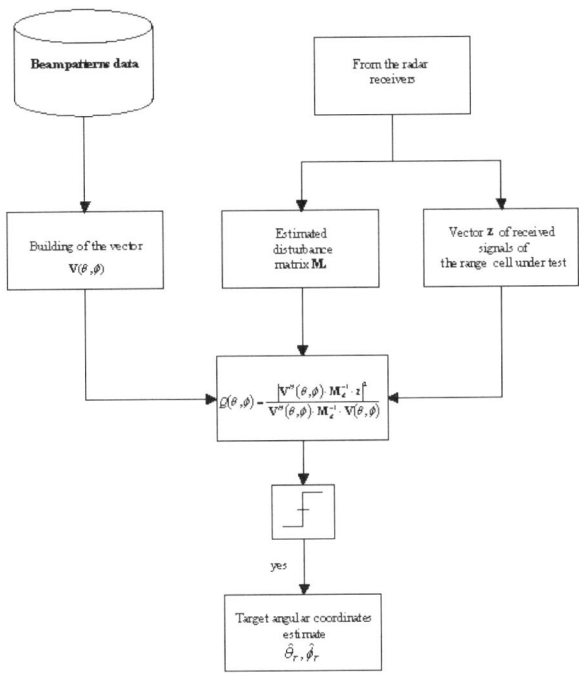

Figure 14.5 Flowchart of processing chain.

$$\mathbf{FIM} = E \left\{ \begin{bmatrix} \frac{\partial^2 F}{\partial \theta^2} & \frac{\partial^2 F}{\partial \theta \partial \phi} & \frac{\partial^2 F}{\partial \theta \partial b_R} & \frac{\partial^2 F}{\partial \theta \partial b_I} \\ \frac{\partial^2 F}{\partial \phi \partial \theta} & \frac{\partial^2 F}{\partial \phi^2} & \frac{\partial^2 F}{\partial \phi \partial b_R} & \frac{\partial^2 F}{\partial \phi \partial b_I} \\ \frac{\partial^2 F}{\partial b_R \partial \theta} & \frac{\partial^2 F}{\partial b_R \partial \theta} & \frac{\partial^2 F}{\partial b_R^2} & \frac{\partial^2 F}{\partial b_R \partial b_I} \\ \frac{\partial^2 F}{\partial b_I \partial \theta} & \frac{\partial^2 F}{\partial b_I \partial \phi} & \frac{\partial^2 F}{\partial b_I \partial b_R} & \frac{\partial^2 F}{\partial b_I^2} \end{bmatrix} \right\} \quad (14.6)$$

where b_R and b_I are the real part and the imaginary parts of the target echo, respectively.

The CRLB values for the standard deviation of the estimates of the azimuth and elevation coordinates of the target are obtained by the following formulas:

$$\sigma_\theta = \sqrt{[\mathbf{FIM}^{-1}]_{1,1}} \quad (14.7)$$

$$\sigma_\phi = \sqrt{[\mathbf{FIM}^{-1}]_{2,2}}$$

In this section, the following scenario is considered for the analysis—one main beam interference with an interference-to-noise ratio (INR) equal to 30 dB set in

$(\theta_I, \phi_I) = (-1.5°, -1.5°)$, and a target with signal-to-noise ratio (SNR) equal to 30 dB, having its angular coordinates a changing parameter.

To simulate the detection process, the false alarm probability has been set to 10^{-6}, and the required detection probability has been set to 80%.

In Figure 14.6, the target angular sectors are shown, in which the requirements of the detection probability and on the CRLB are satisfied using the three beams set. The requirement on the CRLB is satisfied only in a small area around the interference, but in almost the entire sector, the target is not detectable. Therefore, when the target is detectable with the required probability, its position cannot be estimated with satisfactory accuracy.

In Figure 14.7, the target angular sectors are shown, for which the requirements of the detection probability and the CRLB are satisfied using the four beams set. In most parts of the sectors, both requirements are satisfied. Around the interference

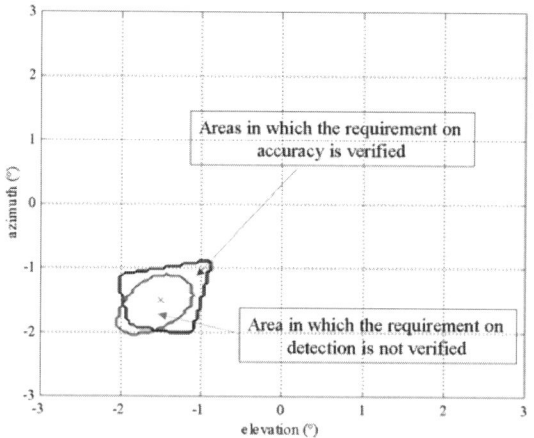

Figure 14.6 Areas where the probability of detection and the CRLB of the angular accuracy satisfy the requirements ($P_d > 0.8$; CRLB $< 0.35°$) for the set of three high-gain beams.

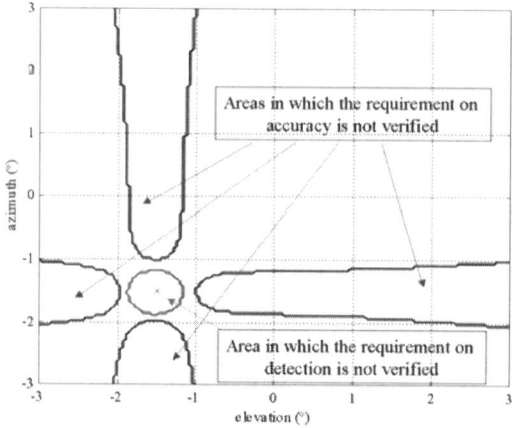

Figure 14.7 Areas where the probability of detection and the CRLB of the angular accuracy satisfy the requirements ($P_d > 0.8$; CRLB $< 0.35°$) for the set of four high-gain beams.

position, it is still not possible to detect the target. The target is detectable with the required probability, and its position can be estimated with the needed accuracy if the separation of the target in azimuth and elevation from the interference is more than 0.5°.

In the following four figures, more details about the CRLB curves are displayed.

In Figures 14.8 and 14.9, the CRLB of the azimuth estimation as a function of the target azimuth is shown for the three-beam set and for the four-beam set, respectively. The minimum of the curves is obtained for the interference azimuth, and it is independent of the target elevation angle and of the number of beams used by the algorithm. For the other target azimuth values, the CRLB of the four-beam set is smaller than the CRLB of the three-beam set. The four-beam curves are much flatter than the three-beam curves.

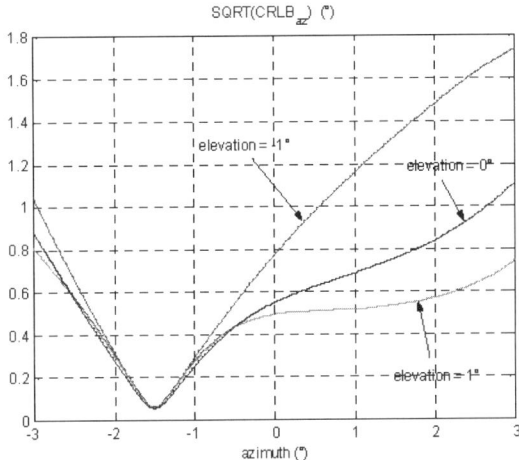

Figure 14.8 CRLB of the azimuth angular accuracy versus the azimuth of the target for the set of three high-gain beams, with different values for the elevation of the target.

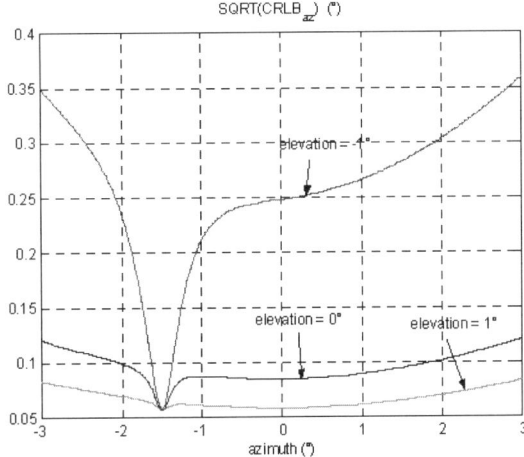

Figure 14.9 CRLB of the angular accuracy versus the azimuth of the target for the set of three high-gain beams, with different values for the elevation of the target.

In Figures 14.10 and 14.11, the CRLB of the azimuth estimation as function of the target elevation is shown for the three-beam set and for the four-beam set, respectively. If the target and interference elevation coincide, then the values of the CRLB are independent of the number of beams used by the algorithm, and they are points of maxima for the four-beam set. For the other target elevation values, the CRLB of the four-beam set is much smaller than the CRLB of the three-beam set.

14.4 MLE Algorithm Simulation Results

The MLE algorithm has been tested for different target positions. The results have been obtained by averaging 100 independent Monte Carlo trials. The results have been summarized and compared to the corresponding CRLB values in Table 14.1. The target has been detected in all the trials.

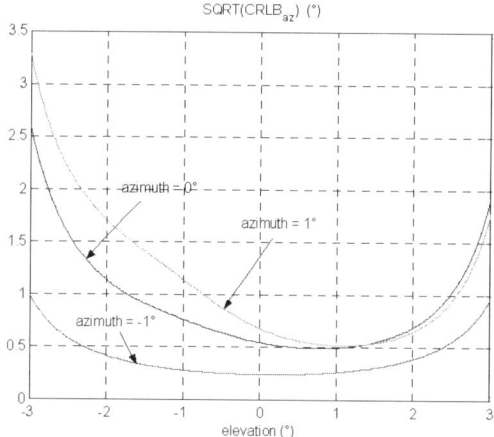

Figure 14.10 CRLB of the azimuth angular accuracy versus the elevation of the target for the set of three high-gain beams, with different values of the azimuth of the target.

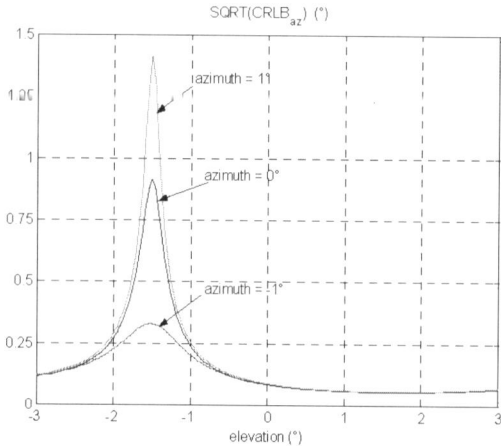

Figure 14.11 CRLB of the azimuth angular accuracy versus the elevation of the target for the set of three high-gain beams, with different values of the azimuth of the target.

14.4 MLE Algorithm Simulation Results

Table 14.1 Results of the MLE Algorithm Simulation

θ_T	ϕ_T	$\theta_T - \hat{\theta}_T$ (3 beams)	$\theta_T - \hat{\theta}_T$ (4 beams)	$\phi_T - \hat{\phi}_T$ (3 beams)	$\phi_T - \hat{\phi}_T$ (4 beams)	σ_θ (3 beams)	$CRLB_\theta$ (3 beams)	σ_θ (4 beams)	$CRLB_\theta$ (4 beams)	σ_ϕ (3 beams)	$CRLB_\theta$ (3 beams)	σ_ϕ (4 beams)	$CRLB_\theta$ (4 beams)
−1°	−1°	−0.01°	−0.01°	−0.08°	0.01°	1.27°	0.28°	0.24°	0.21°	1.31°	0.28°	0.23°	0.21°
−1°	0°	0.15°	0.03°	0.23°	0.03°	0.82°	0.24°	0.10°	0.08°	1.23°	0.77°	0.26°	0.25°
−1°	1°	−0.12°	0.02°	0.30°	0.02°	1.43°	0.26°	0.06°	0.06°	1.89°	1.16°	0.28°	0.26°
0°	−1°	0.14°	0.00°	0.06°	0.03°	1.28°	0.77°	0.27°	0.25°	0.90°	0.24°	0.08°	0.08°
0°	0°	0.18°	0.03°	0.11°	0.02°	0.79°	0.55°	0.09°	0.08°	0.78°	0.55°	0.08°	0.08°
0°	1°	−0.24°	0.02°	0.21°	0.01°	0.90°	0.50°	0.05°	0.05°	1.21°	0.69°	0.09°	0.09°
1°	−1°	0.47°	−0.01°	0.03°	0.03°	1.70°	1.16°	0.30°	0.26°	0.96°	0.26°	0.07°	0.06°
1°	0°	0.48°	0.01°	−0.19°	0.02°	1.50°	0.69°	0.09°	0.09°	1.12°	0.50°	0.06°	0.05°
1°	1°	0.20°	0.01°	0.05°	0.02°	1.21°	0.52°	0.06°	0.06°	1.07°	0.52°	0.06°	0.06°

Note: This scenario consists of a target with angular coordinates (θ_T, ϕ_T) and of an MBI with coordinates $(\theta_T, \phi_T) = (-1.5°, -1.5°)$; SNR = INR = 30 dB; results averaged over 100 Monte Carlo trials; the target has been detected in all the trials.

The standard deviation of the target azimuth and elevation estimated by using three beams always has been unsatisfactory, as expected, by checking the corresponding CRLB values. Moreover, the estimation of the coordinates is not always unbiased, and its standard deviation exceeds the corresponding CRLB value.

The standard deviation of the target azimuth and elevation estimated by using four beams instead fulfills the requirements, and it is practically coincident with the corresponding CRLB values. Moreover, the estimation of the coordinates is always unbiased.

Figures 14.12 and 14.13 show two examples of estimation of the Q function (14.4), on which the MLE algorithm is based. The scenario is the same in both figures [SNR = INR = 30 dB, $(\theta_I, \phi_I) = (-1.5°, -1.5°)$, $(\theta_T, \phi_T) = (+1°, +1°)$]. Figure 14.12 is referred to as the three-beam case, while Figure 14.13 is referred to as the four-beam case.

Both functions have a maximum in the target position, and a "spiky" minimum in the interference position. The width of the maximum is an indicator of the accuracy of the angular estimation. The maximum of the four-beam case is in fact much narrower than in the case of three beams, and indicates a much better accuracy. In the same way, the "spiky" minimum indicates that the MLE algorithm can provide an accurate estimation of the interference position in both cases.

14.5 Test on Recorded Live Data

To test the validity of the MLE approach, several sets of data have been recorded in a site with a ground-based test bed C-band phased array radar (C-PAR) that is equipped with three high-gain beams [8, 9].

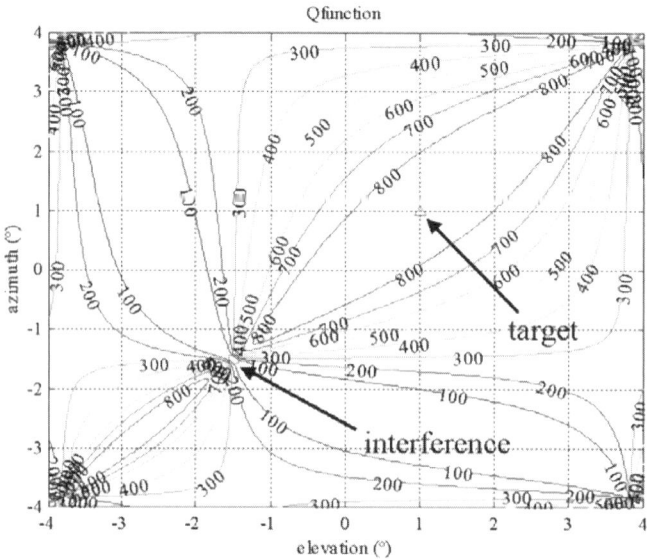

Figure 14.12 Estimation of the Q function, averaged over 100 Monte Carlo trials, with three beams. The target position is marked as a triangle and the interference position as a square.

14.5 Test on Recorded Live Data

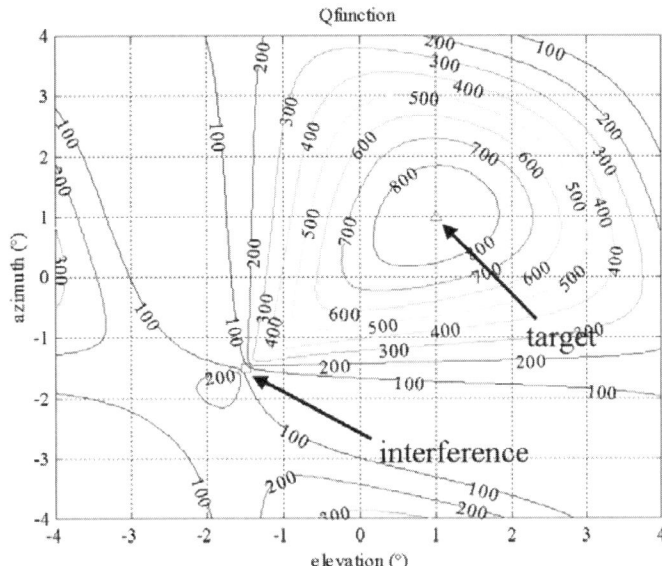

Figure 14.13 Estimation Q function, averaged over 100 Monte Carlo trials, with four beams. The target position is marked as a triangle and the interference position as a square.

The purpose of this section is to analyze the performance of the MLE algorithm in estimating the target angular coordinates in the presence of an MBI on the C-PAR recorded real data. After a brief description of the experimental setup, the subsequent items are discussed.

- A comparison between the beams reconstructed by simulation and the C-PAR real beams is performed.
- A suitable phase compensation on the reconstructed beams is estimated to make them congruent with the real beams. The estimates of the target coordinates, obtained after such compensation, are then compared to the uncompensated beams.
- An estimate of the interference position is performed using the same MLE algorithm, and the estimates with and without phase compensation are compared. The congruence of the reconstructed scenario obtained by the compensated estimates is analyzed and compared to the target and interference real positions.

14.5.1 Description of the Radar for Data Capture

The radar used for the live data capture was a testbed, passive phased array C-band radar equipped with a multichannel digital data recorder (MDDR) to simultaneously acquire the live data from three receiving channels. Its three receiving channels allowed the data to be recorded from the sum, difference in elevation, and difference in azimuth beams. It also was possible to install two low-gain auxiliaries at the extremities of the main antenna to test the sidelobe canceller (SLC) and the sidelobe blanker (SLB) [2, 8, 9]. In one phase of the experiments, it also was possible to acquire data from the sum channel, one of the two difference channels, and one of the two low-gain auxiliary channels, in order to test the radar in a mixture of main beam and sidelobe interference conditions. During the interference acquisition campaign, several recordings were made with the aim of testing the performance of MLE algorithms for TDOA estimation.

14.5.2 Data Capture Experimental Setup

A series of trials were performed to evaluate the performance of the C-PAR under various scenarios. A set of suitably positioned interference emulators and transponders was used to produce the required scenario, and the antenna was either electronically or manually pointed in the direction of the required interference emitter. Figure 14.14 depicts the environmental setup.

Two transponders (TP1 and TP2) emulated the useful targets, and three interference emulators (IE1, IE2, and IE3) tested the performance of the radar in various conditions, to achieve the following goals:

- Characterization of radar receiving channels;
- Cancellation of two contemporaneous continuous noise-like interferences (NLI);
- Impinging of one NLI on the main antenna beam;
- Cancellation of two contemporaneous continuous NLI with angular distance either less than the radar beamwidth or wider;
- Detection of target in presence of two contemporaneous continuous NLIs;
- Blanking probability in presence of coherent repeater interference (CRI) only;

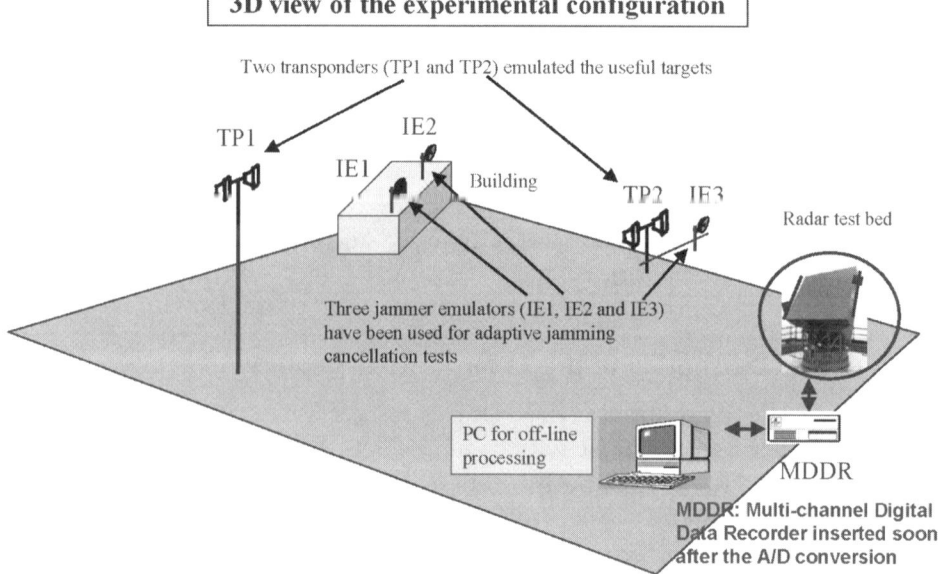

Figure 14.14 Experimental setup.

- Blanking of CRI and cancellation of NLI;
- Contemporaneous cancellation of jamming and clutter via space-time adaptive processing (STAP) algorithms;
- MLE estimation of TDOA under main beam jamming conditions.

Live data was acquired soon after the analog-to-digital (A/D) converters by means of the MDDR, permitting the simultaneous recording on the three receiving channels. Approximately 3 GB of data has been successfully collected.

14.5.3 Comparison Between the Theoretical and Real Radar Beams

A discrepancy has been found between the phase of the "difference and sum C-PAR beams ratio" that has been simulated, and the real C-PAR ratio found during the data recordings. The confirmation of such a hypothesis could negatively influence the accuracy of the angular estimate based on the MLE algorithm described in the Section 14.2. The algorithm can correctly estimate the target angular coordinates from real data, provided that it uses a good approximation of beam patterns from which the data was acquired.

In Figures 14.15 and 14.16, the phase diagrams of the ratio between the simulated difference in azimuth and sum beams, and of the ratio between the simulated difference in elevation and sum beam are shown, respectively, for the 0° elevation and the 0° azimuth planes. These plots are independent, with a good approximation, on the considered elevation and azimuth angles, respectively. They also can be considered valid for the other constant elevation and azimuth planes.

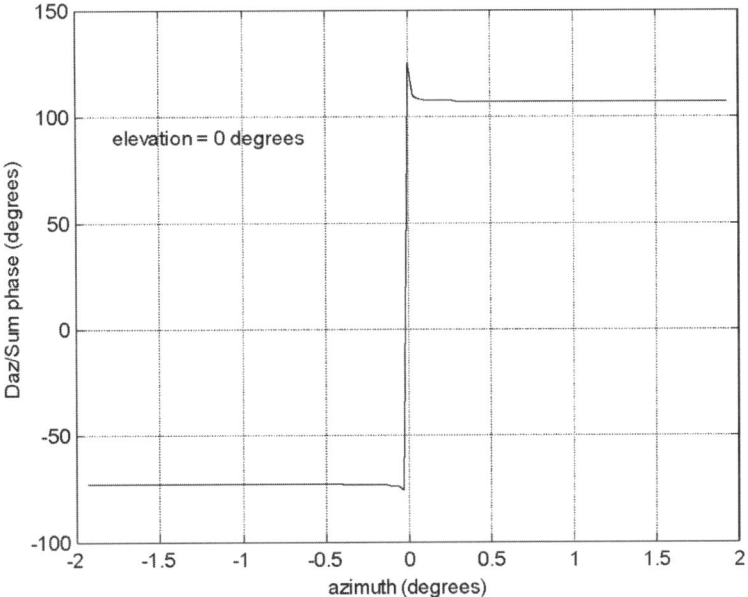

Figure 14.15 Phase diagrams of the simulated ratio between the difference in azimuth and sum beams versus the azimuth angle. The elevation angle is 0°.

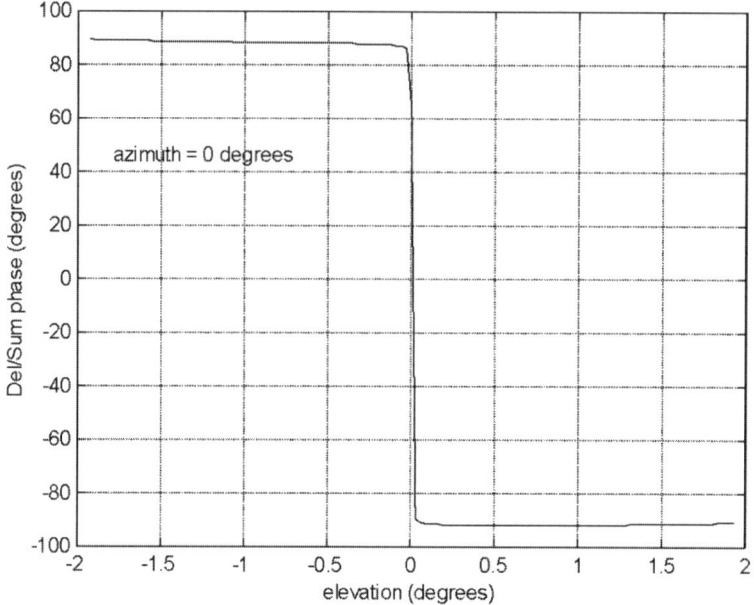

Figure 14.16 Phase diagrams of the ratio between the simulated difference in elevation and sum beams versus the elevation angle. The azimuth angle is 0°.

In Figure 14.15, the phase of the ratio between the difference in azimuth and the sum beams is 107.5° for the positive values of the azimuth, and −72.5° for the negative values. In Figure 14.16, the phase of the ratio between the difference in elevation and the sum beams is −90.8° for the positive values of the elevation, and 89.6° for the negative values. The phase step is in both cases 180°, as the theory requires.

The file P90803k has been considered for a first evaluation of the phases of the ratio between the difference and sum C-PAR beams. In this file, the signals received by the sum, the difference in azimuth, and the difference in elevation beams of the C-PAR have been recorded. The scenario included a target, which is simulated by the transponder TP2 in Figure 14.14, and an MBI, which is simulated by the transmitting antenna IE1 in Figure 14.14. The sweeps corresponding to the first and nineteenth pulse repetition time (PRT) with 1,134 range samples have been considered.

To calculate the phases of the ratio between the difference and the sum beams, the samples from the 600th to the 700th range cell, in which only the interference is present, have been considered, for both the first and the nineteenth PRTs. The results are shown in Figures 14.17 to 14.20, in which the phases are module 2π.

The phases are not constant, but they have some oscillations due to slight variation of the INR from sample to sample. Furthermore, it can be noted that the value of the phase of the ratio between the signal received on the difference in azimuth channel and the signal on the sum changes, passing from the first to the nineteenth PRT (i.e., from approximately −150° to +30°, with a variation of 180°). This can happen if the interference has passed through the null of the difference in azimuth beam during the antenna rotation. The phase of the ratio between the

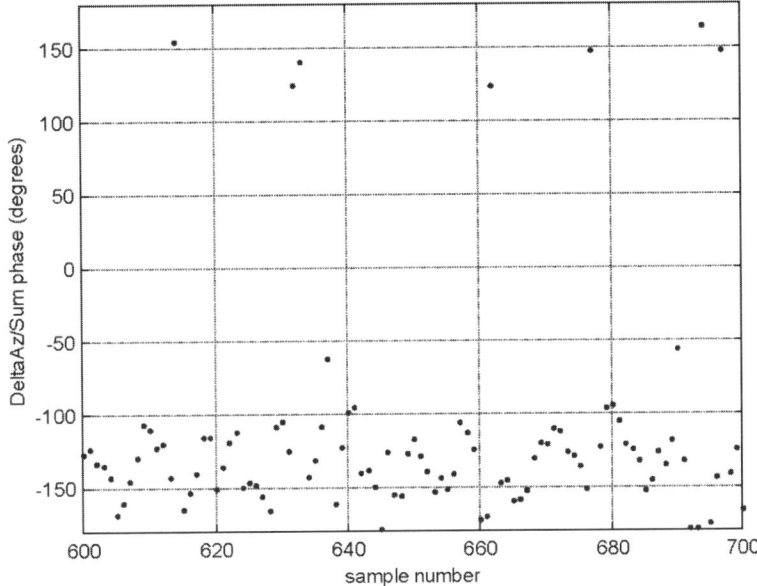

Figure 14.17 Phase of the ratio between the difference in azimuth and the sum beams, using file P90803k, first PRT with 1,134 range samples.

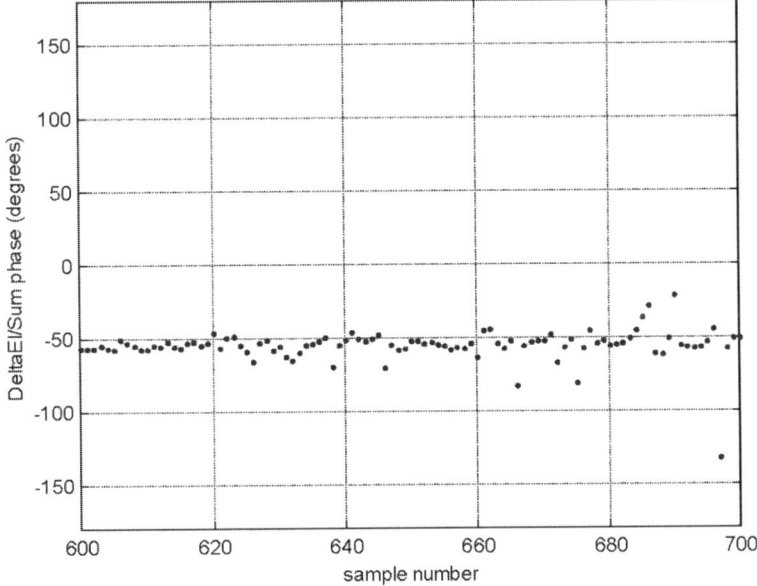

Figure 14.18 Phase of the ratio between the difference in elevation and the sum beams, using file P90803k, first PRT with 1,134 range samples.

signal received by the difference in azimuth channel and the signal received by the sum, therefore, has a 180° step after the interference has passed through the null, as the theory requires. The phase of the ratio between the signal received by the difference in elevation channel and the signal received by the sum remains practically constant, except for a soft ripple due to the noise.

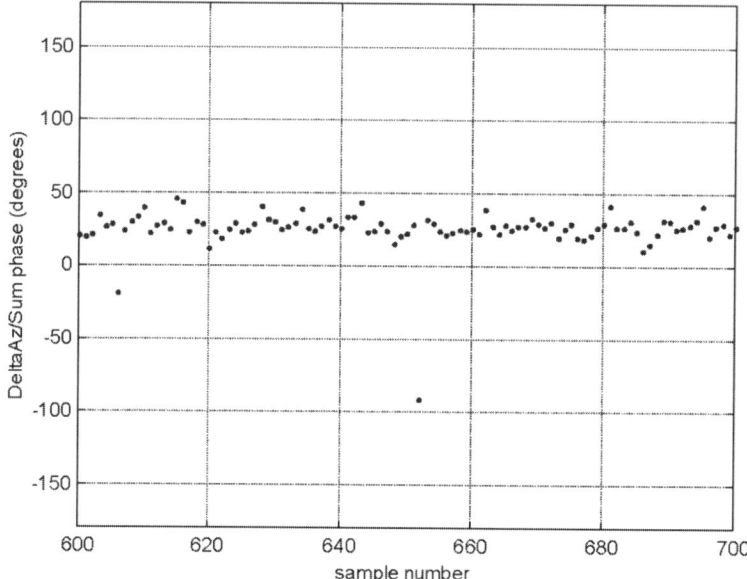

Figure 14.19 Phase of the ratio between the difference in azimuth and the sum beams, using file P90803k, 19th PRT with 1,134 range samples.

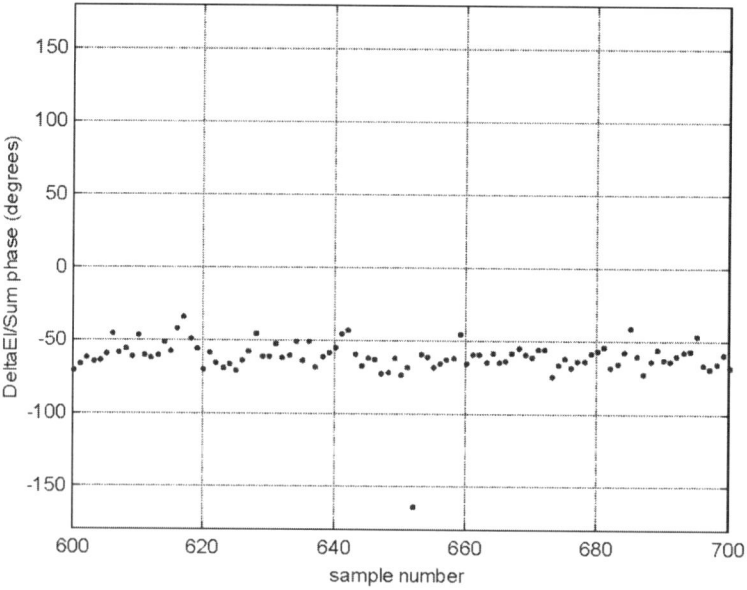

Figure 14.20 Phase of the ratio between the difference in elevation and the sum beams, using file P90803k, 19th PRT with 1,134 range samples.

In conclusion, it appears evident from the results obtained by the recordings that there is some difference between the phase of the simulated beams and the phase of the C-PAR beams. This fact could deteriorate the angle estimation accuracy unless a suitable phase match is made between the simulated beams (i.e., the beams exploited by the MLE algorithm) and the C-PAR beams on the two difference channels.

14.5.4 MLE on Recorded Live Data

Using Figures 14.17 to 14.20, the phase of the ratio between the difference in azimuth and sum beams has been estimated equal to −150°. It has been assumed that the interference had positive azimuth coordinate during the recording under consideration. Thus, a phase shift of −260° has been provided to the reconstructed difference in azimuth beam for matching it to the C-PAR beam. Moreover, the phase of the ratio between the difference in elevation and sum beams in the C-PAR case has been estimated equal to −70°. It has been assumed that the interference had positive elevation during the recording under consideration, thus providing a phase shift of +20° to the reconstructed difference in elevation beam.

Figures 14.21 and 14.22 portray the results obtained with the phase compensation. The estimations obtained with the phase compensation seem more reliable. In fact, the antenna was rotating clockwise in azimuth, with a constant elevation angle.

14.6 Conclusions

The increase of the number of degrees of freedom from three to four greatly improves the accuracy of the MLE of the TDOA, in the presence of one MBI. This also is true when the SNR is comparable to INR. If the target and jammer coordi-

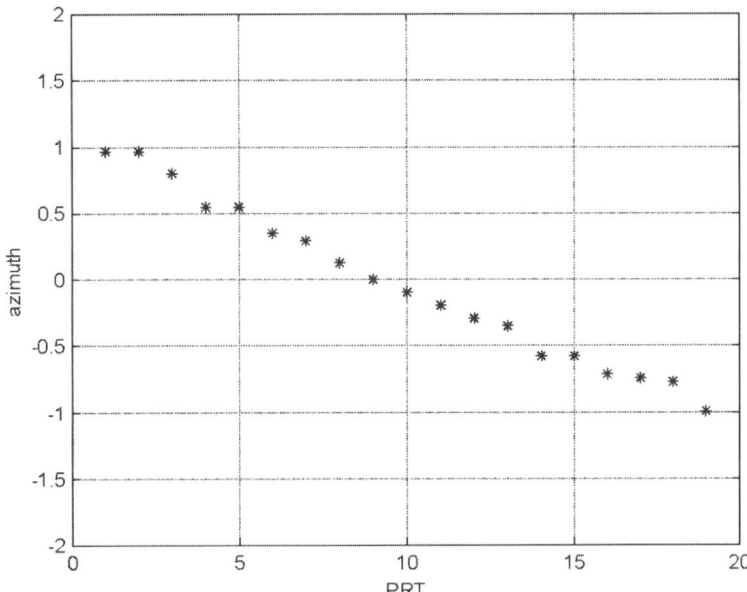

Figure 14.21 Estimate of the target azimuth, using file P90803k, PRT with 1,134 range samples, and phase compensation.

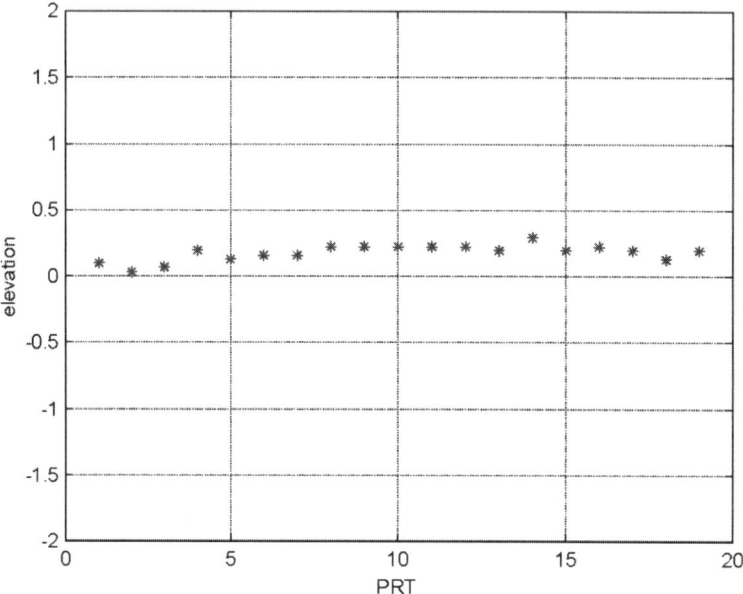

Figure 14.22 Estimate of the target elevation, using file P90803k, PRT with 1,134 range samples, and phase compensation.

nates are not too close, then the MLE with the selected four beams set (i.e., sum, difference in azimuth, difference in elevation, and double difference) is able to estimate the target angular coordinates with the required accuracy. Therefore, this estimate is very close to the correspondent CRLB.

From the test on live data, it has been found that the MLE of TDOA in the presence of MBI needs the knowledge of the amplitude and phase patterns of the antenna beams to properly work. The analysis of the recorded real data has shown that the estimates are very reliable to possible inaccuracies in the reconstructed patterns. This problem has been tackled by obtaining an approximate estimate of the phase characteristics of the antenna patterns. It has been demonstrated that the algorithm provides reliable estimates of the target angular coordinates in the presence of one MBI.

References

[1] Skolnik, M. I., *Radar Handbook*, 2nd ed., New York: McGraw-Hill, 1990, Ch. 18.
[2] Farina, A., *Antenna-Based Signal Processing Techniques for Radar Systems*, Norwood, MA: Artech House, 1992.
[3] Langsford, P., et al., "Monopulse Direction Finding in Presence of Adaptive Nulling," *IEE Colloquium, Advances in Adaptive Beamforming*, Romsey, United Kingdom, June 13, 1995.
[4] Valeri, M., et al., "Monopulse Estimation of Target DOA in External Fields with Adaptive Arrays," *1996 IEEE Symposium of Phased Array Systems and Technology*, Boston, MA, October 15–18, 1996, pp. 386–390.

[5] Farina, A., G. Golino, and L. Timmoneri, "Maximum Likelihood Approach to the Estimate of Target Angular Coordinates Under a Main Beam Interference Condition," *CIE 2001 International Conference on Radar*, Beijing, China, October 15–18, 2001, pp. 834–838.

[6] Farina, A., P. Lombardo, and L. Ortenzi, "A Unified Approach to Adaptive Radar Processing with General Antenna Array Configuration," Special Issue on New Trends and Findings in Antenna Array Processing for Radar, *Signal Processing*, Vol. 84, September 2004, pp. 1593–1623.

[7] Hoffman, J. B., and B. L. Gabelach, "Four-Channel Monopulse for Main Beam Nulling and Tracking," *Proc. of IEEE National Radar Conference NATRAD '97*, Syracuse, NY, May 13–15, 1997, pp. 94–98.

[8] Farina, A., and L. Timmoneri, "Cancellation of Clutter and EM Interference with STAP Algorithms. Application to Live Data Acquired with a Ground-Based Phased-Array Radar Demonstrator," *2004 IEEE Radar Conference*, Philadelphia, PA, pp. 486–491.

[9] Farina, A., et al., "Multichannel Array Processing in Radar: State of the Art, Hot Topics and Way Ahead," *IEEE Sensor and Multichannel Array Processing (SAM04) Workshop*, invited paper, Barcelona, Spain, July 18–21, 2004.

CHAPTER 15
Direction Estimation of Broadband Sources for Auditory Localization and Spatially Selective Listening

Joby Joseph and K. V. S. Hari

15.1 Introduction

DOA estimation of broadband sources and the separation of them using an array of sensors is a challenging problem [1]. This problem has been addressed by nature in many natural organisms. Manmade systems have been built to address this problem in various scenarios, such as communication systems, underwater sensor array processing, and microphone arrays in the audio world. In all these scenarios, there are a number of sources generating broadband signals, and they are received at the sensors propagating through a channel. The models for the source signals and the channel vary with the problem involved.

In this chapter, we deal with the scenario where there are acoustic sources generating sound in the human audible range, and an array of microphones receiving these signals. We specifically deal with the challenge of using only two sensors to achieve the source separation of more than two sources [2–4]. There are two distinct stages.

- Estimating the DOA of the sources;
- Recovering the signals or getting the information in the signals from each of these directions.

Natural organisms have coped with this scenario in their acoustic world by using only two ears, which is against the conventional notion of requiring at least as many sensors as there are sources [5–9]. For example, consider the accuracy with which a barn owl can target a prey prowling on the leaves in near absolute darkness. In the signal processing literature dealing with solutions to this problem, it is also known as a cocktail party problem. We present an algorithm based on [2, 3], which addresses this issue to a limited extent. This does not violate the theoretical limit on the least number of sensors required. Instead, the algorithm requires that the signals from the natural sound sources go to a domain in which the sensors and signals satisfy the conditions for DOA estimation and signal recovery. This also seems to explain how the natural organisms achieve signal processing, and is supported by results from neuroscience [5, 10–12].

15.2 Scenario and Model

The particular scenario addressed in this chapter has M sound sources impinging on K microphones. At the sensors (microphones), the available bandwidth from 20 Hz to 20 kHz is shared by all these sources. If the sources are people speaking, as would be the case with a cocktail party problem, then this bandwidth available to the sources will be reduced to about 8 kHz.

This scenario, as shown in Figure 15.1, has a mathematical model, in which there are M sources generating broadband, possibly quasi-stationary signals, $s_m(t)$. We assume that the signals are quasi-stationary because the parameters of the generation model of speech change with the syllable. These signals for most of the cases propagate through an FIR channel, and are received by K sensors sampling the field. The received signal is denoted by $y_k(t)$. The channel from the mth source to the kth sensor is denoted by $h_{mk}(t)$. Thus, the model is

$$y_k(t) = \sum_{m=1}^{M} h_{mk}(t) * s_m(t) \tag{15.1}$$

Denote the time frequency domain representation of $s_m(t)$ and $y_k(t)$ by $S_m(\omega)$ and $Y_k(\omega)$, respectively. If the channels are stationary, then the channel can be represented by its frequency domain equivalent $H_{mk}(\omega)$. In the time frequency domain, the equation can be written as

$$Y_k(\omega, \tau) = \sum_{m=1}^{M} H_{mk}(\omega) S_m(\omega, \tau) \tag{15.2}$$

When the path from the source to the sensor is a plane wavefront, then the channel from each of the sources to the sensor array can be specified using a single parameter, the DOA of the source signal at the array. Even if the surroundings

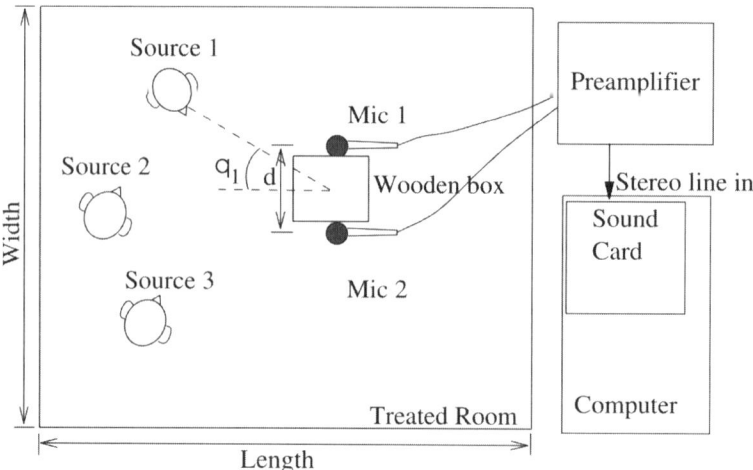

Figure 15.1 Top view of the scenario, in which there are a number of humans producing speech.

are weakly reverberant, and if the angular separation of the sources is reasonably large, then the specification of DOA is sufficient. This is true in many practical scenarios.

There are various approaches described in the literature, in which the number of sources M is less than or equal to the number of sensors K, such as independent component analysis and beamforming [13–17]. We present an algorithm where we can work even if the number of sensors K is restricted to two for number of sources $M > 2$ [2, 3]. This scenario is represented in Figure 15.1.

15.3 Characteristics of Natural Sound Sources

Most natural sound sources are generated by vibration of structures, which are then filtered by a surrounding air column. As a result, within the audible range of frequencies, they tend to restrict themselves to a subset of these frequencies that are characteristic of the physical properties of the structure. Most of the naturally occurring sound-generating structures vary in their physical properties, such as size and density. They tend to produce sounds with significant amounts of nonoverlapping regions of frequencies. One of the sound sources that spans a lot of bandwidth over time is human speech [5, 10, 11]. Even in the case of human speech, there are enough nonoverlapping regions so that we are able to pay selective attention to the desired person's speech when other people are also speaking. Some examples of the spectra of various sound sources at arbitrary times are plotted in Figure 15.2.

A source is recoverable from the mixture by filtering alone if a significant fraction of its spectra is not having interference from other sources in the block

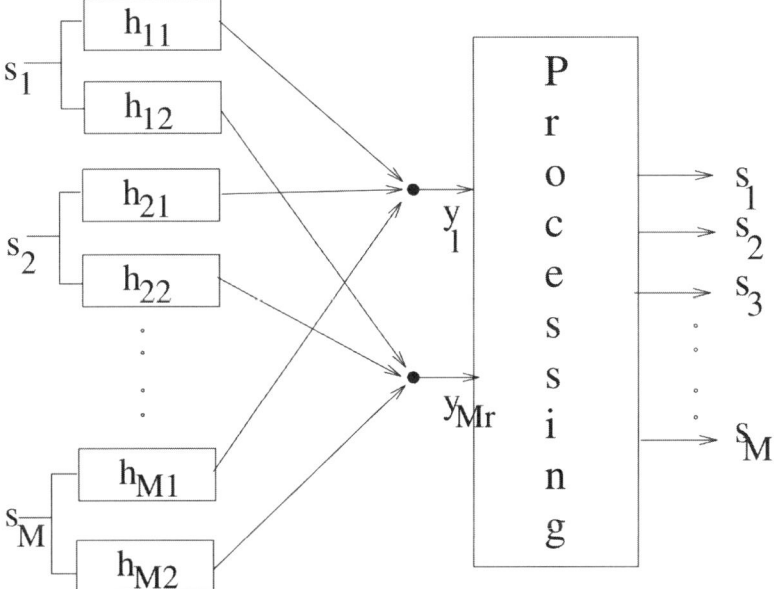

Figure 15.2 Schematic of a system in which source signals are estimated by processing signals at the output of two sensor receiving signals from M sources.

of time being processed. A frequency component is said to be recoverable if the power of the sum of interfering signals at that frequency is less than 20% of the power of the desired source in that frequency. To study the recoverability from mixtures of speech sources, we characterize the percentage of recoverable frequency components in a typical speech segment when various numbers of interfering sources are present.

These observations indicate that we can use some modification of the narrowband schemes for DOA estimation in order to achieve broadband DOA estimation in a scenario in which there are fewer sensors than there are sources.

15.4 Decomposing into Spectral Components

The signals received at the sensors must be decomposed into spectral components. These components then are used for DOA estimation and signal recovery. A popular decomposition uses a DFT filter bank, with the length equaling the average short-time stationary period of speech signals. See Figure 15.3.

15.5 DOA Estimation Based on Time Delays and Relative Attenuation

DOA estimation, using only two sensors when there are more sources than sensors, requires that the sources be broadband. There may be relative attenuation between the sensors for signals from different sources. The broadband nature of speech, which spans over three decades in frequency, also implies that the sensor system may have to be spaced more than the Nyquist spacing (i.e., corresponding to the

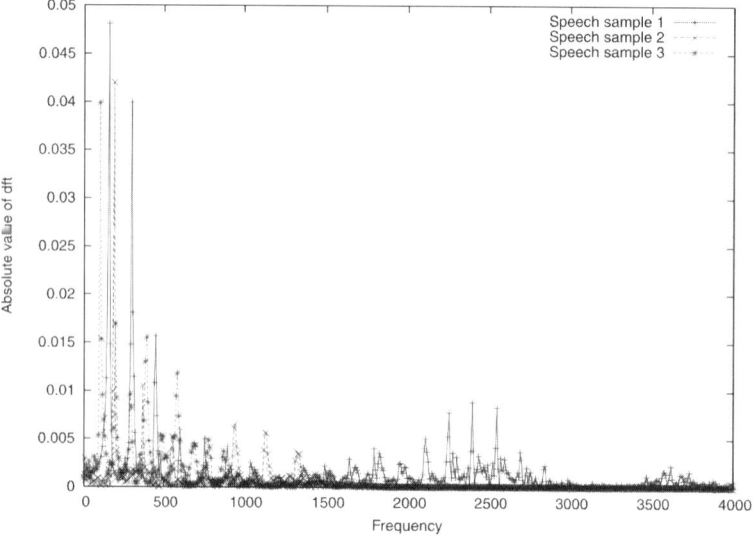

Figure 15.3 Absolute value of the DFT of 1,000 samples of three different speech segments are plotted. The speech was sampled at 8 kHz. As can be seen, there are considerable regions of speech spectra having power for all the sources but are not overlapping in frequency.

15.5 DOA Estimation Based on Time Delays and Relative Attenuation

mid-frequency of speech), in order for some high bandwidth components to achieve reasonable angular resolution. This also implies that we have to use relative attenuation between the sensors to estimate DOA at high frequencies within the bandwidth used. This can be converted to an advantage by introducing relative attenuation between sensors by design, so that we get greater signal-to-interference ratio (SIR) improvement than with only time delay, when we need to recover sources attenuated differently at the two sensors. This also enables multiple sources to have the same frequency, as long as they are attenuated complementarily at different sensors; that is, the sources have no common frequency with appreciable power reaching the same sensor at the same time.

15.5.1 General Approach

The general approach for estimating the DOA of all the speech sources is as follows:

1. Decompose the received signal into short-time spectral components (frequency bins).
2. Use the decomposed signal information at the frequency bins. DOA estimation can be achieved using the relative attenuation between the sensors, at frequencies where the relative attenuation between the two sensors is high because of the shadowing by the headlike structure.
3. Use the decomposed signal information at the frequency bins. DOA estimation can be achieved using the time delays between the sensors at lower frequencies where shadowing is minimal.
4. Combine both DOA estimates from the various frequency bins to identify the number of sources and the DOAs themselves.

A block diagram of showing the steps and signal flow is given in Figure 15.4.

15.5.2 DOA Estimation Using Time Delay Between the Sensors

In order to estimate the DOA using time delays between the sensors, one of the subspace methods, such as MUSIC, or a calculation of the phase difference between

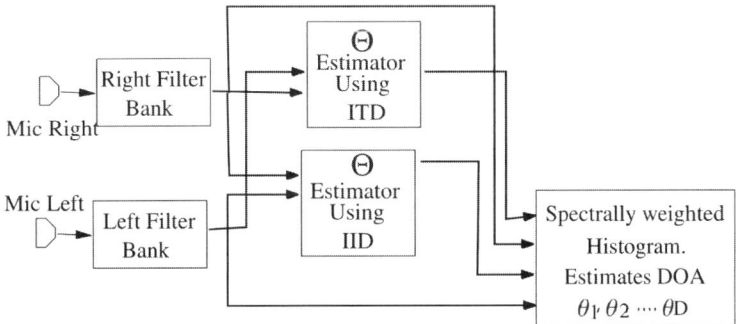

Figure 15.4 A block diagram of the algorithm outline that estimates DOAs of multiple speech sources.

the values at the sensors for the various spectral component at the desired frequency, can be used. The former was used in [2], while the latter was used in [17], as part of a joint phase and attenuation difference estimation.

As pointed out earlier, the speech sources satisfy the nonoverlap of the spectra required by the DOA estimator. This is not exactly true, even though most of the information of speech is in the main lobes of the spectra. There also are sidelobes that are low in amplitude, which interfere with the DOA estimation. Figure 15.5 shows the simulated performance of MUSIC in the presence of interference at various amounts of interfering power and angular separation of the interferer from the desired source. As can be seen, the estimates of the DOA using MUSIC is only perturbed by the interferer as long as the power of the interferer is low compared to the desired source.

15.5.3 Subspace Methods in a Reverberant Room

In an untreated (i.e., reverberant) room, there are many multipaths. At low frequencies, the wavelengths are sufficiently large, so that small changes in the objects in the room do not cause appreciable effect on the propagation of the signal. Thus, the scattering and reflections are less because of this diffraction effect. Hence, for a given constant θ_{mp}, the attenuation of the pth path from the mth source, $\beta_{mp}(n)$, and its phase, $\alpha_{mp}(n)$, may be independent of n. Denote the steering vector from θ_{mp} by $\mathbf{a}_1(\theta_{mp})$, and the wavelength of the spectral component under consideration by λ_1. Thus, the kth element of the sum of all paths,

$$\sum_p \mathbf{a}_1(\theta_{mp}) \beta_{mpl}(n) e^{j\alpha_{mpl}(n)}$$

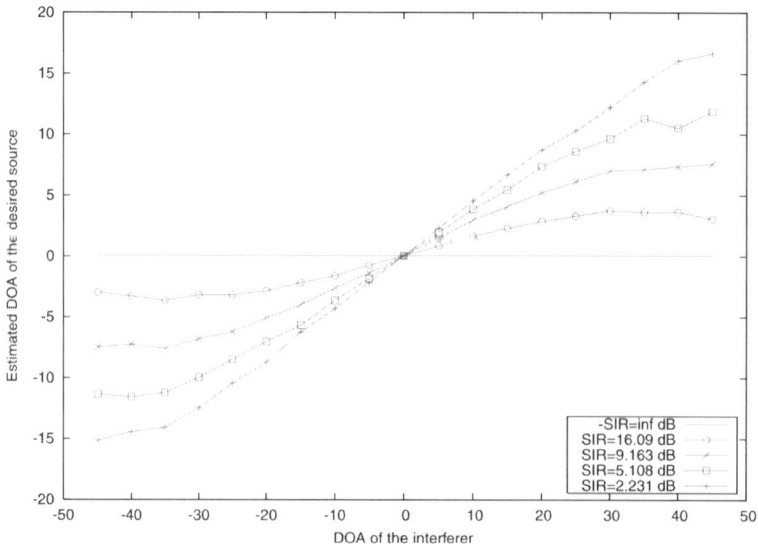

Figure 15.5 Estimate of the DOA using MUSIC for a desired source at 0° while the interferer's DOA and power are varied. Note that the interferer acts as a perturbation, rather than completely spoiling the estimate, and the perturbation increases with decreasing SIR and degree of separation between the sources.

15.5 DOA Estimation Based on Time Delays and Relative Attenuation

would be

$$\sum_p \beta_{mpl} e^{[j2\pi kd \sin(\theta_{mpl})/\lambda_1 + \alpha_{mpl}]} \quad (15.3)$$

Then, assuming $s(n)$ to have unit power, the component due to the mth source in the iqth off-diagonal term in $\mathbf{R}_{xx} = E\{\mathbf{x}_{lm}\mathbf{x}_{lm}^H\}$ would be

$$\sum_p \beta_{mpl} e^{[j2\pi k_i d \sin(\theta_{mp})/\lambda_1 + \alpha_{mpl}]} \sum_{p'} \beta_{mp'l} e^{[-j2\pi k_q d \sin(\theta_{mp'})/\lambda_1 + \alpha_{mp'l}]} \quad (15.4)$$

If the $(p = p_d)$th path is assumed to be the direct path, which has $\beta_{mp_d l} = 1$, and $\beta_{mpl} \ll 1$ for $p \neq p_d$, then (15.4) can be written as $\exp[-j2\pi(k_i - k_j)d \sin(\theta_{mp_d})/\lambda_1] + \rho_m$, where ρ_m is a complex number forming the sum in each element of $\mathbf{R}_{xx}$, except for the term with $p = p' = p_d$. Since $\beta_{mpl} \ll 1$ for the product in the terms of the summation, $|\rho_m| \ll 1$. Therefore, for a weakly reverberant room, we can assume that the reverberation is a perturbation on the direct path at low frequencies, and hence a perturbation on the DOA of the direct path. Then, the estimates of the θ_{mp} may be in error. However, the estimate would be constant and unique for each source in time.

Such a scenario, in which the direct path is much stronger than the reverberant components, occurs if either the room is free from echoes, or the sensors are located away from reflecting bodies, such as walls. This conclusion is corroborated by simulations and experimental results in the work conducted in reverberant rooms.

At high frequencies, $\alpha_{mpl}(n)$ can become independent of $\alpha_{mpl}(n+1)$. Therefore, the terms of $\mathbf{R}_{xx}$ would have an additional averaging over n, thus forming

$$\frac{1}{N} \sum_n \left(\sum_p \beta_{mpl} e^{[j2\pi k_i d \sin(\theta_{mp})/\lambda_1 + \alpha_{mpl}(n)]} \sum_{p'} \beta_{mp'l} e^{[-j2\pi k_j d \sin(\theta_{mp'})/\lambda_1 + \alpha_n]} \right) \quad (15.5)$$

This is a random complex variable that could be changing with time. Thus, for reverberant rooms at higher frequencies, it is not possible to use relative phase information at the sensors for DOA estimation.

15.5.4 DOA Estimation Using Relative Attenuation Between the Sensors

DOA estimation at a given frequency band using relative attenuation depends on the mapping of relative attenuation between the sensors to the DOA. This depends on the physical dimension, material, and shape of the object between the microphones. See Figure 15.6 for the specifications of the structure used in the experiments. We have a box with dimensions as given in Figure 15.6, and we rely on measurements made to estimate the relative attenuation at various frequency bins from various DOAs. The measured power ratios at various angles and frequency are given in Figure 15.7. The estimate $\theta(1)$ is a function of the ratio of the absolute values—$\theta(1) = f[|X_2(1)|/|X_1(1)|]$. Analogous to the human head-related transfer

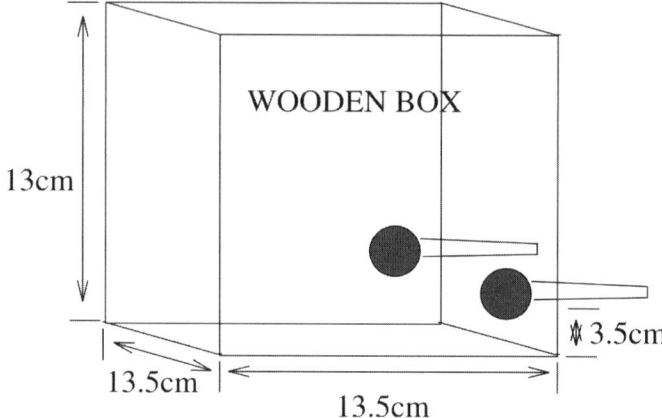

Figure 15.6 The physical dimensions of the object between the microphones used in the experiments in this chapter. These dimensions are of the order of a human head, since this broadband DOA estimation method was developed as an attempt to model the binaural hearing process to apply to source separation.

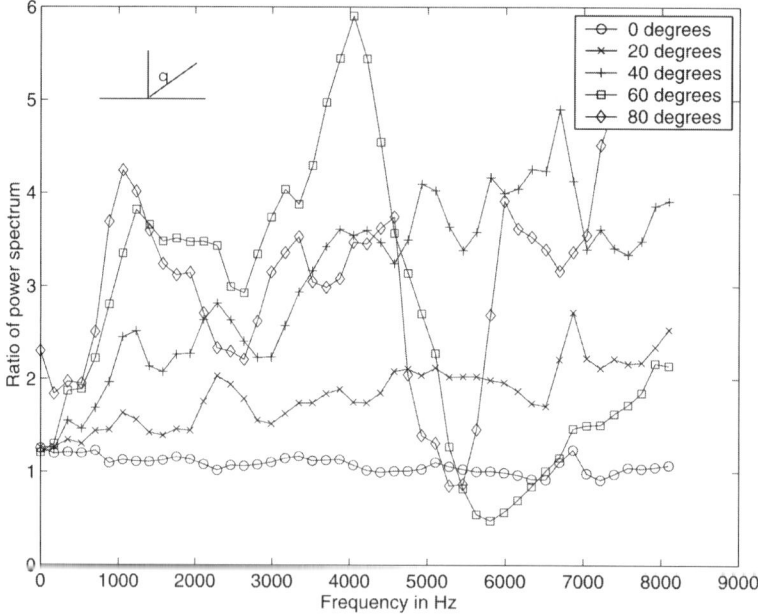

Figure 15.7 The ratio of the magnitude spectrum of the signals at the two microphones placed as in Figure 15.6. The sound source was a speaker playing white noise at various angles to the broadside 1m away from the microphone and box setup. At angles greater that 60°, the ratio is not reliable. Therefore, in the algorithm, the sources are restricted to be within 60° for bandwidth ranging from 1 to 4 kHz.

function, we propose to use box-related transfer function (BRTF) for this box structure between the microphones. This gives the ratio of the magnitude spectrum of signals at the two microphones placed as in Figure 15.7. The sound source was a speaker playing white noise at various angles to the broadside 1m away from the microphone and box setup. It can be observed that:

1. There is uniqueness in the transfer function at various DOAs.
2. At higher frequencies, there is no monotonicity of increase in the ratio with DOA.
3. The functions are smooth and thus will be tolerant to errors in the estimation of the magnitude spectrum.
4. The DOA estimates are reliable in the range from 1 to 4 kHz, for angles in the range from +60° to −60°.

15.6 Spectrally Weighted DOA Estimate

After obtaining the DOA estimates using time delays and relative attenuation at each frequency bin, we give weight to the DOA estimates using the spectral power at each bin. Then, we form a histogram of the weighted DOA estimates to detect the number of sources that are active while the block is being processed. Let $\mathbf{P}(\theta)$ denote the histogram before weighting, and θ_h denote the interval of the histogram, so that the weighted histogram is

$$\mathbf{P}_w(\theta) = \mathbf{P}(\theta) \sum_{l:|(\theta_1 - \theta < \theta_h/2)|} |\mathbf{X}(l)|^2 \qquad (15.6)$$

The number of prominent peaks of this weighted histogram is taken as the number of sources, and the corresponding bins of the histogram give the DOAs. The block diagram of the process is given in Figure 15.8. The processing steps on the output of the DFT filter bank for estimating the DOA of the broadband sources are as follows:

1. Estimate the DOA using narrowband MUSIC at each of the frequency bins, at frequencies where relative attention between the sensors is negligible and where the condition number is large.
2. Estimate the DOA using the relative attenuation at each of the frequency bins, at frequencies where relative attenuation between the sensors is appreciable.
3. Estimate the power in each frequency bin.
4. Use (15.6) to compute the weighted histogram.
5. Pick the peaks of $\mathbf{P}_w(\theta)$ to find the DOA of the sources.

An example for the sequence of steps is given in Figure 15.9. Here, only the plots up to 4 kHz are shown, because there is no appreciable power in the higher

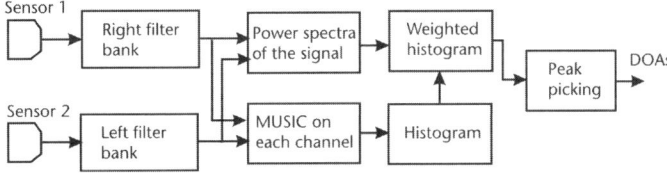

Figure 15.8 Steps involved in the estimation of the DOA of the broadband source-like speech, using only two sensors.

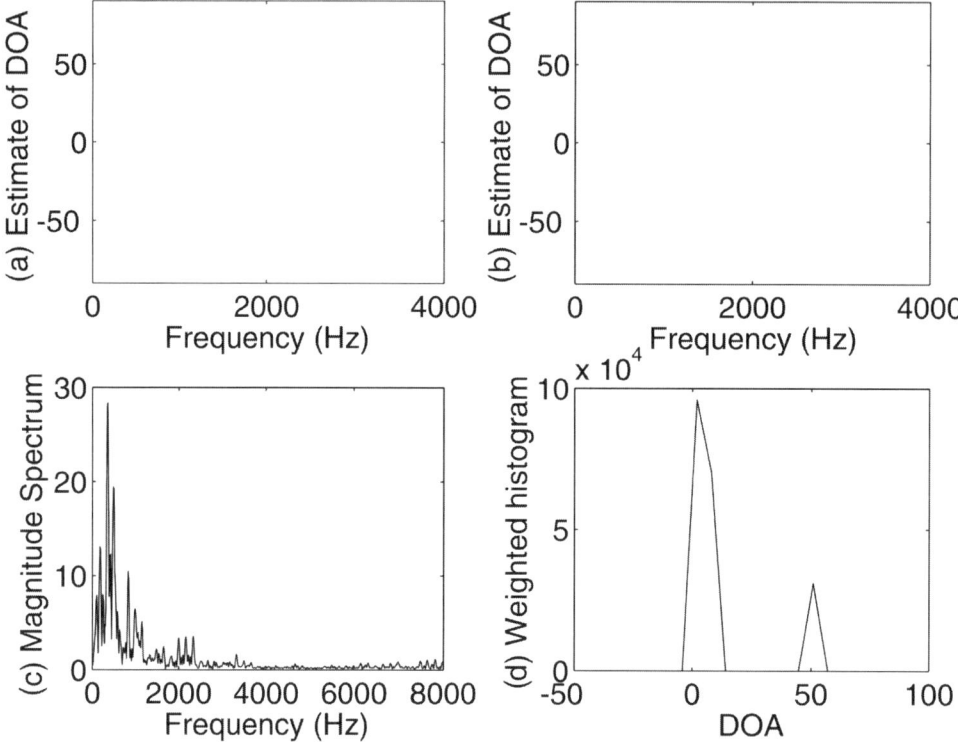

Figure 15.9 Steps involved in obtaining the spectrally weighted estimate of DOAs of sources, from a segment where there were two females and one male reading out text. They were located approximately 2m from a structure, as in Figure 15.1, and at angles −40°, +40°, and 0° in azimuth, respectively. In (a) and (b), those frequencies for which the DOA could not be estimated, either due to lack of power or because of small condition number, are represented as −90°. (a) Estimate of the DOA at each frequency, using MUSIC. (b) Estimate of DOA at each frequency, using MUSIC and relative attenuation. (c) Estimate of magnitude spectrum for a given block. (d) Weighted histogram of DOA with minor peaks removed, estimated from (b) and with weights from (c).

frequencies for speech signals. It is seen from the weighted histogram that the detector missed the DOA of −40°, because the power in that source is less for this segment. It is detected in other segments, where power of this source is larger. This also shows that this detector is probably not optimal.

15.7 Source Separation Algorithm

The steps involved in the source separation algorithm, using the DOA estimation algorithm, is given below, and a block schematic is given in Figure 15.10.

We choose a DFT filter bank as the equivalent of the tonotopic mapping in the cochlea and cochlear nuclei. The algorithm takes the present sample vector from the microphones into the DFT filter bank and temporarily stores the result. Although a Mel spectral-based filter bank might provide perceptually better results, we have not used it in the present implementation.

15.7 Source Separation Algorithm

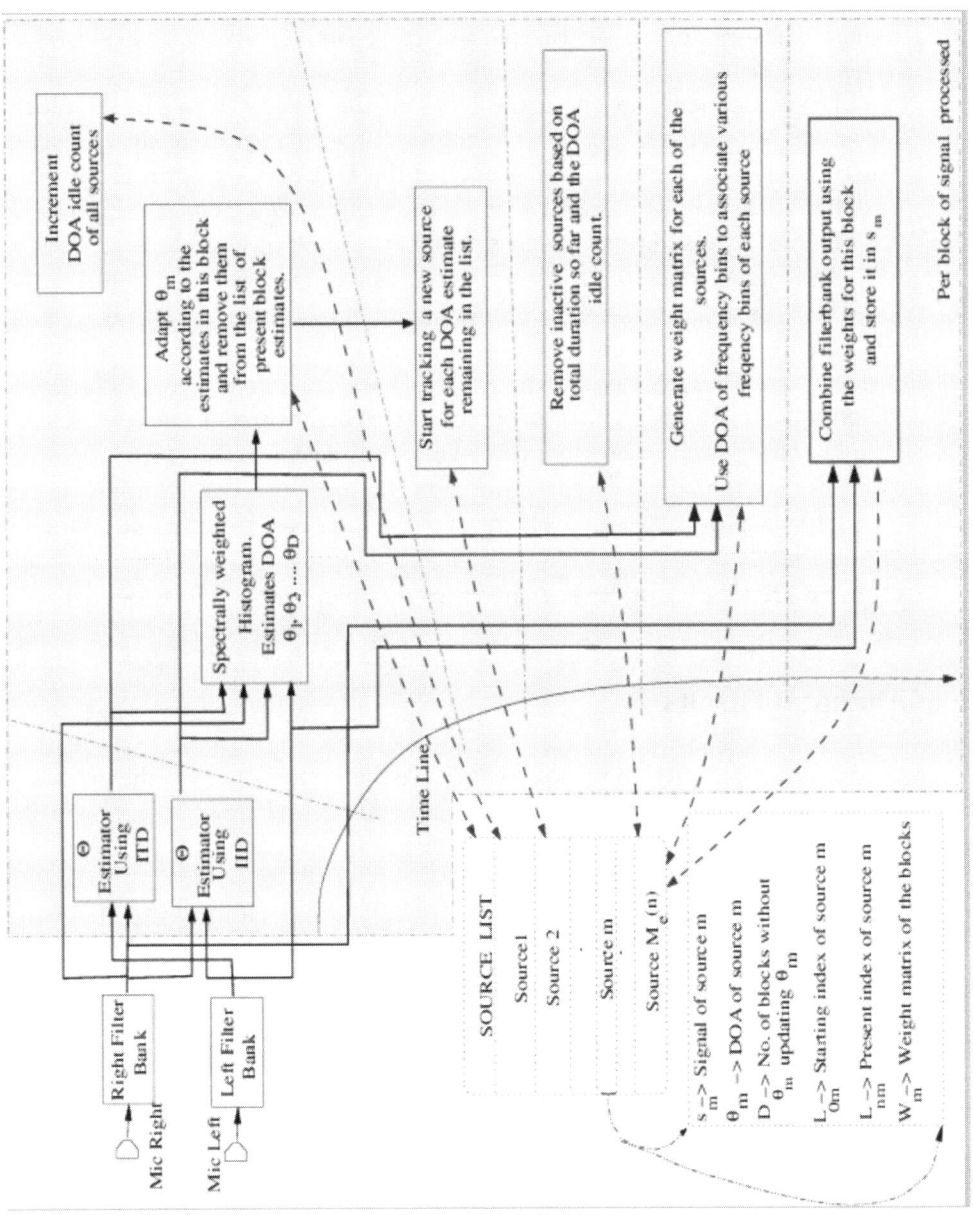

Figure 15.10 Details of the algorithm when DOA is estimated, also using relative delay between the sensors. This represents the algorithm when only DOA cue is used for separation. This can be applied to measurements when a headlike structure is present.

When the number of samples input to the DFT filter bank crosses a multiple of block length L, we process the DFT filter bank output for that block. This involves the following steps.

1. Estimate the DOA at each frequency bin l, $\hat{\theta}(l)$. If it was not possible to estimate the DOA, then the DOA is assigned the value of $-90°$.
2. Make decisions based on the parameters estimated in the previous step. This involves
   a. Draw a conclusion about the DOA of the sources active in the block, using a spectrally weighted histogram of $\hat{\theta}(l)$, as previously described.
   b. Initiate tracking of a new source, if the DOAs detected in the present block cannot be associated with the sources detected in the previous blocks; that is, if those sources present in the set of sources being tracked up to that time, $\{\theta_m\}$.
3. Associate the frequency bins with the sources already detected, thus forming a mask to be used for each source. This involves associating the frequency bins using DOAs; that is, setting $W_m(l) = 1$, if $|\hat{\theta}(l) - \hat{\theta}_m| < \theta_{\text{thresh}}$ for some fixed θ_{thresh}. This θ_{thresh} can depend on reverberation and noise, which are the factors limiting performance of DOA estimation.
4. Reconstruct the signal of each source in the present block by applying the mask formed for each source ($W\_\{m\}$) to the filter bank output and summing these components.
5. If there are more samples coming from the microphone, go to Step 1.

15.8 Experimental Results

Next, we present experimental results to evaluate the performance of the proposed algorithm, done in two steps. Two sets of experiments are done to evaluate the performance of the algorithm.

1. The first set of experiments is aimed at studying the DOA-based algorithm under controlled conditions to validate the principle. Therefore, in these experiments, the array data is synthesized using speech and sinusoid signals. The speech data used is from the TIMIT database. Since we have the unmixed signals in these cases, performance is quantified numerically.
2. The second set of experiments uses real multitalker recordings, using a setup similar to that shown in Figure 15.11. Performance is evaluated using listening tests.

15.8.1 Parameter Values Used in the Implementation

The block length used for processing consists of 1,024 samples. There are 20 bins in the histogram for estimating DOA. A particular frequency is associated to a particular DOA if they are closer than 20°. DOA is estimated using MUSIC, up to 1 kHz. Beyond 1 kHz and up to 4 kHz, the estimate is done using the relative

15.8 Experimental Results

Figure 15.11 Photograph of the setup used for recording with a headlike structure.

delay between the sensors from measurements. Adaptation of DOA for the mth source was $\theta_m(k+1) = 0.9\theta_m(k) + 0.1\hat{\theta}_m(k)$, where $\hat{\theta}_m(k)$ is the estimate of DOA of the mth source using the current block.

Performance for Synthesized Array Data

This section studies the performance of the sound source separation algorithm under various simulated scenarios. It examines the performance when sources are close and when the number of sources is increased. To study this, we synthesize the array output under nonreverberant conditions. In these experiments, we have the original unmixed signal available, so that the performance measure can be computed. Performance has to be measured in the following terms.

1. The similarity of the recovered signal to the original signal, as compared to the case in which the mixture itself is used as an estimate.
2. The suppression of the interfering signal in each of the recovered sequences, as compared to the case in which the mixture itself is used as an estimate.

If $\mathbf{s}(n) = [s_1(n), s_2(n), \ldots, s_M(n)]$, where the original signals $y_1(n)$ is the mixture output at the first sensor and $\hat{\mathbf{s}}(n) = [\hat{s}_1(n), \hat{s}_2(n), \ldots, \hat{s}_M(n)]$ are the estimates obtained by the algorithm, then to observe the performance, we can do the following.

1. Matrix $\mathbb{Q}$ whose ijth element is

$$\mathbb{Q} = E[s_i(n)s_j(n)]/\{E[s_i^2(n)]E[s_j^2(n)]\}^{0.5}$$

2. Matrix $\hat{\mathbb{Q}}$ whose ijth element is

$$\hat{\mathbb{Q}}_{ij} = E[s_i(n)\hat{s}_j(n)]/\{E[s_i^2(n)]E[\hat{s}_j^2(n)]\}^{0.5}$$

3. Matrix $\hat{\mathbb{Q}}^o$ whose ijth element is

$$\hat{\mathbb{Q}}_{ij}^o = E[s_i(n)y_1(n)]/\{E[s_i^2(n)]E[y_1^2(n)]\}^{0.5}$$

In the ideal case, $\hat{\mathbb{Q}} = \mathbb{Q}$ for perfect separation. When the sources are uncorrelated, $\mathbb{Q}$ is identity. $\hat{\mathbb{Q}}$ indicates the quality when the mixture itself is used as the estimate. Then the closeness of $\hat{\mathbb{Q}}$ to $\mathbb{Q}$, as compared to closeness of $\hat{\mathbb{Q}}^o$ to $\mathbb{Q}$, is a measure of the performance of the algorithm. Let $\|\cdot\|_F$ denote the Frobenius norm of the enclosed matrix. To reduce the above mentioned similarity quality to a single number, we can use the measure $\mathbf{Q}_e = |\mathbb{Q} - \hat{\mathbb{Q}}|_F$ and $\mathbf{Q}_o = |\mathbb{Q} - \hat{\mathbb{Q}}^o|_F$, respectively. In this measure, the algorithm has performed best when $\mathbf{Q}_e$ attains zero. $\mathbf{Q}_o$ would be the performance if the mixture itself was used as the estimate. It is possible to compare them to see to what extent the algorithm has performed source recovery and interference suppression. Ideally we would like $\mathbf{Q}_e$ to go to zero even when $\mathbf{Q}_o$ is large.

Conditions Used in the Simulations

In all the cases of synthesized data, the spacing between the two microphones was 0.05 cm. The sampling rate was assumed to be 44.1 kHz. Speech segments from TIMIT database that had a sampling rate of 8 kHz was chosen as data. They were resampled to 44.1 kHz for compatibility with the code. Time averages are used as estimates for expectation.

15.8.2 Performance of the Algorithm When the Sources Are Sinusoidal

To check if the algorithm works for sinusoidal sources, we generate a mixture output of two sensors with three sources. The three sources are $\sin(2\pi \cdot 1{,}200t)$, $\sin(2\pi \cdot 650t)$, and $\sin(2\pi \cdot 300t)$, and at angles $-40°$, $0°$, and $+40°$, respectively. The performance of the algorithm is given in Table 15.1. The performance of the algorithm is seen from the lesser value of $\mathbf{Q}_e$ compared to that of $\mathbf{Q}_o$.

15.8.3 Performance of the Algorithm with Increasing Number of Sources

This simulation studies how well the algorithm can perform when the number of sources is increased from two to five. Performance in terms of $\mathbf{Q}_e$ and $\mathbf{Q}_o$ is plotted in Figure 15.12. $\mathbf{Q}_e$ increases as the number of sources increases. This means that

15.8 Experimental Results

Table 15.1 Performance of Source Separation on Sinusoidal Sources

$$\hat{Q} = \begin{bmatrix} 0.9419 & 0.3072 & 0.0840 \\ 0.1230 & 0.9150 & 0.3699 \\ 0.1053 & 0.2234 & 0.9639 \end{bmatrix} \quad \hat{Q}^o = \begin{bmatrix} 0.5766 & 0.5770 & 0.5770 \\ 0.5766 & 0.5770 & 0.5770 \\ 0.5766 & 0.5770 & 0.5770 \end{bmatrix}$$

Performance measures $Q_e = 0.5707$, $Q_o = 1.5915$

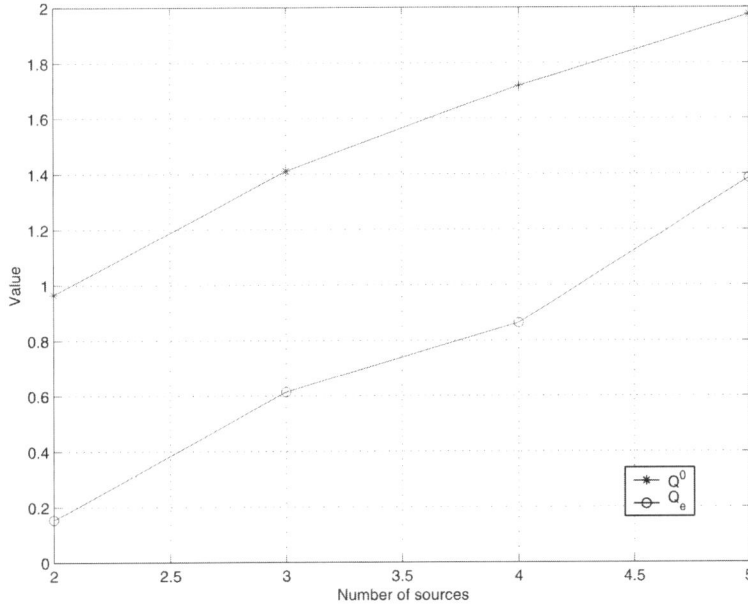

Figure 15.12 Performance of the source separation algorithm as the number of sources is increased. Q_e is the quality of the separated signals, and Q_o is the quality of the mixture. The difference in the height for the mixture and separation is an indicator of the degree of improvement that the algorithm has made.

the recovery of desired sources is influenced by other sources that act like interferers. Similar trends can be observed with respect to Q_o. This is due to the fact that the mixture signal has more interferers. The ratio Q_o/Q_e indicates the improvement factor of using the algorithm. It is observed that this ratio decreases with the number of sources.

15.8.4 Performance of the Algorithm When the Sensor Spacing Is Varied

The sensor spacing has a direct effect on the resolution of DOA estimation and the maximum usable frequency at which MUSIC can be used. To study these effects, we take the scenario of three speech sources and check the performance of the algorithm when the sensor spacing is varied. The sources were fixed at $-50°$, $+5°$, and $+45°$, respectively. Sensor spacing is varied from $d = 0$m to $d = 0.4$m. The values for Q_e and Q_o are plotted in Figure 15.13.

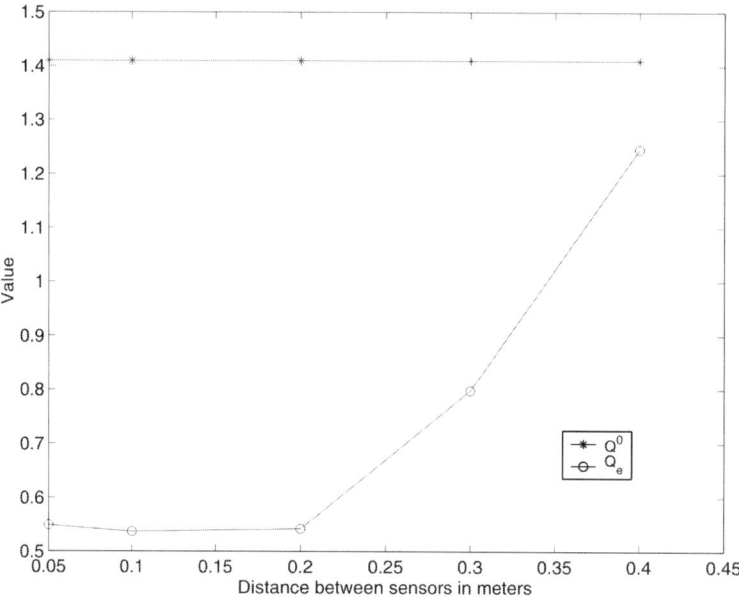

Figure 15.13 Performance of the algorithm with sensor spacing. There is slight improvement in performance in increasing the sensor separation when they are close together. Then as the spacing is increased, the frequency up to which MUSIC can be used decreases, and severely affects the performance.

15.8.5 Resolvability of the Algorithm When Sources Are Closely Spaced

This simulation studies the performance of the sound source separation algorithm when the difference in the DOA of the sources is varied. One source is fixed at −30°, while the other is changed in steps of 5°, up to +5°. The performance measures Q_e and Q_o, given in Section 15.8.1, are plotted in Figure 15.14. Performance improves with angular separation. In the implementation of the algorithm, the permissible deviation in DOA, within which a frequency component is associated with a particular source, is 20°. As seen from the plot, there is a jump in performance corresponding to this angle. The performance improves with Q_e, changing from 0.95 to 0.15 as angular separation increases.

15.8.6 Performance of the Sound Source Separation from Real Two Sensor Array Recordings in Multitalker Scenario

The Setup

Experiments have been done to verify the cases where there is no headlike structure, and where there is a headlike structure between the microphones. A symbolic top view of these two cases is shown in Figure 15.1. Experiments are done in a treated hall. The dimensions of the room were as follows: length = 8m, width = 8m, and height = 4m. The recorded signal mixture used for the experiment and the separated signals are available at http://www.dsp.ece.iisc.ernet.in/~joby/audpro/.

Two matched-pair Sennheizer MBC 660 condenser microphones were used as sensors. A Yamaha MX 12/4 mixing console acted as stereo preamplifier. A Yamaha

15.8 Experimental Results

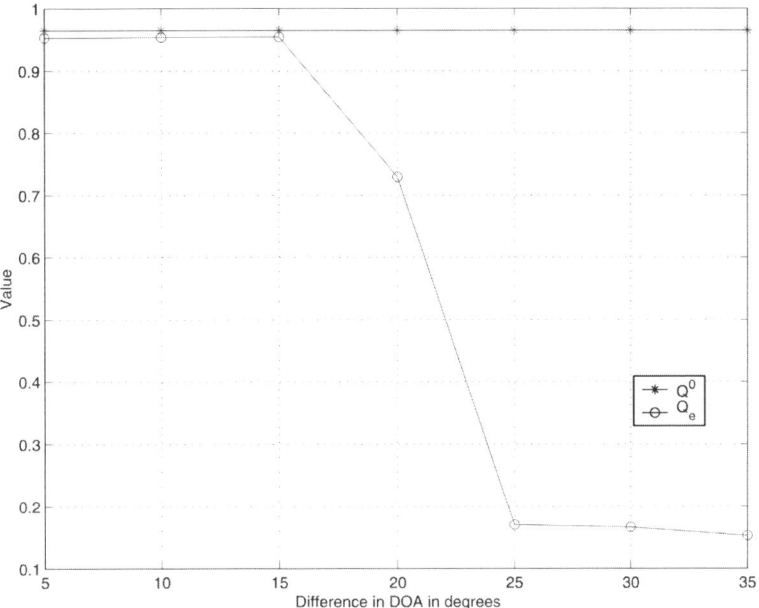

Figure 15.14 Plot giving performance of the source separation algorithm when separation between the two sources is increased. Q_e is the quality of the separated signals, and Q_o is the quality of the mixture. Performance increases with separation. In the implementation of the algorithm, the permissible deviation in DOA, within which a frequency component is associated with a particular source, is 20°. As seen from the plot, there is a jump in performance corresponding to this. The performance improves from approximately 0.95 to 0.15.

Corporation YMF-724 sound card in a 1.6-GHz Pentium 4 computer, with 512-MB RAM, was used as the system for recording and processing. SAA software that was developed in the Statistical Signal Processing Lab, Indian Institute of Science, available at http://www.dsp.ece.iisc.ernet.in/~joby/saajoby/, running on Redhat 9 Linux, was used as the recording and processing tool. This SAA software incorporates the algorithms presented in this chapter. All recordings were done with 16 bits per sample. In the experiment were BRTF was measured, the white noise was played out through a Creative Sound Blaster SBS320 loudspeaker. The processing blocks were of 23.2-ms duration.

All recordings were done with a set of people reading text in front of a pair of microphones placed approximately at the height of the speakers. Experiments were done in a treated room. The distance and angle measurements are approximate, since the speakers were allowed the freedom to shift themselves in the seat. When there was no headlike structure, the spacing between the microphones was $d = 13$ cm. When there was a headlike structure, the spacing between the microphones was $d = 17$ cm.

15.8.7 Performance Measure for Listening Tests

There are several listening tests available for testing speech and audio quality for compression, watermarking, audio effects, and synthesized speech. These tests are not directly suitable for experiments presented in this work. This is because the objective of the listening test in this work is to assess the performance of the algorithm in separating a mixture of simultaneous speech signals, which is not the objective in the existing listening tests. Thus, we propose the following listening test methodology, which is similar to existing methodology. The measures used to judge the performance of the algorithm while doing the listening tests are as follows.

- *Intelligibility:* Intelligibility is a measure of the understandability of the words and sentences forming the speech.
- *Fidelity:* Fidelity of the estimate of a signal is proportional to the exactness of reproduction of the signal. Even if the source signal is intelligible, it may be distorted.
- *Suppression:* Suppression is a measure of the attenuation of the unwanted signals.

In each case, the listeners were asked to listen to the mixture in stereo mode, so that the comparison for separability was more realistic. In the mixture, they were asked to pay attention to each source in turn, and then listen to the recovered signal. The following questions were answered by the persons who took the listening test to grade the performance of the algorithm on each of the source signals in the mixture.

1. How intelligible is the separated source signal, as compared to the same source while listening to the mixture?
2. To what extent is the fidelity of the source signals in the mixture preserved in the separated source signal?
3. By what percentage are the other source signals suppressed in the recovered signals, as compared to the desired signal in the mixture?

15.8.8 Performance of the Algorithm with Headlike Structure

This experiment tries to see the performance of the algorithm with headlike structure when the sound sources are from different gender combinations. Results of the listening tests are given in Table 15.2. The four cases given in the table are as follows:

Table 15.2 Results of the Listening Test (Experiment 5) on the Output of the Algorithm in the Four Cases, Obtained by Averaging the Gradings Given by 10 People

Case	Quality of Recovered Source 1 (%)			Quality of Recovered Source 2 (%)			Quality of Recovered Source 3 (%)		
	Intelligibility	Fidelity	Suppression	Intelligibility	Fidelity	Suppression	Intelligibility	Fidelity	Suppression
1	75.5	54	62.5	50.5	41	47			
2	73.8	57.5	59.5	71	54.5	54			
3	43.2	37.5	44	64	54	53			
4	71.8	53.5	52	68.8	54.5	50.5	65.5	52.5	46

- *Case 1:* One male and one female reading text, while seated in front of a setup approximately 2m away, at angles −40° and +40°.
- *Case 2:* Two females reading text, while seated in front of a setup approximately 2m away, at angles −40° and +40°.
- *Case 3:* Two males reading text, while seated in front of a setup approximately 2m away, at angles −40° and +40°.
- *Case 4:* Two females and one male reading text, while seated in front of a setup approximately 2m away, at angles −40°, +40°, and 0°, respectively.

We can infer from the table that the performance is reasonably good. An observation is that the performance of Case 3 is worse than the others. One possible reason is that both are male voices having large overlaps in spectrum at low frequencies for large numbers of processing blocks. Note: This experiment demonstrates the potential of the proposed algorithm in separating a mixture of speech signals recorded in a real scenario using a two-microphone array.

15.9 Conclusions

In this chapter, we give an algorithm to estimate the DOA, and use it in a source separation algorithm, in a scenario where there are acoustic sources generating sound and an array of only two microphones receiving these signals.

References

[1] Laheld, B., and J. F. Cardoso, "Adaptive Source Separation with Uniform Performance," *Proc. EUSIPCO,* Edinburgh, U.K., September 1994, pp. 183–186.

[2] Joseph, J., (Ph.D. thesis supervised by K. V. S. Hari), "Why Only Two Ears? Some Indicators from the Study of Source Separation Using Two Sensors," Ph.D dissertation, Indian Institute of Science, Bangalore, India, 2004.

[3] Yilmaz, Ö, and S. Rickard, "Blind Separation of Speech Mixtures via Time-Frequency Masking," *IEEE Trans. on Signal Processing*, Vol. 52, No. 7, July 2004, pp. 1830–1847.

[4] Shamsunder, S., and G. B. Giannakis, "Modeling of Non-Gaussian Array Data Using Cumulants: DOA Estimation of More Sources with Less Sensors," *Signal Processing*, Vol. 30, 1993, pp. 279–297.

[5] Bregman, A. S., *Auditory Scene Analysis*, Cambridge, MA: MIT Press, 1999.

[6] Carr, C. E., and D. Soares, "Evolutionary Convergence and Shared Computational Principles in the Auditory System," *Brain Behav. Evol.*, 2002, pp. 294–311.

[7] Moore, D. R., J. W. H. Schnupp, and A. J. King, "Coding the Temporal Structure of Sounds in Auditory Cortex," *Nature Neuroscience*, Vol. 4, 2001, pp. 1055–1056.

[8] Shinn-Cunningham, B., and K. Kawakyu, "Neural Representation of Source Direction in Reverberant Space," *Proc. of the 2003 IEEE Workshop on Applications of Signal Processing to Audio and Acoustics*, 2003, pp. 79–82.

[9] Goldstein, E. B., *Sensation and Perception*, 6th ed., Pacific Grove, CA: Wadsworth, 2002.

[10] Cooke, M., "A Glimpsing Model of Speech Perception," *ICPhS*, 2003.

[11] Cooke, M., and D. P. Ellis, "The Auditory Organization of Speech and Other Sources in Listeners and Computational Models," *Speech Communication*, Vol. 35, No. 3–4, October 2001, pp. 141–177.

[12] Moore, B. C. J., *An Introduction to the Psychology of Hearing*, New York: Academic Press, 1989.

[13] Buckley, K. M., "Broadband Beamforming and the Generalized Sidelobe Canceler," *IEEE Trans. on Acoustics, Speech, and Signal Processing*, Vol. 34, 1986, pp. 1322–1323.

[14] Sydow, C., "Broadband Beamforming for a Microphone Array," *JASA*, Vol. 96, 1994, pp. 845–849.

[15] Wu, Q., and K. M. Wong, "Blind Adaptive Beamforming for Cyclostationary Signals," *IEEE Trans. on Speech*, Vol. 44, 1996, pp. 2757–2767.

[16] Torkkola, K., "Blind Separation for Audio Signals—Are We There Yet," *Proc. Workshop on Independent Component Analysis and Blind Signal Separation*, 1999, p. 44.

[17] Mansour, A., A. K. Barros, and N. Ohnishi, "Blind Separation of Sources: Methods and Applications," *IEICE Trans. Fundamentals*, E88-A, 2000, pp. 1498–1512.

Selected Bibliography

Applebaum, S. P., "Adaptive Arrays with Main Beam Constraints," *IEEE Trans. on Acoustics and Processing*, Vol. AP24, 1976, pp. 650–662.

Brandstein, M. S., "On the Use of Explicit Speech Modeling in Microphone Array Application," *ICASSP*, 1998, pp. 3613–3616.

Buckley, K. M., and L. J. Griffiths, "Broadband Signal Subspace Spatial-Spectrum (Bass-Ale) Estimation," *IEEE Trans. on Signal Processing*, Vol. 36, 1988, pp. 953–964.

Chocheyras, Y., and L. Kopp, "Limitations of Joint Space and Time Processing for Moving Source Localization with a Few Sensors," *Proc. ICASSP*, 1995, pp. 3559–3562.

Cooke, M., and D. P. Ellis, *The Auditory Organization of Speech in Listeners and Machines*, Tech. Rep. TR-98-016, Berkeley, CA, 1998.

Cosi, P., and E. Zovato, "Lyon's Auditory Model Inversion: A Tool for Sound Separation and Speech Enhancement," *Proc. of ESCA Workshop on the Auditory Basis of Speech Perception*, Keele University, 1996, pp. 194–197.

Emile, B., and P. Comon, "Estimation of Time Delays Between Unknown Colored Signals," *Signal Processing*, Vol. 68, 1998, pp. 93–100.

Emile, B., P. Comon, and J. L. Roux, "Estimation of Time Delay with Few Sensors Than Sources," *IEEE Trans. on Signal Processing*, Vol. 46, 1998, pp. 2012–2015.

Gardner, W. A., and C. K. Chen, "Signal-Selective Time Difference of Arrival Estimation for Passive Location of Manmade Signal Sources in Highly Corruptive Environments," *IEEE Trans. on Signal Processing*, Vol. 40, 1992, pp. 1168–1184.

Gold, B., and N. Morgan, *Speech and Audio Signal Processing*, 1st ed., New York: John Wiley & Sons, 2002.

Haykin, S., *Adaptive Filter Theory*, 3rd ed., Upper Saddle River, NJ: Prentice Hall, 1996.

Ho, K. C., and Y. T. Chan, "Optimum Discrete Wavelet Scaling and Its Application to Delay and Doppler Estimation," *IEEE Trans. on Signal Processing*, 1998, pp. 2285–2290

Hoang-Lan, N. Thi, and C. Jutten, "Blind Source Separation for Convolutive Mixtures," *Signal Processing*, Vol. 45, 1995, pp. 209–222.

Hu, G., and D. Wong, "Monoaural Speech Segregation Based on Pitch Tracking and Amplitude Modulation," Technical Report: OSU-CISR-3/02-TR6, 2002.

Jan, E. E., and J. Flanagan, "Sound Capture from Spatial Volumes: Matched-Field Processing of Microphone Arrays Having Random Distributed Sensors," *Proc. ICASSP*, 1996, pp. 917–920.

Johnson, D. H., "The Application of Spectral Estimation Methods to Bearing Estimation Problems," *IEEE Proc.*, Vol. 70, 1982, pp. 1018–1027.

Johnson, D. H., *Array Processing*, 1st ed., Upper Saddle River, NJ: Prentice Hall, 1993.

Kandel, E. R., J. H. Schwartz, and T. M. Jessel, *Principles of Neural Science*, 4th ed., New York: McGraw-Hill, 2000.

Karjalainen, M., and T. Tolonen, "Multipitch and Periodicity Analysis Model for Sound Separation and Auditory Scene Analysis," *International Conference on Acoustics, Speech and Signal Processing*, II, 1999, pp. 929–932.

Kellerman, W., "Strategies for Combining Acoustic Echo Cancellation and Adaptive Beamforming Microphone," *ICASSP97*, I, 1997, pp. 219–222.

Kennedy, R. A., T. D. Abhayapala, and D. B. Ward, "Broadband Near Field Beamforming Using a Radial Beampattern Transformation," *IEEE Trans. on Signal Processing*, Vol. 46, 1998, pp. 2147–2156.

Klapuri, A. P., "Multipitch Estimation and Sound Separation by the Spectral Smoothness Principle," *Proc. of ICASSP2001*, 2001.

Knapp, C. H., and G. C. Carter, "The Generalized Correlation Method for Estimation of Time Delay," *IEEE Trans. on Acoustics, Speech, and Signal Processing*, Vol. 24, 1976, pp. 320–327.

Krolik, J., and D. Swingler, "Multiple Broadband Source Location Using Steered Covariance Matrices," *IEEE Trans. on Signal Processing*, Vol. 37, 1989, pp. 1481–1494.

Landone, C., and M. Sandler, "Issues in Performance Prediction of Surround Systems in Sound Reinforcement Applications," *Proc. 2nd COST G-6 Workshop on Digital Audio Effects (DAFX99)*, NTNU, Trondheim, Norway, December 1999.

Meddis, R., and L. O'Mard, "A Unitary Model of Pitch Perception," *J. Acoust. Soc. Am.*, Vol. 102, 1997, pp. 1811–1820.

Michael, H. F. S., and S. Brandstein, "A Robust Method for Speech Signal Time-Delay Estimation in Reverberant Rooms," *Proc. ICASSP*, 1997, pp. 375–378.

Nakadai, K., H. G. Okuno, and H. Kitano, "Auditory Fovea Based Speech Separation and Its Application to Dialog System," *Proc. of IEEE/RSJ IROS-2002*, 2002, pp. 1314–1319.

Oppenheim, A. V., and R. W. Shaffer, *Discrete Time Signal Processing*, Upper Saddle River, NJ: Prentice Hall, 1996.

Pham, D.-T., and J.-F. Cardoso, "Blind Separation of Instantaneous Mixtures of Nonstationary Sources," *IEEE Trans. on Signal Processing*, Vol. 49, 2000, pp. 1837–1848.

Read, H., J. Winer, and C. S. Schreiner, "Functional Architecture of Primary Auditory Cortex," *Current Opinion in Neurobiology*, Vol. 12, pp. 433–440.

Roweis, S. T., "One Microphone Source Separation," *Neural Information Processing Systems*, Vol. 13, 2000, pp. 889–896.

Sahlin, H., *Blind Separation by Second Order Statistics*, Department of Signals and Systems, Chalmers University of Technology, 1998.

Schmidt, R. H., "A Signal Subspace Approach to Multiple Emitter Location and Spectral Estimation," Ph.D. dissertation, Stanford University, Stanford, CA, 1981.

Seifritz, E., et al., "Spatio-Temporal Pattern of Neural Processing in the Human Auditory Cortex," *Science*, Vol. 297, 2002, pp. 1706–1708.

Shackleton, T., et al., "Interaural Time Difference Discrimination Thresholds for Single Neurons in the Inferior Colliculus of Guinea Pigs," *J. Neuroscience*, 2003, pp. 716–724.

Shinn-Cunningham, B., S. Santarelli, and N. Kopco, "Tori of Confusion: Binaural Cue for Sources Within Reach of a Listener," *JASA*, Vol. 107, 2000, pp. 1627–1636.

Sofiene, S. G., and Y. Grenier, "An Algorithm for Multisource Beamforming and Multitarget Tracking," *IEEE Trans. on Signal Processing*, Vol. 44, 1996, pp. 1512–1522.

Vale, C., and D. H. Sanes, "The Effect of Bilateral Deafness on Excitatory Synaptic Strength in the Auditory Midbrain," *European Journal of Neuroscience*, Vol. 16, 2002, pp. 2394–2404.

Veprek, P., and M. S. Scordilis, "Analysis, Enhancement and Evaluation of Five Pitch Determination Techniques," *Speech Communication*, 2002.

Wang, D. L., and G. J. Brown, "Separation of Speech from Interfering Sounds Based on Oscillatory Correlation," *IEEE Trans. on Neural Networks*, Vol. 10, 1999.

Weinstein, E., M. Feder, and A. V. Oppenheim, "Multichannel Signal Separation by Decorrelation," *IEEE Trans. on Speech and Audio Processing*, Vol. 1, 1993, pp. 405–413.

CHAPTER 16
Multiple Target Estimation in a Surveillance Radar System with a Mechanically Rotating Antenna

Fulvio Gini, Maria S. Greco, and Alfonso Farina

16.1 Introduction

The subject of this chapter is the estimation of multiple targets present in the same range-azimuth resolution cell of a surveillance coherent radar system with a mechanically rotating antenna. In most modern radar systems, the target DOA is estimated by the monopulse technique [1], which in principle can work with just a single pulse. When one angular coordinate is needed, the price to pay is the need for two tightly matched receiving channels: the sum (Σ) and the difference (Δ). DOA estimate is a function of the ratio of the Δ and Σ channel outputs. When multiple targets are present in the range-azimuth cell under test, the monopulse system provides an erroneous DOA measure [2, 3], and it is not useful to track any of the targets in the antenna main beam. A few methods are proposed to reduce the deleterious effects of target multiplicity. Blair et al. developed a DOA estimator using the real and imaginary parts of the monopulse ratio and the observed signal strengths for two unresolved Swerling I or Swerling III targets [4]. Sinha et al. presented a single pulse ML estimator that uses both in-phase and quadrature components of the sum and difference channels [5]. The approach is considered for no more than two Swerling I or Swerling III targets, and it cannot be easily extended to the case of more than two targets. To derive a computationally efficient detection algorithm, we derived an estimator termed AML-RELAX, based on the ML technique and on a RELAXation method, which exploits knowledge of the antenna main beam pattern and the fact that the mechanically scanning antenna impresses an amplitude modulation on the signals backscattered by the targets [6–8].

The rest of the chapter is organized as follows. The data model and the problem statement are briefly introduced in Section 16.2. Section 16.3 concerns the estimation problem. The ML and the asymptotic (i.e., large sample size), ML (AML) estimators are derived in Section 16.3.1. In Section 16.3.2, we describe an efficient way to implement the AML, based on a RELAXation method. It decouples the nonlinear multidimensional AML problem into simpler estimation problems, where the DOA and Doppler frequency of each target are separately and sequentially estimated. In Section 16.4, the performance of the AML-RELAX algorithm is numerically investigated through Monte Carlo simulation and compared to the CRLB. Some concluding remarks are reported in Section 16.5.

16.2 Data Model and Problem Statement

Consider a radar antenna that rotates mechanically with constant angular velocity ω_R rad/s, and denote the one-way antenna beam pattern by $h(\theta)$, the maximum gain by G_0, and the −3 dB azimuth beam width by θ_B; that is, the angle such that $h^2 = (\pm\theta_B/2) = G_0/2$. The number of pulses N collected during the time-on-target (ToT) by the radar within the −3 dB points of the rotating antenna is given by $N = \theta_B/(\omega_R T)$, where $T = 1/\text{PRF}$ is the radar pulse repetition time (PRT), and the pulse repetition frequency (PRF) is in the denominator. Assume that M pointlike targets, with direction of arrivals $\{\theta_{TGi}\}_{i=1}^M$ and Doppler frequencies $\{f_{Di}\}_{i=1}^M$, are present in the range-azimuth resolution cell under test. The data vector $\mathbf{z}$ is composed of the collection of the N echoes received during the ToT. The nth element of $\mathbf{z}$ is given by

$$z(n) = \sum_{i=1}^{M} b_i G(n, \theta_{TGi}) e^{j2\pi f_{Di} n} + d(n)$$

where b_i is the unknown complex amplitude of the ith target, $G(n, \theta_{TGi}) = h^2(\theta_{TGi} - n\omega_R T)$ is the two-way antenna gain for the nth pulse from the ith DOA, $\theta_{TGi} \in [0, \theta_B)$, and $f_{Di} \in [-0.5, 0.5)$ is the Doppler frequency of the ith target normalized to the PRF. The assumption that $\theta_{TGi} \in [0, \theta_B)$ is equivalent to the assumption that no contaminating targets are present in adjacent azimuth cells. The relaxation of this assumption is discussed in [8]. The term $d(n)$ models the disturbance, which is composed of the superposition of clutter and thermal noise. The problem is investigated here under the following two assumptions on $d(n)$:

1. $d(n)$ is a stationary complex-valued process independent of the signal components;
2. $d(n)$ satisfies the so-called *mixing condition* [9], which states that the kth-order cumulant of $d(n)$ at lag $l = (l_1, l_2, \ldots, l_{k-1})$, which is denoted by $c_{kd}(l)$, is absolutely summable:

$$\sum_l |c_{kd}(l)| < \infty, \forall k$$

For $k = 2$, the disturbance autocovariance sequence $c_{2d}(l)$ is absolutely summable. The mixing condition requires that sufficiently separated samples are approximately independent and is satisfied by all finite memory signals in practice. The second assumption was used in [8] for establishing consistency of the proposed estimator.

In vector notation, the data model for M targets is given by $\mathbf{z} = \mathbf{A}(\boldsymbol{\theta})\mathbf{b} + \mathbf{d}$, where $\mathbf{z} = [z(0) \ \ldots \ z(N-1)]^T$ is the $N \times 1$ complex data vector, $(\cdot)^T$ denotes the transpose operation, $\mathbf{b} = [b_1 \ b_2 \ \ldots \ b_M]^T$ is the $M \times 1$ vector of the unknown complex amplitudes, $\mathbf{A}(\boldsymbol{\theta}) = [\mathbf{a}(\theta_{TG1}, f_{D1}) \ \ldots \ \mathbf{a}(\theta_{TGM}, f_{DM})]$ is the $N \times M$ steering matrix, $\boldsymbol{\theta} = [\theta_{TG1} \ \ldots \ \theta_{TGM} \ f_{D1} \ \ldots \ f_{DM}]^T$ is the $2M \times 1$ vector of the unknown DOAs and Doppler frequencies, and $\mathbf{a}(\theta_{TGi}, f_{Di})$ is an $N \times 1$ vector that can be factored as $\mathbf{a}(\theta_{TGi}, f_{Di}) = \mathbf{g}(\theta_{TGi}) \odot \mathbf{p}(f_{Di})$, whose elements are given by

$$[\mathbf{a}(\theta_{TGi}, f_{Di})]_n = [\mathbf{g}(\theta_{TGi})]_n \cdot [\mathbf{p}(f_{Di})]_n = G(n-1, \theta_{TGi}) e^{j2\pi f_{Di}(n-1)} \tag{16.1}$$

where $\odot$ represents the Hadamard product or element-wise multiplication [10], $[\mathbf{g}(\theta_{TGi})]_n = G(n-1, \theta_{TGi})$ and $[\mathbf{p}(f_{Di})]_n = \exp[j2\pi f_{Di}(n-1)]$. Note that $\mathbf{g}(\theta_{TGi})$ is a function of only θ_{TGi}, whereas $\mathbf{p}(f_{Di})$ is function of only f_{Di}. The $N \times 1$ disturbance vector $\mathbf{d}$ is composed of the sum of thermal noise $\mathbf{n}$ and clutter $\mathbf{c}$. The thermal noise $\mathbf{n}$ is modeled as a complex zero-mean white Gaussian vector. In shorthand notation, we write $\mathbf{n} \sim CN(0, \sigma_n^2 \mathbf{I})$, where σ_n^2 is the variance of each noise component, and $\mathbf{I}$ is the $N \times N$ identity matrix. Clutter $\mathbf{c}$ is modeled as a complex Gaussian distributed random vector having zero-mean and covariance matrix $E\{\mathbf{cc}^H\} = \sigma_c^2 \mathbf{M}_c$, where $(\cdot)^H$ is the conjugate-transpose operator, σ_c^2 is the variance of each clutter component, and $\mathbf{M}_c$ is the normalized covariance matrix; that is, $[\mathbf{M}_c]_{i,i} = 1$, for $i = 1, 2, \ldots, N$. The disturbance covariance matrix is $\mathbf{M}_d = E\{\mathbf{dd}^H\} = \sigma_c^2 \mathbf{M}_c + \sigma_n^2 \mathbf{I} = \sigma_d^2 \mathbf{M}$, where $\sigma_d^2 = \sigma_c^2 + \sigma_n^2$ is the total disturbance power, and $\mathbf{M}$ is the normalized disturbance covariance matrix, which is given by

$$\mathbf{M} = \frac{CNR}{CNR + 1} \mathbf{M}_c + \frac{1}{CNR + 1} \mathbf{I}$$

where CNR is the clutter-to-noise power ratio defined as $CNR = \sigma_c^2/\sigma_n^2$. In shorthand notation, $\mathbf{d} \sim CN(0, \sigma_d^2 \mathbf{M})$. Note that Assumption 1 implies that $\mathbf{M}$ is a Toeplitz matrix. In summary, the goal here is to estimate jointly target parameters $\mathbf{b}$ and $\boldsymbol{\theta}$ (i.e., target complex amplitudes), Doppler frequencies, and directions of arrival, based on the observation of N consecutive samples $\{z(n)\}_{n=0}^{N-1}$. Here we assume that the number of targets M is a priori known, or already has been estimated. An algorithm for estimating M is described in [11].

16.3 The Estimation Algorithm

The goal is to exploit knowledge of the antenna main beam pattern and the consequent amplitude modulation impressed on the signal backscattered by each target to estimate jointly the target parameters. This basic idea was originally described in [12] for a single target scenario. In [6], a linear algorithm for the estimation of a single target DOA was proposed and analyzed. The method was extended in [7, 8] to consider the presence of multiple targets in the same range-azimuth resolution cell.

16.3.1 Asymptotic Maximum Likelihood Estimation

Conditioned to a given $\boldsymbol{\theta}$ and $\mathbf{b}$, the data vector $\mathbf{z}$ is complex Gaussian distributed with probability density function (PDF) given by:

$$p_{z|b}(z|b; \theta) = \frac{1}{(\pi\sigma_d^2)^N |M|} \exp\left[-\frac{[z - A(\theta)b]^H M^{-1}[z - A(\theta)b]}{\sigma_d^2}\right] \quad (16.2)$$

The PDF depends on the assumptions on vectors θ and b. No a priori information is assumed for θ and b, and they are modeled as a vectors of unknown deterministic constants, as in [13, 14]. The ML estimate is obtained by maximizing the LF $p_{z|b}(z|b; \theta)$ with respect to θ and b. Maximization of the LF is equivalent to minimization of the quadratic form $[z - A(\theta)b]^H M^{-1}[z - A(\theta)b]$. After straightforward manipulations, we find [15]:

$$\hat{\theta}_{ML} = \arg\max_{\theta} z^H M^{-1} A (A^H M^{-1} A)^{-1} A^H M^{-1} z \quad (16.3)$$

$$\hat{b}_{ML} = (A^H M^{-1} A)^{-1} A^H M^{-1} z \quad (16.4)$$

where, for ease of notation, we omitted the dependence of $A(\theta)$ on θ. Equations (16.3) to (16.4) imply that first we derive $\hat{\theta}_{ML}$ and then $\hat{b}_{ML}$, using $\hat{\theta}_{ML}$ previously obtained [i.e., calculating $A(\hat{\theta}_{ML})$]. Calculation of $\hat{\theta}_{ML}$ requires the $2M$-dimensional ($2M$-D) nonlinear maximization of the functional $F_N(\theta) = z^H M^{-1} A (A^H M^{-1} A)^{-1} A^H M^{-1} z$, where $\theta = [\theta_1 \ldots \theta_M \ f_1 \ldots f_M]^T$ denotes the generic parameter vector, and the subscript N points out the fact that the functional also depends on the number N of collected samples. When only one target is present in the range-azimuth cell under test, θ is a 2×1 vector, and b is a scalar, and the estimators in (16.3) and (16.4) become:

$$(\hat{\theta}_{ML}, \hat{f}_{ML}) = \arg\max_{\theta, f} \frac{|z^H M^{-1} a(\theta, f)|^2}{a^H(\theta, f) M^{-1} a(\theta, f)} \quad (16.5)$$

$$\hat{b}_{ML} = \frac{a^H(\hat{\theta}_{ML}, \hat{f}_{ML}) M^{-1} z}{a^H(\hat{\theta}_{ML}, \hat{f}_{ML}) M^{-1} a(\hat{\theta}_{ML}, \hat{f}_{ML})} \quad (16.6)$$

where $a(\theta, f) = g(\theta) \odot p(f)$ is the overall target steering vector introduced in Section 16.2. It is useful for future developments to define the following quantity:

$$\Gamma_N(\theta, f) = \frac{|z^H M^{-1} [g(\theta) \odot p(f)]|^2}{[g(\theta) \odot p(f)]^H M^{-1} [g(\theta) \odot p(f)]} \quad (16.7)$$

which is the LF for the single target case. Therefore,

$$(\hat{\theta}_{ML}, \hat{f}_{ML}) = \arg\max_{\theta, f} \Gamma_N(\theta, f)$$

is the ML estimator for the single target case. The ML estimator (16.3) requires a nonlinear $2M$-D maximization. Generally, this maximization is computationally

16.3 The Estimation Algorithm

cumbersome and may not be feasible in real time. Therefore, it is necessary to find a suboptimum algorithm that trades off good performance with computational complexity. Estimator (16.3) is used as a starting point to derive a suboptimal algorithm based on M 2-D maximizations, instead of one $2M$-D maximization of $F_N(\boldsymbol{\theta})$. Under the hypothesis that the Doppler frequencies are sufficiently separated, that is $|f_{Di} - f_{Dj}| \geq 1/N$ for $i \neq j$, in [8] we proved that

$$\lim_{N \to \infty} N^{-1} F_N(\boldsymbol{\theta}) = \lim_{N \to \infty} N^{-1} G_N(\boldsymbol{\theta}) = \lim_{N \to \infty} \sum_{i=1}^{M} N^{-1} \Gamma_N(\theta_i, f_i)$$

Based on this observation, we propose to estimate θ by solving the following problem:

$$\hat{\boldsymbol{\theta}}_{AML} = \underset{(\theta_1, \theta_2, \ldots, \theta_M, f_1, f_2, \ldots, f_M)}{\arg \max} \sum_{i=1}^{M} \Gamma_N(\theta_i, f_i) \qquad (16.8)$$

We denote (16.8) as the AML estimator. It calculates an estimate of $\boldsymbol{\theta}$ from the locations of the M highest peaks of $\Gamma_N(\theta_i, f_i)$. It is less computationally heavy than the ML, and replaces the $2M$-D nonlinear search required by the ML of (16.3) with the search of the locations of the M highest peaks of a two-dimensional functional. Consistency of the AML estimator is established in [8]. The complex amplitudes are estimated by making use of $\hat{\mathbf{b}}_{ML}$ of (16.3). In Figure 16.1, we plot $\Gamma_N(\theta, f)$ as a function of θ and f.

The signal parameters are $\boldsymbol{\theta}_{TG1} = 0.9°$, $\boldsymbol{\theta}_{TG2} = 1.5°$, $N = 16$, $\theta_B = 2°$, $G_0 = 1$, $f_{D1} = -0.3$, $f_{D2} = -0.3$, $SDR_1 = SDR_2 = 20$ dB, and CNR $\to \infty$ (i.e., $\mathbf{M} = \mathbf{I}$). The

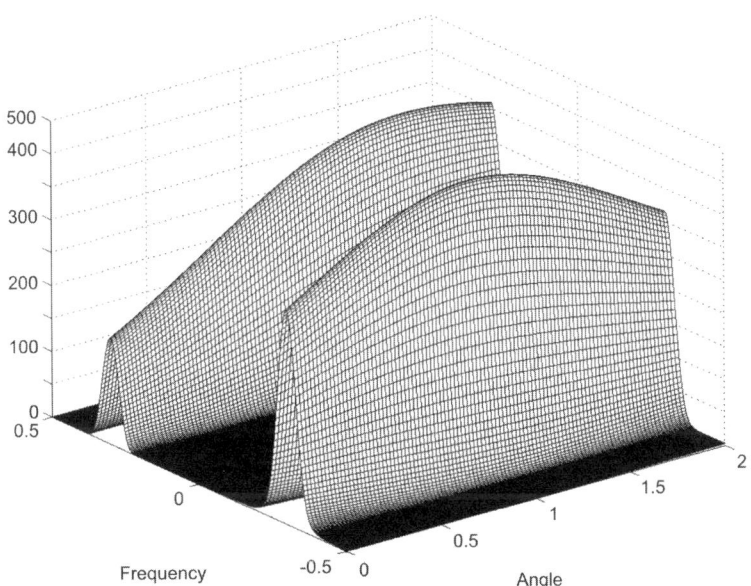

Figure 16.1 Plot of $\Gamma(\theta, f)$ for $|f_{D1} - f_{D2}| \gg 1/N$.

plot for $\rho = 0.9$ and CNR $\to \infty$ is quite similar, and is not reported here. A problem arises when the frequencies are not well separated; that is, when $|f_{Di} - f_{Dj}| \geq 1/N$ for some $i \neq j$. In this case, $\Gamma_N(\theta, f)$ presents only K maxima, where $K < M$ is the number of different Doppler frequencies.

16.3.2 The AML-RELAX Estimator

To further reduce the computational complexity, we propose an approach based on the RELAX method, which provides an efficient method to implement the AML of (16.8). In general, maximizations with respect to $(\theta_1, f_1), (\theta_2, f_2), \ldots, (\theta_M, f_M)$ are coupled. However, when N goes to infinity, $N^{-1}\Gamma_N(\theta, f)$ approaches nonzero values only for $f = \{f_{Di}\}_{i=1}^{M}$, whatever the values of θ and $\{\theta_{TGi}\}_{i=1}^{M}$ are. This suggests that it is possible to estimate $(\theta_{TG1}, f_{D1})_{i=1}^{M}$ by first estimating all the Doppler frequencies from the locations of the highest peaks of the Fourier transform of the whitened data. Then $\{\theta_{TGi}\}_{i=1}^{M}$ is estimated by plugging $\{\hat{f}_{Di}\}_{i=1}^{M}$ in the AML or in the ML estimator for known Doppler frequencies [7]. In [8], it is shown that the AML for known Doppler frequencies estimates sequentially the M DOAs performing M one-dimensional nonlinear maximizations of $\Gamma_N(\theta, f)$ in (16.7), calculated sequentially for $f = \hat{f}_{D1}, \hat{f}_{D2}, \ldots, \hat{f}_{DM}$. These ideas are exploited to derive the following relaxation-based estimation algorithm. The most important feature of the RELAX algorithm is its ability to decouple the search of the locations of the M highest peaks in the two-dimensional functional $\Gamma_N(\theta, f)$ into a sequence of two-dimensional nonlinear maximization problems. Actually, as explained in the following, we perform only one-dimensional maximizations. Roughly speaking, the RELAX algorithm first estimates the parameters of the strongest component; then it removes the contribution of the strongest component from the data and proceeds with the second strongest component; and so on, up to the Mth component. Then, it iteratively refines the estimates of each pair (θ_{TGi}, f_{Di}), working again on a component-by-component basis. Without loss of generality, assume that $|b_1| \geq |b_2| \geq \ldots \geq |b_M|$; that is, the target components are ordered so that the strongest component has index "1," and the weakest index "M." The steps of the AML-RELAX algorithm can be summarized as follows.

Step 0. Rough Estimate of the Parameters of the Strongest Component

1. Obtain the estimate $\hat{f}_{Di}$ from the location of the highest peak of the periodogram of the whitened data $\mathbf{w} = \mathbf{L}^H \mathbf{z}$, where $\mathbf{M}^{-1} = \mathbf{L}\mathbf{L}^H$,

$$\hat{f}_{D1} = \arg\max_f |\mathbf{w}^H \mathbf{p}(f)|^2 \quad \text{with } f \in (-0.5, 0.5] \qquad (16.9)$$

2. Estimate the complex amplitude b_1 and the DOA θ_{TG1} of the strongest component, using the ML estimator (16.5) to (16.6) for the single target scenario. Then, go to Step 2. Note that $\hat{f}_{Di}$ is used to get the first rough estimate of the frequency of the mth component when the strongest $m-1$ have been estimated and removed from the data. In this case, once a first rough estimate of $\{\hat{f}_{Dm}, \hat{\theta}_{TGm}, \hat{b}_m\}$ is obtained, we go back to Step m to derive iteratively improved estimates of $\{\hat{f}_{Dk}, \hat{\theta}_{TGk}, \hat{b}_k\}_{k=1}^{m}$.

16.3 The Estimation Algorithm

Step 1

1. Set $l = 1$, and compute the data sequence $z_1(n)$, defined as follows:

$$z_l(n) = z(n) - \sum_{\substack{k=1 \\ k \neq l}}^{P} \hat{b}_k G(n, \hat{\theta}_{TGk}) e^{j2\pi \hat{f}_{Dk} n}, \quad n = 0, 1, \ldots, N-1$$

(16.10)

To calculate (16.10), we use $\{\hat{f}_{Dk}, \hat{\theta}_{TGk}, \hat{b}_k\}_{k=1}^{P}$ as previously derived, for $k \neq 1$. P is the number of components we have already estimated, so it is a number that assumes different values at different steps.[1]

2. Estimate f_{D1} using the ML estimator (16.5) with $\theta = \hat{\theta}_{TG1}$. See also (16.11). Use the available DOA estimate $\hat{\theta}_{TG1}$ for this purpose, which was obtained from the previous step, in place of the unknown θ_{TG1}. Finally, again estimate the DOA θ_{TG1} and the complex amplitude b_1 using the single target ML estimator (16.6).

Step 2

1. Set $l = 2$ and compute the data vector $\mathbf{z}_2$ using (16.10) and $\{\hat{f}_{Dk}, \hat{\theta}_{TGk}, \hat{b}_k\}_{k=1}^{P}$, as previously derived. P is the number of components that we have already estimated. Calculate $\{\hat{f}_{D2}, \hat{\theta}_{TG2}, \hat{b}_2\}$ by processing the data vector $\mathbf{z}_2$ in the same way as $\mathbf{z}_1$. See Step 0, and the second part of Step 1.

2. Set $l = 1$ and compute $\mathbf{z}_1$ from (16.10) by using $\{\hat{f}_{Dk}, \hat{\theta}_{TGk}, \hat{b}_k\}_{k=1}^{P}$, as previously derived, and again determine $\{\hat{f}_{D1}, \hat{\theta}_{TG1}, \hat{b}_1\}$ from the data $\mathbf{z}_1$, as in Step 1.

3. Iterate the previous two substeps until "practical convergence" is achieved. This convergence is measured on the cost function:

$$CF(\{\hat{f}_{Dk}, \hat{\theta}_{TGk}, \hat{b}_k\}_{k=1}^{P}) = \sum_{n=0}^{N-1} \left| z(n) - \sum_{k=1}^{P} \hat{b}_k G(n, \hat{\theta}_{TGk}) e^{j2\pi \hat{f}_{Dk} n} \right|^2$$

(16.11)

Convergence is determined by checking the relative change of the cost function $CF(\cdot)$ in (16.11), between the jth and $(j + 1)$th iterations. In our numerical simulations, we terminated the iterations when the relative change is lower than $\epsilon = 10^{-4}$, as in [16].

---

1. Note that the first time we end up with Step m, only the estimates $\{\hat{f}_{Dk}, \hat{\theta}_{TGk}, \hat{b}_k\}_{k=1}^{m}$ are available. Therefore, we can remove only the contribution of the first m components. Once we end up for the first time with Step M, all the estimates $\{\hat{f}_{Dk}, \hat{\theta}_{TGk}, \hat{b}_k\}_{k=1}^{M}$ become available. Therefore, in the iterations that follow, we can remove all the components to refine the estimates, except the one we want to estimate.

Step m

1. Set $l = m$ and compute $\mathbf{z}_m$ from (16.10) by using $\{\hat{f}_{Dk}, \hat{\theta}_{TGk}, \hat{b}_k\}_{k=1}^{P}$, as previously derived. Calculate $\{\hat{f}_{Dm}, \hat{\theta}_{TGm}, \hat{b}_m\}$ from $\mathbf{z}_m$, as in Step 1.
2. Set $l = 1$ and compute $\mathbf{z}_1$ from (16.10) by using $\{\hat{f}_{Dk}, \hat{\theta}_{TGk}, \hat{b}_k\}_{k=1}^{P}$, as previously derived, and again determine $\{\hat{f}_{D1}, \hat{\theta}_{TG1}, \hat{b}_1\}$ from the data $\mathbf{z}_1$, as in Step 1.
3. Set $l = 2$ and compute $\mathbf{z}_2$ from (16.10) by using $\{\hat{f}_{Dk}, \hat{\theta}_{TGk}, \hat{b}_k\}_{k=1}^{P}$, as previously derived, and again determine $\{\hat{f}_{D2}, \hat{\theta}_{TG2}, \hat{b}_2\}$ from the data $\mathbf{z}_2$, as in Step 1.
m. Set $l = m - 1$ and compute $\mathbf{z}_{m-1}$ from (16.10) by using $\{\hat{f}_{Dk}, \hat{\theta}_{TGk}, \hat{b}_k\}_{k=1}^{P}$, as previously derived, and again determine $\{\hat{f}_{Dm-1}, \hat{\theta}_{TGm-1}, \hat{b}_{m-1}\}$ from the data $\mathbf{z}_{m-1}$, as in Step 1.
m + 1. Iterate the previous m substeps until practical convergence is achieved.

Step M

This is the last step, which is composed of $M + 1$ substeps. The details are immediately obtained from the explanation of Step m by setting $m = M$.

All the nonlinear searches involved in the calculation of the above steps are one-dimensional. This guarantees a reduction of the computational complexity over the ML and AML algorithms. In particular, we replaced the two-dimensional maximization of the AML algorithm with two separate one-dimensional searches, where the Doppler frequency is estimated first, and then used to separately estimate the DOA.

16.4 Numerical Performance Analysis

We now investigate numerically the performance of the AML-RELAX algorithm. The RMSE is derived by Monte Carlo simulation, and compared with the square root of the Cramér-Rao lower bound (RCRLB). Calculation of the CRLB is outlined in [8]. The RMSE is derived by running 10⁴ independent Monte Carlo runs. The RMSE of the estimation of the normalized frequency is obviously dimensionless, and $\{f_{Di}\}$ are the target Doppler frequencies normalized to the radar PRF. The RMSE of the estimation of the DOA is measured in degrees. In all the runs, the azimuth beam width is set at $\theta_B = 2°$, $G_0 = 1$, and the one-way antenna beam pattern was assumed to be Gaussian shaped [12]. The autocovariance function (ACF) is assumed to have an exponential shape, so that $[\mathbf{M}_c]_{ij} = \rho^{|i-j|}$, with $0 \leq \rho \leq 1$, where ρ is the clutter one-lag correlation coefficient. We investigate the scenario in which two targets are present in the range-azimuth cell under test, with unknown deterministic amplitude $\mathbf{b}$. The reference scenario is related to the following set of parameters:

- Azimuth –3 dB beamwidth: $\theta_B = 2°$;
- Number of integrated pulses: $N = 16$;
- Number of targets: $M = 2$;
- Target DOAs: $[\theta_{TG1} \quad \theta_{TG2}] = [0.9° \quad 1.5°]$;

16.4 Numerical Performance Analysis

- Doppler frequencies: $[f_{D1} \quad f_{D2}] = [-0.3 \quad 0.3]$;
- Signal-to-disturbance power ratio: $SDR = SDR_1 = SDR_2 = 20$ dB;
- Clutter one-lag correlation coefficient: $\rho = 0.9$.

Performance is investigated as a function of N, SDR (where $SDR = SDR_1 = SDR_2$), SDR_2 (while keeping SDR_1 constant), CNR, θ_{TG2}, and f_{D2}.

A first rough estimate of the target Doppler frequencies is obtained using the periodogram of the whitened data, as described in Step 0 of the AML-RELAX algorithm. The periodogram has two peaks for $f \cong \{f_{Dk}\}_{k=1,2}$. Therefore, it is successfully used to get a first rough estimate of the target Doppler frequencies. The maximum of the functional (16.9) is found by using the FFT. The signal is zero-padded to $N_{zp} = 2^{14} = 16{,}384$ points. The final frequency step-size of the nonlinear search obtained in this manner is approximately $N_{zp}^{-1} = 6.1 \cdot 10^{-5}$. The same is done for the single target ML frequency estimator that we get from (16.5) when the DOA is known. Concerning DOA estimation, once the estimate $\hat{f}_{Dk}$ is obtained, $\hat{\theta}_{TGk}$ is derived, and the maximum of the single target ML functional (16.5) [i.e., the calculation of $\hat{\theta}_{TGk} = \arg \max_\theta \Gamma(\theta, \hat{f}_{Dk})$], is found by two successive one-dimensional searches. The first coarse search is carried out on a grid of 100 points. Then, a fine search is performed in an identical manner around the maximum found by the previous coarse search. The final DOA step-size of the nonlinear search obtained in this manner is approximately 10^{-4}.

In Figures 16.2 and 16.3, we investigate how the clutter-to-noise power ratio (CNR) affects estimation accuracy.

The power of the overall disturbance is kept constant at $\sigma_d^2 = 1$, and the clutter one-lag correlation coefficient is set at $\rho = 0.9$. The worst-case scenario is when

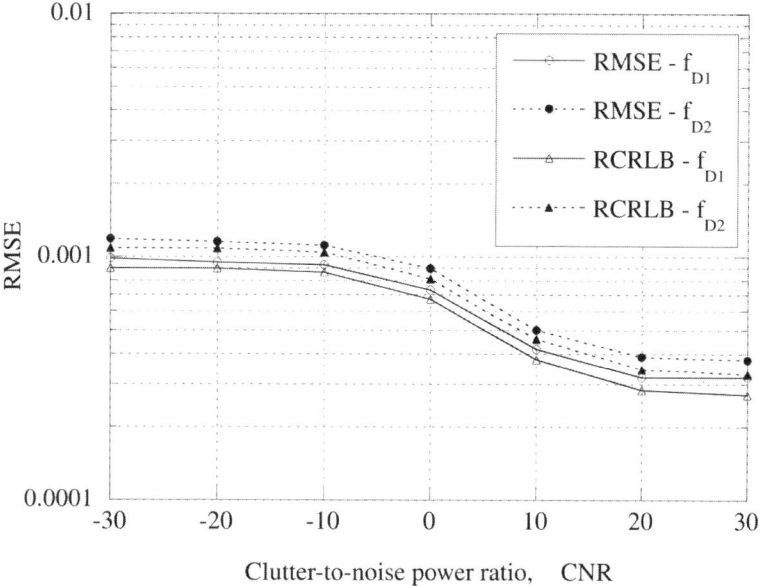

Figure 16.2 RCRLB and RMSE of the Doppler frequency estimator versus the CNR.

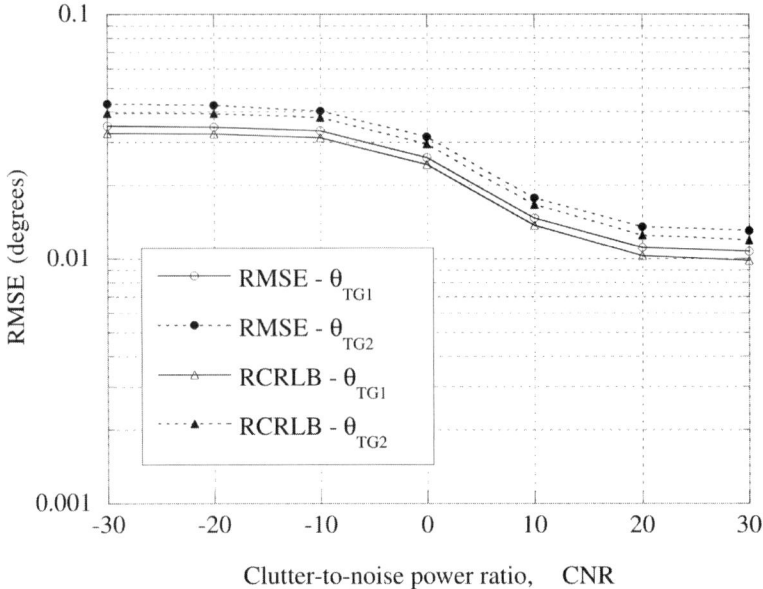

Figure 16.3 RCRLB and RMSE of the DOA estimator versus the CNR.

the disturbance is composed of only thermal noise (i.e., CNR $\to \infty$) [7]. Figures 16.4 to 16.13 were derived for the worst case scenario.

In Figures 16.4 to 16.5, we investigate the behavior of RMSE($\hat{f}_{Dk}$), RMSE($\hat{\theta}_{TGk}$), $\sqrt{\text{CRLB}(f_{Dk})}$, and $\sqrt{\text{CRLB}(\theta_{TGk})}$ as a function of the separation between the two targets in the Doppler domain.

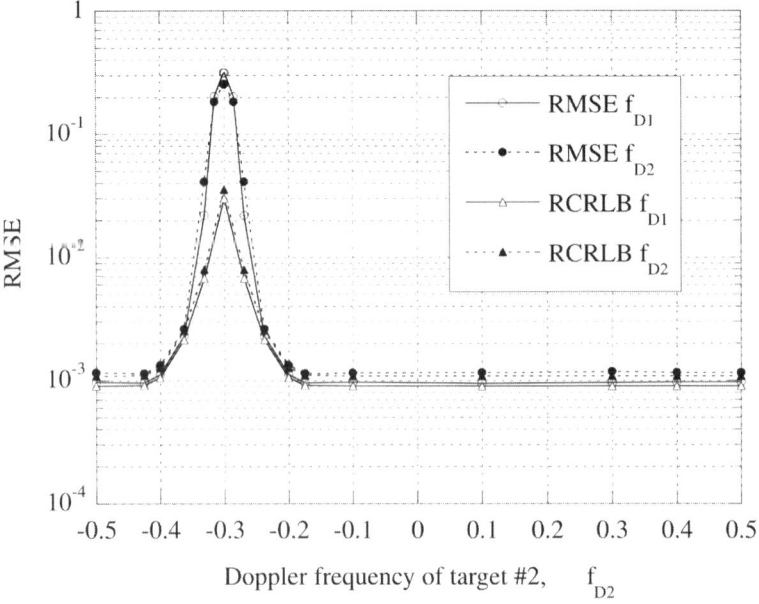

Figure 16.4 RCRLB and RMSE of the Doppler frequency estimator versus the Doppler frequency of target #2, f_{D2}.

16.4 Numerical Performance Analysis

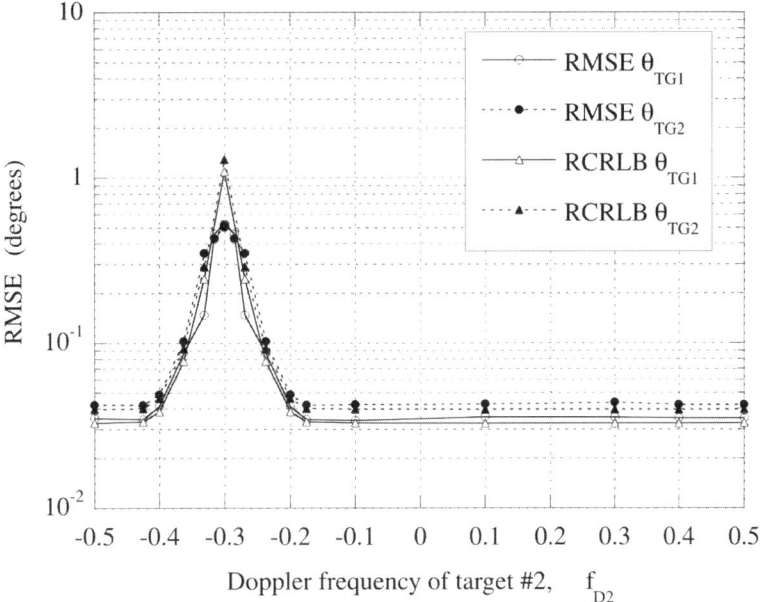

Figure 16.5 RCRLB and RMSE of the DOA estimator versus the Doppler frequency of target #2, f_{D2}.

Note that we calculated the RMSE for values of $\mathbf{f}_{D2}$ increasingly dense around f_{D1}. The numerical results show that when $|f_{D1} - f_{D2}| < 1/N = 0.0625$, RMSE($\hat{f}_{Dk}$) departs from the RCRLB. In this case, we would say that AML-RELAX is not able to resolve the two targets. Concerning the DOA estimates, the fact that their RMSE is lower than the CRLB is only due to the fact that the CRLB does not take into account that the estimates are confined into the limited interval $I \in [0°, \theta_B)$. As a consequence, the RMSE is upper bounded by that of a random variable uniformly distributed in I.

In Figures 16.6 and 16.7, we change the DOA of target #2, while keeping constant the DOA of target #1.

The results show that the accuracy of the estimation of the parameters of target #1 is not affected by the actual value of θ_{TG2}, not even if $\theta_{TG1} = \theta_{TG2}$. This means that targets exhibiting an arbitrarily small difference in azimuth can be separated, as long as their Doppler frequencies differ by at least $1/N$. Obviously, RMSE($\hat{\theta}_{TG2}$) is the lowest when $\theta_{TG2} = \theta_B/2 = 1°$, since when θ_{TG2} deviates from this value, the output SDR becomes increasingly lower, due to amplitude modulation effects.

Performance as a function of the number N of integrated pulses is investigated in Figures 16.8 to 16.9.

For $N \geq 16$, the estimation accuracy is very close to the RCRLB, and as N increases, the RMSE becomes increasingly closer to the RCRLB. In Figure 16.10, we compare the RMSE($\hat{f}_{Dk}$) of the AML-RELAX algorithm with the square root of the CRLB (RCRLB) as a function of SDR = $SDR_1 = SDR_2$.

The threshold effect, which is typical of nonlinear estimators, is well evident for SDR < 5 dB. For SDR $\geq$ 5 dB, the RMSE is quite close to the RCRLB. The

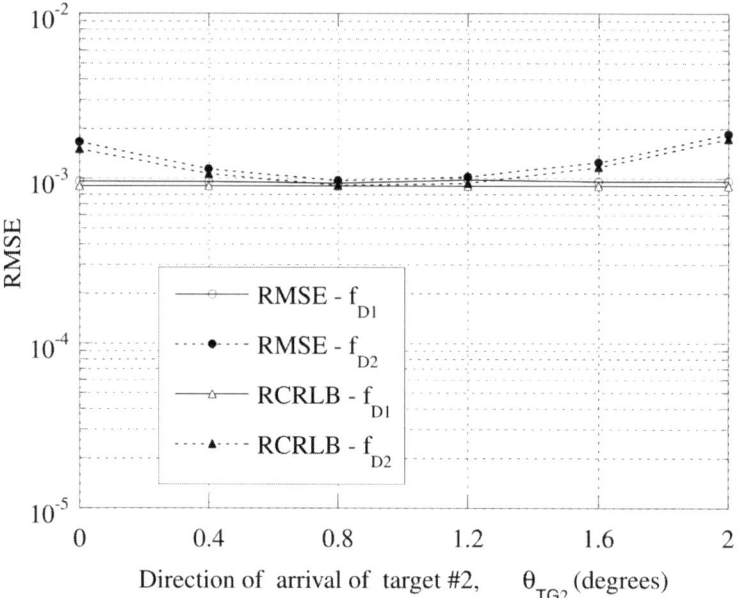

Figure 16.6 RCRLB and RMSE of the Doppler frequency estimator versus the DOA of target #2, θ_{TG2}.

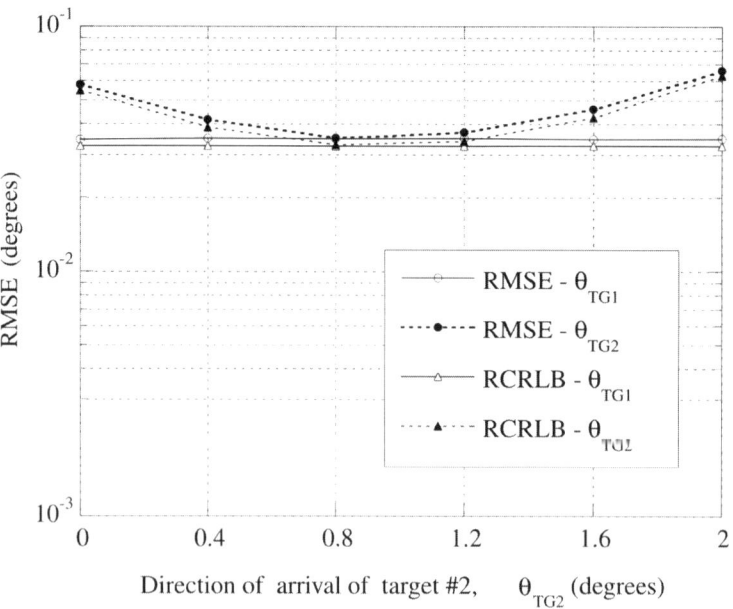

Figure 16.7 RCRLB and RMSE of the DOA estimator versus the DOA of target #2, θ_{TG2}.

curves in Figure 16.11 suggest that AML-RELAX provides performance quite close to the CRLB. For SDR ≥ 5, we get RMSE($\hat{\theta}_{TGk}$) ≤ $\theta_B/10 = 0.2°$.

As a rule of thumb, the standard deviation by monopulse estimation is one-tenth of the beamwidth for the usual value of SDR = 13 dB [14]. In Figures 16.12

16.4 Numerical Performance Analysis

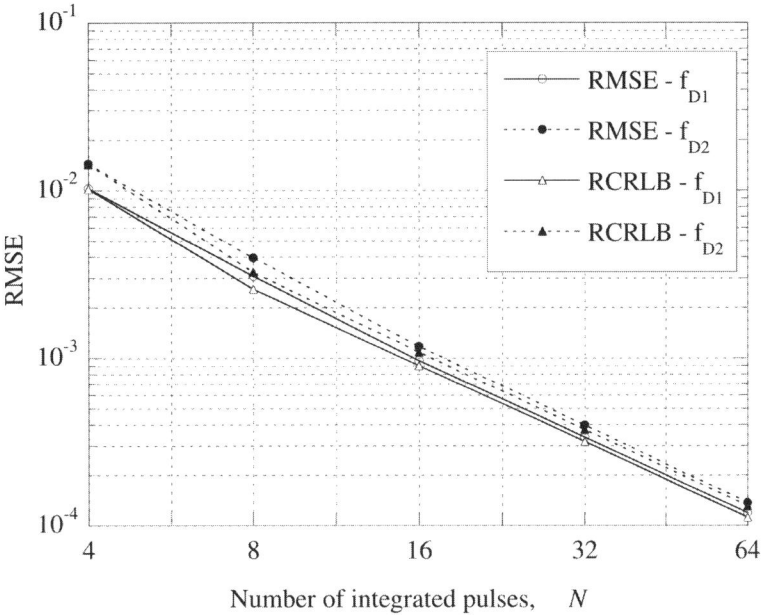

Figure 16.8 CRLB and RMSE of the Doppler frequency estimator versus the number N of integrated pulses.

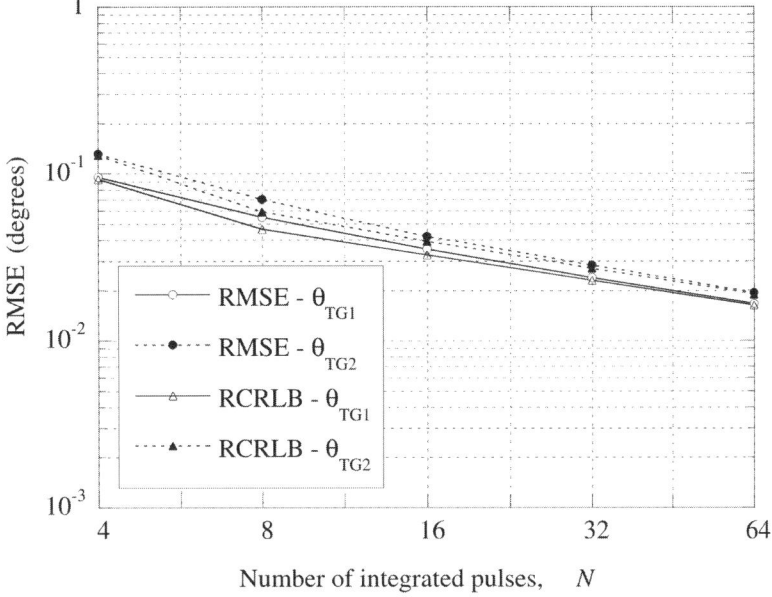

Figure 16.9 RCRLB and RMSE of the DOA estimator versus the number N of integrated pulses.

and 16.13, we change the SDR of target #2, while keeping constant the SDR of target #1.

The results shown here are quite interesting, since they tell us that already for SDR ≥ 5, the presence of the second target signal does not affect the estimation

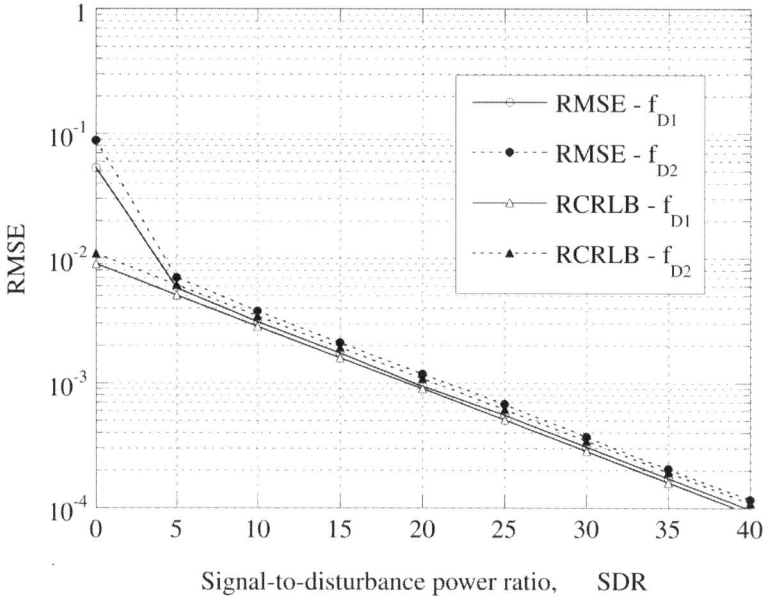

Figure 16.10 RCRLB and RMSE of the Doppler frequency estimator versus SDR.

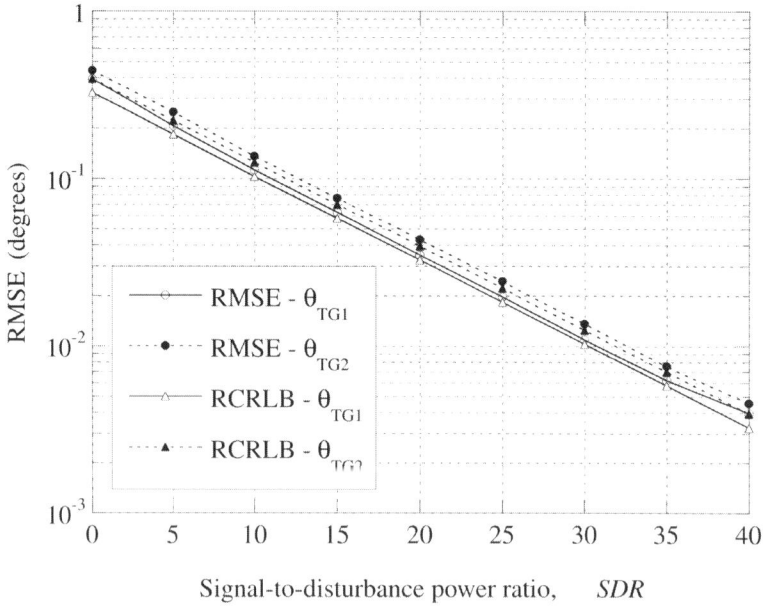

Figure 16.11 RCRLB and RMSE of the DOA estimator versus SDR.

accuracy of the parameters of the first target. This "decoupling" effect is obtained from the use of the RELAX approach.

In summary, the numerical results suggest that the RMSEs of Doppler frequency and DOA estimators are close to the RCRLB, except for very critical circumstances, such as the case in which SDR < 5 dB (i.e., the well-known threshold effect of any

16.4 Numerical Performance Analysis

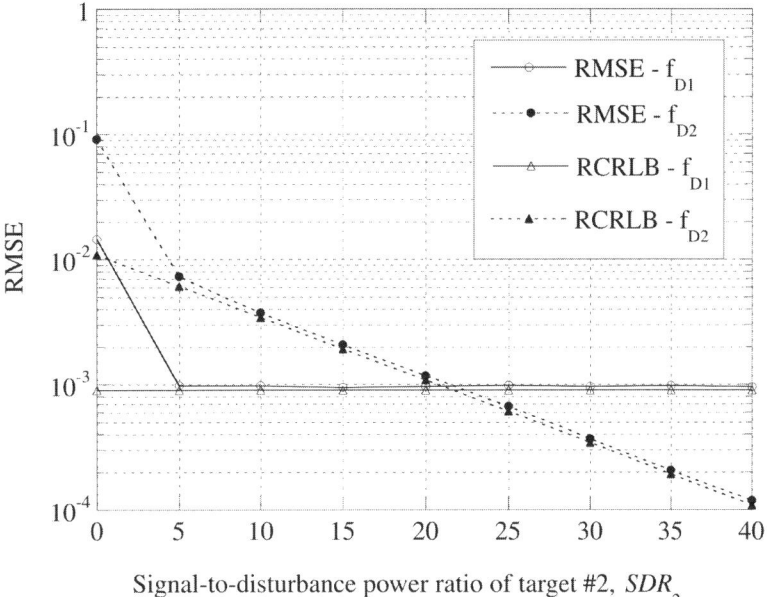

Figure 16.12 RCRLB and RMSE of the Doppler frequency estimator versus SDR_2.

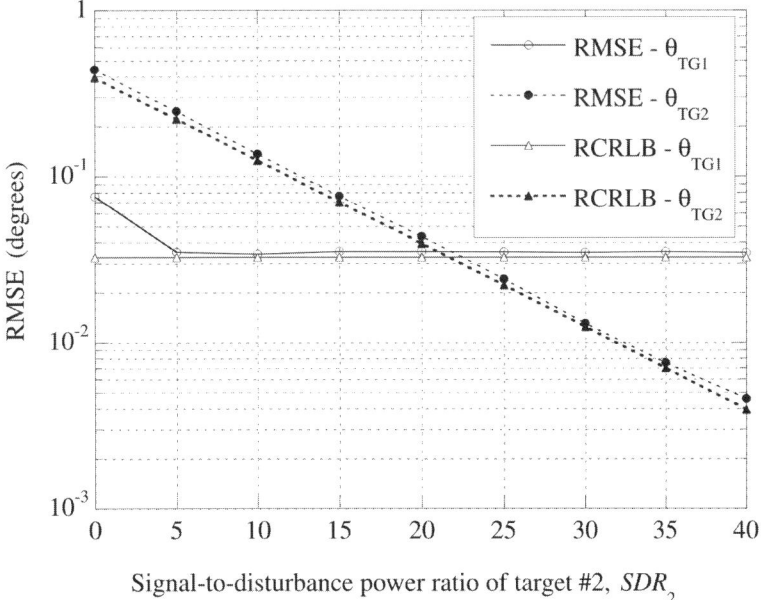

Figure 16.13 RCRLB and RMSE of the DOA estimator versus SDR_2.

nonlinear estimator), or when the target Doppler frequencies are close (i.e., $|f_{D1} - f_{D2}| < 1/N$). To evaluate the computational complexity of the AML-RELAX estimator, we calculate the average number of iterations required to achieve convergence. When two targets are present, on the average three to four iterations are enough to guarantee convergence, whatever the value of SDR. Obviously, it makes sense to evaluate estimation accuracy only once the multiple targets have been detected by the "detection" algorithm, which is described in [11].

16.5 Conclusions

In this work, we deal with the problem of estimating multiple targets present in the same range-azimuth resolution cell of surveillance radar with a mechanically rotating antenna. AML-RELAX is a computationally efficient method to calculate the asymptotical ML estimates based on the RELAX method. It decouples the problem of jointly estimating the parameters of all signal components into a sequence of simpler problems, in which we separately and recursively estimate the parameters of each component. Performance of the algorithm is numerically investigated through Monte Carlo simulation. Different scenarios are considered to demonstrate the ability of the proposed method to correctly estimate multiple target parameters in a typical surveillance radar scenario.

References

[1] Skolnik, M., *Introduction to Radar Systems*, 3rd ed., New York: McGraw-Hill, 2001.
[2] Kanter, I., "Multiple Gaussian Targets: The Track-on Jam Problem," *IEEE Trans. on Aerospace and Electronics Systems*, Vol. 13, No. 6, 1977, pp. 620–623.
[3] Sherman, S. M., *Monopulse Principles and Techniques*, Dedham, MA: Artech House, 1984.
[4] Blair, W. D., and M. Brandt-Pearce, "Monopulse DOA Estimation of Two Unresolved Rayleigh Targets," *IEEE Trans. on Aerospace and Electronics Systems*, Vol. 37, No. 2, April 2001, pp. 452–468.
[5] Sinha, A., T. Kirubarajan, and Y. Bar-Shalom, "Maximum Likelihood Angle Extractor for Two Closely Spaced Targets," *IEEE Trans. on Aerospace and Electronics Systems*, Vol. 38, No. 1, January 2002, pp. 183–203.
[6] Farina, A., G. Gabatel, and R. Sanzullo, "Estimation of Target Direction by Pseudo-Monopulse Algorithm," *Signal Processing*, Vol. 80, 2000, pp. 295–310.
[7] Farina, A., F. Gini, and M. Greco, "DOA Estimation by Exploiting the Amplitude Modulation Induced by Antenna Scanning," *IEEE Trans. on Aerospace and Electronic Systems*, Vol. 38, No. 4, October 2002, pp. 1276–1286.
[8] Gini, F., M. Greco, and A. Farina, "Multiple Radar Targets Estimation by Exploiting Induced Amplitude Modulation," *IEEE Trans. on Aerospace and Electronic Systems*, Vol. 39, No. 4, October 2003, pp. 1316–1332.
[9] Brillinger, D. R., *Time Series: Data Analysis and Theory*, San Francisco, CA: Holden-day, 1981.
[10] Stoica, P., and R. Moses, *Introduction to Spectral Analysis*, Upper Saddle River, NJ: Prentice Hall, 1997.
[11] Gini, F., F. Bordoni, and A. Farina, "Multiple Radar Targets Detection by Exploiting the Induced Amplitude Modulation," *IEEE Trans. on Signal Processing*, Vol. 52, No. 4, April 2004, pp. 903–913.
[12] Swerling, P., "Maximum Angular Accuracy of a Pulsed Search Radar," *IRE Proc.*, Vol. 44, No. 9, September 1956, pp. 1146–1155.
[13] Ziskind, I., and M. Wax, "Maximum Likelihood Localization of Multiple Sources by Alternating Projection," *IEEE Trans. on Acoustics, Speech and Signal Processing*, Vol. 36, No. 10, October 1988, pp. 1553–1560.

[14] Wirth, W. -D., *Radar Techniques Using Array Antennas*, London, England: The Institution of Electrical Engineers, 2001.

[15] Kay, S. M., *Fundamentals of Statistical Signal Processing, Estimation Theory*, Upper Saddle River, NJ: Prentice Hall, 1993.

[16] Li, J., and P. Stoica, "Efficient Mixed-Spectrum Estimation with Applications to Target Featured Extraction," *IEEE Trans. on Signal Processing*, Vol. 44, No. 2, February 1996, pp. 281–295.

CHAPTER 17
Joint DOA/DOD/DTOA Estimation System for UWB Double Directional Channel Modeling

Jun-ichi Takada, Katsuyuki Haneda, and Hiroaki Tsuchiya

17.1 Introduction

High data rate short range wireless links have been intensively investigated. The UWB communication system has drawn attention as a promising technique to satisfy this demand. To construct an effective and efficient UWB system, the propagation channel must be investigated. Although the channel model standards for the high- and low-rate personal area networks have already been standardized in IEEE 802.15.3a and 4a task groups [1, 2], they include the transmitter and receiver antennas into the channel. However, in the UWB systems, the distortions of the waveform and radiation pattern are obvious, due to the frequency characteristics of the antennas. In other words, the channel response may drastically change by changing the antennas. Therefore, the propagation channels should be separated from the antennas, especially for the performance evaluation of the antennas. This concept is known as a double directional channel [3, 4], and the concept has been extensively deployed into the standard channel models for MIMO systems [5, 6]. Meanwhile, this concept has not been seriously taken into account in the UWB systems, since they are regarded as the classical single-input single-output (SISO) system.

The double directional channel is modeled as the sum of the paths. Each of the paths is characterized by the DOA, the direction-of-departure (DOD), and the delay-time-of-arrival (DTOA). The complex amplitude of the path is approximated as constant in the conventional wideband modeling, but its frequency dependence is considered to model the path loss of the UWB signal more accurately.

Theoretical and experimental schemes of the double directional channel sounding for wideband channels have already been constructed [7–11]. Nonparametric spectral estimation has been deployed in [7], where the robust beamforming technique is used at the sacrifice of resolution. Since the path model is a parametric model, the high resolution parameter estimation techniques can be deployed. Multi-dimensional unitary ESPRIT has been initially used [8, 9]. However, the smoothing preprocessing is necessary after the compensation of the mutual coupling among antennas, and the array geometry is limited to either linear or rectangular arrays. Therefore, the ML-based estimation techniques have been getting more popular [10, 11].

Comparing the conventional wideband sounding and the UWB sounding, the former satisfies the following relation between the signal bandwidth B and the antenna aperture size A.

$$B\frac{A}{c} \ll 1 \qquad (17.1)$$

where c is the velocity of the light. Obviously, the UWB sounding does not satisfy the condition.

For UWB channels, the Sensor-CLEAN algorithm has been used to estimate DOAs and TDOAs, as well as the incident waveforms [12, 13]. Their aims are very similar to the authors' aims, but they have not yet considered the double directional models.

In this chapter, the authors describe the joint DOA/DOD/DTOA estimation system for the UWB double directional channel model. The system consists of a vector network analyzer (VNA), and two antenna positioners with ultrawideband antennas. As the postprocessing, an ML-based estimator extracts the parameters of the paths. Special consideration is necessary for the short range scenario, where the plane wave model is not applicable.

This chapter consists of the following sections. In Section 17.2, the sounding system is described. The parametric multipath modeling for the UWB is described in Section 17.3, and an ML-based channel parameter estimation scheme is presented in Section 17.4. Experimental results are presented in Section 17.5, and Section 17.6 concludes this chapter.

17.2 UWB Double Directional Channel Sounding System

To model the UWB double directional channel, it is necessary to measure the spatial transfer function distribution over the ultrawide bandwidth. Figure 17.1 depicts the proposed system for the UWB double directional channel sounding [14, 15]. The authors have chosen a VNA to measure the ultrawideband transfer function, since it has a very sophisticated function of calibration. For the spatial sampling, it is difficult to deploy an array antenna, since a UWB antenna is large relative to the wavelength, even at the lowest frequency [16, 17], and the grating lobes cannot be suppressed. Moreover, the mutual coupling between the antennas shall not be negligible, and its compensation is needed. Therefore, the authors decided to use the antenna positioners in both the transmitter (Tx) and the receiver (Rx) sides, and the synthetic antenna arrays are constructed. A low-noise preamplifier is inserted between the Rx antenna and the input port of the VNA to obtain a sufficient SNR.

In the present system, each of the antenna positioners has three scanning axes. It takes a few hours for the scanning to realize a synthetic array with sufficient size. Thus, the system is not applicable for the evaluation of the time-varying channels. On the other hand, the array configuration (i.e., the shape of the array) and the number of the elements can be flexibly configured.

A personal computer controls the antenna positioners, and collects the transfer function from the VNA via the general-purpose interface bus (GPIB).

17.3 Parametric Multipath Modeling for UWB

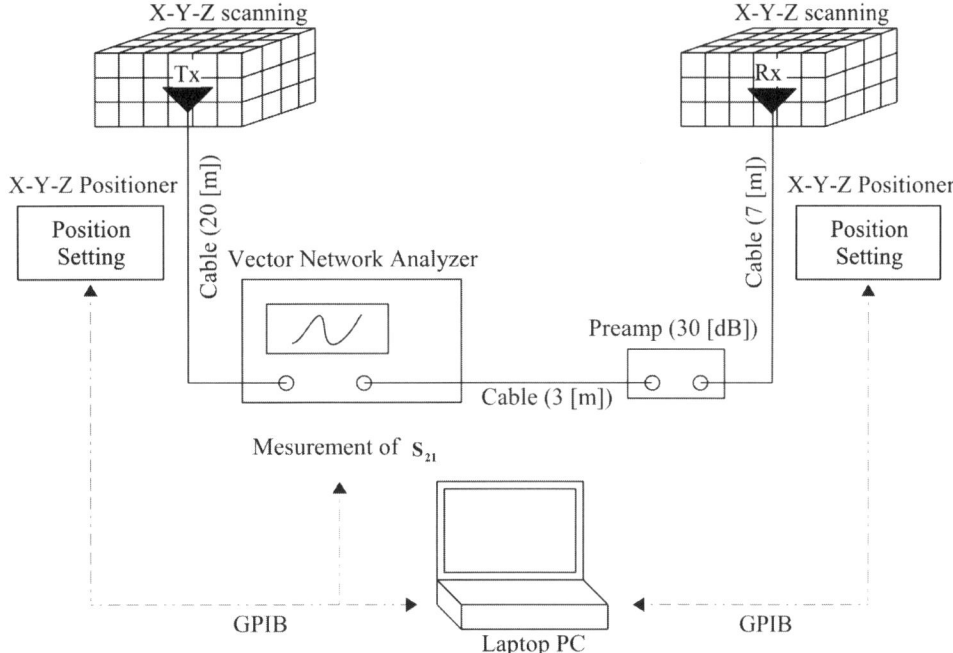

Figure 17.1 UWB double directional channel sounding system.

The calibration of the system is done twice. First, the port-to-port calibration is done by using the internal functionality of the VNA. Next, the back-to-back calibration is done in the open space by connecting Tx and Rx antennas. The latter step can be skipped if the antenna input impedance is well matched to the cable, and the complex gain of the antenna is known in advance.

17.3 Parametric Multipath Modeling for UWB

17.3.1 Multipath Model

The ray path model is based on a high frequency approximation. It is applicable when the scattering objects are much bigger than the wavelength. In the model, each propagation path has individual DOA, DOD, and DTOA that do not depend on frequency, and has frequency-dependent complex amplitude.

Figure 17.2 shows the definition of the coordinates. z is the vertical axis, while x and y are the horizontal axes. $0 \leq \phi < 2\pi$ is the azimuth angle, and $-\pi < \psi < \pi$ is the elevation angle. The direction vector points from the origin toward the DOD at the Tx side, whereas it points toward the origin from the DOA at the Rx side. $D_{\alpha\beta}(f, \Omega_\alpha)$ denotes the complex gain of the antenna at the frequency f toward the $\Omega_\alpha = [\psi_\alpha, \phi_\alpha]$ direction. The subscript $\alpha = $ t or r corresponds to either the transmitter or the receiver, and the subscript $\beta = \psi$ or ϕ denotes either the vertical or horizontal polarization component. Note that the reciprocity guarantees

$$D_{t\beta}(f, \Omega) = D_{r\beta}^{*}(f, \Omega) \tag{17.2}$$

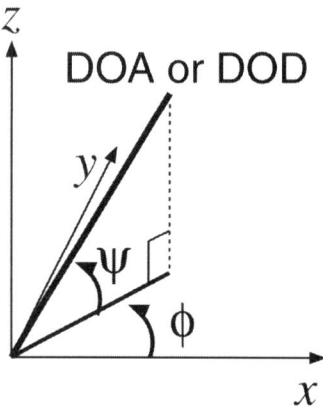

Figure 17.2 Definition of the coordinates.

The transfer function of the lth ray path, including Tx and Rx antennas $h_l(f)$ is expressed as

$$h_l(f) = h_0(f, \tau_l) \sum_{\beta_r = \psi, \phi} \sum_{\beta_t = \psi, \phi} \gamma_{\beta_r \beta_t l}(f) D_{r\beta_r}(f, \mathbf{\Omega}_{rl}) D_{t\beta_t}(f, \mathbf{\Omega}_{tl}) \quad (17.3)$$

where τ_l is the DTOA, $\mathbf{\Omega}_{tl}$ is the DOD, $\mathbf{\Omega}_{rl}$ is the DOA, and $\gamma_{\beta_r \beta_t l}(f)$ is the complex polarimetric scattering gain of lth path, respectively. $h_0(f, \tau)$ is the free-space complex path gain, given as the complex form of Friis' transmission formula.

$$h_0(f, \tau) = \frac{1}{4\pi f \tau} \exp(-j2\pi f \tau) \quad (17.4)$$

Let us assume the synthetic array realized by the parallel translation of the antenna. The phase center of the m_rth Rx antenna is spatially shifted along the geometrical vector $\mathbf{r}_{rm_r}$ as

$$\mathbf{r}_{rm_r} = \hat{\mathbf{x}} x_{rm_r} + \hat{\mathbf{y}} y_{rm_r} + \hat{\mathbf{z}} z_{rm_r} \quad (17.5)$$

where $\hat{\mathbf{x}}$, $\hat{\mathbf{y}}$, and $\hat{\mathbf{z}}$ are geometrical unit vectors pointing to x, y, and z directions, respectively. Then, the complex gain of the Rx antenna $D_{r\beta m_r}(f, \mathbf{\Omega}_r)$ is expressed as

$$D_{r\beta m_r}(f, \mathbf{\Omega}_r) = D_{r\beta}(f, \mathbf{\Omega}_r) \exp\left(j \frac{2\pi f}{c} \mathbf{r}_{rm_r} \cdot \hat{\boldsymbol{\omega}}_r\right) \quad (17.6)$$

In (17.6), $\hat{\boldsymbol{\omega}}_r$ is the geometrical unit vector pointing toward DOA,

$$\hat{\boldsymbol{\omega}}_r = \hat{\mathbf{x}} \cos \psi_r \cos \phi_r + \hat{\mathbf{y}} \cos \psi_r \sin \phi_r + \hat{\mathbf{z}} \sin \psi_r \quad (17.7)$$

The complex gain of the Tx antenna also can be derived from the reciprocity (17.2).

17.3.2 Subband Model

In practice, the scattering coefficient $\gamma_{\beta_r\beta_t l}(f)$ is not drastically changing within a sufficient bandwidth. Therefore, the entire UWB channel is divided into I subbands, and $\gamma_{\beta_r\beta_t l}(f)$ within each of the subbands is modeled to be constant.

17.3.3 Spherical Wave Model

Equation (17.3) implicitly assumes that the propagation path can be modeled as the plane wave at both Tx and Rx sides. However, the plane wave model is not appropriate when the distance between Tx and Rx is small, or the scattering point is close to either end. For simplicity, let us assume that both Tx and Rx antennas are purely polarized, and the summation in (17.3) is reduced to a single term. If the spherical wave model [18] is deployed only at the Rx side, it shall be represented by the curvature radius from the scattering center R_{rl} in addition to the DOA $\Omega_{rl} = [\psi_{rl}, \phi_{rl}]$, and DTOA τ_l, as shown in Figure 17.3. Therefore, the position vector from the origin of the coordinates to the scattering center $\mathbf{R}_{rl}$ is expressed as

$$\mathbf{R}_{rl} = R_{rl}\,\hat{\boldsymbol{\omega}}_{rl} \tag{17.8}$$

In this case, (17.3) for m_tth Tx antenna and m_rth Rx antenna is modified as

$$h_{lm_t m_r}(f) = h_0(f, \tau_l)\gamma_l(f)D_r(f, \Omega_{rl})D_t(f, \Omega_{tl})\exp\left[j\frac{2\pi f}{c}\left(\|\mathbf{R}_{rl} - \mathbf{r}_{rm_r}\| - R_{rl}\right)\right]$$
$$\exp\left(-j\frac{2\pi f}{c}\mathbf{r}_{tm_t}\cdot\hat{\boldsymbol{\omega}}_{tl}\right) \tag{17.9}$$

Although it seems to be straightforward to extend the model (17.9) to the Tx side, a contradiction may occur. Figures 17.4 and 17.5 compare two different mechanisms. Figure 17.4 is the case when the scattering center seen from the Rx is independent of the Tx antenna position. Physically, the corner diffraction can be classified in the category. The model can be easily extended to the Tx side.

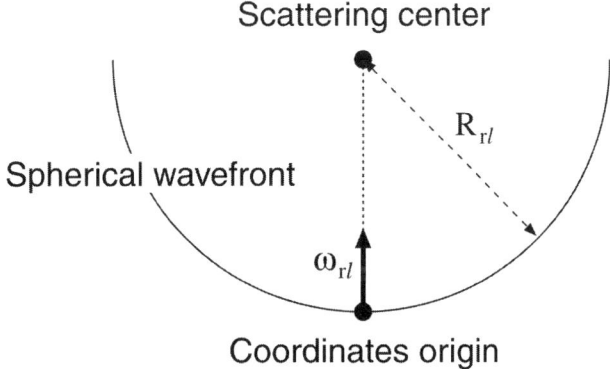

Figure 17.3 Spherical wave model and its parameters.

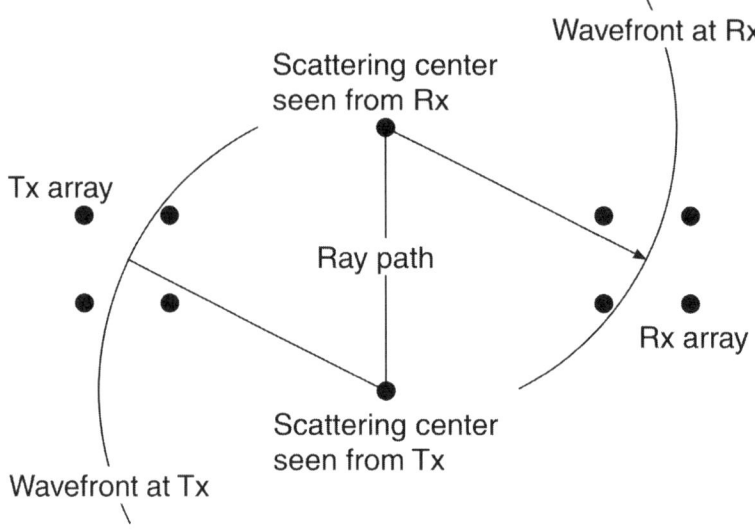

Figure 17.4 The case when double directional spherical wave model is applicable.

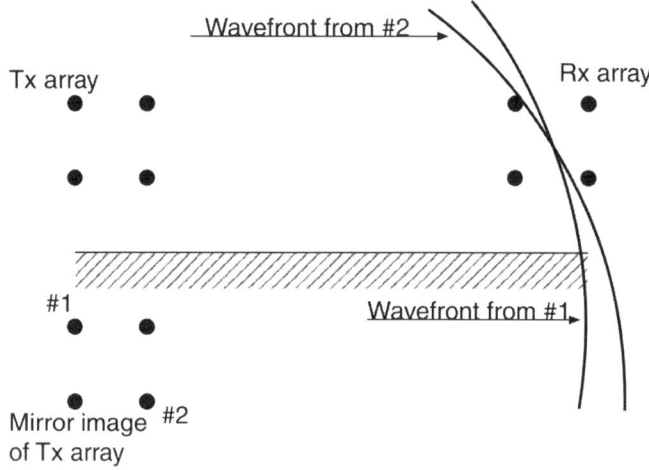

Figure 17.5 The case when double directional spherical wave model is not applicable.

Figure 17.5 shows that the scattering center changes according to the Tx antenna position. In this example, the specular reflection is considered. Unfortunately, we cannot know in advance which path satisfies the former and which the latter in the estimation process. Therefore, the joint estimation of DOD and DOA is impossible when the spherical wave model is deployed.

17.3.4 Two Single Directional Measurements, Pairing, and Clusterization

To introduce the spherical wave model, the authors propose a two-step scheme. DODs are measured by the Tx array and the single Rx antenna, while DOAs are measured by the single Tx antenna and the Rx array.

Since the delay time can be accurately estimated due to the nature of the UWB, DOD and DOA can be related by using the delay time. In some cases, however, more than one path is estimated at the same delay time. Figure 17.6 shows the case of two paths that are detected at the same DTOA. The ray tracing helps to solve this problem. Possible ray paths are traced for each pair of DODs and DOAs. The TDOA is evaluated from the length of the ray path, and it can be compared with the estimated value to select the appropriate DOD-DOA pair.

In reality, several paths experience the similar propagation histories, and the set of these paths is called a cluster [19]. However, it is more difficult to pair the paths within a single DOD-delay cluster and a single DOA-delay cluster. Therefore, the ray tracing is applied to the clusters, not to the individual rays. In the present study, the heuristic approach is applied for the clusterization.

For the single directional model at the Rx side, the transfer function of the multipath channel is expressed as

$$H_{m_r}(f) = \sum_{l=1}^{L} h_{lm_r}(f) \tag{17.10}$$

In the measurement procedure, receiver noise perturbs the output. Therefore, the measured transfer function at the kth frequency is modeled as

$$y_{m_r k} = H_{m_r k} + n_{m_r k} \tag{17.11}$$

where $H_{m_r k} = H_{m_r}(f_k)$ and $n_{m_r k}$ is the independent and identically distributed Gaussian random variable with the variance σ^2.

17.4 ML-Based Parameter Estimation for UWB Directional Channel

As described in the previous section, the spherical wave model cannot be fully compatible with the double directional model. Therefore, the parameter estimation for the single directional model only at the Rx side is described in this section. However, the single directional model at the Tx side can be derived straightforwardly.

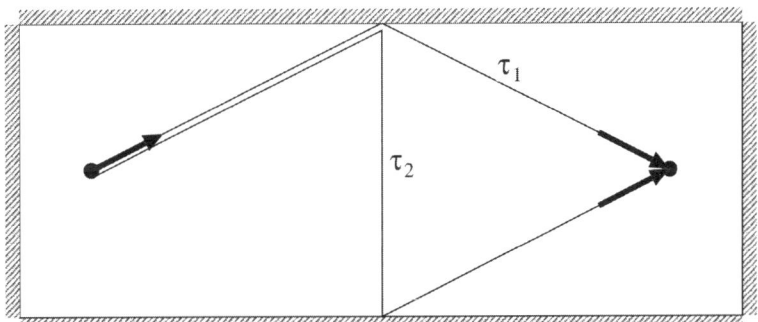

Figure 17.6 Help of ray tracing to relate DOD and DOA for specific DTOA.

17.4.1 Maximum Likelihood Parameter Estimation

An ML estimation is a procedure that finds the set of the parameters that maximize the likelihood function. The likelihood function is the conditional probability of the observation of the measured data sample, assuming the PDF is predetermined with respect to the parameters.

Let us define the parameter set of the lth path $\boldsymbol{\mu}_l$ as

$$\boldsymbol{\mu}_l = \left[(\gamma_{li})_{i=1}^{I}, \psi_{rl}, \phi_{rl}, R_l, \tau_l \right] \tag{17.12}$$

where γ_{li} is the value of $\gamma_l(f)$ in the ith subband. The parameter set of the whole channel $\boldsymbol{\mu}$ is then defined as

$$\boldsymbol{\mu} = \bigcup_{l=1}^{L} \boldsymbol{\mu}_l \tag{17.13}$$

According to the measured transfer function model in (17.11), the likelihood function of the measured data

$$\mathbf{y} = \{ y_{m_r k} | 1 \leq m_r \leq M_r, 1 \leq k \leq K \} \tag{17.14}$$

under the condition of the parameter set $\boldsymbol{\mu}$ is given as

$$p(\mathbf{y}|\boldsymbol{\mu}) = \prod_{k=1}^{K} \prod_{m_r=1}^{M_r} \left[\frac{1}{\pi \sigma} \exp\left(-\frac{|y_{m_r k} - H_{m_r k}(\boldsymbol{\mu})|^2}{\sigma^2} \right) \right] \tag{17.15}$$

By taking the logarithm of (17.15), the maximum likelihood condition is simplified to the following minimization problem.

$$\boldsymbol{\mu} = \arg \max_{\boldsymbol{\mu}} \ln p(\mathbf{y}|\boldsymbol{\mu}) = \arg \min_{\boldsymbol{\mu}} \| \mathbf{y} - \mathbf{H}(\boldsymbol{\mu}) \|^2 \tag{17.16}$$

However, (17.16) requires a simultaneous $(2I+4)L$-dimensional search, which is computationally prohibitive.

17.4.2 Expectation Maximization Algorithm

The EM algorithm has been proposed to reduce the multidimensional simultaneous search of ML estimation [20]. The EM algorithm estimates "the complete data" $\mathbf{x}_l$ from "the incomplete data" $\mathbf{y}$ as

$$\mathbf{x}_l = \mathbf{h}_l + b_l (\mathbf{y} - \mathbf{H}) \tag{17.17}$$

In (17.17), the nature of the additive white Gaussian noise (i.e., that the sum of the Gaussian variables is also Gaussian distributed) is considered in the derivation. b_l is a positive number that has a constraint

$$\sum_{l=1}^{L} b_l = 1 \qquad (17.18)$$

that is, in practice, set to $1/L$ for all l. This makes it possible to maximize the conditional Fisher information of complete data $\mathbf{x}_l$ [21]. In the present case, the incomplete data is the measured data, and the complete data corresponds to each of the paths that are not directly measured in the experiment. The log-likelihood function for the complete data (17.17) is analogous to (17.16) as

$$\arg\max_{\boldsymbol{\mu}} p(\mathbf{x}_l | \boldsymbol{\mu}) = \arg\min_{\boldsymbol{\mu}_l} \| \mathbf{x}_l - \mathbf{h}_l(\boldsymbol{\mu}_l) \|^2 \qquad (17.19)$$

The simultaneous search dimension is reduced to $(2I + 4)$. The right-hand side of (17.19) is a least-square problem, and is solved by the matched filtering. Let us define the steering vector

$$a_{m_r k}(\boldsymbol{\mu}_l) = b_0(f_k, \tau_l) D_r(f_k, \Omega_{rl}) \exp\left[j \frac{2\pi f}{c} \left(\| \mathbf{R}_{rl} - \mathbf{r}_{rm_r} \| - R_{rl} \right) \right] \qquad (17.20)$$

where

$$\boldsymbol{\mu}_l = \{ \psi_{rl}, \phi_{rl}, R_l, \tau_l \} \qquad (17.21)$$

is the subset of the parameters of lth path, excluding the complex gain. Then, the minimization problem of the right-hand side of (17.19) is rewritten as

$$\boldsymbol{\mu}_l = \arg\max_{\boldsymbol{\mu}_l} \frac{|\mathbf{a}^H(\boldsymbol{\mu}_l) \mathbf{x}_l|}{\sqrt{\mathbf{a}^H(\boldsymbol{\mu}_l) \mathbf{a}(\boldsymbol{\mu}_l)}} \qquad (17.22)$$

The estimates of the complex gain γ_{li} are equivalent to the output of the matched filter, and is given as

$$\hat{\gamma}_{li} = \frac{\mathbf{a}_i^H(\boldsymbol{\mu}_l) \mathbf{x}_{li}}{\mathbf{a}_i^H(\boldsymbol{\mu}_l) \mathbf{a}_i(\boldsymbol{\mu}_l)} \qquad (17.23)$$

where $\mathbf{a}_i(\boldsymbol{\mu}_l)$ and $\mathbf{x}_{li}$ are the subset vectors of $\mathbf{a}(\boldsymbol{\mu}_l)$ and $\mathbf{x}_l$ corresponding to ith subband, respectively.

17.4.3 SAGE Algorithm

Equation (17.22) still requires a simultaneous four-dimensional search. The space alternating generalized EM (SAGE) algorithm further divides the whole search space of the EM algorithm into hidden data spaces [21, 22]. In the present case, the search is carried out sequentially in the following order.

$$\hat{\psi}_{rl} = \arg\max_{\psi_{rl}} \frac{|\mathbf{a}^H(\psi_{rl}, \phi_{rl}, R_l, \tau_l)\mathbf{x}_l|}{\sqrt{\mathbf{a}^H(\psi_{rl}, \phi_{rl}, R_l, \tau_l)\mathbf{a}(\psi_{rl}, \phi_{rl}, R_l, \tau_l)}} \qquad (17.24)$$

$$\hat{\phi}_{rl} = \arg\max_{\phi_{rl}} \frac{|\mathbf{a}^H(\hat{\psi}_{rl}, \phi_{rl}, R_l, \tau_l)\mathbf{x}_l|}{\sqrt{\mathbf{a}^H(\hat{\psi}_{rl}, \phi_{rl}, R_l, \tau_l)\mathbf{a}(\hat{\psi}_{rl}, \phi_{rl}, R_l, \tau_l)}} \qquad (17.25)$$

$$\hat{R}_l = \arg\max_{R_l} \frac{|\mathbf{a}^H(\hat{\psi}_{rl}, \hat{\phi}_{rl}, R_l, \tau_l)\mathbf{x}_l|}{\sqrt{\mathbf{a}^H(\hat{\psi}_{rl}, \hat{\phi}_{rl}, R_l, \tau_l)\mathbf{a}(\hat{\psi}_{rl}, \hat{\phi}_{rl}, R_l, \tau_l)}} \qquad (17.26)$$

$$\hat{\tau}_l = \arg\max_{\tau_l} \frac{|\mathbf{a}^H(\hat{\psi}_{rl}, \hat{\phi}_{rl}, \hat{R}_l, \tau_l)\mathbf{x}_l|}{\sqrt{\mathbf{a}^H(\hat{\psi}_{rl}, \hat{\phi}_{rl}, \hat{R}_l, \tau_l)\mathbf{a}(\hat{\psi}_{rl}, \hat{\phi}_{rl}, \hat{R}_l, \tau_l)}} \qquad (17.27)$$

Finally, $\hat{\gamma}_{li}$ is obtained by (17.23). Note that initial parameters μ_l are sufficiently accurate. We can update the estimates of parameters by repeating the procedure above for all L paths until the convergence of estimated parameters or log-likelihood.

17.4.4 Successive Interference Cancellation-Type Procedure

In the maximization of the log-likelihood function, the successive interference cancellation (SIC)-type procedure is preferable when the number of samples is sufficiently large [10, 23]. In this procedure, the strongest wave is detected and removed from the measured data in a sequential manner. The process is equivalent to the EM algorithm when the number of waves equals one. Therefore, the complete data is exactly the same as incomplete data, according to (17.17). An example of the search implementation is depicted in Figure 17.7. This scheme is a combination of the coarse global mesh search based on the EM algorithm, and the fine local search based on the SAGE algorithm. In the coarse global mesh search, we find the region that may include a maximum point of the log-likelihood function. After the region is specified, the fine local search is carried out inside the selected region to find an accurate peak.

The SIC iteration continues until the number of waves reaches a predefined number, or the residual power is below the threshold, although determination of the number of paths is still an open issue. The former implementation gives the stable result of parameters and spectra without being affected by the estimation of the number of the waves.

The error propagation is the biggest drawback of the SIC scheme, especially when a large number of paths are detected. Fortunately, the high resolution property of the UWB channel in the time domain can make it less significant, since the detection and removal of the paths are conducted in quite an accurate manner.

On the other hand, a careful search is necessary to avoid the false detection of the paths due to the sidelobes. The coarser the global mesh search in the Figure

17.4 ML-Based Parameter Estimation for UWB Directional Channel

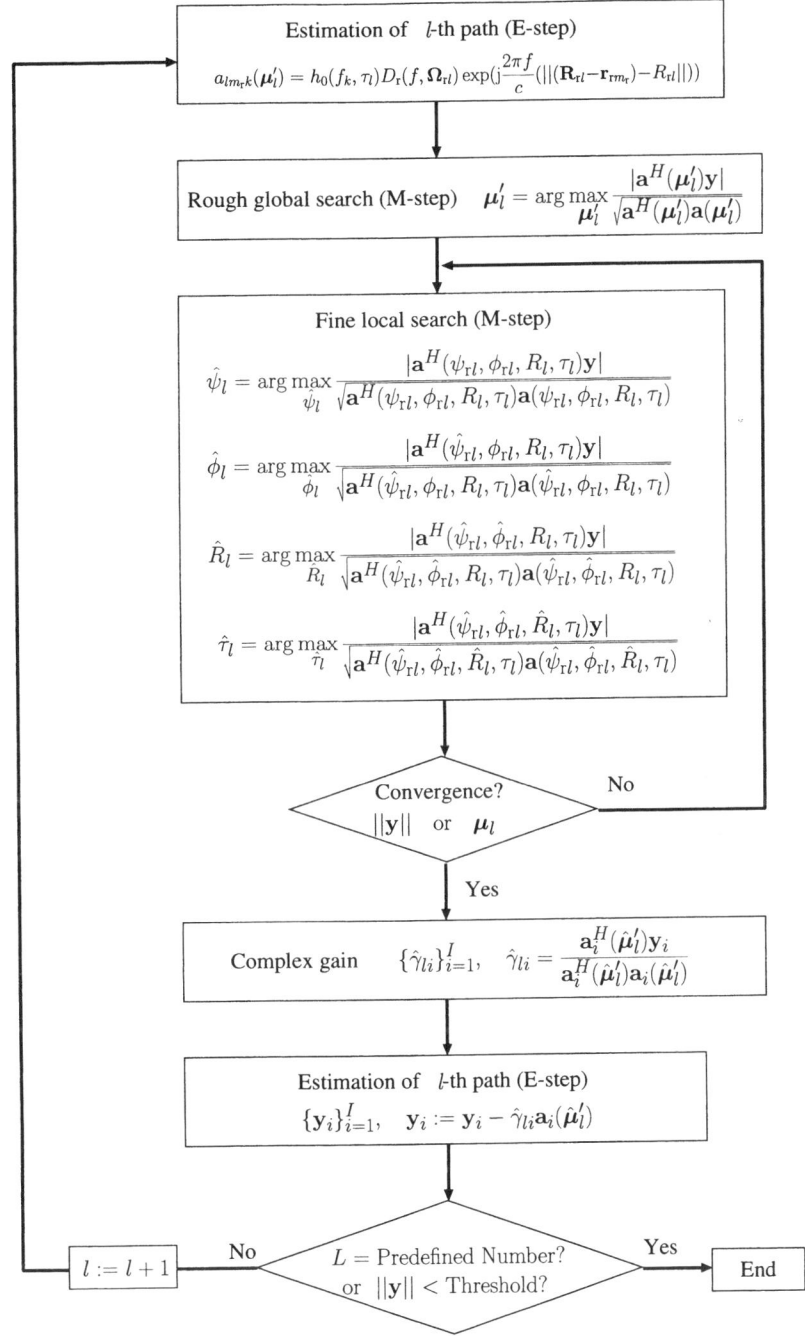

Figure 17.7 An SIC-type procedure of SAGE algorithm for UWB channel parameter estimation.

17.7, the more probable it is to detect the false paths. Taking into account the trade-off between the computational cost of the search and the accuracy of detection, the global mesh size shall be less than one-half of the beamwidth of the matched filter in both delay and angular domains.

17.4.5 Consideration About the Subband

As mentioned earlier, the whole bandwidth is divided into several subbands to estimate a frequency-dependent spectrum for each path. The log-likelihood function of the whole bandwidth is expressed as a sum of the log-likelihood functions in the subbands.

There is a trade-off between the estimation error of the spectrum and the resolution of the DTOA, in the choice of the bandwidth of the subband. Specifically, when two paths are closely separated in the delay domain and are within the beamwidth, they cannot be resolved when the subband width Δf is less then the inverse of the delay difference $1/\Delta t$.

17.4.6 Deconvolution of Antenna Directivity

It is necessary to deconvolve the antenna characteristics from the measurement results for an antenna-independent channel model. We shall include the antenna characteristics in the steering vector (17.20) for the parameter search. However, the antenna directivity $D(f, \Omega)$ is common among all the elements in the synthetic array. If the group delay of the antenna directivity is almost constant over the bandwidth, then $D(f, \Omega)$ may not be included in the steering vector for the EM parameter search, but is later subtracted from the frequency-dependent complex gain of the path.

17.5 Experiments

In this section, some experimental results are presented as an example of the data processing.

17.5.1 Specification of Measurement Equipment

Table 17.1 lists the specifications of the measurement equipment. The FCC-compliant frequency band has been chosen for the measurements [24]. The element spacing of the synthesized array is chosen to be equal to 0.49λ at the lowest frequency to satisfy the sampling theorem. Vertical biconical antennas are used for both Tx and Rx, and only the vertical-vertical polarization pair is considered.

Table 17.1 Specification of the Measurement Equipment

Bandwidth	3.1 to 10.6 GHz
Frequency samples	751
Frequency interval	10 MHz
Spatial samples	10 × 10 × 7 at both ends
Spatial interval	48 mm
Type of antennas	Biconical [17]
Polarization	Vertical-Vertical
IF bandwidth of VNA	100 Hz
SNR at VNA Rx port	Approximately 30 dB
Subband	3 GHz

17.5 Experiments

The resolution in delay, elevation, and azimuth are estimated by the following equations:

$$\Delta \tau = \frac{1}{f_{\max} - f_{\min}} \tag{17.28}$$

$$\Delta \psi = \sin^{-1} \frac{\lambda_{\max}}{L_z} \tag{17.29}$$

$$\Delta \phi = \sin^{-1} \frac{\lambda_{\max}}{L_x} \tag{17.30}$$

where L_x and L_z are the horizontal and the vertical size of the array, respectively,

$$\lambda_{\max} = \frac{c}{f_{\min}} \tag{17.31}$$

is the longest wavelength in the band, and $f_{\min}$ and $f_{\max}$ are the lowest and the highest frequencies, respectively. In case of Table 17.1, they are 0.13 ns, 13.1°, and 19.9°, respectively.

17.5.2 Measurement Environment

The measurement was performed in an empty meeting room, with no furniture. Figure 17.8 shows the layout of the room. Note that the azimuth angle is defined with respect to the line of sight (LOS), and different for the Tx and Rx sides.

17.5.3 Results

From the measurement data, the parameters and spectrum of the paths are estimated by using the SIC-based EM algorithm, as described in the previous section. A total of 120 paths are extracted at Tx and Rx positions, respectively. After the extraction, approximately 30% of the total power still remains. The residuals are considered as diffuse components that cannot be modeled as (17.20), and the whole channel model (17.10) is modified to include these components. The similar components also have been found in other measurements [11, 25] that propose different models for the diffuse components. The modeling is still an open issue.

Figures 17.9 and 17.10 show the extracted path angular-delay domains at Tx and Rx positions, respectively. It is obvious that the detected paths form clusters in the angular-delay domains. The heuristic clusterization has been done based on DOD, DOA, and DTOA, but these clusters are related to the physical scattering objects in the environment. Clusters A and D consist of scattering from the pillars. They are the strongest clusters. Clusters B, C, and E consist of scattering from the pillars, metal doors, and window frames. Clusters F and J consist of scattering from the walls. Clusters G, H, I, K, L, and M consist of scattering from the pillars, metal doors, window frames, and walls. They have long DTOAs, due to multiple

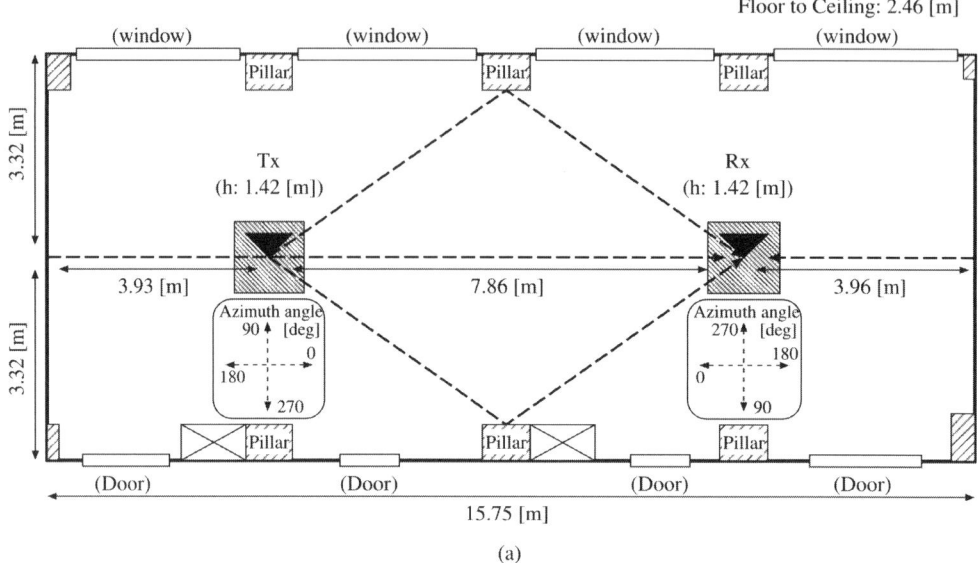

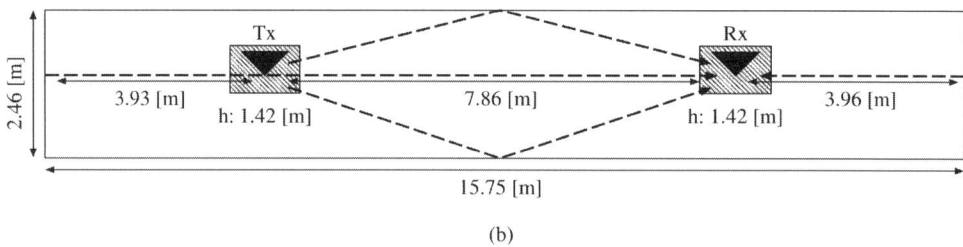

Figure 17.8 Measurement environment: (a) top view and (b) side view.

scattering. Clusters N and P consist of scattering waves at the ceiling and floor, respectively. Cluster N is also one of the strongest clusters. More detailed analysis about the clusters is presented in [17].

17.6 Conclusions

In this chapter, the authors describe the joint DOA/DOD/DTOA estimation system for the UWB double directional channel model. The system consists of VNA and synthetic array antennas at both Tx and Rx sides. An SIC-type EM algorithm has been implemented for the parameter extraction. Spherical wave model has been introduced to model the short range path. It requires the joint processing of two single-directional measurements. The clusterization concept has been deployed, in conjunction with the utilization of the fine DTOA resolution to pair DOD and DOA. Experimental results in an empty office have been presented to demonstrate the proposed system.

17.6 Conclusions

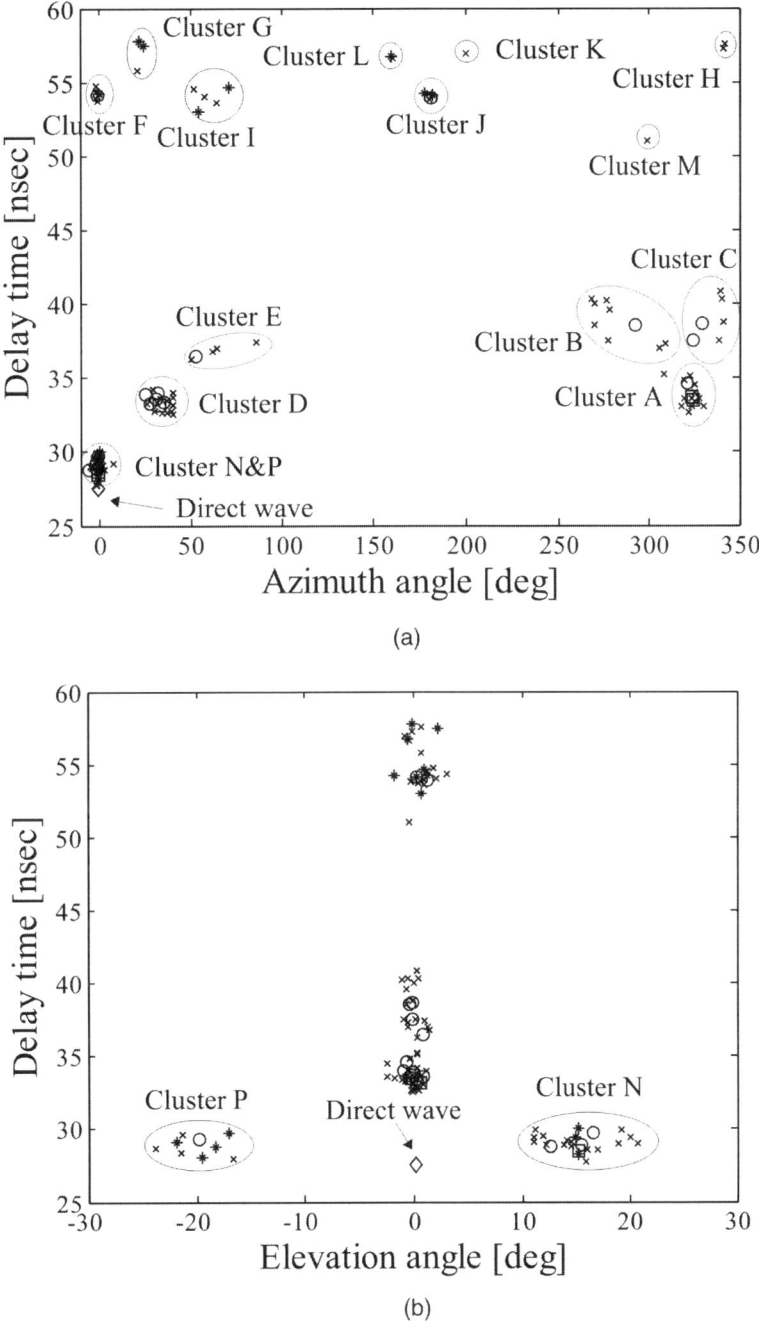

Figure 17.9 Clusterization of paths at Tx position: (a) azimuth delay, and (b) elevation delay.

This system has been extensively used for the MIMO channel modeling of the indoor home environments, with European cooperation in the field of scientific and technical research (COST) action 273 "Towards mobile broadband multimedia networks" [26].

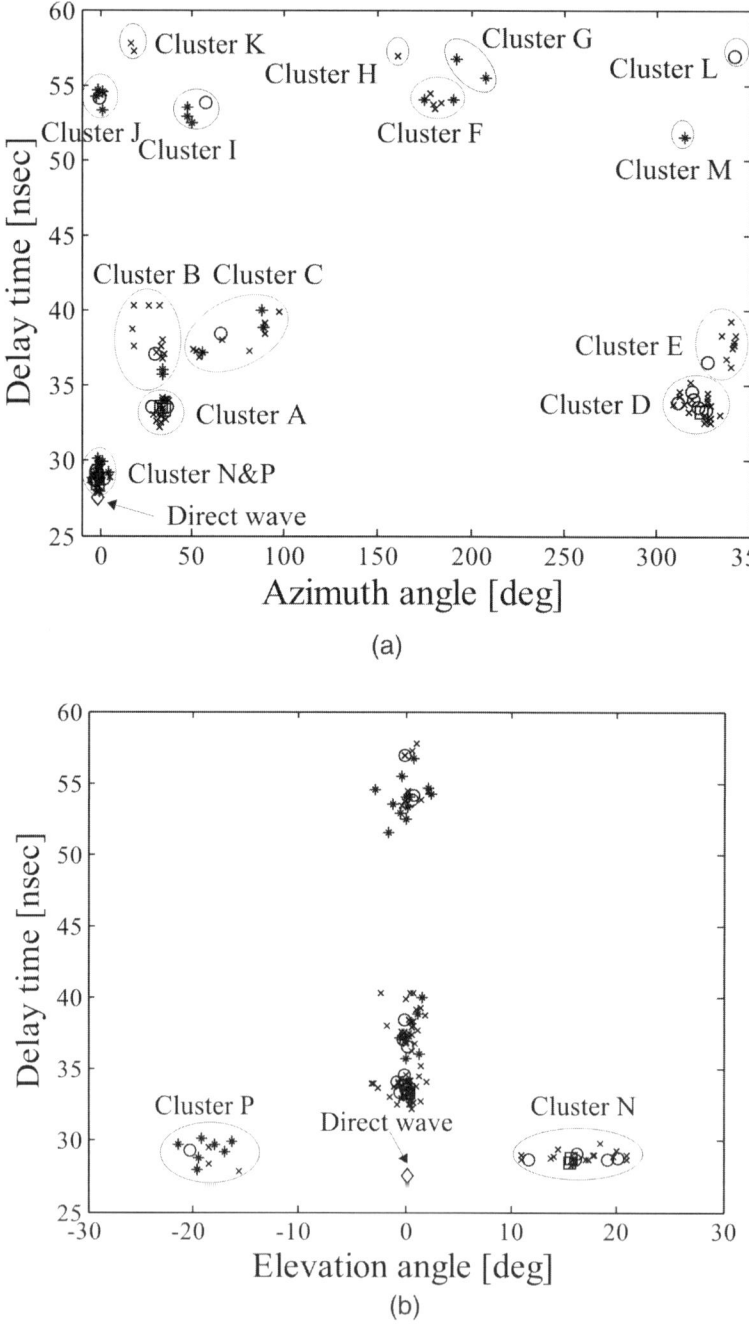

Figure 17.10 Clusterization of paths at Rx position: (a) azimuth delay, and (b) elevation delay.

References

[1] Foerster, J., "Channel Modeling Sub-Committee Report Final," *IEEE 802.15 Working Group Document*, IEEE P802.15-03/490r0, 2002.

[2] Molisch, A. F., et al., "IEEE 802.15.4a Channel Model—Final Report," *IEEE 802.15 Working Group Document*, IEEE P802.15-04/662r0, 2004.

[3] Zwick, T., et al., "A Stochastic Spatial Channel Model Based on Wave-Propagation Modeling," *IEEE J. of Selected Areas in Communications*, Vol. 18, No. 1, 2000, pp. 6–15.

[4] Steinbauer, M., A. F. Molisch, and E. Bonek, "The Double Directional Mobile Radio Channel," *IEEE Antennas and Propagation Magazine*, Vol. 43, No. 4, 2001, pp. 51–63.

[5] Erceg, V., et al., "TGn Channel Models," *IEEE 802.11 Working Group Document*, IEEE 802.11-03/940r4, 2004.

[6] Technical Specification Group Radio Access Network, "Spatial Channel Model for Multiple Input Multiple Output (MIMO) Simulations (Release 6)," *3GPP Specifications*, 3GPP TR 25.996 V6.1.0, 2003.

[7] Kalliola, K., et al., "3-D Double-Directional Radio Channel Characterization for Urban Macrocellular Applications," *IEEE Trans. on Antennas and Propagation*, Vol. 51, No. 11, 2003, pp. 3122–3133.

[8] Richter, A., et al., "Joint Estimation of DOD, Time-Delay, and DOA for High-Resolution Channel Sounding," *Proc. of 2000 Spring IEEE Vehicular Technology Conference*, Tokyo, Japan, May 2000, pp. 1045–1049.

[9] Sakaguchi, K., J. Takada, and K. Araki, "A Novel Architecture for MIMO Spatio-Temporal Channel Sounder," *IEICE Trans. on Electronics*, Vol. E85-C, No. 3, 2002, pp. 436–441.

[10] Fleury, B. H., P. Jourdan, and A. Stucki, "High-Resolution Channel Parameter Estimation for MIMO Applications Using the SAGE Algorithm," *Proc. of 2002 International Zurich Seminar on Broadband Communications*, Zurich, Switzerland, February 2002, pp. 30-1–30-9.

[11] Thomae, R. S., et al., "Multidimensional High-Resolution Channel Sounding Measurement," *Proc. of 2004 IEEE Instrumentation and Measurements Technical Conference*, Como, Italy, May 2004, pp. 257–262.

[12] Cramer, R. J., R. A. Scholtz, and M. Z. Win, "Evaluation of an Ultra-Wide-Band Propagation Channel," *IEEE Trans. on Antennas and Propagation*, Vol. 50, No. 5, 2002, pp. 561–570.

[13] Poon, A. S. Y., and M. Ho, "Indoor Multiple Antenna Channel Characterization from 2 to 8 GHz," *Proc. of 2003 IEEE International Conference on Communications*, Anchorage, AK, May 2003, pp. 3519–3523.

[14] Haneda, K., J. Takada, and T. Kobayashi, "Double Directional LOS Channel Characterization in a Home Environment with Ultrawideband Signal," *Proc. of 7th International Symposium on Wireless Personal Multimedia Communications*, Abano Terme, Italy, September 2004, pp. 214–218.

[15] Tsuchiya, H., K. Haneda, and J. Takada, "UWB Indoor Double-Directional Channel Sounding for Understanding the Microscopic Propagation Mechanisms," *Proc. of 7th International Symposium on Wireless Personal Multimedia Communications*, Abano Terme, Italy, September 2004, pp. 95–99.

[16] Taniguchi, T., and T. Kobayashi, "An Omni-Directional and Low-VSWR Antenna for the FCC-Approved UWB Frequency Band," *Proc. of 2003 IEEE AP-S International Symposium*, OH, June 2003, pp. 460–463.

[17] Promwong, S., W. Hachitani, and J. Takada, "Free Space Link Budget Evaluation of UWB-IR Systems," *Proc. of 2004 International Workshop on Ultra Wideband Systems Joint with Conference on Ultra Wideband Systems and Technologies*, Kyoto, Japan, May 2004, pp. 312–316.

[18] Ohmae, A., M. Takahashi, and T. Uno, *Localization of Sources in the Finite Distance Using MUSIC Algorithm by the Spherical Mode Vector*, Technical Report of IEICE, AP2003-64/SAT2003-56/MW2003-70/OPE2003-57, July 2003.

[19] Saleh, A. A. M., and R. A. Valenzuela, "A Statistical Model for Indoor Multipath Propagation," *IEEE J. of Selected Areas in Communications*, Vol. 5, No. 2, 1987, pp. 128–137.

[20] Dempster, A. P., N. M. Laird, and D. B. Rubin, "Maximum Likelihood from Incomplete Data Via the EM Algorithm," *J. of Royal Statistical Society, Series B*, Vol. 39, No. 1, 1977, pp. 1–38.

[21] Fleury, B. H., et al, "Channel Parameter Estimation in Mobile Radio Environments Using the SAGE Algorithm," *IEEE J. of Selected Areas in Communications*, Vol. 17, No. 3, 1999, pp. 434–449.

[22] Fessler, J. A., and A. O. Hero, "Space-Alternating Generalized Expectation Maximization Algorithm," *IEEE Trans. on Signal Processing*, Vol. 42, No. 10, 1994, pp. 2664–2677.

[23] Haneda, K., and J. Takada, "High-Resolution Estimation of NLOS Indoor MIMO Channel with Network Analyzer Based System," *Proc. of 14th Personal Indoor and Mobile Radio Communications*, Beijing, China, September 2003, pp. 675–679.

[24] "Radio Frequency Devices," *Federal Communications Commission Rules*, Part 15, 2003.

[25] Molisch, A. F., "A Generic Model for MIMO Wireless Propagation Channels in Macro- and Microcells," *IEEE Trans. on Signal Processing*, Vol. 52, No. 1, 2004, pp. 61–71.

[26] COST 273 Web site, http://www.lx.it.pt/cost273/.

CHAPTER 18
Conventional Method Improvements by Signal Property Exploitation
Pascal Chargé and Yide Wang

Array processing techniques basically rely on the spatial properties of the signals impinging on the array of sensors. In applications such as radar, sonar, or telecommunications, the signals of interest may have rich properties that are exploited to minimize interferences and background noise. Modulated signals are frequently used in these systems. Most modulated communication signals exhibit a cyclostationarity or periodic correlation property, corresponding to the underlying periodicity arising from carrier frequencies or baud rates. The aim of this chapter is the estimation of the DOA of impinging signals in the telecommunications systems area, where almost all signals exhibit the cyclostationarity property [1].

In this chapter, the cyclostationarity is briefly presented, and then an array data model that allows the exploitation of this property of incoming signals is given. This chapter presents several direction-finding algorithms that exploit cyclostationarity in order to improve the direction-of-arrival estimation. Some simulations are also proposed, in order to illustrate the benefits provided by these techniques.

18.1 Cyclostationarity

The cyclostationarity was first introduced into array processing by Gardner [2]. We can find in the literature several algorithms that exploit cyclostationarity to improve the performances of the conventional methods [3–5]. Instead of using the correlation matrix of the antenna output as in the conventional methods, these algorithms require the estimation of the cyclic correlation matrix that reflects the cyclostationarity of incoming signals. This assumes that they have baud rates and/or are carrier modulated signals, since they would be in radar and radio communication applications.

The cyclic autocorrelation function and the cyclic conjugate autocorrelation function of a signal $s(t)$ are defined as the following infinite-time averages [2]:

$$R_{ss}^{\alpha}(\tau) = \left\langle s\left(t+\frac{\tau}{2}\right)s^*\left(t-\frac{\tau}{2}\right)e^{-j2\pi\alpha t} \right\rangle \quad (18.1)$$

and

$$R_{ss^*}^{\alpha}(\tau) = \left\langle s\left(t+\frac{\tau}{2}\right) s\left(t-\frac{\tau}{2}\right) e^{-j2\pi\alpha t} \right\rangle \quad (18.2)$$

respectively. $s(t)$ is said to be cyclostationary if $R_{ss}^{\alpha}(\tau)$ or $R_{ss^*}^{\alpha}(\tau)$ does not equal zero at cycle frequency α for some lag parameter τ. Depending on the type of modulation used, the cycle frequency α is usually equal to two times the carrier frequency, multiples of the baud rate, spreading codes repetition rate, chip rate, or some combinations of these. Moreover, some signals have both nonzero cyclic correlation and nonzero conjugate cyclic correlation; for example, binary phase shift keying (BPSK), offset quaternary phase shift keying (OQPSK) or minimum shift keying (MSK).

Let us now consider the vector $s(t)$ of cyclostationary signals. The cyclic autocorrelation matrix and the cyclic conjugate autocorrelation matrix are given by:

$$\mathbf{R}_{ss}^{\alpha}(\tau) = \left\langle \mathbf{s}\left(t+\frac{\tau}{2}\right) \mathbf{s}^{H}\left(t-\frac{\tau}{2}\right) e^{-j2\pi\alpha t} \right\rangle \quad (18.3)$$

and

$$\mathbf{R}_{ss^*}^{\alpha}(\tau) = \left\langle \mathbf{s}\left(t+\frac{\tau}{2}\right) \mathbf{s}^{T}\left(t-\frac{\tau}{2}\right) e^{-j2\pi\alpha t} \right\rangle \quad (18.4)$$

respectively. Note that in practice, the finite-time average operator estimates these correlation matrices.

18.2 Data Model

Let us consider an array of M antennas. Suppose K electromagnetic waves impinging on the array from angular directions θ_k, with $k = 1, \ldots, K$. The incident waves are assumed to be plane waves, as generated from farfield point sources. Furthermore, the signals are assumed to be narrowband. Let us assume that K_α sources emit cyclostationary signals with cycle frequency α, with $K_\alpha \leq K$. In the following, the vector $s(t)$ contains only the K_α signals that exhibit cycle frequency α, and all of the remaining $K - K_\alpha$ signals (i.e., those that do not have cycle frequency α), along with any noise, are lumped into a vector $i(t)$. Using this assumption, the signal received by the array from the emitting narrowband sources is written as:

$$\mathbf{z}(t) = \mathbf{A}\mathbf{s}(t) + \mathbf{i}(t) \quad (18.5)$$

where the vector $\mathbf{s}(t) = [s_1(t), \ldots, s_{K_\alpha}(t)]^T$ contains the temporal signals that have cycle frequency α, and the vector $\mathbf{i}(t)$ represents interfering sources and noise. The matrix $\mathbf{A} = [\mathbf{a}(\theta_1), \ldots, \mathbf{a}(\theta_{K_\alpha})]$ contains the steering vectors of the impinging signals of interest (SOI). The received signals are sampled at N distinct times t_n, with $n = 1, \ldots, N$.

The cyclic autocorrelation matrix and the cyclic conjugate autocorrelation matrix at cycle frequency α for some lag parameter τ are then nonzero, and are estimated by:

$$\mathbf{R}_{zz}^{\alpha}(\tau) = \frac{1}{N} \sum_{n=1}^{N} \mathbf{z}(t_n + \tau/2) \mathbf{z}^H(t_n - \tau/2) e^{-j2\pi\alpha t_n} \tag{18.6}$$

and

$$\mathbf{R}_{zz^*}^{\alpha}(\tau) = \frac{1}{N} \sum_{n=1}^{N} \mathbf{z}(t_n + \tau/2) \mathbf{z}^T(t_n - \tau/2) e^{-j2\pi\alpha t_n} \tag{18.7}$$

18.3 High-Resolution Techniques

18.3.1 Cyclic MUSIC Algorithm

This section deals with a signal selective direction-finding method. This algorithm allows us to select desired signals and to ignore interferences and background noise, by exploiting the cyclostationarity property of the signals of interest. The method only requires some basic a priori information about the signals, such as the modulation type, the baud rate, and the carrier frequency. This method relies on the MUSIC method principle, and the cyclostationarity is taken into account with the cyclic correlation matrix $\mathbf{R}_{zz}^{\alpha}(\tau)$ [or $\mathbf{R}_{zz^*}^{\alpha}(\tau)$]. Contrary to the covariance matrix exploited by the MUSIC algorithm [6], the cyclic correlation matrix (18.6) [or (18.7)] exploited by the Cyclic MUSIC method [3] is generally not hermitian. Instead of using the eigenvalue decomposition (EVD), Cyclic MUSIC uses the singular value decomposition (SVD) of the cyclic correlation matrix, and relies on the principle that the left null space of $\mathbf{R}_{zz}^{\alpha}(\tau)$ [or $\mathbf{R}_{zz^*}^{\alpha}(\tau)$] is orthogonal to the steering vectors of source. For a finite number of time samples, the algorithm is implemented as follows:

- Estimate matrix $\mathbf{R}_{zz}^{\alpha}(\tau)$ by using (18.6) [or $\mathbf{R}_{zz^*}^{\alpha}(\tau)$ by using (18.7)].
- Compute SVD of the estimated matrix:

$$[\mathbf{E}_S \quad \mathbf{E}_N] \begin{bmatrix} \mathbf{\Sigma}_S & 0 \\ 0 & \mathbf{\Sigma}_N \end{bmatrix} [\mathbf{V}_S \quad \mathbf{V}_N]^H$$

where $[\mathbf{E}_S \quad \mathbf{E}_N]$ and $[\mathbf{V}_S \quad \mathbf{V}_N]$ are unitary matrices, and the diagonal elements of the diagonal matrices $\mathbf{\Sigma}_S$ and $\mathbf{\Sigma}_N$ are arranged in decreasing order. $\mathbf{\Sigma}_N$ tends toward zero as the number of time samples tends to infinity.
- Find the minima of $\|\mathbf{E}_N^H \mathbf{a}(\theta)\|^2$ or the maxima of $\|\mathbf{E}_S^H \mathbf{a}(\theta)\|^2$.

18.3.2 Extended Cyclic MUSIC Algorithm

Another signal selective direction-finding method is presented, which is an extension of the Cyclic MUSIC algorithm [3].

Extended Data Model

By concatenating the conventional array data model and its conjugate expression in a single extended data vector, the method can exploit simultaneously the information contained in both the cyclic correlation matrix (18.6) and the cyclic conjugate correlation matrix (18.7). This so-called extended data vector is given by:

$$\mathbf{z}_{CE}(t) = \begin{bmatrix} \mathbf{z}(t) \\ \mathbf{z}^*(t) \end{bmatrix} \tag{18.8}$$

Extended Autocorrelation Matrix

For this method, we generate a cyclic correlation matrix that contains both the cyclic autocorrelation (18.6) and cyclic conjugate autocorrelation (18.7) matrices. This so-called extended cyclic correlation matrix is expressed as:

$$\mathbf{R}_{CE}^{\alpha}(\tau) = \frac{1}{N} \sum_{n=1}^{N} \mathbf{I}_{2M}^{\alpha}(t_n) \mathbf{z}_{CE}(t_n + \tau/2) \mathbf{z}_{CE}^{H}(t_n - \tau/2) \tag{18.9}$$

where the time dependent matrix $\mathbf{I}_{2M}^{\alpha}(t)$ is defined by:

$$\mathbf{I}_{2M}^{\alpha}(t) = \begin{bmatrix} \mathbf{I}_M e^{-j2\pi\alpha t} & 0 \\ 0 & \mathbf{I}_M e^{+j2\pi\alpha t} \end{bmatrix} \tag{18.10}$$

where $\mathbf{I}_M$ is the M-dimensional identity matrix. The extended cyclic correlation matrix is developed as:

$$\mathbf{R}_{CE}^{\alpha}(\tau) = \begin{bmatrix} \mathbf{R}_{zz}^{\alpha}(\tau) & \mathbf{R}_{zz^*}^{\alpha}(\tau) \\ \mathbf{R}_{zz^*}^{\alpha *}(\tau) & \mathbf{R}_{zz}^{\alpha *}(\tau) \end{bmatrix} \tag{18.11}$$

Extended-Cyclic-MUSIC

By computing the SVD of $\mathbf{R}_{CE}^{\alpha}(\tau)$ similarly to the Cyclic MUSIC algorithm, a left signal subspace is defined by the K'_{α} left singular vectors associated with the K'_{α} nonzero singular values. These singular vectors form the column vectors of the matrix $\mathbf{U}_S$. In the same way, a left null-space is spanned by the remaining $2M - K'_{\alpha}$ singular vectors associated with the zero singular values of $\mathbf{R}_{CE}^{\alpha}(\tau)$, and these singular vectors are the column vectors of the matrix $\mathbf{U}_N$. Note that in practice, there are no zero singular values, but only small singular values, and the dimension on the signal subspace is estimated by the MDL criterion [7]. Some cyclostationary signals have a rank two in the signal subspace spanned by column vectors of the matrix $\mathbf{U}_S$, and others have only a rank one, so that the dimension of the signal subspace is equal to K'_{α} with $K_{\alpha} \leq K'_{\alpha} \leq 2K_{\alpha}$ [8]. Then, for a fixed set of parameters (α, τ), the number of sources that is estimated is N_S, with $(M - 1) \leq N_S \leq 2(M - 1)$. This extended data model allows handling more sources than the number of sensors.

18.3 High-Resolution Techniques

According to the extended data model (18.8), the following extended normalized steering vector is proposed [8]:

$$\mathbf{b}(\theta, \mathbf{c}) = \begin{bmatrix} \mathbf{a}(\theta) & 0 \\ 0 & \mathbf{a}^*(\theta) \end{bmatrix} \frac{\mathbf{c}}{\|\mathbf{c}\|} \quad (18.12)$$

where the (2×1) vector $\mathbf{c}$ contains unknown coefficients. This vector forms a single steering vector from $\mathbf{a}(\theta)$ and $\mathbf{a}^*(\theta)$, so that we can exploit simultaneously the information provided by both the cyclic autocorrelation matrix and the cyclic conjugate autocorrelation matrix.

According to the subspace-based methods principle, and by using this extended steering vector, the DOA of the SOI are given by the values of θ and $\mathbf{c}$, which minimizes the following function:

$$\overline{P}(\theta, \mathbf{c}) = \left\| \mathbf{U}_N^H \mathbf{b}(\theta, \mathbf{c}) \right\|^2 \quad (18.13)$$

This two-dimensional minimization problem is turned into a one-dimensional maximization search [8]. It is easily shown that the spatial spectrum of the proposed Extended Cyclic MUSIC (ECM) method is given by:

$$P(\theta) = \frac{1}{\mathbf{a}^H(\theta)\mathbf{U}_{N1}\mathbf{U}_{N1}^H\mathbf{a}(\theta) - \left| \mathbf{a}^T(\theta)\mathbf{U}_{N2}\mathbf{U}_{N1}^H\mathbf{a}(\theta) \right|} \quad (18.14)$$

where $\mathbf{U}_{N1}$ and $\mathbf{U}_{N2}$ are two submatrices of the same dimension, defined by:

$$\mathbf{U}_N = \begin{bmatrix} \mathbf{U}_{N1}^T & \mathbf{U}_{N2}^T \end{bmatrix}^T \quad (18.15)$$

Both cyclic and conjugate cyclic correlation matrices are used in this cyclic method, so that it is considered to be an extension of the Cyclic MUSIC algorithm [3]. This extended cyclic method allows an increase in the resolution power and the noise robustness.

The final expression of this method, which is rather similar to the one given in [9], estimates the DOA of noncircular sources. However, this method is dedicated to cyclostationary signals. Moreover, the direction-finding method presented in [9] has no signal-selectivity capability, and is limited to a special kind of noncircular signals, whereas this extended cyclic method is a signal-selective method that has no particular limitation.

It should be emphasized that this extended cyclic method has better estimation performances than the Cyclic MUSIC method [3], when both the cyclic correlation and the cyclic conjugate correlation matrices are nonzero. This occurs for many digitally modulated signals. If one of these two correlation matrices is equal to zero, then our method is strictly equivalent to the Cyclic MUSIC method.

Simulations

Some simulation results are presented here, in order to show the behavior of the ECM method and to compare it with the MUSIC algorithm [6] and with the classical Cyclic MUSIC algorithm [3].

Consider here a linear uniformly spaced array with $M = 6$ sensors spaced by one-half wavelength of the incoming signals. Incoming BPSK cyclostationary signals are generated with noise, and the SNR is 10 dB for each signal. The bit rate of the BPSK SOI is 4 Mbps. Other signals are considered as interferers, which are BPSK modulated signals with a 3.2-Mbps bit rate. In order to choose correctly the parameters α and τ, we have estimated the magnitude of cyclic correlation function in Figure 18.1 for a 4 Mbps BPSK modulated signal, sampled with the frequency 32 MHz during 25 μs. It is noted that the magnitude of the cyclic autocorrelation function and that of the conjugate cyclic autocorrelation function are equal for a BPSK signal. According to this result, the proposed method and the Cyclic MUSIC algorithm are simulated with $\alpha = 4$ MHz and $\tau = 0.125$ μs. In the next simulation, the averaging time is equal to 25 μs and the sample frequency is 32 MHz. The contributions of both the interferer signals and the noise are theoretically zeroed in the two cyclic correlation matrices.

In this simulation, the signal selectivity and the resolution power of the cyclic algorithms are tested. Two BPSK SOI arrive from 5° and 10°, and one interferer from 30° DOA. The resulting spatial spectra are shown in Figure 18.2. The rank of the signal subspace provided by the MDL criterion is 3 for the MUSIC method and 2 for the cyclic methods. The ECM method allows us to perfectly select the two SOI, and ignore the interferer signal. Due to its signal selectivity, the Cyclic MUSIC algorithm also ignores the interferer signal, but does not separate the two SOI. The ECM algorithm performs better than the Cyclic MUSIC, due to the exploitation of both cyclic correlation matrices. More information about sources is used, and the observation dimension space [i.e., the size of the extended covariance

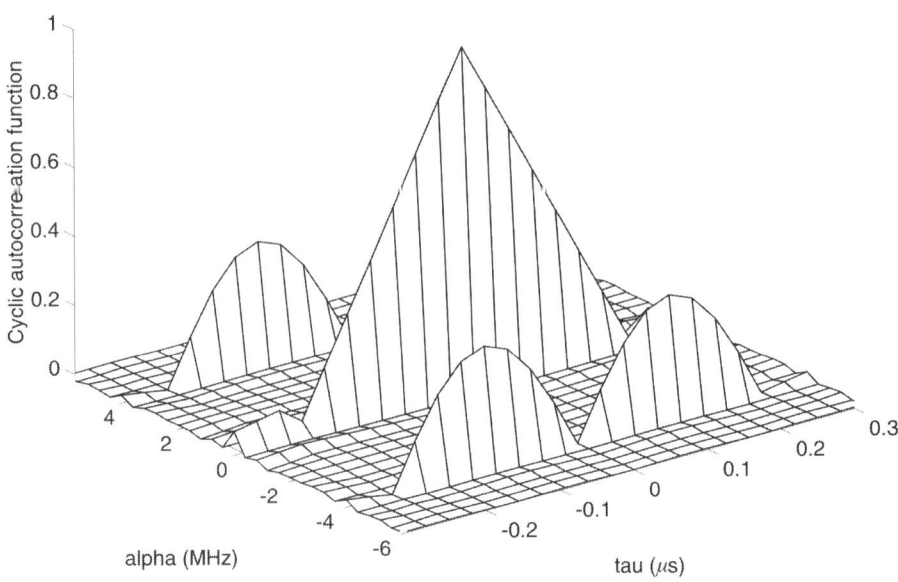

Figure 18.1 Magnitude of the BPSK cyclic autocorrelation function versus α and τ.

18.3 High-Resolution Techniques

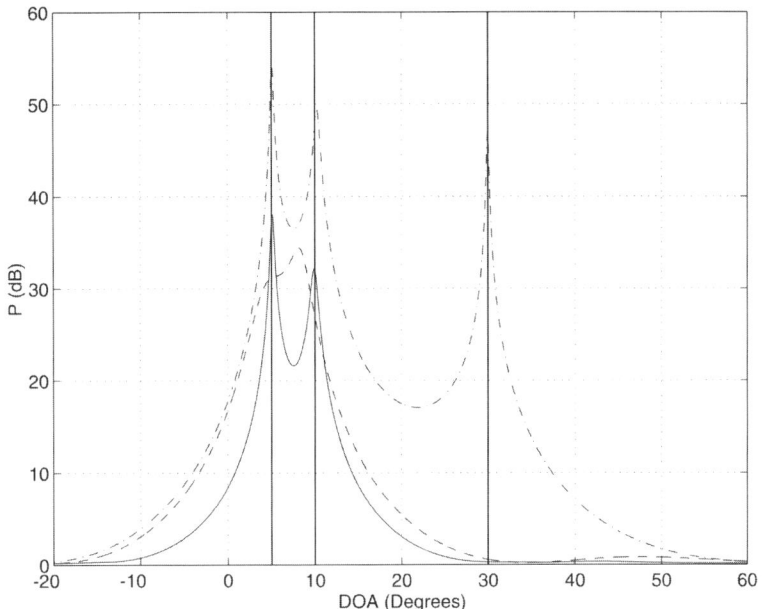

Figure 18.2 Spatial spectra for environment containing two SOI with 5° and 10° DOA and one interferer with 30° DOA: with the proposed method (solid line), with the Cyclic MUSIC method (dashed line), and with the MUSIC method (dash-dot line).

matrix $\mathbf{R}_{CE}^{\alpha}(\tau)]$ is doubled. The conventional MUSIC algorithm allows localizing the three signals.

In the next simulation, the fact that the ECM method handles more SOI than the number of sensors is emphasized. Recalling that the linear uniform array is made of six sensors spaced $\lambda/2$ apart, seven 4-Mbps BPSK SOI are generated, with −70°, −50°, −30°, −10°, +20°, +40°, and +60° DOA. Three 3.2-Mbps BPSK interferers are also generated, with −20°, 0°, and +30° DOA. The SNR is 10 dB for each signal. Figure 18.3 represents 20 independent simulations with the ECM method. For each of the simulations, the MDL criterion is used to estimate the rank of the signal subspace, and the estimated rank is always 7. Figure 18.3 shows that the ECM algorithm allows the estimation of seven SOI with a six-sensor array. As expected, the presence of the three interferers has little effect on the estimation of the seven SOI.

18.3.3 Root Cyclic MUSIC Algorithm

The Root Cyclic MUSIC (RCM) method relies on exactly the same data model as that used in the ECM method. This method is inspired from the Root-MUSIC method [10–12], and is restricted to linear uniformly spaced arrays. However, it has the distinct advantage over the ECM, since it does not require a search over parameter space. Instead, the RCM algorithm requires calculation of the roots of a polynomial, which is a simple process with low computation cost.

As previously mentioned, the extended cyclic covariance matrix $\mathbf{R}_{CE}^{\alpha}(\tau)$ needs to be estimated. Signal and noise subspaces are determined by computing the SVD

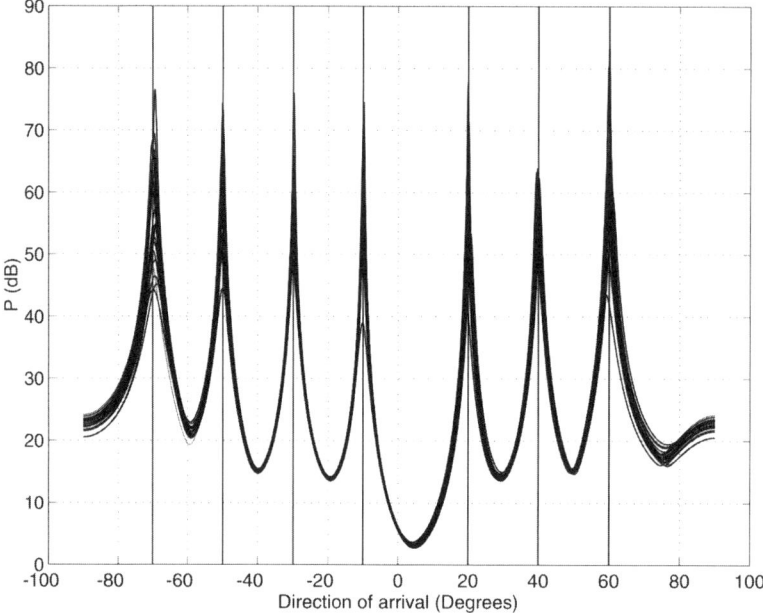

Figure 18.3 Spatial spectrum provided by the Extended Cyclic MUSIC method for an environment containing seven SOI and three interferers. The DOAs of interferers are −20°, 0°, and +30°.

of this matrix. The dimension of the signal subspace also is estimated by the MDL criterion.

According to the subspace processing principle, and by using the previously defined extended steering vector $\mathbf{b}(\theta, \mathbf{c})$ (18.12), the DOA of the SOI are given by the minima of the following function:

$$\overline{P}(\theta, \mathbf{c}) = \left\| \mathbf{U}_N^H \mathbf{b}(\theta, \mathbf{c}) \right\|^2 = \mathbf{c}^H \mathbf{M} \mathbf{c} \tag{18.16}$$

where $\mathbf{M}$ is a (2×2) matrix

$$\mathbf{M} = \begin{bmatrix} \mathbf{a}^H(\theta) \mathbf{U}_{N1} \mathbf{U}_{N1}^H \mathbf{a}(\theta) & \mathbf{a}^H(\theta) \mathbf{U}_{N1} \mathbf{U}_{N2}^H \mathbf{a}^*(\theta) \\ \mathbf{a}^T(\theta) \mathbf{U}_{N2} \mathbf{U}_{N1}^H \mathbf{a}(\theta) & \mathbf{a}^H(\theta) \mathbf{U}_{N1} \mathbf{U}_{N1}^H \mathbf{a}(\theta) \end{bmatrix} \tag{18.17}$$

The minimum of this quadratic form for a particular θ is given by the smallest eigenvalue of the matrix $\mathbf{M}$. This eigenvalue is always nonnegative, since the quadratic form is nonnegative.

When θ corresponds exactly to the value of one of the true DOA, the function $\overline{P}(\theta, \mathbf{c})$ equals zero. In this case, the smallest eigenvalue of $\mathbf{M}$ is equal to zero, and the determinant of the matrix $\mathbf{M}$ also equals zero.

A complex variable

$$z = e^{j \frac{2\pi \delta}{\lambda} \sin(\theta)}$$

needs to be defined here, where δ denotes the distance between two adjacent antennas, and λ the wavelength of impinging SOI. The steering vector $\mathbf{a}(\theta)$ is written as $\mathbf{a}(z) = [1, z, z^2, \ldots, z^{M-1}]^T$, and the matrix $\mathbf{M}$ is a function of z. We estimate the DOAs by finding the values of z, such that:

$$\det\{\mathbf{M}\} = 0 \qquad (18.18)$$

The left side of this last equation is a polynomial of z. The DOA estimation problem is then transformed into a polynomial rooting problem, which is solved using computationally efficient root-solving algorithms. The roots of the polynomial $\det\{\mathbf{M}\}$ are computed using any polynomial rooting algorithm.

Due to the symmetry property of the polynomial coefficients, roots appear in reciprocal conjugate pairs, z_i and $1/z_i^*$. In each pair, one root is inside the unit circle while the other is outside the unit circle, and the two roots coincide if they are on the unit circle. Either one of the two is used for DOA estimation, since they have the same angle in the complex plane. The roots inside the unit circle are chosen. According to the defined complex variable

$$z = e^{j\frac{2\pi\delta}{\lambda}\sin(\theta)}$$

the modulus of the roots corresponding to incoming SOI should be equal to one. In the presence of noise, the modulus is not necessary one, but is expected to be close to one. We then select the K_α roots (z_k, with $k = 1, 2, \ldots, K_\alpha$) that are nearest to the unit circle as being the roots corresponding to the DOA estimates. The DOA estimates are then:

$$\theta_k = \arcsin\left[\frac{\lambda}{2\pi\delta}\arg(z_k)\right] \qquad (18.19)$$

The degree of the polynomial $\det\{\mathbf{M}\}$ is $4M - 4$. Hence, the number of roots is $4M - 4$. Since roots appear in reciprocal pairs, the proposed procedure allows, in some conditions (e.g., BPSK, or AM signals), to determine $2(M - 1)$ signals. This point should be emphasized, since the number of signals estimated by the proposed method may be larger than the number of sensors.

Simulations

The behavior of the RCM and the classical Root-MUSIC methods are compared by drawing the selected roots by these two methods in the polar plane.

Figure 18.4 shows the distribution of the selected roots by the RCM method obtained from 500 Monte Carlo simulations when two BPSK SOI arrive from $-3°$ and $+3°$, with one interferer BPSK source from $15°$ DOA. BPSK SOI and the interferer are generated as previously explained.

Figure 18.5 shows the distribution of the selected roots by the classical Root-MUSIC method obtained from 500 Monte Carlo simulations in the same situation.

These two figures show that the RCM method perfectly selects the two SOI, and ignores the interferer signal. The polynomial roots selected by the proposed

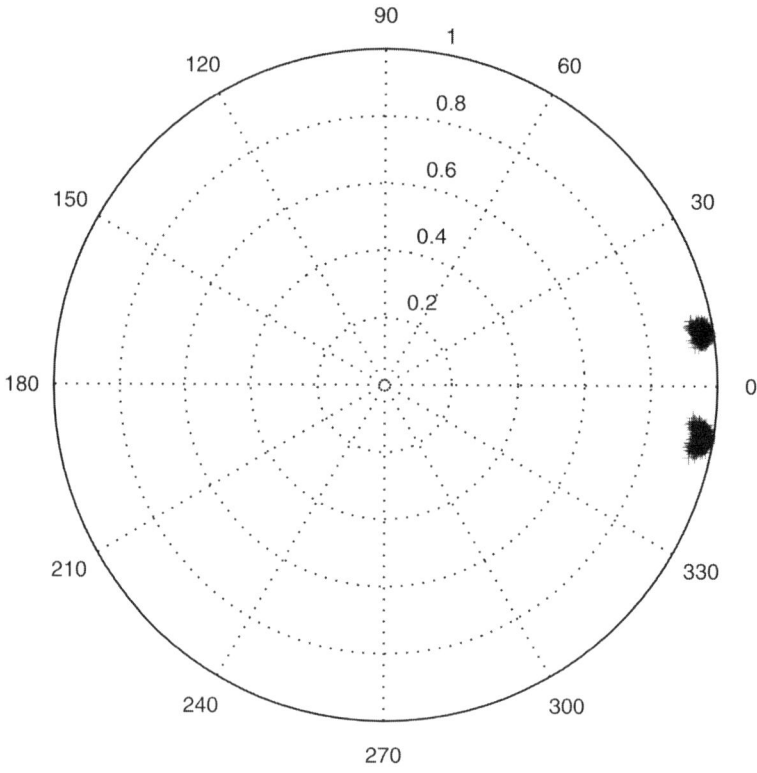

Figure 18.4 Selected polynomial roots by the Root Cyclic MUSIC method.

cyclic method are all located in two small areas. Those obtained from the classical Root MUSIC method are more scattered. It is deduced that the estimation provided by the proposed method is more accurate.

The cyclic algorithm performs better than the classical Root-MUSIC method, due to the exploitation of both cyclic correlation matrices. More information about sources is used, and the observation dimension space [i.e., the size of the extended covariance matrix $\mathbf{R}_{CE}^{\alpha}(\tau)$] is doubled.

18.4 Nonselective Beamforming Techniques

Two direction-finding procedures that exploit cyclostationarity of incoming signals are presented in this section. Both methods are based on the conventional beamforming principle. They can accommodate perfect correlation among signals, and they offer the advantage of simple implementation and computational efficiency, compared with the eigenstructure-based methods [3–5, 8, 13].

The two methods presented here require only knowledge of the baud rate, carrier frequency, or other frequencies that characterize the underlying periodicity exhibited by the desired signals. Contrary to these techniques, in these procedures the *prior* knowledge of cyclostationarity is not used to discriminate a desired signal

18.4 Nonselective Beamforming Techniques

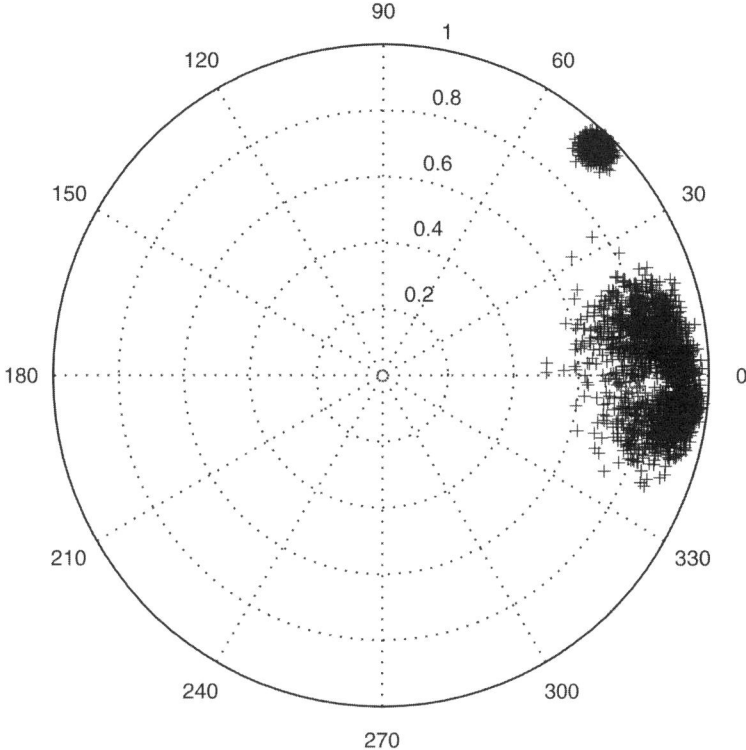

Figure 18.5 Selected polynomial roots by the classical Root MUSIC method.

from undesired signals, but it is exploited to outperform conventional methods without significant growth in complexity.

18.4.1 Data Model

In order to exploit cyclostationarity of incoming signals, another extended observation vector is formed. The so-called $\alpha\tau$-data observation vector is then:

$$\mathbf{z}_{\alpha\tau}(t) = \begin{bmatrix} \mathbf{z}(t) \\ \mathbf{z}^*(t-\tau)e^{j2\pi\alpha(t-\tau/2)} \end{bmatrix} \quad (18.20)$$

The corresponding $\alpha\tau$-covariance matrix is:

$$\mathbf{R}_{\alpha\tau} = \langle \mathbf{z}_{\alpha\tau}(t)\mathbf{z}_{\alpha\tau}^H(t) \rangle = \begin{bmatrix} \mathbf{R}_{zz} & \mathbf{R}_{zz^*}^{\alpha}(\tau) \\ \mathbf{R}_{zz^*}^{\alpha H}(\tau) & \mathbf{R}_{zz}^* \end{bmatrix} \quad (18.21)$$

where

$$\mathbf{R}_{zz^*}^{\alpha}(\tau) = \langle \mathbf{z}(t)\mathbf{z}^T(t-\tau)e^{-j2\pi\alpha(t-\tau/2)} \rangle \quad (18.22)$$

and

$$\mathbf{R}_{zz} = \langle \mathbf{z}(t)\mathbf{z}^H(t) \rangle = \langle \mathbf{z}(t-\tau)\mathbf{z}^H(t-\tau) \rangle \qquad (18.23)$$

This model allows simultaneous exploitation of the information provided in both the covariance matrix (18.21) and the cyclic conjugate correlation matrix (18.20).

18.4.2 Beamforming for Cyclostationary Signals

Let the beamformer output $y(t)$ be expressed as the inner product of a $2M$-dimensional weight vector $\boldsymbol{\omega}$ and the extended data vector $\mathbf{z}_{\alpha\tau}(t)$:

$$y(t) = \boldsymbol{\omega}^H \mathbf{z}_{\alpha\tau}(t) \qquad (18.24)$$

The output power is written as:

$$\langle |y(t)|^2 \rangle = \boldsymbol{\omega}^H \mathbf{R}_{\alpha\tau} \boldsymbol{\omega} \qquad (18.25)$$

In order to estimate the DOA, we have to maximize this beamformer output power over θ. For any angle θ, it is shown [14] that the mean power output of this beamformer can take the following form:

$$P_{BF\alpha\tau}(\theta) = \mathbf{a}^H(\theta)\mathbf{R}_{zz}\mathbf{a}(\theta) + \left| \mathbf{a}^T(\theta)\mathbf{R}_{zz^*}^{\alpha H}(\tau)\mathbf{a}(\theta) \right| \qquad (18.26)$$

It should be noted that, in the case where signals are not cyclostationary with cycle frequency α and lag parameter τ, the matrix $\mathbf{R}_{zz^*}^{\alpha}(\tau)$ equals zero, and this last function is reduced to the conventional beamforming spectrum.

18.4.3 Minimum Variance Beamforming for Cyclostationary Signals

Another direction-finding method based on the linearly constrained minimum variance principle also exists. The basic idea of this method is to constrain the response of the beamformer, so that the contribution of the two signal components (i.e., nonconjugate and conjugate-delayed components) from the direction of interest are passed with a specified gain; that is, the whole contribution will be normalized. The weight vector $\boldsymbol{\omega}$ is chosen to minimize output variance or power subject to the response constraints. This has the effect of preserving the contribution of both the nonconjugate and conjugate parts of the desired signal, while minimizing contributions from noise and interfering signals that come from directions other than the direction of interest.

According to this principle, the minimum variance beamformer output power spectrum is defined as [14]:

$$P_{MV\alpha\tau}(\theta) = \left[\min_{\mathbf{c}} \mathbf{b}(\theta, \mathbf{c})^H \mathbf{R}_{\alpha\tau}^{-1} \mathbf{b}(\theta, \mathbf{c}) \right]^{-1} \qquad (18.27)$$

where $\mathbf{b}(\theta, \mathbf{c})$ is the $(2M \times 1)$ vector defined in (18.12). Then, it is shown in [14] that for a fixed θ, the minimum value within the brackets is given by the minimum eigenvalue of the matrix $\mathbf{B}^H \mathbf{R}_{\alpha\tau}^{-1} \mathbf{B}$ with

$$\mathbf{B} = \begin{bmatrix} \mathbf{a}(\theta) & 0 \\ 0 & \mathbf{a}^*(\theta) \end{bmatrix}$$

$\mathbf{B}^H \mathbf{R}_{\alpha\tau}^{-1} \mathbf{B}$ is a (2×2)-dimensional hermitian matrix, a low computational cost is required to determine this minimum eigenvalue. DOA are given by the maxima of this minimum variance beamformer output power spectrum.

Similar to the previous cyclic beamforming method, in the case where signals are not cyclostationary with cycle frequency α and lag parameter τ, the matrix $\mathbf{R}_{zz^*}^{\alpha}(\tau)$ equals zero. In this situation, it is shown that the spectrum obtained by this cyclic minimum variance beamforming technique is reduced to the classical Capon's method spectrum [15].

18.4.4 Discussion

The two algorithms presented above are compared with each other. Both are designed to estimate efficiently the DOA of cyclostationary signal sources. The cyclostationarity-exploiting minimum variance beamforming ($MV\alpha\tau$) method has a better resolution power than the cyclostationarity-exploiting beamforming ($BF\alpha\tau$) method. However, the computation cost for the $MV\alpha\tau$ algorithm is less attractive, since this method requires an inversion of a $(2M \times 2M)$-dimensional matrix. The $BF\alpha\tau$ method is basically a beamforming method with cyclostationarity assumption, comparable to the conventional beamforming method. The $MV\alpha\tau$ is a cyclostationarity-exploiting minimum variance beamforming technique, comparable to the minimum variance Capon's method [15].

One common point between the two proposed methods is that being optimum, they both require the conjugate cyclic correlation to be nonzero. So $BF\alpha\tau$ and $MV\alpha\tau$ methods perform optimally for those signals that exhibit conjugate cyclostationarity, such as BPSK, OQPSK, MSK, or GMSK modulated signals. That kind of modulation occurs essentially in terrestrial and/or satellite telecommunication systems.

18.4.5 Simulations

In this section, we will compare the simulation results provided by the $BF\alpha\tau$ and $MV\alpha\tau$ cyclic methods with results obtained by the conventional beamforming technique and the Capon's method (which do not take cyclostationarity into account), respectively. These results are obtained by simulating a linear uniformly spaced array with six sensors spaced $\lambda/2$ apart. The number of data samples generated at each sensor output is 100. In the simulations, the receiving signals are 8-Mbps OQPSK modulated signals. The cyclic beamforming methods $BF\alpha\tau$ and $MV\alpha\tau$ are simulated with $\alpha = 4$ MHz and $\tau = 0$.

Figure 18.6 shows the power spectrum of the $BF\alpha\tau$ beamforming method when two OQPSK modulated sources are impinging on the array, with SNR = 10 dB.

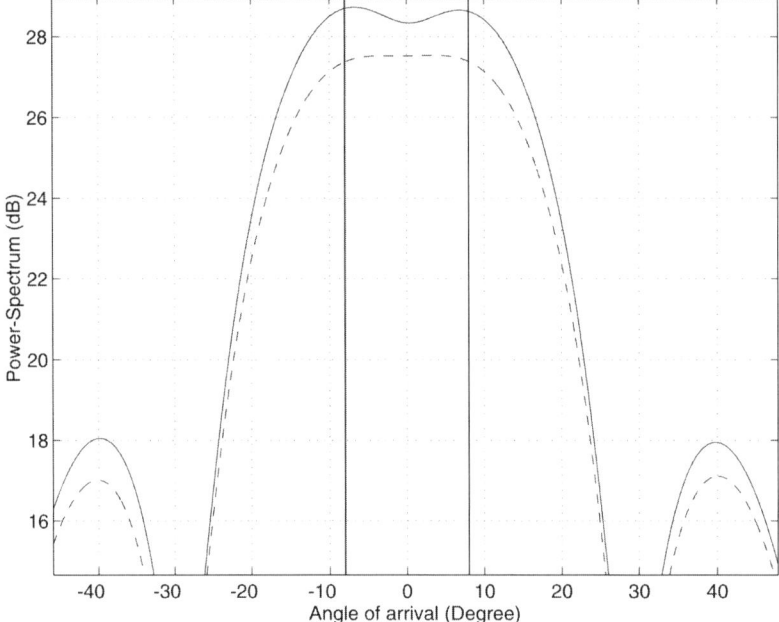

Figure 18.6 DOA estimation, with the proposed $BF\alpha\tau$ method (solid line), and with conventional beamforming method (dashed line).

The DOA of the sources are −8° and +8°. Since the half-power beamwidth of the array used for these simulations is approximately 20°, the estimation provided by the $BF\alpha\tau$ method is good. The $BF\alpha\tau$ method allows separation of the two sources, while the conventional beamforming method fails.

Figure 18.7 shows the power spectrum of the $MV\alpha\tau$ beamforming method when two OQPSK sources are impinging on the array, with a DOA of −6° and +6°, and with SNR = 0 dB. The $MV\alpha\tau$ method allows separation of the two sources with a better accuracy than the Capon's method.

18.5 Conclusions

In this chapter, several signal properties exploitation direction-finding techniques are presented. These techniques allow a greater performance than the estimation performance of conventional techniques, due to the exploitation of the a priori known cyclostationarity property of the incoming signals. Some other improvements to the conventional methods are proposed in the literature that exploit the noncircularity property of signals [9, 16, 17]. In [9], the MUSIC algorithm is extended to noncircular signals; in [16], the Root-MUSIC technique is adapted to exploit the noncircular property; and in [17], the ESPRIT technique is improved for noncircular sources.

18.5 Conclusions

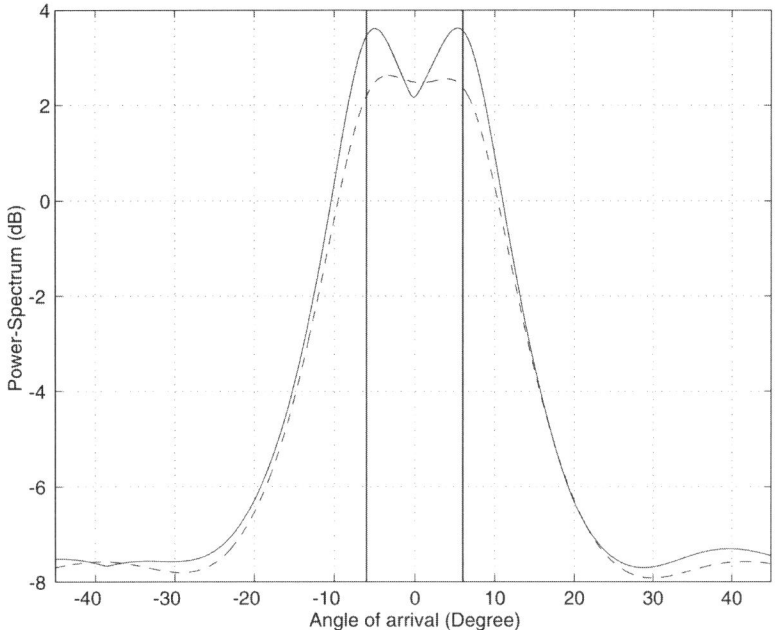

Figure 18.7 DOA estimation, with the proposed $MV\alpha\tau$ method (solid line), and with Capon's method (dashed line).

References

[1] Gardner, W. A., et al., *Cyclostationarity in Communications and Signal Processing*, New York: IEEE Press, 1993.

[2] Gardner, W. A., "Simplification of MUSIC and ESPRIT by Exploitation of Cyclostationarity," *IEEE Proc.*, Vol. 76, No. 7, July 1988, pp. 845–847.

[3] Schell, S. V., et al., "Cyclic MUSIC Algorithms for Signal-Selective DOA Estimation," *Proc. IEEE ICASSP*, Glasgow, Scotland, May 1989, pp. 2278–2281.

[4] Izzo, L., L. Paura, and G. Poggi, "An Interference-Tolerant Algorithm for Localization of Cyclostationary-Signal Sources," *IEEE Trans. on Signal Processing*, Vol. 40, July 1992, pp. 1682–1686.

[5] Xu, G., and T. Kailath, "Direction of Arrival Estimation Via Exploitation of Cyclostationarity—A Combination of Temporal and Spatial Processing," *IEEE Trans. on Signal Processing*, Vol. 40, July 1992, pp. 1775–1786.

[6] Schmit, R. O., "Multiple Emitter Location and Signal Parameters Estimation," *IEEE Trans. on Antennas and Propagation*, Vol. 34, March 1986, pp. 276–280.

[7] Wax, M., and T. Kailath, "Detection of Signals by Information Theoretic Criteria," *IEEE Trans. on Acoustics, Speech, and Signal Processing*, Vol. ASSP-33, No. 2, April 1985, pp. 387–392.

[8] Chargé, P., Y. Wang, and J. Saillard, "An Extended Cyclic MUSIC Algorithm," *IEEE Trans. on Signal Processing*, Vol. 51, No. 7, July 2003, pp. 1695–1701.

[9] Galy, J., "Antenne Adaptative: Du Second Ordre aux Ordres Supérieurs, Applications aux Signaux de Télécommunications," Toulouse, Ph.D. thesis, 1998.

[10] Barabell, A. J., "Improving the Resolution Performance of Eigenstructure-Based Direction Finding Algorithms," *Proc. IEEE ICASSP*, 1983, pp. 336–339.

[11] Wong, K. T., and M. D. Zoltowski, "Root-MUSIC-Based Azimuth-Elevation Angle-of-Arrival Estimation with Uniformly Spaced but Arbitrarily Orientated Velocity Hydrophones," *IEEE Trans. on Signal Processing*, Vol. 47, No. 12, December 1999, pp. 3250–3260.

[12] Swindlehurst, A. L., P. Stoica, and M. Jansson, "Application of MUSIC to Arrays with Multiple Invariances," *Proc. IEEE International Conference on Acoustics, Speech, and Signal Processing*, June 2000.

[13] Chargé, P., and Y. Wang, "A Root-MUSIC-Like Direction Finding Method for Cyclostationary Signals," *EURASIP J. on Applied Signal Processing*, No. 1, January 2005, pp. 69–73.

[14] Chargé, P., Y. Wang, and J. Saillard, "Cyclostationarity-Exploiting Direction Finding Algorithms," *IEEE Trans. on Aerospace and Electronic Systems*, Vol. 39, No. 3, July 2003, pp. 1051–1056.

[15] Capon, J., "High Resolution Frequency-Wavenumber Spectrum Analysis," *IEEE Proc.*, Vol. 57, August 1969, pp. 1408–1418.

[16] Chargé, P., Y. Wang, and J. Saillard, "A Noncircular Sources Direction Finding Method Using Polynomial Rooting," *Elsevier Signal Processing*, No. 81, 2001, pp. 1765–1770.

[17] Zoubir, A., P. Chargé, and Y. Wang, "Noncircular Sources Localization with ESPRIT," *ECWT*, Munich, Germany, October 2003.

PART V
Experimental Setup and Results

CHAPTER 19
DOA Antenna Array Measurement and System Calibration

Manuel Sierra-Pérez and Manuel Sierra-Castañer

This chapter presents the antenna array calibration process for the DOA applications. The first part of the chapter studies the calibration of unknown phase and unknown gain in the antenna elements, and the effects of the coupling between the antenna elements (i.e., the coupling in the radio frequency circuits and the free space between the antenna elements). The last part of the chapter shows one example of the application of this calibration procedure.

19.1 Introduction

Antenna array calibration has been a very active and interesting area of research in antenna array processing for the past several years [1–8]. For DOA applications, calibration of high accuracy antenna arrays becomes necessary to achieve adequate results, while the assumption of knowing a priori antenna array manifold is rarely satisfied [9]. Therefore, the combined effects of the unknown gain, the unknown phase, and the mutual coupling, as well as errors in the sensor positions, will degrade the performances of the antenna array processing algorithms [10]. The effects of mutual coupling become particularly important in antenna array calibration. These effects are widely studied for the DOA estimation in [11, 12]. The problems of the DOA and the steering vector estimation using the partially calibrated antenna arrays, and the problem of self-calibration of antenna arrays are also studied [13, 14].

This chapter deals with the calibration process of the phase, the gain, and the effects due to the mutual coupling. The emphasis in the first part of the chapter is given to the mutual coupling model and the calibration. In this model, the input impedance is computed through an N-port network model described by the impedance (Z) or the scattering (S) matrix. The radiated field is computed if the active radiated field from each element that takes into account the complete array influence of each element radiated field is known, which is presented by Mailloux [15]. For the receiving antennas, and for the DOA applications, the model is the same, giving an equivalent radiation pattern and an active output impedance that depends on the loading circuit. In large antenna arrays, most antenna elements present similar conditions: their classification, position, and load. This situation almost provides equal active impedance for most antenna elements in the antenna array, except for the extreme elements. In [16], Wasylkiwskij shows the relation between the input impedance and the radiated field for the minimum scattering antennas (MSA). On many occasions, the real antennas are assumed to be the MSA with good results [17]. In other cases, that assumption becomes invalid, and the relation between the input coupling matrix (Z) and the active radiated field becomes uncertain. One of the most important questions in the design of the active or adaptive antenna arrays is whether it is possible to model the coupling by an N × N coupling matrix (C), as described in [18]. In this chapter, a matrix model [19] is presented for the antenna arrays fed via the linear networks. This description is independent of the feeding network, the feeding distribution, or the transmission-reception applications [12]. For many printed patches, only one resonant radiation mode is used to describe the radiated field of each antenna element. When this happens, a 2N × 2N matrix describes the complete antenna array model. Based on this model and an application of the feeding network parameters, a coupling C matrix is obtained to characterize the influence of the antenna element coupling in the structure.

This chapter is divided into four sections. Section 19.2 explains the electromagnetic model that characterizes the coupling among the different antenna elements included in the antenna array. This model is based on the system that is described in Chapter 13. Section 19.3 explains the calibration process to determine the parameters of the previous model, and shows one example. Section 19.4 concludes the main ideas that are developed in the chapter.

19.2 Steering Vector Model

The antenna array is characterized by several individual antenna elements, placed in specific points in the space (r_k) that receive the incoming signal. The signals received by each antenna array element usually are processed, assuming the ideal behavior of antennas as ideal electromagnetic sensors. This leads to the classical definition of the steering vector, which mentions the behavior of the antenna array as a function of the direction of arrival, and takes into account only the position of the antenna elements.

$$\mathbf{a}_s = \begin{bmatrix} e^{(-j\omega_c \tau_{s1})} \\ \cdots \\ e^{(-j\omega_c \tau_{sP})} \end{bmatrix} \quad (19.1)$$

where $\tau_{sk} = (|\vec{r}_s - \vec{r}_k| - |\vec{r}_s|)/c$ is the differential time propagation from the source to the antenna element k ($k = 1$ to P); r_s and r_k are the sources; k is the antenna element position; $\omega_c = 2\pi f_c$, where fc is the carrier frequency; and c is the speed of light.

The snapshot or input signal vector $\mathbf{x}(t)$ is then usually written as:

$$\mathbf{x}(t) = f_s(t)\mathbf{a}_s \quad (19.2)$$

where $f_s(t)$ is the lowpass complex time function or the source.

19.2 Steering Vector Model

This formulation is appropriate when the entire antenna elements are of equal lengths and are equally orientated, and connected to ideally matched networks with the same gain and phase behaviors.

However, there are several reasons why the signals from the sensor not ideal. One of these reasons is the unbalance between the different RF chains that feed the antennas (i.e., a difference in the amplitude and the phase). This unbalance is corrected in any calibration process, and only requires N complex numbers to characterize the different antenna elements of the antenna array. $N - 1$ complex numbers are required if one of the elements acts as reference antenna element. The steering vector is then written as the product of the position phase vector and a diagonal matrix of complex gain factors.

$$\begin{bmatrix} \rho_1 e^{(-j\omega_c \tau_{s1})} \\ \cdots \\ \rho_P e^{(-j\omega_c \tau_{sP})} \end{bmatrix} = \begin{bmatrix} \rho_1 & 0 & 0 \\ 0 & \ddots & 0 \\ 0 & 0 & \rho_P \end{bmatrix} \begin{bmatrix} e^{(-j\omega_c \tau_{s1})} \\ \cdots \\ e^{(-j\omega_c \tau_{sP})} \end{bmatrix} = \mathbf{G}\mathbf{a_s} \quad (19.3)$$

The second factor is the mutual coupling among the different antenna elements of the antenna array, either in the RF feeding circuit, or in the nearfield free space. The correct electromagnetic full wave characterization of these factors is normally difficult, and requires a complete electromagnetic analysis of the antenna structure. However, as shown in Chapter 13, when the antenna elements are characterized by a fundamental radiating mode, a coupling matrix is obtained to compensate for the steering vector. The most important factors of the coupling matrix are determined through the fundamental circuit and the radiation pattern measurements. If we include the RF gain, or even an RF coupling problem, then we can finally write a steering vector as a product of an overall coupling matrix and a position phase vector.

$$\begin{bmatrix} c_{11} & \cdots & c_{1P} \\ \vdots & \ddots & \vdots \\ c_{P1} & \cdots & c_{PP} \end{bmatrix} \begin{bmatrix} e^{(-j\omega_c \tau_{s1})} \\ \cdots \\ e^{(-j\omega_c \tau_{sP})} \end{bmatrix} = \mathbf{C}\mathbf{a_s} \quad (19.4)$$

In this section, a matrix model is presented for the antenna arrays fed through the linear networks. The model is based on the assumption that there is only one fundamental mode field radiation by each antenna element of the antenna array. This model is valid for many antennas as printed patches, and for other resonant antennas. A more general situation requires the formulation of the active or the embedded antenna element patterns, leading to a formulation where the steering vector correction depends on the direction of arrival, not only in the position phase factor, but in the amplitude and the phase due to the pattern variations, as well [15].

In this simple case, a $2P$ square matrix will describe the antenna array behavior. This description is independent of the feeding network, the feeding distribution, or even the transmission-reception application [12]. Based on this model and an application of the feeding network parameters, a coupling matrix $\mathbf{C}$ is obtained to characterize the influence of the antenna element coupling on the signal structure.

The parameters of this model are obtained through the measurements of the radiation pattern for each individual antenna element in the final antenna array disposition, as shown in Section 19.3. These measurements are used for the antenna calibration process.

If we define the input wave vector as $[\mathbf{a} = (a_1, a_2, \ldots, a_P)^t]$ and the received or reflected wave vector as $[\mathbf{b} = (b_1, b_2, \ldots, b_P)^t]$, then we can write the relation through the coupled antenna array elements as:

$$\mathbf{b} = \mathbf{S}_a \mathbf{a} \quad (19.5)$$

where $\mathbf{S}_a$ is the scattering parameters square matrix of $P \times P$ antenna elements.

As shown in Chapter 13, the radiated field of an antenna array is obtained as a sum of the fundamental fields radiated by each antenna element. If we assume that all the antenna elements radiate in the fundamental mode, and the pattern is the same for all the antenna elements, then the field is written as:

$$\vec{E}(\theta, \phi) = \sqrt{2\eta_0} \frac{e^{(-jk_0 r)}}{r} \hat{e}_A(\theta, \phi) F_A(\theta, \phi) \sum_{k=1}^{P} b_k^e e^{[-j\omega_c \tau_k(\theta, \phi)]} \quad (19.6)$$

where b_k^e is a coefficient of the main mode excitation for each antenna element.

Assuming linear behavior and some coupling between antenna elements, the vector or radiation coefficients $\mathbf{b}_e = (b_1^e, b_2^e, \ldots, b_N^e)^t$ must be related to the input voltage wave vector a through a coupling matrix $\mathbf{S}_e$:

$$\mathbf{b}_e = \mathbf{S}_e \mathbf{a} \quad (19.7)$$

In the same way for the receiving antennas, the equivalent generator for each antenna array element when the antenna receives a field wave from a source s depends on the antenna radiation pattern. This pattern is similarly expressed as a function of the fundamental mode pattern, giving an output vector at the terminal (**b**), proportional to the input field coefficients $\mathbf{a}_e = (a_1^e, a_2^e, \ldots, a_N^e)^t$:

$$\mathbf{b} = \mathbf{S}_r \mathbf{a}_e \quad (19.8)$$

$$a_k^v = \frac{\lambda}{\sqrt{2\eta_0}} \vec{E}_s(t) \hat{e}_A(\theta_s, \phi_s) F_A(\theta_s, \phi_s) e^{(-j\omega_c t_{sk})} = f_s(t) e^{(-j\omega_c t_{sk})} \quad (19.9)$$

If an antenna array of P antenna elements is considered, then it is represented by a $2P$-port network, as shown in Figure 19.1. The input ports at the left represent the antenna element input terminals, while the ports at the right represent the radiating modes. The outgoing waves to the load ($\mathbf{b}_e$) describe the amplitude and the phase of the main mode radiated field, and the generators ($\mathbf{a}_e$) describe the amplitude and the phase of the incoming field to the antenna [19]. The matrix equation relating previous variables is given as:

$$\begin{pmatrix} \mathbf{b} \\ \mathbf{b}_e \end{pmatrix}_{(2N \times 1)} = \mathbf{S}_{(2N \times 2N)} \cdot \begin{pmatrix} \mathbf{a} \\ \mathbf{a}_e \end{pmatrix}_{(2N \times 1)} \Rightarrow \begin{matrix} \mathbf{b} = \mathbf{S}_a \cdot \mathbf{a} + \mathbf{S}_r \cdot \mathbf{a}_e \\ \mathbf{b}_e = \mathbf{S}_e \cdot \mathbf{a} + \mathbf{S}_s \cdot \mathbf{a}_e \end{matrix} \quad (19.10)$$

19.2 Steering Vector Model

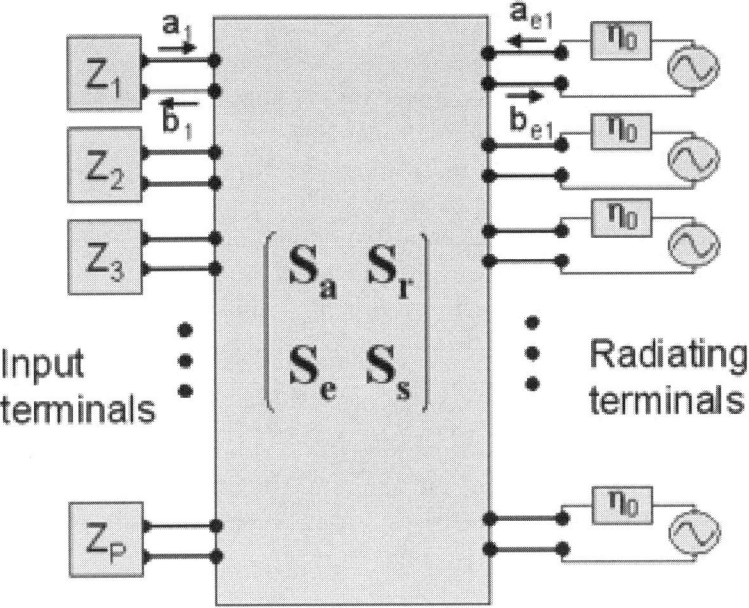

Figure 19.1 Scattering matrix for the antenna array.

where S_a represents the reflection coefficient, S_e represents the transmission parameter for the fundamental mode, S_r represents reception coupling for that mode, and S_s indicates the scattered field.

In this scattering parameter definition, the input terminals are referenced to the standard impedance Z_0, and the radiating terminals are referenced to the free space impedance η_0.

When operating as a receiver, which is typical for the DOA applications, (19.10) is simplified, since we are only interested in the first line. The terminals are connected to the receivers with load impedance Z_k or equivalent input reflection coefficient $\Gamma_k = (Z_k - Z_0)/(Z_k + Z_0)$, in such a way that we can write a second relation as:

$$\mathbf{a} = \mathbf{\Gamma}_L \mathbf{b} \qquad (19.11)$$

with

$$\mathbf{\Gamma}_L = \begin{bmatrix} \Gamma_1 & \cdots & 0 \\ 0 & \cdots & 0 \\ 0 & 0 & \Gamma_P \end{bmatrix}$$

The total received signal at each antenna terminal is expressed as the sum of the contributions from any of the radiating antenna elements:

$$\mathbf{b} = \mathbf{S}_a \mathbf{a} + \mathbf{S}_r \mathbf{a}_e = (\mathbf{I} - \mathbf{S}_a \mathbf{\Gamma}_L) \mathbf{S}_r \mathbf{a}_e \qquad (19.12)$$

The received signal vector is then expressed as a function of the source field function in a similar way (19.12), with the only difference being that the steering vector is multiplied by a coupling matrix $\mathbf{C}$.

$$\mathbf{b} = f_s(t)(\mathbf{I} - \mathbf{S}_a \mathbf{\Gamma}_L)^{-1} \mathbf{S}_r \mathbf{a}_s = f_s(t) \mathbf{C} \mathbf{a}_s \qquad (19.13)$$

A reception coupling matrix (**C**) is obtained during the calibration process.

When the feeding network is included, this problem is studied like two cascaded multipoles—the first for the feeding network, and the second for the coupling among the radiating antenna elements. The process for the complete structure (i.e., the feeding network with the radiating antenna elements) becomes similar to the previous process, but results in a new coupling matrix. Assuming the loads at the receiving point are perfectly matched to the reference impedance, the coupling matrix as a function of the multipole matrices is:

$$\mathbf{C} = \mathbf{S}_g (\mathbf{I} - \mathbf{S}_a \mathbf{S}_c)^{-1} \mathbf{S}_r \qquad (19.14)$$

where $\mathbf{S}_g$ and $\mathbf{S}_c$ represent the gain and the reflection, including the coupling, in the receiver RF circuits, respectively, as shown in Figure 19.2.

19.3 Calibration Process

In this section, the calibration process is discussed. In the simplest model, it is considered that all the antenna elements are ideal sensors, and that all the receiving patterns are equal for all radiators, with no coupling between them. The calibration process is then reduced to a measurement of the relative amplitude and phase of the receiving patterns between all the receiver terminals. In this situation, (19.3) is applied.

The measurement of the active or the embedded antenna element patterns is always necessary to detect the malfunctions in each radiator. If the errors between the radiation patterns of antenna element are comparable, then the matrix $\mathbf{S}_r$ is written as a nearly diagonal matrix, and only a gain factor is considered. In this case, the measurement of the $\mathbf{S}_a$ matrix is done with a vector network analyzer, and a common antenna element pattern is used to obtain the output power as a function of the incoming field. If the coupling between the RF receivers is considered,

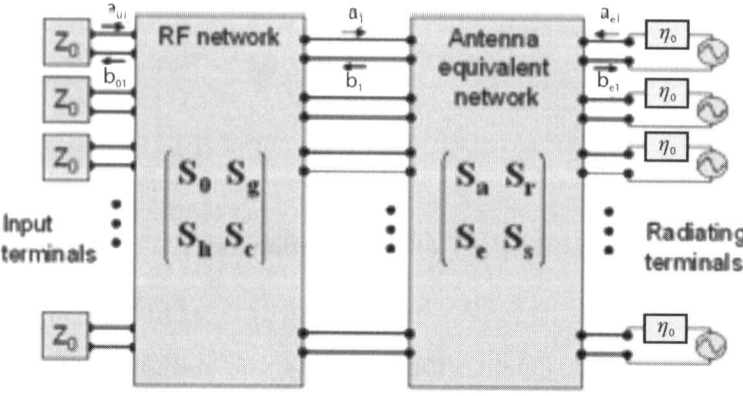

Figure 19.2 Matrix representation of coupling in antennas and RF subsystem.

then this system is characterized by the measurement of matrices S_h and S_c. These parameters are again obtained through the vector network analyzer measurements.

However, if the antenna patterns are not the same for all the antenna elements, then the pattern of each antenna element included in the steering vector is considered individually. If there are major changes in the patterns due to the main mode coupling matrix S_r, then the model described in the previous section is used for the calibration of the antenna array. Adequate compensations are provided through the general coupling matrix C (19.3). In this case, we need to measure the radiation pattern for each antenna element in an anechoic system, and to compute a coupling matrix C_r. To ensure that there are no other radiating modes created in the coupling process, the radiation patterns modeled with the above matrix theory are checked.

19.3.1 Coupling Through Terminals

The first measurement of coupling is to find the antenna terminal coupling matrix S_a. To measure this matrix, it is necessary to connect a network analyzer at each pair of terminals, keeping the rest loaded with the nominal impedance, as shown in Figure 19.3. As a general rule, the coupling parameters are almost the same for the antenna elements placed in the same direction. This means that for a linear, equally spaced antenna array, matrix S_a has almost the same parameters for all the antenna elements in the same direction. In a large, linear, or planar antenna array, this approximation is done for the antenna elements far from the edge of

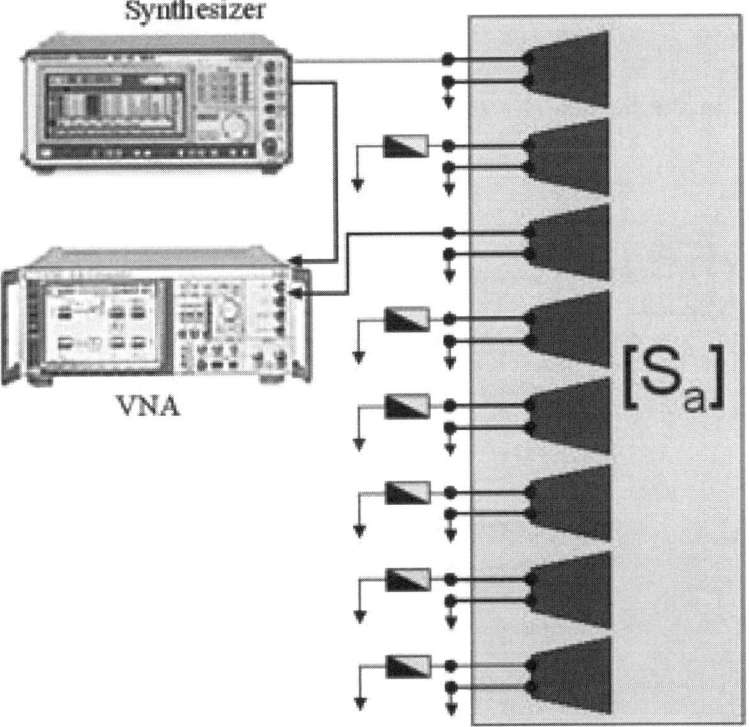

Figure 19.3 Coupling measurement through terminals.

the antenna array. For small antenna arrays (i.e., less than 10 antenna elements), it is recommended to measure all the coupling parameters involved.

19.3.2 Radiation Pattern Measurements

In this section, the measurements of the radiated field of the antenna array in an anechoic environment are discussed. Each antenna array element is in the farfield pattern of the measurement probe antenna. The system allows the measurement of the relative amplitude and phase of the received signals from both the reference antenna element and the antenna array under consideration. The antenna array being tested is installed on an azimuth positioner in such a way that the measurements are taken in all possible angles of directions of arrival (θ_n), as shown in Figure 19.4.

Once the main mode antenna element pattern is measured ($\hat{e}_A$ and F_A) for the DOA of interest, then the antenna array measurements are carried out for the same DOA. Interpolation is done to get the appropriate values, and to get the equivalent wave vector $\mathbf{a}_e(\theta_n)$ associated with the angular position θ_n. The vector components are written as shown in (19.9) for the particular direction and the incoming field:

$$a_{k,n}^e = \lambda b_0 \langle \hat{e}_S(\theta_{k,n}^S) \hat{e}_A(\theta_{k,n}) \rangle F_S(\theta_{k,n}^S) F_A(\theta_{k,n}) \frac{e^{(-jk_0 r_{k,n})}}{r_{k,n}} \qquad (19.15)$$

where $\hat{e}_S$ is the polarization vector of the feeding antenna, F_S is the relative field pattern, and $\theta_{k,n}^S$ is the angular position in the feeding antenna coordinate system for the antenna element k at position n. Constant b_0 represents the amplitude and the phase of the transmitter.

In the far-field system measurements, all the angular positions are the same for all the antenna elements $\theta_{k,n} = \theta_n$ and $\theta_{k,n}^S = 0$. It is important to point out

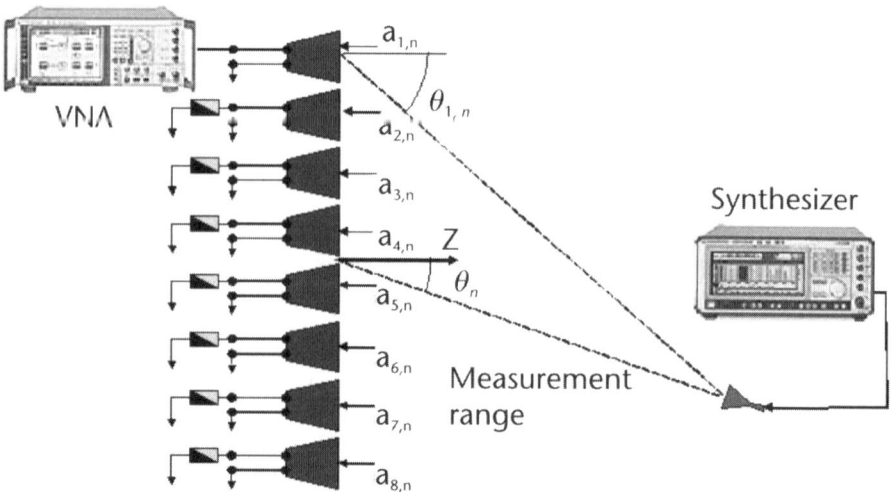

Figure 19.4 Farfield antenna calibration system.

that (19.15) also considers the polarization coupling between the radiation antenna element and the radiation probe.

For each angular position, the transmitted signals $\mathbf{b}(\theta_n)$ are measured, and by performing the same number of independent measurements on the antenna array elements, the coupling matrix is obtained as, using (19.8):

$$[\mathbf{b}_1 \quad \mathbf{b}_2 \quad \ldots \quad \mathbf{b}_N] = \mathbf{B} = \mathbf{S}_r[\mathbf{a}_{e,1} \quad \mathbf{a}_{e,2} \quad \ldots \quad \mathbf{a}_{e,N}] = \mathbf{S}_r \mathbf{A}_e \quad (19.16)$$

The selection of the different angular positions is critical. It is necessary to have independent situations to solve this problem. This implies that there is no symmetry in the DOA, and avoids the case where the antenna array factor is periodic. Normally, the number N of measurement points is greater than the P antenna elements, so an optimization process is developed to obtain the coupling matrix $\mathbf{C}_r$.

Using all the equations, in order to get a minimum root square error for the determination of the norm of the difference:

$$\|\mathbf{B} - \mathbf{S}_r \mathbf{A}_e\| \quad (19.17)$$

As a solution, we can write:

$$\mathbf{S}_r = \left(\mathbf{A}_e \mathbf{A}_e^H\right)^{-1} \mathbf{B} \mathbf{A}_e^H \quad (19.18)$$

Once the parameters of $\mathbf{S}_r$ matrix are computed, the quality of the model is validated through the radiation patterns of the antenna elements. One mathematical evaluation is by (19.17), but a good agreement between the measured and the modeled pattern is done in (19.6), with $\mathbf{b}_e$ obtained using (19.7) for unity vectors (**a**) as the feeding excitation.

19.3.3 Calibration Process for the RF Subsystem

The calibration of the RF network is done with a network analyzer. In many situations for the DOA systems, the frequency conversion and the A/D conversion is usually done in the receiver scheme. It is better to measure directly with the digital processor to get the overall influence of the receiver channels. Figure 19.5 shows a schematic arrangement of the measurement setup. A signal generator or synthesizer excites one of the inputs, and the output is then measured at all the N terminals. This setup also characterizes the difference of the gain and the phase for each of the RF channels, which usually is the most important parameter.

In many active systems, the frequency conversion to a baseband digital signal makes the measurement of the absolute value of the phase transmission parameter ($s_{g(k,k)}$) difficult. In fact, there is a phase uncertainty among these parameters, but the important point of interest is the relative phase between them. If the RF circuit is subjected for the measurements of the relative phase, as shown in Figure 19.5, then the output vector for a specific input at terminal k is:

$$e^{j\Phi_k} \mathbf{x}_k = \mathbf{S}_g \mathbf{b}_k \quad (19.19)$$

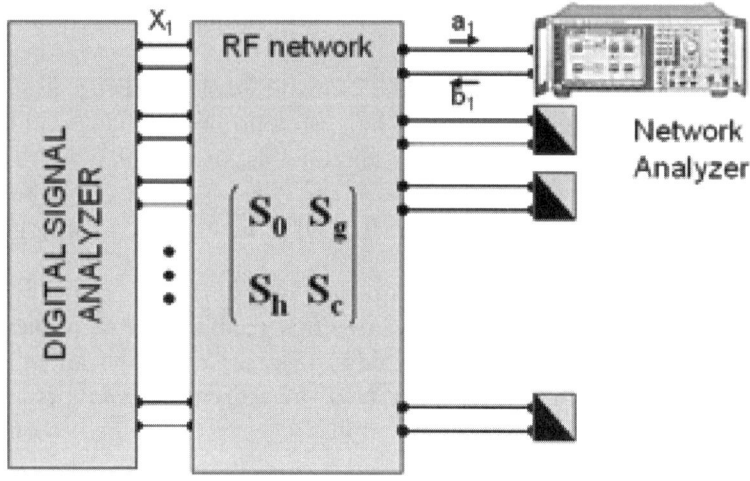

Figure 19.5 Calibration scheme for coupling in RF subsystem.

where $\mathbf{b}_k$ represents the unit vector $\mathbf{b}_k = (0, 0, \ldots, 1, \ldots, 0)^t$, and Φ_k is the unknown phase reference that we can only measure if we lock the RF signal and the local conversion oscillators. Even in this case, if there is any unlocking process between one measure and the next one, then the measurements are not guaranteed.

If we measure the RF circuit by placing the synthesizer outputs in all the input terminals, then the final matrix equation does not allow a solution for the $\mathbf{S}_g$ matrix, because of the unknown Φ diagonal phase matrix:

$$\mathbf{X}\Phi = \mathbf{S}_g \mathbf{B} \tag{19.20}$$

To gather a better understanding of the RF receiver, we need to have data from a large number of measurements. In fact, the unknowns are $P^2 + P - 1$. If an extra measurement of the system is done, then the required number of independent equations is obtained to solve the problem. The kind of measurement needed is shown in Figure 19.6, where the same RF signal is applied at all the input terminals, creating an extra term,

$$\mathbf{b}_{P+1} = \sqrt{\frac{1}{P}} (1, 1, \ldots, 1, \ldots, 1)^t$$

With all the extra measurements, applying a minimum-square error algorithm solves (19.20). However, to solve the phase uncertainty, the combination measurements must be done with all the input terminals. For instance, we can use a two-way power divider, and measure $P - 1$ times the combination of each two consecutive terminals, as shown in Figure 19.7. It is important to use the matched power dividers as the Wilkinson or the magic-T dividers, to avoid the influence of reflection matrix ($\mathbf{S}_c$) in the transmission measurement.

Finally, the matrices $\mathbf{S}_g$ and Φ are obtained through a minimum function problem as:

19.3 Calibration Process

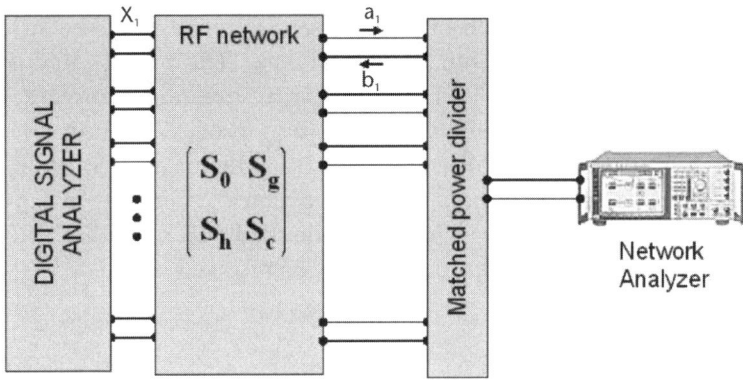

Figure 19.6 Calibration scheme for coupling in RF subsystem, with extra phase measurement.

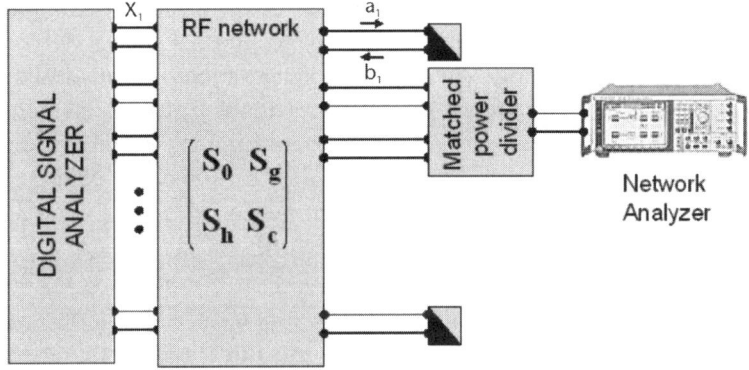

Figure 19.7 Extra phase measurement in consecutive input ports.

$$\min_{\Phi, S_g} \|X\Phi - S_g B\| \qquad (19.21)$$

The same kind of equations are obtained if the full antenna array and the RF circuit are measured directly to get the snapshot vector for several antenna positions in the anechoic chamber relative to the probe. In this case, a similar minimum-square equation must be solved to get the coupling matrix given in (19.14).

$$\min_{\Phi, C} \|X\Phi - CA_e\| \qquad (19.22)$$

19.3.4 Application Example

As an example, this calibration process has been applied to a similar antenna structure, which is an array of four elements separated by one wavelength, working at 2 GHz. The array is excited with one signal coming in with a 0° direction of arrival, generated from one antenna placed at a distance of 6.5m, in a farfield anechoic chamber. In order to analyze the effect of the calibration process, the signals received from each one of the four antennas have been processed using an LMS algorithm, as in a smart antenna case. The process has been repeated with and without calibration, and the results are shown in Figure 19.8. It is shown that the radiation pattern after the calibration process is much more similar to the theoretical result (i.e., uniform excitation antenna pattern) than is the result from the noncalibrated structure. This improvement comes from the calibration of the coupling among the four elements, while the LMS algorithm can correct the differences in phase and amplitude of the four received signals.

19.4 Conclusions

This chapter has presented several DOA system calibration methods. The steering vector model that characterizes the antenna-RF system considers an array of real antenna elements, including the mutual coupling among them, and the random errors in the different RF chains. In order to simplify the model, only the fundamental electromagnetic mode for each antenna element is considered. A good approximation for printed patches or other resonant antenna elements is shown. The calibration process, which consists of obtaining the different antenna elements of the coupling matrix that characterizes the DOA system, is also illustrated. Several cases are studied, depending on the influence of factors such as coupling in free space, coupling inside RF circuits, and so forth.

This chapter did not study the self-calibration process that consists of getting a consistent calibration while the antenna is used. The changes in the performance of the antenna are mainly due to variations in the receiver gain, due to temperature changes, input impedance of the receivers, or small changes in the radiator position. Friedlander and Weiss [11] present a self-calibration process that requires an initial

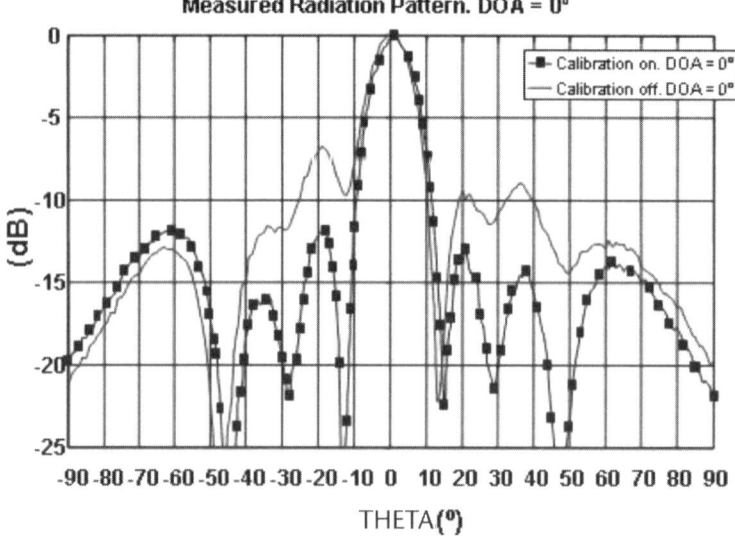

Figure 19.8 Radiation pattern with and without calibration.

calibration matrix C, and the DOA process, such as MUSIC, to receive the signal arriving at the antenna array.

References

[1] Solomon, I. S. D., et al., "Receiver Array Calibration Using Disparate Sources," *IEEE Trans. on Antennas and Propagation*, Vol. AP-47, March 1999, pp. 496–505.

[2] Chong Ng, B., and C. M. Samson See, "Sensor-Array Calibration Using a Maximum-Likelihood Approach," *IEEE Trans. on Antennas and Propagation*, Vol. AP-44, June 1996, pp. 827–835.

[3] Rockah, Y., and P. M. Schultheiss, "Array Shape Calibration Using Sources in Unknown Locations—Part I: Far-Field Sources," *IEEE Trans. on Acoustics, Speech, and Signal Processing*, Vol. ASSP-35, March 1987, pp. 286–299.

[4] Rockah, Y., and P. M. Schultheiss, "Array Shape Calibration Using Sources in Unknown Locations—Part II: Near-Field Sources and Estimator Implementator," *IEEE Trans. on Acoustics, Speech, and Signal Processing*, Vol. ASSP-35, June 1987, pp. 724–735.

[5] Steinberg, B. D., *Microwave Imaging with Large Antenna Arrays*, New York: John Wiley & Sons, 1983.

[6] Steinberg, B. D., and H. Subbaram, *Microwave Imaging Techniques*, New York: John Wiley & Sons, 1991.

[7] See, C. M. S., "Sensor Array Calibration in the Presence of Mutual Coupling and Unknown Sensor Gain and Phases," *Electronic Lett.*, Vol. 30, March 1994, pp. 373–374.

[8] See, C. M. S., "Method for Array Calibration in High-Resolution Sensor Array Processing," *Inst. Elect. Eng. Proc. Radar, Sonar, Navig.*, Vol. 142, June 1995, pp. 90–96.

[9] Porat, B., and B. Friedlander, "Accuracy Requirements in Off-Line Array Calibration," *IEEE Trans. on Aerospace and Electronic Systems*, Vol. AES-33, April 1997, pp. 545–556.

[10] Weiss, A. J., and B. Friedlander, "Comparison of Signal Estimation Using Calibrated and Uncalibrated Arrays," *IEEE Trans. on Aerospace and Electronic Systems*, Vol. AES-33, January 1997, pp. 241–249.

[11] Friedlander, B., and A. J. Weiss. "Direction Finding in Presence of Mutual Coupling," *IEEE Trans. on Antennas and Propagation*, Vol. AP-39, March 1991, pp. 273–284.

[12] Segovia-Vargas, D., R. Martín-Cuerdo, and M. Sierra-Pérez, "Mutual Coupling Effects Correction in Microstrip Arrays for Direction of Arrival (DOA) Estimation," *IEE Proc. on Microwave, Antennas, and Propagation*, Vol. 149, April 2002, pp. 113–118.

[13] Weiss, A. J., and B. Friedlander, "DOA and Steering Vector Estimation Using a Partially Calibrated Array," *IEEE Trans. on Aerospace and Electronic Systems*, Vol. AES-32, July 1996, pp. 1047–1057.

[14] Ng, B. P., "Array Shape Self-Calibration Technique for Direction Finding Problems," *IEE Proc.-H*, Vol. 139, December 1992, pp. 521–525.

[15] Mailloux, J. R., *Phased Array Antenna Handbook*, Norwood, MA: Artech House, 1994.

[16] Wasylkiwskyj, W., and W. K. Kahn, "Theory of Mutual Coupling Among Minimum-Sacttering Antennas," *IEEE Trans. on Antennas and Propagation*, Vol. AP-18, March 1970, pp. 204–216.

[17] Gupta, I. J., and A. A. Ksienski, "Effect of Mutual Coupling on the Performance of Adaptive Arrays," *IEEE Trans. on Antennas and Propagation*, Vol. AP-31, September 1983, pp. 785–791.

[18] Steiskal, H., and J. S. Herd, "Mutual Coupling Compensation in Small Array Antennas," *IEEE Trans. on Antennas and Propagation*, Vol. AP-38, December 1990, pp. 1971–1975.

[19] Fernández, J. M., et al., "Estimation of Patch Array Coupling Model Through Radiated Field Measurements," *Microwave and Optical Technology Letters*, Vol. 43, October 2004, pp. 59–64.

CHAPTER 20
ESPAR Antenna Signal Processing for DOA Estimation

Jun Cheng and Takashi Ohira

High-precision direction finding for mobile user terminals is investigated by employing a single-port antenna, the electronically steerable parasitic array radiator (ESPAR) antenna. The structure of the ESPAR antenna is introduced, along with formulation of its output. The main feature of the antenna is that it only requires one receiver chain. This significantly reduces the cost, size, complexity, and power consumption. Another feature is that it carries out signal combination in space by electromagnetic mutual coupling among array elements. This implies that a compact implementation of the antenna is possible. These features make the ESPAR antenna applicable to mobile user terminals. However, because only the single-port output of the antenna is observed, the direct application of conventional array antenna algorithms is ineffective and impractical. It is thus important to investigate a signal processing algorithm specialized for the ESPAR antenna. This work proposes ESPAR antenna signal processing for DOA estimation. The antenna correlation matrix is estimated as it switches over a set of beam patterns by a set of reactances. The MUSIC algorithm is then applied to the correlation matrix to estimate the DOAs of impinging signals. Computer simulation and an experiment are successfully carried out, which verified the reactance domain MUSIC algorithm for high precision DOA estimation. We also describe three practical applications of the ESPAR antenna: satellite attitude control, a wireless locator and homing system, and a wireless ad hoc network. Our results show that estimating DOA by employing the ESPAR is a promising alternative to the conventional array antenna, especially for mobile user terminals.

20.1 Introduction

Direction finding of electromagnetic waves is of great importance in a variety of applications, such as satellite control, personal locating service, and wireless communications. In the last two decades, direction finding with antenna signal processing has attracted considerable interest in the areas of signal processing and wireless communication. The theory and sophistical algorithms for high precision DOA estimation have already been established. For example, the MUSIC algorithms provide an asymptotically unbiased estimation of the characteristics of electromagnetic waves, such as the number of signals and their DOAs [1]. Although the performance of these algorithms is excellent, there is a high cost in employing one receiver chain per branch of antenna [2]. For a two-element array antenna, for instance, this doubles the receiver hardware, and it is not an easy trade-off to make when it is essential to reduce system complexity. System complexity is precisely what prevents the array antenna technique from being applicable to mobile user terminals.

Analog smart antennas, such as the ESPAR antenna [3–9], recently have shown the potential for application to mobile user terminals, since the antenna only uses a single active radio receiver, which significantly reduces the antenna's cost, size, complexity, and power consumption. For the ESPAR antenna, only a single radiator is connected to the receiver. This active radiator is surrounded by parasitic radiators loaded with variable reactors. The radiation directivity of the antenna is controlled by changing the reactance values. In the ESPAR antenna, the signal combination is carried out in space by electromagnetic mutual coupling among array elements, not in circuits. This permits compact implementation of the antenna.

The concept of the ESPAR antenna dates back to the early work of Harrington's model [10], where the "electric length" of the element is adjusted by changing the element's loaded reactance. This causes a change in the radiation pattern. Dinger demonstrated a reactive approach that uses planar parasitic elements for antijamming [11, 12]. Another single-port approach related to the ESPAR antenna is the switched parasitic antenna [13–18]. Details on direction finding by using the switched parasitic antenna can are found in [13–16].

For the ESPAR antenna, however, the signal on the surrounding parasitic elements is not observed. Only the single-port output is observed. This differs from the conventional array antenna, where the received signal on each element is observed. This characteristic prevents the conventional algorithms for array antennas from being applicable to the ESPAR antenna.

In this chapter, we describe ESPAR antenna signal processing for DOA estimation with high precision and high resolution. The key idea is to estimate the array correlation matrix from the signal received through the single-port output of the ESPAR antenna as it switches over a set of antenna patterns [19, 20]. This is based on the angular diversity of the ESPAR antenna. The goal is to recreate the spatial diversity of a conventional array based on the variable reactances connected to the parasitic elements, under the assumption that the signals are periodically sent. Once the correlation matrix is available, the conventional MUSIC algorithm can be applied to this correlation matrix, and used to estimate the DOA of multiple incident signals. The proposed DOA estimation is called the "reactance-domain MUSIC algorithm."

The organization of this chapter is as follows. In the next section, we describe the basic structure and output formulation of the ESPAR antenna, and then present a signal model for ESPAR antenna signal processing. Section 20.3 explains how to obtain a correlation matrix from the single-port output of the ESPAR antenna, and presents a reactance-domain MUSIC algorithm that allows the correlation matrix to estimate the DOA of radio waves. A computer simulation and an experiment are carried out to verify the proposed algorithm. In Section 20.4, we give three examples of ESPAR antenna applications in practical systems: a satellite

20.2 ESPAR Antenna Fundamentals

attitude control, a wireless locator and homing system, and a wireless ad hoc network.

20.2 ESPAR Antenna Fundamentals

This section describes the structure of the ESPAR antenna and gives a formulation of the antenna output, which is a nonlinear function of loaded reactance. The signal model used for ESPAR antenna signal processing is also given.

20.2.1 Structure and Formulation

An $(M + 1)$-element ESPAR antenna with $M = 6$ is depicted in Figure 20.1. The 0th element is an active radiator located at the center of a circular ground plane. It is a 0.25λ-length monopole, where λ is the wavelength, and is excited from the bottom in a coaxial fashion. The remaining M elements of 0.25λ-length monopoles are passive radiators surrounding the active radiator symmetrically, with the circle's radius $R = 0.25\lambda$. Each of these M elements is terminated by a variable reactance x_m, $(m = 1, 2, \ldots, M)$. In practical applications, the value of reactance x_m may be constrained in certain ranges (e.g., from -300Ω to $+300\Omega$). The vector denoted by

$$\mathbf{x} = [x_1, x_2, \ldots, x_M]^T \tag{20.1}$$

is called the reactance vector, which is variable, and is used to form the specified beam. The superscript T is the transpose of the vector or matrix. Note that in a

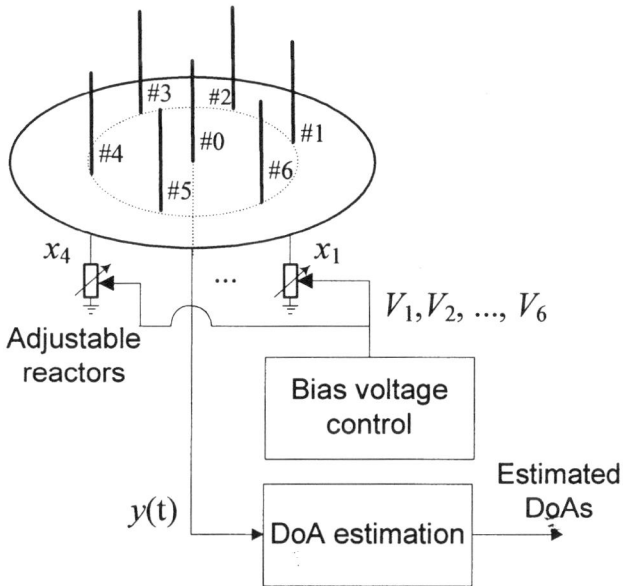

Figure 20.1 ESPAR antenna applied to DOA estimation.

fabricated ESPAR antenna, a bias voltage V_m on the value x_m adjusts it. The variable beam pattern is provided by controlling the M bias voltages, and thus the value of reactances.

We first briefly formulate the output of the ESPAR antenna as a function of these M reactances [3, 8, 21]. Denote $\mathbf{s}(t) = [s_0(t), s_1(t), \ldots, s_M(t)]^T$, where the component $s_m(t)$ is the RF signal impinging on the mth element of the ESPAR antenna. Then, the single-port RF output $y(t)$ of the antenna is given by

$$y(t) = \mathbf{i}^T \mathbf{s}(t) + n(t) \tag{20.2}$$

where $\mathbf{i} = [i_0, i_1, \ldots, i_M]^T$ is the RF current vector with the component i_m appearing on the mth element. The notation $n(t)$ is the additive white Gaussian noise. Precisely speaking, the current vector $\mathbf{i}$ should be normalized to match the physical dimension, as described in [8], but $\mathbf{i}$ is used directly for simplicity in this chapter.

Next, we give a representation of the current vector as a function of the reactance vector, as shown in (20.7). For easier understanding, we derive the representation under the assumption that the ESPAR antenna is operating in a transmit mode. The theorem of reciprocity tells us that the receive-mode radiation pattern of an antenna is equal to that of the transmit-mode radiation pattern. Therefore, the representation of (20.7) is also available in the receive mode.

Denote the RF voltage vector by $\mathbf{v} = [v_0, v_1, v_2, \ldots, v_M]^T$, where

$$v_m = -jx_m i_m \quad m = 1, 2, \ldots, M \tag{20.3}$$

is the RF voltage imposed on the reactance x_m, and

$$v_0 = V_s - Z_0 i_0 \tag{20.4}$$

is the RF voltage at the central element. The notation Z_0 is the output impedance of the transmitter, or the input impedance of the receiver in the receive mode, and is not influenced by the mutual coupling of elements. The value of Z_0 is a constant; for example, $Z_0 = 50\Omega$. In (20.4), V_s is the internal source RF voltage of the transmitter. According to the theorem of reciprocity, the receive mode radiation pattern or array factor of the ESPAR antenna is the same as that of the transmit-mode radiation pattern. Once obtained, the array factor is not affected by V_s. Arranging (20.3) and (20.4) in a vector form yields

$$\mathbf{v} = \begin{bmatrix} V_s - Z_0 i_0 \\ -jx_1 i_1 \\ -jx_2 i_2 \\ \vdots \\ -jx_1 i_1 \end{bmatrix} = V_s \mathbf{u}_0 - \mathbf{X}\mathbf{i} \tag{20.5}$$

where $\mathbf{u}_0 = [1, 0, 0, \ldots, 0]^T$, and the diagonal matrix

20.2 ESPAR Antenna Fundamentals

$$\mathbf{X} = diag[Z_0, jx_1, jx_2, \ldots, jx_M]$$

is called the reactance matrix.

On the other hand, the RF current vector $\mathbf{i}$ and RF voltage vector $\mathbf{v}$ have the relationship

$$\mathbf{v} = \mathbf{Z}\mathbf{i} \qquad (20.6)$$

where $\mathbf{Z} = [z_{kl}]_{(M+1)\times(M+1)}$ is referred to as the impedance matrix, with z_{kl} expressing the mutual impedance between the elements k and l ($0 \leq k, l \leq M$). Substituting (20.5) into (20.6) yields

$$\mathbf{i} = V_s(\mathbf{Z} + \mathbf{X})^{-1}\mathbf{u}_0 \qquad (20.7)$$

For the $(M + 1)$-element ESPAR antenna with $M = 6$, the impedance matrix $\mathbf{Z}$ is determined by only six independent components of the mutual impedances, explained as follows. By the reciprocity theorem, and similarly to conventional array antennas, it holds that

$$z_{kl} = z_{lk} \qquad (20.8)$$

Furthermore, the cyclic symmetry of the elements of the ESPAR antenna implies

$$\begin{aligned} z_{11} &= z_{22} = z_{33} = z_{44} = z_{55} = z_{66} \\ z_{01} &= z_{02} = z_{03} = z_{04} = z_{05} = z_{06} \\ z_{12} &= z_{23} = z_{34} = z_{45} = z_{56} = z_{61} \\ z_{13} &= z_{24} = z_{35} = z_{46} = z_{51} = z_{62} \\ z_{14} &= z_{25} = z_{36} \end{aligned} \qquad (20.9)$$

Equations (20.8) and (20.9) imply that the impedance matrix $\mathbf{Z} = [z_{kl}]_{(M+1)\times(M+1)}$ is determined by only the six components of the mutual impedance: z_{00}, z_{10}, z_{11}, z_{21}, z_{31}, and z_{41}. The values of the six components depend on the physical structure of the antenna (e.g., radius, space intervals and lengths of the elements), and therefore are constant after the antenna is fabricated.

The formulation of (20.7) is derived under the assumption that the ESPAR antenna is operating in the transmit mode. As stated above, the theorem of reciprocity means that the formulation of (20.7) is also valid in the receive mode. However, in the receive mode, the vector i of (20.7) does not mean a current vector, but can be seen as a weight vector, since it plays the role of weight vector like that in the conventional adaptive array antenna [2]. This is why the formulation of (20.7) is called an "equivalent vector model" [21]. An experimental verification of the equivalent vector model with actual far-field measurement is found in [22]. It reports correlation coefficients from 0.933 to 0.988, between the radiation patterns of the equivalent weight vector model and measured far-field results [22]. In addition to the equivalent weight vector model, an equivalent steering vector model is reported in [23]. This model includes the current distribution and the effects of dielectric material.

It should be emphasized that the signal vector $\mathbf{s}(t)$ in (20.2) impinging on the elements of the ESPAR antenna is not directly measurable. This differs from the conventional adaptive array antenna, where the received signal vector on the elements can be observed. For the ESPAR antenna, only the single-port output $y(t)$ is measured. More problematically, as shown in (20.7), the single-port output $y(t)$ is a highly nonlinear function of reactances x_m, $(m = 1, 2, \ldots, M)$. It includes an intractable matrix inverse, which makes it difficult to produce an analytical expression of signal processing performance. It is also worth noting that, from (20.7), each component of equivalent weight vector $\mathbf{i}$, unlike the weight coefficient vector of the conventional adaptive array, is not independent, but rather is mutually coupled with each other. The discussion above implies that it is impractical to directly apply most of the algorithms devised for the conventional adaptive array to the ESPAR antenna. It is necessary to develop single-port–oriented signal processing algorithms specifically for the ESPAR antenna.

20.2.2 Signal Model

As background to the signal processing used for the ESPAR antenna, a model of the received signal is described in this section.

Before proceeding, we describe the steering vector of the ESPAR antenna. Consider the $(M + 1)$-element ESPAR antenna, as shown in Figure 20.2. The mth element is placed at an angle

$$\varphi_m = \frac{2\pi}{M}(m-1) \quad m = 1, 2, \ldots, M$$

relative to an arbitrary axis. When an incoming wavefront is shown impinging on the antenna from a DOA of θ relative to the same reference axis, there is a spatial

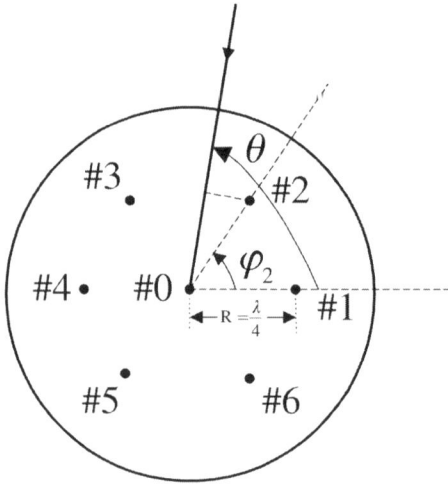

Figure 20.2 ESPAR antenna geometry.

delay of $R\cos(\theta - \varphi_m)$ between the signals received at the pair of the mth element and 0th element. With a wavelength of λ, this spatial delay is translated into an electrical angular difference defined by

$$\frac{2\pi}{\lambda} R \cos(\theta - \varphi_m)$$

Consequently, when the radius $R = \lambda/4$, the steering vector of the ESPAR antenna on the DOA of θ is defined by

$$\mathbf{a}(\theta) = \left[1, e^{j\frac{\pi}{2}\cos(\theta - \varphi_1)}, e^{j\frac{\pi}{2}\cos(\theta - \varphi_2)}, \ldots, e^{j\frac{\pi}{2}\cos(\theta - \varphi_M)} \right]^T \quad (20.10)$$

We could extend the above simple case to a more general case. Suppose there are a total number of Q signals $u_q(t)$, with DOA θ_q, $(q = 1, 2, \ldots, Q)$. Let $s_m(t)$, $(m = 0, 1, \ldots, M)$ denote the signal impinging on the mth element of the antenna, and let $\mathbf{s}(t)$ be the column vector with mth component $s_m(t)$. The signal $s_m(t)$ is a superposition of all Q signals

$$s_m(t) = \sum_{q=1}^{Q} a_m(\theta_q) u_q(t) \quad m = 0, 1, \ldots, M \quad (20.11)$$

where $a_m(\theta_q)$, $(m = 0, 1, \ldots, M)$ is the mth component of (20.10), with θ_q instead of θ. Then, the column vector $\mathbf{s}(t)$ appearing at the elements of the antenna may be expressed as

$$\mathbf{s}(t) = \sum_{q=1}^{Q} \mathbf{a}(\theta_q) u_q(t) \quad (20.12)$$

where $\mathbf{a}(\theta_q) = [a_0(\theta_q), a_1(\theta_q), \ldots, a_M(\theta_q)]^T$ is the steering vector defined in (20.10) with θ_q instead of θ. From (20.2), the output of the ESPAR antenna can be written as

$$y(t) = \mathbf{i}^T \mathbf{s}(t) + n(t) = \sum_{q=1}^{Q} \mathbf{i}^T \mathbf{a}(\theta_q) u_q(t) + n(t) \quad (20.13)$$

Note that the current vector $\mathbf{i}$, and thus antenna output $y(t)$, is a function of the reactances x_m, $(m = 1, 2, \ldots, M)$ in (20.1).

20.3 High-Precision DOA Estimation

This section explains how to obtain a correlation matrix of the single-port output of the ESPAR antenna, and presents a reactance-domain MUSIC algorithm from the correlation matrix to estimate the DOA.

20.3.1 Correlation Matrix for Single-Port Antennas

In conventional array antennas, measuring the signals on each element of the antenna creates the correlation matrix. For all kinds of single-port output antennas (e.g., the ESPAR antenna), the problem is how to create a correlation matrix with only one output port. The goal is to recreate, with a single-port output ESPAR antenna, the spatial diversity of a conventional array antenna. First, we consider a conventional array antenna. It is well-known that signals in the M elements of the conventional array antenna are monitored in the same time, and thus their correlation matrix is computed. This is the normal case, as shown in Figure 20.3(a).

Now we consider only a single-element antenna. If we want to recreate the spatial diversity available in a conventional array antenna with M elements, then we must change the position of the single-element antenna and measure the output

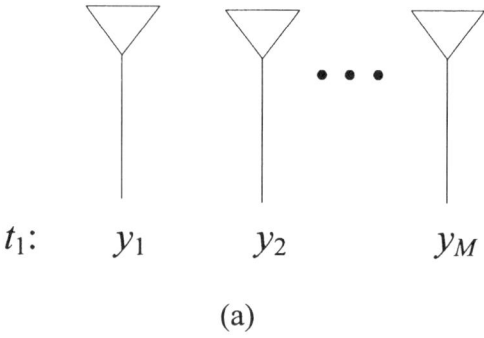

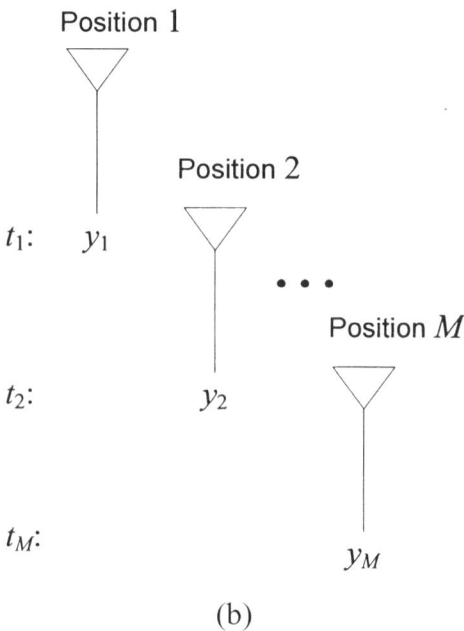

Figure 20.3 Correlation matrix: (a) conventional array antenna, and space domain with $\mathbf{y} = [y_1, y_2, \ldots, y_M]^T$, $\mathbf{R} = E[\mathbf{yy}^H]$; and (b) single-element antenna, and decomposition in time domain with $\mathbf{y} = [y_1, y_2, \ldots, y_M]^T$, $\mathbf{R} = E[\mathbf{yy}^H]$.

M times by sending the same signal each time, as shown in Figure 20.3(b). Then, we can obtain the M output values as if we were using an M-element antenna. In fact, we use decomposition in time to recreate the spatial diversity of the M-element antenna. We now have all of the outputs needed to compute the correlation matrix. We can apply the same reasoning with the single-port output ESPAR antenna, but instead will use the reactance domain to recreate the spatial diversity.

Although we can only measure the ESPAR antenna's single-port output, its M variable reactances make it possible to recreate a spatial diversity. First, we set a reactance vector and measure the output. Then, we change the loaded reactances, and thus the antenna pattern, and measure the output by sending the same signals. By repeating this process M times, we can get the M outputs needed to create a correlation matrix, as shown in Figure 20.4 [19].

20.3.2 Reactance-Domain Signal Processing

The waveforms received by the ESPAR antenna elements are combinations of the incident wavefronts and noise. Accordingly, for a given set of reactances, we can compute the equivalent current vector $\mathbf{i}$, which is a function of the reactances. Then, we can get the output of the ESPAR antenna, as expressed in (20.13).

For the ESPAR antenna, we are in the reactance domain [3, 7, 8, 19], so we use the set $\mathbf{x}^m$ of reactances to compute the current vector $\mathbf{i}^m$ and to obtain the output y_m (see Figure 20.4). By changing the values of the reactances, and thus the antenna pattern and the weight vector, M times, we can recreate the spatial diversity available in the conventional case, and finally obtain the vector $\mathbf{y}$ used to compute the correlation matrix. The M measured outputs can be expressed as

$$y_m(t_m) = \sum_{q=1}^{Q} \mathbf{i}_m^T \mathbf{a}(\theta_q) u_q(t_m) + n(t_m) \quad m = 1, 2, \ldots, M \quad (20.14)$$

under the assumption that

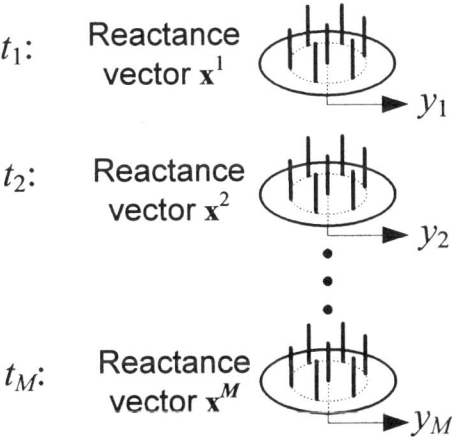

Figure 20.4 Correlation matrix in reactance domain with $\mathbf{y} = [y_1, y_2, \ldots, y_M]^T$, $\mathbf{R} = E[\mathbf{yy}^H]$.

$$u_q(t_1) = u_q(t_1) = \ldots = u_q(t_M) \tag{20.15}$$

for $q = 1, 2, \ldots, Q$. This assumption implies that the proposed reactance-domain MUSIC algorithm requires that the source waveform remain constant throughout the M samplings. By putting a periodic signal $u_q(t)$ in a pilot channel, this assumption can be satisfied in practice. This leads to the output vector

$$\begin{bmatrix} y_1 \\ y_2 \\ \vdots \\ y_M \end{bmatrix} = \begin{bmatrix} \mathbf{i}_1^T \\ \mathbf{i}_2^T \\ \vdots \\ \mathbf{i}_M^T \end{bmatrix} [\mathbf{a}(\theta_1) \quad \mathbf{a}(\theta_2) \quad \ldots \quad \mathbf{a}(\theta_M)] \begin{bmatrix} u_1 \\ u_2 \\ \vdots \\ u_Q \end{bmatrix} + \begin{bmatrix} n_1 \\ n_2 \\ \vdots \\ n_M \end{bmatrix} \tag{20.16}$$

or

$$\mathbf{y} = \mathbf{I}^T \mathbf{A} \mathbf{u} + \mathbf{n} \tag{20.17}$$

where the time indices are omitted for the sake of simplicity. Denote

$$\mathbf{a}_{rac}(\theta_q) = \mathbf{I}^T \mathbf{a}(\theta_q) \tag{20.18}$$

as the reactance-domain steering vector for the DOA estimation by the reactance-domain MUSIC algorithm.

We can now rewrite (20.16) as

$$\mathbf{y} = [\mathbf{a}_{rac}(\theta_1) \quad \mathbf{a}_{rac}(\theta_2) \quad \ldots \quad \mathbf{a}_{rac}(\theta_M)] \mathbf{u} + \mathbf{n} \tag{20.19}$$

In this shape, the equation is similar to the equation used in the conventional MUSIC algorithm [1].

Now that we know how to compute a correlation matrix with the ESPAR antenna, we can give some explanation of the reactance-domain MUSIC algorithm. The correlation matrix of the ESPAR antenna is given by

$$\mathbf{R} = E[\mathbf{y}\mathbf{y}^H] \tag{20.20}$$

where the superscript H is the transpose and conjugate operation. After making the eigendecomposition of the correlation matrix, we can estimate the number of impinging signals by choosing from the eigenvalues the highest values λ_i ($i = 1, 2, \ldots, Q$), determined by a threshold T_Q. The number of eigenvalues that are chosen determines the number of impinging signals Q. Then, the number of eigenvalues corresponding to the noise subspace is $(M - Q)$. Thus, the MUSIC spectrum [1] for the ESPAR antenna becomes

$$P_{MU}^{ESPAR}(\theta) = \frac{1}{\mathbf{a}_{rac}^H(\theta) \mathbf{E}_N \mathbf{E}_N^H \mathbf{a}_{rac}(\theta)} \tag{20.21}$$

for $0° \leq \theta \leq 360°$, where $\mathbf{E}_N$ is an $M \times (M - Q)$ matrix, whose columns are the $(M - Q)$ noise eigenvectors, $\mathbf{e}_{Q+1}, \mathbf{e}_{Q+2}, \ldots, \mathbf{e}_M$. It is now easy to estimate the DOA of the Q impinging signals by finding on the x-axis the values of the MUSIC spectrum maxima. A simulation evaluation of the reactance-domain MUSIC algorithm is carried out in Section 20.3.3, and an experimental verification is shown in Section 20.3.4. The statistical performance on estimation error variance of the reactance-domain MUSIC estimation with Cramér-Rao lower bound can be found in [19, 24].

Now we can determine the number of impinging signals and their estimated DOA, and use them to learn the power of the incident signals. It is attractive to determine the SOI, under the assumption that it is the more powerful signal. Since we know the estimated DOA, the matrix $\mathbf{A}$ in (20.17) becomes available, and may be used to compute the parameters of the incident signals contained in the $Q \times Q$ matrix $\mathbf{P}$ [1] of cross and auto powers of the incident signals, as follows.

$$\mathbf{P} = (\mathbf{A}^H \mathbf{I}^* \mathbf{I}^T \mathbf{A})^{-1} \mathbf{A}^H \mathbf{I}^* (\mathbf{R} - \lambda_{\min} \mathbf{I}_d) \mathbf{I}^T \mathbf{A} (\mathbf{A}^H \mathbf{I}^* \mathbf{I}^T \mathbf{A})^{-1} \quad (20.22)$$

where $\mathbf{I}_d$ is an $M \times M$ unit matrix, and $\lambda_{\min}$ is the smallest eigenvalue of the correlation matrix. The superscript * is the conjugate operation. This is nearly the same expression as the conventional MUSIC algorithm [1], with the only modification being that we replace the matrix $\mathbf{A}$ by the matrix $\mathbf{I}^T \mathbf{A}$. The diagonal of this matrix $\mathbf{P}$ contains the estimated power of each incident signal, so it is easy to find the SOI, under the assumption that it is the more powerful.

20.3.3 Computer Simulations

The reactance-domain MUSIC algorithm described above is evaluated in this section. In all of the following simulations, it is assumed that all of the impinging signals have the same carrier frequency, and that the data clock synchronization is perfect. Note also that the number Q of impinging signals is a priori known. The size M of the correlation matrix is set to 6, due to the particular configuration chosen for the ESPAR antenna. In most of the simulations, a total number of $6 \times 1,000$ samples, without specific declaration, are used to compute this correlation matrix in (20.20) and the MUSIC spectrum in (20.21). It is also assumed that all of the source signals are periodic, in accordance with (20.15).

In the simulations, we use a shifted reactance method to determine the value of sets of reactances for practical application. For the reactance-domain MUSIC algorithm described in (20.21), each set of reactances is chosen independently. However, the structure of the ESPAR antenna is symmetric, and because of its calibration [21, 25–27], it is easier in practice to use one set of reactances (e.g., $\mathbf{x} = [193.3, 18.7, 162.9, 262.3, 260.6, 91.2]^T$), and to cyclically shift the values of the reactances each time we want to measure an output. The shifted version of the reactances gives the remaining five antenna patterns. It is clear that each pattern is also the shifted version of the others, due to the symmetry of the ESPAR antenna.

The first simulation is an example showing concretely how to get the estimated DOA from the MUSIC spectrum in (20.21). Figure 20.5 shows the MUSIC spectrum and the values of the estimated DOA for four incident signals, each with an SNR

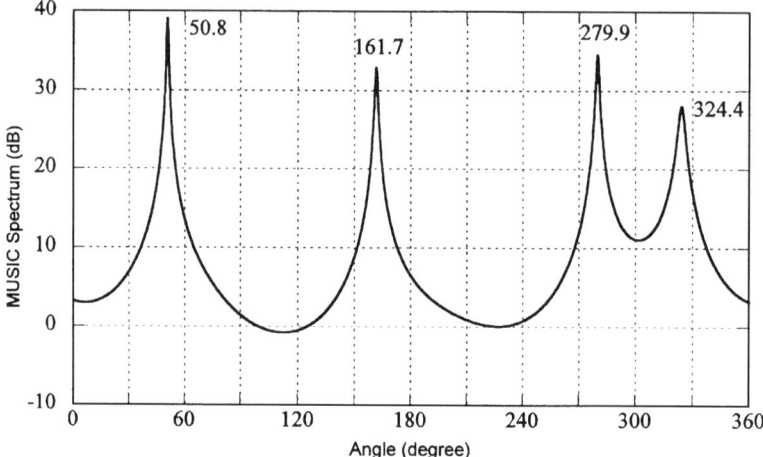

Figure 20.5 MUSIC spectrum with the estimated DOA = 50.8°, 161.7°, 279.9°, and 324.4°. The DOA of incident signals are 50°, 162°, 280°, and 325°, respectively, with SNR = 5 dB for each signal.

of 5 dB. In fact, the x-axis values of the four maximum peaks give the estimation of the DOA. The average precision of the reactance-domain MUSIC algorithm is less than 0.5°. In addition, Figure 20.6 illustrates the case of two signals impinging on the antenna with only 10° of angular distance, each with an SNR of 5 dB. The result implies that the reactance-domain MUSIC estimator has a resolution capability of 10° at an SNR of 5 dB.

In the second simulation, we want to show the influence of the SNR of incident signals on the MUSIC spectrum. Therefore, we show and discuss the results for SNRs of 20 and 0 dB for each incident signal. Figure 20.7 shows the respective MUSIC spectra for these two SNR values. It is clear by looking at the curves that

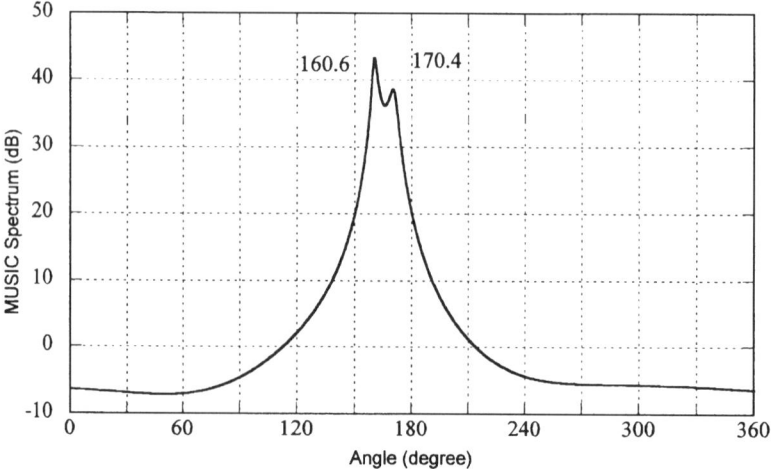

Figure 20.6 MUSIC spectrum with the estimated DOA = 160.6° and 170.4°. The incident signals are from DOA = 160° and 170°, respectively, with SNR = 5 dB for each signal.

20.3 High-Precision DOA Estimation

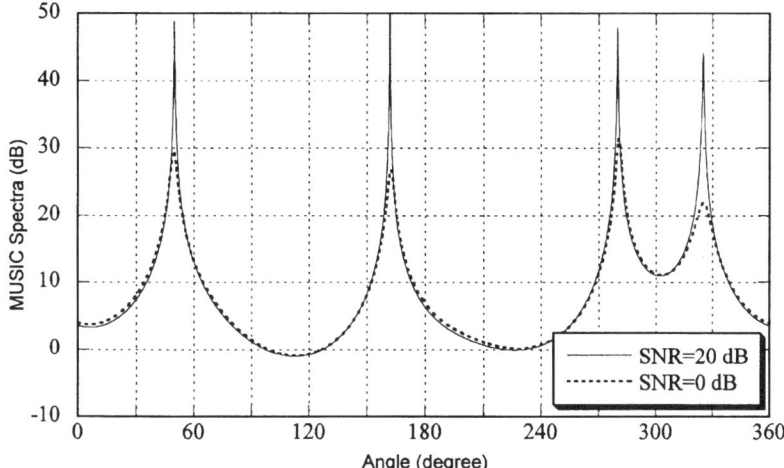

Figure 20.7 MUSIC spectra for SNR = 20 and 0 dB for each signal. The incident signals are from DOA = 50°, 162°, 280°, and 325°. The estimated DOA = 49.9°, 162.0°, 280.0°, and 325.1°, respectively, for SNR = 20 dB; and DOA = 49.5°, 162.5°, 280.7°, and 325.3°, respectively, for SNR = 0 dB.

when the SNR increases, the peaks become sharper and higher, and thus the precision becomes better. The estimated precision is 0.1° with an SNR of 20 dB.

In the third simulation, the goal is to determine the SOI, which is assumed to be more powerful than the other incident signals. This assumption is frequently made in practice. Since the MUSIC spectrum does not give enough information on the power of the incident signals, we can estimate the power of each signal by computing the matrix **P** of the cross and auto power from (20.22). The diagonal of the matrix **P** gives the estimated power of each impinging signal. Accordingly, the position of the highest value in the diagonal of **P** indicates the SOI. The matrix **P** computed for the simulation parameters in Table 20.1 is

$$\mathbf{P} = \begin{bmatrix} 4.02 & 0.01 & 0.02 & 0.01 \\ 0.02 & 3.15 & 0.02 & 0.02 \\ 0.02 & 0.02 & 8.01 & 0.01 \\ 0.01 & 0.02 & 0.01 & 5.00 \end{bmatrix}$$

The highest power is 8.01 in the third entry of the diagonal. Thus, the SOI is the third incident signal with an estimated DOA of 180.3°.

Table 20.1 DOA Estimation for Signal of Interest

	Impinging Signals			
Signal Number	1	2	3	4
SNR (dB)	3	2	6	4
DOA	10°	110°	180°	300°
Estimated DOA	10.2°	109.6°	180.3°	300.1°

In the above simulations, according to the reactance-domain MUSIC algorithm, all of the incident signals are sent periodically, in order to compute the correlation matrix **R** of (20.20). In the fourth simulation, we consider the case where only one signal is sent periodically to the ESPAR antenna. The other signals are interferences that are assumed to come from unknown sources, and so are not sent periodically. The MUSIC spectrum (see Figure 20.8) shows that even if we cannot estimate the DOA of the interferences, we can estimate the DOA of the periodically sent signal with the same precision as in the previous simulations. This property is used to estimate the (periodic) signal of interest, even in the presence of (nonperiodic) interference signals.

The aim of the fifth simulation is to investigate the influence of the range of reactances from (20.1). For the above simulations, the range of reactances was set over [−300Ω, +300Ω]. However, in practical cases, the available range of reactances is set to, for example, [−93.6Ω, −4.8Ω] because of the hardware design. In this simulation, each set of reactances used is randomly chosen over its corresponding range. Here, we want to see whether or not the available range of the reactance-domain MUSIC algorithm can maintain the performance without degradation, even with a narrower range of reactances. Figure 20.9 shows the MUSIC spectra obtained for the previous narrow range of reactances. This shows that the narrow range of reactances also gives an estimation of DOA without much degradation in estimated precision.

20.3.4 Experimental Verification

In Section 20.3.3, the reactance-domain MUSIC algorithm was evaluated in simulations. In this section, experiments and results of the algorithm are reported, which verify the algorithm in a radio anechoic room.

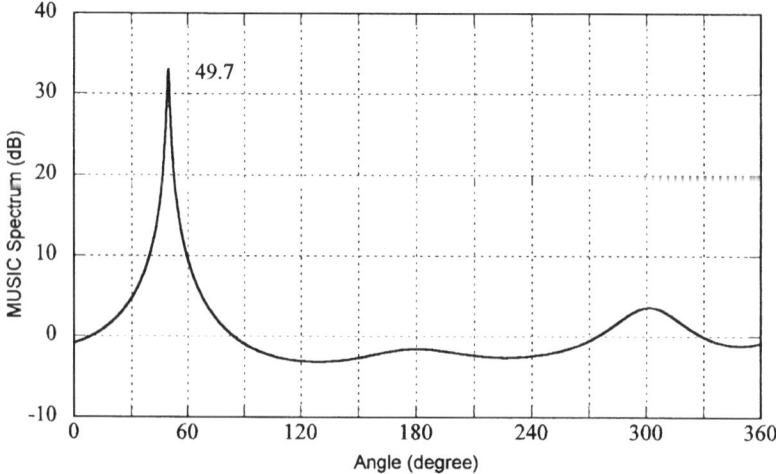

Figure 20.8 MUSIC spectrum with the estimated DOA = 49.7° for one periodic signal (DOA = 50°), and DOA = 162°, 280°, and 325° for three nonperiodic interferences. The SNR = 5 dB for the periodic signal, and the interference to noise power ratio is 5 dB for each interference.

20.3 High-Precision DOA Estimation

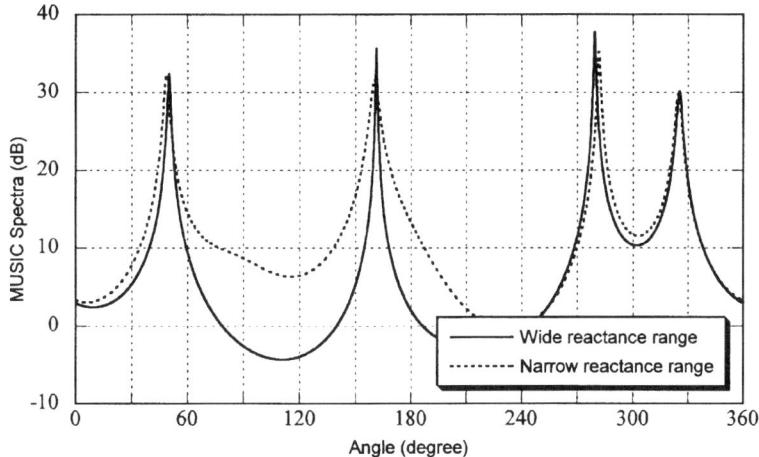

Figure 20.9 MUSIC spectra for wide and narrow range of reactances. The incident signals are from DOA = 50°, 162°, 280°, and 325°, with SNR = 5 dB for each signal. The estimated DOA = 48.8°, 160.4°, 281.9°, and 324.8°, for a narrow range of reactances over [−93.6Ω, −4.8Ω]; and the estimated DOA = 50.2°, 161.4°, 279.6°, and 325.4°, for a wide range of reactances over [−300Ω, +300Ω].

20.3.4.1 Antenna Structure and Measurement Configuration

Before proceeding, the fabricated ESPAR antenna used in the experiment is described, and the measurement configuration is given.

A 2.4-GHz ESPAR antenna [28] is illustrated in Figures 20.10 and 20.11. The 7-element monopole elements are arranged on a finite circular ground structure. The center element is the feed element. The remaining elements are parasitic, and make a 0.25λ ring around the center element, where λ = 120 mm is a free-space wavelength corresponding to the operation frequency of the 2.4-GHz band. The bottom of each parasitic element is loaded with a variable reactance. A bias voltage V_m on the element adjusts the value of the reactance. Variable beamforming is carried out by controlling the six bias voltages (control voltages) V_m, (m = 1, 2, . . . , 6), and thus the values of the reactances. A skirting is used on the ground

Figure 20.10 A fabricated 7-element ESPAR antenna.

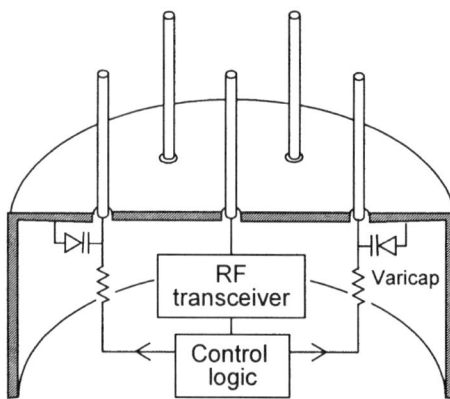

Figure 20.11 Cross-sectional view of a 7-element ESPAR antenna.

plane to reduce the main lobe elevation. The radius of the circular ground plane is 0.5λ, and the skirting height is 0.25λ, which provides maximum gain of the antenna's radiation in its horizontal direction. The reactive-loaded design gives a variable range of reactance at the bias voltage, varying from +20V to −0.5V for the frequency of 2.4 GHz [28].

An experimental setup is needed to verify the reactance-domain DOA estimation algorithm. The two transmission antennas and the ESPAR antenna are installed on the same horizontal plane. The transmission antennas are the Horn antenna and the Yagi antenna. The data, which consists of different m-sequences, are modulated in BPSK using signal generators. The modulated signals are transmitted at a frequency of 2.4 GHz from each transmission antenna.

In the receiver, only the center element of the ESPAR antenna is connected to a lower noise amplifier module. The 2.4-GHz RF signal is directly converted to a 70-MHz intermediate frequency (IF) by a downconverter and local oscillators. The IF signal is routed into a bandpass filter, and demodulated into orthogonal baseband signals in the in-phase and quadrature-phase channels. The baseband signals are then digitized individually with 12-bit resolution by analog-to-digital converters at the rate of 500 kHz. A PC-based ESPAR controller is used to set six bias voltages and to feed the voltages to the ESPAR antenna. The bias voltages change the values of reactances and thus the antenna pattern. A sequence of N samples of antenna output was obtained by setting a set of bias voltages, for example, $\mathbf{V} = [-0.5, 20, 20, 20, 20, 20]$V. The other sequences were obtained by shifting the set of bias voltages. The accumulated sequences were used to calculate the correlation matrix $\mathbf{R}$ of (20.20) off-line.

20.3.4.2 Calibration and DOA Estimation Experiments

To calculate the MUSIC spectrum of (20.21), we want to know the value of the reactance-domain steering vector $\mathbf{a}_{rec}(\theta)$ for $0° \leq \theta < 360°$. From (20.18) and (20.7), it is seen that $\mathbf{a}_{rec}(\theta)$ is a function of impedance matrix $\mathbf{Z}$ and related to reactances (i.e., the bias voltages). This implies that the accuracy of the estimated

MUSIC spectrum strongly depends on the estimation of **Z** for a practical ESPAR antenna. As it is known, it is difficult to estimate the value of **Z**.

We now consider a calibration scheme to directly obtain a practical estimated value of $\mathbf{a}_{rec}(\theta)$ [20, 29]. Assume that noise is free, and only a (constant) signal $u(t)$ [e.g., $\forall t$, $u(t) = 1$, and $Q = 1$] impinges on the antenna from the direction of θ. For a given bias voltage vector $\mathbf{V}^m = \left[V_1^m, V_2^m, \ldots, V_M^m \right]$ (i.e., a vector $\mathbf{x}^m$ of reactances), we observe from (20.13) that the antenna output y_m, at an angle of θ, is

$$y_m = \mathbf{i}_m^T \mathbf{a}(\theta)$$

Note that the value of y_m is obtained by measuring the phase and power of the antenna output. It follows that

$$\mathbf{a}_{rec}(\theta) = \left[\mathbf{i}_1^T, \mathbf{i}_2^T, \ldots, \mathbf{i}_M^T \right] \mathbf{a}(\theta) = [y_1, y_2, \ldots, y_M]^T$$

for a given θ. This means that the vector $\mathbf{a}_{rec}(\theta)$ for $0° \leq \theta < 360°$ can be estimated a priori by measurement. For the DOA estimation, we can calculate the correlation matrix from samples of antenna output, and obtain the matrix $\mathbf{E}_N$ by eigendecomposition. Then, from the matrix $\mathbf{E}_N$ and the measured vector $\mathbf{a}_{rec}(\theta)$, $0° \leq \theta < 360°$, we can calculate the MUSIC spectrum of (20.21) and estimate the DOA. The MUSIC spectra calculated from the correlation matrices, which are obtained by experiment [30], are shown in Figures 20.12 and 20.13. Figure 20.12 illustrates that the estimated DOAs are −133° and +1°, when two incident signals are from −135° and 0°, respectively. In Figure 20.13, the MUSIC spectrum shows two peaks at −46° and +2° when the two impinging signals are from −45° and 0°, respectively. The experimental results verify the reactance-domain MUSIC algorithm, and show that the two impinging signals can be estimated at a precision of 2°. Note that the

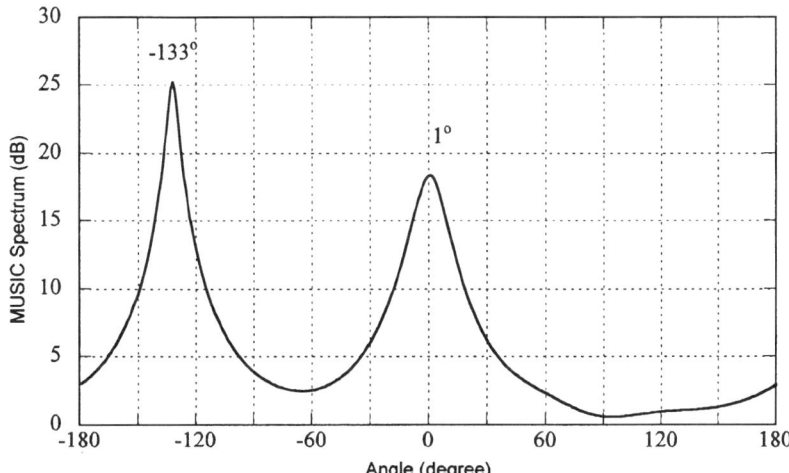

Figure 20.12 Experimental MUSIC spectrum with estimated DOA = −133° and +1°. The incident signals are from −135° and 0°.

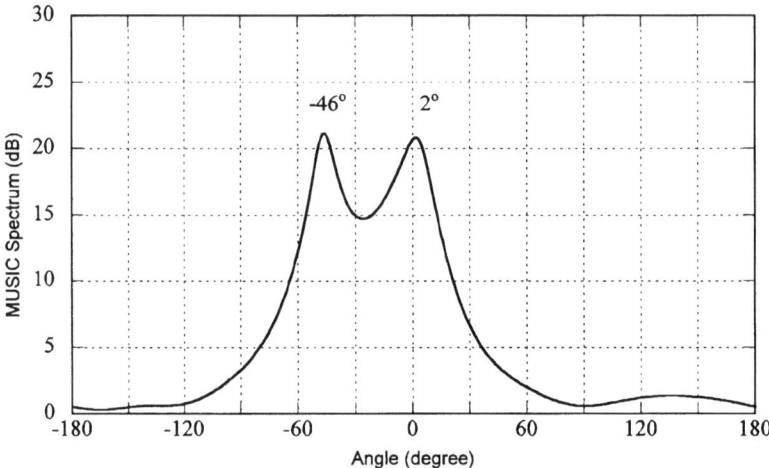

Figure 20.13 Experimental MUSIC spectrum with estimated DOA = −46° and +2°. The incident signals are from −45° and 0°.

waves for the DOA estimation are noncoherent here. For the DOA estimation of coherent waves by the ESPAR antenna, see [30–34].

20.4 Practical System Applications

Since the ESPAR antenna offers smart functionalities without using any beamforming networks or multiple RF circuits, it is possible to apply at low cost to various radio systems. Here, we discuss practical examples of ESPAR antenna applications.

20.4.1 Satellite Attitude and Direction Control

The attitude (i.e., roll, yaw, and pitch) of a satellite can be controlled by exploiting the direction-finding function of the ESPAR antenna. Figure 20.14 shows a prototype engineering model of the earthquake prediction satellite developed by the Tokyo Metropolitan College of Aeronautical Engineering. The satellite is loaded

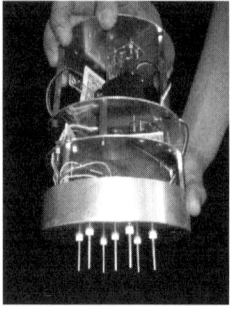

Figure 20.14 Prototype engineering model of an earthquake prediction satellite.

with an ESPAR antenna that senses radio waves transmitted from an Earth station. The Earth station sends a constant-amplitude wave to the satellite, and the satellite receives the wave while rotating the beam direction of the ESPAR antenna. Based on the time series of received amplitude, the satellite computes its direction. This means the satellite detects its relative angle to the station. Thus, the satellite knows its attitude deviation, and feeds this information to the attitude control mechanism for adjustment. The actual flight model will carry an FM transmitter for ionospheric sounding that is believed to sense a prognostic of earthquakes.

20.4.2 Wireless Locator and Homing System

Another application of the ESPAR antenna's direction-finding function is a wireless homing system [35]. Figure 20.15 shows a wireless locator applied to the homing vehicle developed by the Kobe University International Rescue System Laboratory [36] which tracks the signal emitted from a cellular telephone or wireless PDA. In a disaster area, the rescue device calls the cellular telephone of a victim. The cell phone emits a constant amplitude wave, and the locator receives the wave while rotating the beam direction of the ESPAR antenna mounted on the locator. Based on the time series of received amplitude, the locator computes the direction of the victim's cell phone. This means that the locator detects its relative angle to the cell phone. Thus, the locator feeds this information to the caterpillar control mechanism for optimal movement. Even if the victim cannot answer the call, the telephone emits an acknowledgment of receipt. The wireless locator picks it up and detects the direction of arrival. The vehicle moves a certain distance toward the detected direction. It again calls the phone, picks up the acknowledgment signal, adjusts its

Figure 20.15 Wireless locator and homing vehicle.

direction, and moves on. This cycle is iterated until the vehicle reaches the signal source (i.e., the victim).

20.4.3 Wireless Ad Hoc Network

A wireless ad hoc network offers the advantage of direct communication between nodes without any infrastructure, such as a base station. In a practical application of the standard communication environment, it is necessary that the node itself independently functions as a router in order to act separately and harmoniously between other nodes, since a node can move easily during a dynamic change.

However, if direct communication is not possible between nodes due to microwave signal attenuation, then a multihop communication method is used toward a terminal node, by passing through another node as a relay station. Finding the best relay station is a key routing task in a wireless ad hoc network. The traditional approach, which is to search for a route by flooding packets using a nondirectional antenna, results in a lower level of performance by heavily loading the network. Therefore, the proposed approach suggests improved network performance by checking the differential interference in the same channel. That is, the directional beam is regularly stirred at a set angle unit, and the received signal intensity of each node is measured to create Angle-SINR-Table (AST) information. The system searches for the relay station that offers the most suitable microwave signal conditions for communication with the directional beam, based on the above information. Figure 20.16 shows the developed test system, which consists of an ESPAR antenna, a wireless module, and a notebook PC [37].

20.5 Conclusions

High-precision direction finding for mobile user terminals is investigated by employing a single-port antenna, the ESPAR antenna. The antenna has only a single radiator connected to a receiver or transmitter, and the surrounding parasitic

Figure 20.16 Prototype networking test consisting of ESPAR-equipped wireless terminals.

radiators are loaded with variable reactors. A set of beam patterns is obtained by varying the loaded reactance, and this set is used to estimate the DOA of the impinging signals. The main feature of the antenna is that it only requires one receiver chain. This significantly reduces the cost, size, complexity, and power consumption of an antenna. Another feature is that its signal combination is carried out in space by electromagnetic mutual coupling among array elements, rather than in circuits. This makes possible a compact implementation of the antenna.

Here, we propose ESPAR antenna signal processing for DOA estimation. We estimate the received signal correlation matrix from the antenna as it switches over a set of beam patterns by setting a set of reactances, and then we apply the MUSIC algorithm to the correlation matrix to estimate the DOA of impinging signals. Computer simulation and an experiment have been successfully carried out to verify the effectiveness of the reactance-domain MUSIC algorithm for high precision DOA estimation. Three practical applications of the ESPAR antenna are described: satellite attitude control, a wireless locator and homing system, and a wireless ad hoc network. This work shows that estimating DOA by employing the ESPAR antenna is a promising alternative to the conventional array antenna, especially for mobile user terminals.

In this chapter, we introduce ESPAR antenna signal processing for DOA estimation and its applications. Further issues involving the ESPAR antenna include adaptive beamforming and ESPRIT-based techniques. For these advanced applications of the ESPAR antenna, interested readers can refer to [3, 5, 8, 28, 38–40].

Acknowledgments

The authors would like to thank Professor Takamori of Kobe University International Rescue System Laboratory, Dr. Kawakami of Antenna Giken, and Professor Wakabayashi of Tokyo Metropolitan College of Aeronautical Engineering, for providing pictures of ESPAR antennas and applied devices.

References

[1] Schmidt, R. O., "Multiple Emitter Location and Signal Parameter Estimation," *IEEE Trans. on Antennas and Propagation*, Vol. AP-34, No. 3, March 1986, pp. 276–280.

[2] Nicolau, E., and D. Zaharia, *Adaptive Arrays*, New York: Elsevier, 1989.

[3] Cheng, J., Y. Kamiya, and T. Ohira, "Adaptive Beamforming of ESPAR Antenna Based on Steepest Gradient Algorithm," *IEICE Trans. on Communications*, Vol. E84-B, No. 7, July 2001, pp. 1790–1800.

[4] Ohira, T., and K. Gyoda, "Handheld Microwave Direction-of-Arrival Finder Based on Varactor-Tuned Analog Aerial Beamforming," *Proc. 2001 Asia-Pacific Microwave Conf.*, Taipei, December 2001, pp. 585–588.

[5] Ohira, T., "Analog Smart Antennas: An Overview (Invited)," *IEEE Intl. Symp. Personal Indoor Mobile Radio Commun. (PIMRC2002)*, 4, Lisbon, Portugal, September 2002, pp. 1502–1506.

[6] Ohira, T., "Analog Renaissance in Adaptive Antennas," *Progress in Electromagnetics Research Symp. (PIERS2003)*, Singapore, January 2003, p. 270.

[7] Ohira, T., "Reactance Domain Signal Processing in Parasite-Array Antennas (Invited)," *Asia-Pacific Microwave Conf. (APMC2003)*, WD6-1, 1, Seoul, Korea, November 2003, pp. 290–293.

[8] Ohira, T., and J. Cheng, "Analog Smart Antennas," *Adaptive Antenna Arrays: Trends and Applications*, Berlin: Springer Verlag, June 2004, pp. 184–204.

[9] Ohira, T., and K. Iigusa, "Electronically Steerable Parasitic Array Radiator Antenna," *Electronics and Communications in Japan (Part 2: Electronics), ECJB*, Wiley Periodicals, Vol. 87, No. 10, 2004, pp. 25–45.

[10] Harrington, R. F., "Reactively Controlled Directive Arrays," *IEEE Trans. on Antennas and Propagation*, Vol. AP-26, No. 3, May 1978, pp. 390–395.

[11] Dinger, R. J., "Reactively Steered Adaptive Array Using Microstrip Patch Elements at 4 GHz," *IEEE Trans. on Antennas and Propagation*, Vol. AP-32, No. 8, August 1984, pp. 848–856.

[12] Dinger, R. J., "A Planar Version of a 4.0 GHz Reactively Steered Adaptive Array," *IEEE Trans. on Antennas and Propagation*, Vol. AP-34, No. 3, March 1986, pp. 427–431.

[13] Preston, S. T., et al., "Base-Station Tracking in Mobile Communications Using a Switch Parasitic Antenna Array," *IEEE Trans. on Antennas and Propagation*, Vol. AP-46, No. 6, June 1998, pp. 841–844.

[14] Preston, S. T., D. V. Thiel, and J. W. Lu, "A Multibeam Antenna Using Switched Parasitic and Switched Active Elements for Space-Division Multiple Access Applications," *IEICE Trans. Electron.*, Vol. E82-C, No. 7, July 1999, pp. 1202–1210.

[15] Scott, N. L., M. O. Leonard-Taylor, and R. G. Vaughan, "Diversity Gain from a Single-Port Adaptive Antenna Using Switched Parasitic Elements Illustrated with a Wire and Monopole Prototype," *IEEE Trans. on Antennas and Propagation*, Vol. AP-47, No. 6, June 1999, pp. 1066–1070.

[16] Svantesson, T., and M. Wennstrom, "High-Resolution Direction Finding Using a Switched Parasitic Antenna," *Proc. of 11th IEEE Signal Processing Workshop on Statistical Signal Processing*, Singapore, August 2001, pp. 508–511.

[17] Thiel, D. V., and S. Smith, *Switched Parasitic Antennas for Cellular Communications*, Norwood, MA: Artech House, 2001.

[18] Vaughan, R., "Switched Parasitic Elements for Antenna Diversity," *IEEE Trans. on Antennas and Propagation*, Vol. AP-47, No. 2, February 1999, pp. 399–405.

[19] Plapous, C., et al., "Reactance Domain MUSIC Algorithm for Electronically Steerable Parasitic Array Radiator," *IEEE Trans. on Antennas Propagation*, Vol. AP-52, No. 12, December 2004, pp. 3257–3264.

[20] Taillefer, E., et al., "Reactance-Domain MUSIC for ESPAR Antennas (Experiment)," *IEEE Wireless Communications and Networking Conference*, Vol. 1, New Orleans, LA, March 2003, pp. 98–102.

[21] Ohira, T., et al., *Equivalent Weight Vector and Array Factor Formulation for ESPAR Antennas*, IEICE Technical Report, Vol. AP2000-44 (SAT2000-41, NW2000-41), July 2000 (in Japanese).

[22] Han, Q., K. Inagaki, and T. Ohira, "Array Antenna Characterization Technique Based on Evanescent Reactive-Nearfield Probing in an Ultrasmall Anechoic Box," *IEEE MTT-S International Microwave Symp., IMS2003*, Vol. 3/3, TH4C-3, June 2003, pp. 1841–1844.

[23] Iigusa, K., T. Ohira, and B. Komiyama, "Equivalent Steering Vector for ESPAR Antennas and Its Derivation by Using Structure Parameters of Vector Effective Length," *Proc. of the 2004 Int. Symp. on Antennas and Propagation*, Sendai, Japan, August 17–21, 2004, pp. 1297–1300.

[24] Taillefer, E., et al., "Fisher-Cramer-Rao Lower Bound and MUSIC Standard Deviation Formulation for ESPAR Antennas," *International Symp. on Antennas and Propagation*, ISAP2004, 4D2-2, Sendai, Japan, August 2004, pp. 1301–1304.

[25] Chiba, K., H. Yamada, and Y. Yamaguchi, "Experimental Verification of Antenna Array Calibration Using Known Sources," *IEICE Technical Report*, Vol. AP2002-41, July 2002, pp. 7–12 (in Japanese).

[26] Iigusa, K., et al., "ESPAR Antenna Parameters Fitting Based on Measured Data," *IEICE Technical Report*, Vol. AP2001-104 (RCS2001-143), October 2001, pp. 93–100 (in Japanese).

[27] Iigusa, K., and T. Ohira, "Adjustable Impedance Matching for ESPAR Antenna," *IEICE Technical Report*, Vol. SCE2002-14 (MW2002-14), April 2002, pp. 71–76 (in Japanese).

[28] Cheng, J., et al., "Electronically Steerable Parasitic Array Radiator Antenna for Omni- and Sector-Pattern Forming Applications to Wireless Ad Hoc Networks," *IEE Proc. on Microwaves, Antennas, and Propagation*, Vol. 150, No. 4, August 2003.

[29] Taillefer, E., A. Hirata, and T. Ohira, "Direction-of-Arrival Estimation Using Radiation Power Pattern with an ESPAR Antenna," *IEEE Trans. on Antennas and Propagation*, Vol. AP-53, No. 2, February 2005, pp. 678–668.

[30] Hirata, A., et al., "Reactance-Domain SSP MUSIC for an ESPAR Antenna to Estimate the DOAs of Coherent Waves," *Proc. on the Wireless Personal Multimedia Conference (WPMC'03)*, Vol. 3, Yokosuka, Japan, October 2003, pp. 242–246.

[31] Hirata, A., H. Yamada, and T. Ohira, "Reactance-Domain MUSIC DOA Estimation Using Calibrated Equivalent Weight Matrix of ESPAR Antenna," *2003 IEEE AP-S International Symp.*, Columbus, OH, June 2003, pp. 252–255.

[32] Hirata, A., et al., "Correlation Suppression Performance for Coherent Signals in RD-SSP-MUSIC with a 7-Element ESPAR Antenna," *European Conf. Wireless Tech., ECWT2004*, Amsterdam, October 2004, pp. 149–152.

[33] Ogawa, Y., et al., "Experiment of DOA Estimation with RD-CUBA-MUSIC Using 7-Element ESPAR Antennas," *International Symp. on Antennas and Propagation, ISAP2004*, 4D2-4, Sendai, Japan, August 2004, pp. 1309–1312.

[34] Ikeda, K., et al., "DOA Estimation by Using MUSIC Algorithm with a 9-Element Rectangular ESPAR Antenna," *International Symp. on Antennas and Propagation, ISAP2004*, 1A4-5, Sendai, Japan, August 2004, pp. 45–48.

[35] Shimizu, T., S. Tawara, and T. Ohira, "A Proposal of Portable Wireless Locator and Foxhunting System: Handheld DOA Finder and Public Mobile Communication Infrastructure," *International Symp. Wireless Personal Multimedia Communications (WPMC2003)*, Yokosuka, Japan, October 2003, pp. 372–376.

[36] Takamori, T., et al., "Development of UMRS (Utility Mobile Robot for Search) and Searching System for Sufferers with Cell Phone," *SICE/RSJ Int'l Symp. Systems and Human Science (SSR2003)*, Osaka, Japan, November 2003, pp. 47–52.

[37] Watanabe, M., and S. Tanaka, "Directional Beam MAC for Node Direction Measurement in Wireless Ad Hoc Network," *EuMW2003*, No. ECWT9-14, Munich, Germany, October 2003, pp. 155–158.

[38] Cheng, J., et al., "Adaptive ESPAR Antenna Experiments Using MCCC and MMC Criteria," *IEICE General Conference 2002*, Vol. B-1-117, Nagoya, Japan, March 2002, p. 133.

[39] Cheng, J., et al., "Blind Aerial Beamforming Based on a Higher-Order Maximum Moment Criterion (Part II: Experiments)," *Asia-Pacific Microwave Conference (APMC2002)*, Vol. 1, Kyoto, Japan, November 19–22, 2002, pp. 185–188.

[40] Taillefer, E., E. Chu, and T. Ohira, "ESPRIT Algorithm for a Seven-Element Regular-Hexagonal Shaped ESPAR," *European Conf. Wireless Tech. (ECWT2004)*, Amsterdam, October 2004.

CHAPTER 21
A New DOA Estimation Method for Pulsed Doppler Radar

Jian Wang, R. Lynn Kirlin, and Xiaoli Lu

In this chapter, we propose a new algorithm to estimate the DOA of far-field targets in radar echoes. The improvement provided by this algorithm is based on the use of a state-space model and a double stage two-dimensional prefilter, which is finally combined with the high-resolution (MUSIC) method for DOA estimation. A simplified theoretical resolution threshold is derived, and both the theory and simulations verify the effectiveness of our algorithm. Results from an experiment based on real HF radar sea clutter also confirm the algorithm.

21.1 Introduction

In sensor array signal processing, especially in radar applications, there is a growing interest in the detection and estimation of the DOA of multiple targets in strong background noise/clutter or interference. High resolution methods, such as MUSIC [1], Minimum Norm (Min-Norm) [2], and ESPRIT [3, 4], are attractive candidates, which give resolution far beyond the classical Fourier limit when the signal model is sufficiently accurate, and these methods require less computation compared to the optimal maximum likelihood method [5]. Additionally, only a second-order statistical characterization is explored, since the estimation procedure utilizes the eigenvectors associated with either the smallest eigenvalues (noise subspace) or the largest eigenvalues (signal subspace) of the received signal's covariance matrix. These so-called subspace-based methods typically provide asymptotically unbiased estimation, and have proven effective in a great variety of applications. In this chapter, we focus on MUSIC, which we suggest is the most promising algorithm within this class [6].

The first detailed analysis of resolvability of MUSIC and Min-Norm is reported by Kaveh and Barabell [7]. A closed form of resolution threshold, derived under the assumption of white Gaussian background noise, is expressed as a function of the number of snapshots, the number of sensors, and the angular separation. Stoica and Nehorai present a more general performance analysis of MUSIC in [8], where the explicit expression of DOA estimation error covariance and CRB are derived and compared. Other performance analyses due to model error or interference can be found in [9, 10]. In practice, there are some limiting scenarios, such as low

SNR, highly correlated signals, limited number of snapshots, or model error, under which the conventional MUSIC will seriously degrade or even fail [8–12]. A large number of efforts have been proposed to improve or enhance the performance of MUSIC in different scenarios. A large part of them can be classified under spatial smoothing [13–18]. Among them, Kirlin and Du [16–18] incorporate the cross-correlation information into the estimation of the covariance matrix and spatial smoothing, resulting in better angular resolution.

Another notable class of methods couples MUSIC with spatial prefiltering, leading to the so-called beamspace MUSIC (BS-MUSIC) [19–22], whose most obvious advantage is lower computational complexity, since a small dimensionality usually will result. Xu and Buckley [21] show a lower estimate bias obtained by BS-MUSIC, compared to element space MUSIC, and Lee and Wengrovitz [22] show that a lower resolution threshold is achieved by BS-MUSIC.

In this chapter, we consider the special characteristics of radar application, and propose a new algorithm to estimate the DOA of superimposed cisoidal radar echoes from far-field targets. Due to the dominating clutter in radar echoes, long dwelling time and Doppler processing [23, 24] are routinely applied on the demodulated radar echo, in order to enhance the signal-to-noise-and-clutter ratio (SNCR), since targets possess a neat Doppler structure, while noise and clutter do not. The only successes for subspace methods in HF surveillance radar have been those applied to Doppler processed data (frequency domain) in strong clutter [25, 26]. The performances of these methods are mainly inhibited by the limited number of available snapshots (i.e., usually only one snapshot after Doppler processing). Another direct disadvantage is computational expense.

Notice that the Doppler processing is just a very narrow bandpass temporal filtering, as generally provided by the FFT. In this chapter, we incorporate a state space realization to more accurately model the received array signal, and we propose a two-stage, two-dimensional (spatial and temporal) prefilter, whose output is subsequently combined with the high resolution (MUSIC) method for DOA estimation. More specifically, we perform a linear transform (mapping) on a particular group of signals within a chosen space-frequency (sector-Doppler) area, and thereafter eliminate or significantly attenuate other existing groups of signals from both spatial and temporal domains. The main advantage of this algorithm is that it reduces the interference from groups of signals, noise, and clutter having spatial or temporal spectra outside the space-frequency region of interest. Another advantage of the prefiltering is that it tends to prewhiten the noise and clutter when their spectra are not spatially or temporally white. Our proposal is based on the MUSIC-like algorithm, but the extension to other high resolution methods, such as Minimum Norm or ESPRIT, is straightforward. After briefly reviewing the MUSIC and BS-MUSIC algorithms and establishing our notation expression, we specify the characteristics of a Doppler radar echo, present its state space model, and produce our proposed two-dimensional spatial-temporal prefiltering-based MUSIC (2DP-MUSIC) algorithm. We then give a simplified theoretical derivation for the resolution threshold of our new algorithm that is consistent with a more general conclusion of [22]. Finally, simulation and experimental results are presented to support our algorithm.

21.2 Review of MUSIC and BS-MUSIC

In this section, we give a brief description of two fundamental array processing algorithms: sensor domain traditional MUSIC, or so-called spectral MUSIC, and BS-MUSIC. The convention adopted in this chapter is that bold lowercase letters denote vectors, and bold uppercase letters denote matrices. The superscripts T and H denote transpose and conjugate transpose, respectively. Assuming that the receiving array snapshot at time instant t ($t = 1, 2, \ldots, T$) is $\mathbf{y}(t) \in \mathbb{C}^m$, and a ULA is deployed, then the traditional narrowband array model for far-field planar incident waves is:

$$\mathbf{y}(t) = \mathbf{A}\mathbf{s}(t) + \mathbf{n}(t) \quad (21.1)$$

where $\mathbf{n}(t) \in \mathbb{C}^m$ is the additive spatially white noise vector, and $\mathbf{s}(t) \in \mathbb{C}^k$ contains temporal signals reflected by the k independent scattered sources, and is uncorrelated with $\mathbf{n}(t)$. Matrix $\mathbf{A} = [\mathbf{a}(\theta_1), \ldots, \mathbf{a}(\theta_k)]$ contains the direction vectors of the sources, and the ith column is:

$$\mathbf{a}(\theta_i) = \left[1, e^{-j2\pi \frac{d\sin(\theta_i)}{\lambda}}, \ldots, e^{-j2\pi \frac{d\sin(\theta_i)}{\lambda}(m-1)}\right]^T \quad (21.2)$$

where m is the number of sensors, θ_i is the DOA of the ith source, and λ and d are the radar carrier wavelength and sensor spacing, respectively. The covariance matrix $\mathbf{R}_{yy} \in \mathbb{C}^{m \times m}$ of the sensor array output vector $\mathbf{y}(t)$ is

$$\mathbf{R}_{yy} = E\{\mathbf{y}(t)\mathbf{y}^H(t)\} = \mathbf{A}\mathbf{R}_{ss}\mathbf{A}^H + \sigma^2 \mathbf{I}_{m \times m} \quad (21.3)$$

where σ^2 is the common variance of noise $\mathbf{n}(t)$, $\mathbf{R}_{ss}$ is the signal covariance matrix with full rank, and $E\{\cdot\}$ is the expectation operator. $\mathbf{I}_{m \times m}$ is the identity matrix with dimension m. The subspace decomposition of $\mathbf{R}_{yy}$ is:

$$\mathbf{R}_{yy} = \sum_{i=1}^{m} \lambda_i \mathbf{h}_i \mathbf{h}_i^H = \mathbf{U}_s \mathbf{\Sigma}_s \mathbf{U}_s^H + \mathbf{U}_n \mathbf{\Sigma}_n \mathbf{U}_n^H \quad (21.4)$$

where $\lambda_1 \geq \ldots \geq \lambda_k > \lambda_{k+1} = \ldots = \lambda_m = \sigma^2$ are the eigenvalues of $\mathbf{R}_{yy}$, and $\mathbf{h}_i$ are the corresponding orthonormal eigenvectors. $\mathbf{U}_s = [\mathbf{h}_1 \mathbf{h}_2 \ldots \mathbf{h}_k]$ with rank k spans the signal subspace, and $\mathbf{U}_n = [\mathbf{h}_{k+1} \mathbf{h}_{k+2} \ldots \mathbf{h}_m]$ spans the noise subspace. $\mathbf{\Sigma}_s$ and $\mathbf{\Sigma}_n$ are diagonal matrices, with the corresponding eigenvalues as diagonal. It is shown [1, 12] that the subspace spanned by the columns of $\mathbf{U}_n$ is orthogonal to $\{\mathbf{a}(\theta_1), \mathbf{a}(\theta_2), \ldots, \mathbf{a}(\theta_k)\}$. Hence, we obtain the MUSIC spatial spectrum (pseudospectrum) as

$$P_{MU}(\theta) = \frac{1}{\mathbf{v}^H(\theta)\mathbf{U}_n \mathbf{U}_n^H \mathbf{v}(\theta)} = \frac{1}{D(\theta)} \quad (21.5)$$

where $D(\theta)$ is defined as the null spectrum, and the steering vector $\mathbf{v}(\theta)$ is given by

$$\mathbf{v}(\theta) = \frac{1}{\sqrt{m}} \left[1, e^{-j2\pi \frac{d \sin(\theta)}{\lambda}}, \ldots, e^{-j2\pi \frac{d \sin(\theta)}{\lambda}(m-1)} \right]^T \quad (21.6)$$

When $\mathbf{v}(\theta)$ coincides with a source at DOA θ_i, it is orthogonal to the noise subspace $\mathbf{U}_n$, and P_{MU} theoretically takes on the value of infinity. MUSIC estimates the source arrival directions by determining the θs associated with those k highest peaks in the spectrum P_{MU}.

BS-MUSIC takes advantage of some a priori knowledge of the target direction, thus first passing the sensor snapshots through a beamforming preprocessor prior to the application of MUSIC algorithm. By so doing, the target information from a desired region (sector) is kept intact, while interference and unwanted clutter from other directions are suppressed, resulting in an enhanced SNCR. The preprocessor performs the following data transformation:

$$\mathbf{x}(t) = \mathbf{B}^H \mathbf{y}(t) \quad (21.7)$$

where $\mathbf{B}$ is an $m \times m'$ beamforming matrix, which maps the vector $\mathbf{y}(t)$ from an m-dimension sensor space into an m'-dimension beamspace ($m > m'$).

In this chapter, we adopt discrete prolate spheroidal sequence (DPSS) [27, 28] beamformer, with the following property:

$$\mathbf{B}^H \mathbf{B} = \mathbf{I}_{m' \times m'} \quad (21.8)$$

where columns of $\mathbf{B}$ are orthonormal set. The covariance matrix $\mathbf{R}_{xx} \in \mathbb{C}^{m' \times m'}$ of the $\mathbf{x}(t)$ is:

$$\mathbf{R}_{xx} = E\{\mathbf{x}(t)\mathbf{x}^H(t)\} = \mathbf{B}^H \mathbf{A} \mathbf{R}_{ss} \mathbf{A}^H \mathbf{B} + \sigma^2 \mathbf{I}_{m' \times m'} \quad (21.9)$$

which has a subspace decomposition similar to that in (21.3):

$$\mathbf{R}_{xx} = \sum_{i=1}^{m'} \lambda_{xi} \mathbf{h}_{xi} \mathbf{h}_{xi}^H = \mathbf{U}_{xs} \mathbf{\Sigma}_{xs} \mathbf{U}_{xs}^H + \mathbf{U}_{xn} \mathbf{\Sigma}_{xn} \mathbf{U}_{xn}^H \quad (21.10)$$

where $\lambda_{x1} \geq \ldots \geq \lambda_{xk} > \lambda_{x(k+1)} = \ldots = \lambda_{xm'} = \sigma^2$ are the eigenvalues of $\mathbf{R}_{xx}$, and $\mathbf{h}_{xi}$ are the corresponding orthonormal eigenvectors. Again $\mathbf{U}_{xs} = [\mathbf{h}_{x1} \mathbf{h}_{x2} \ldots \mathbf{h}_{xk}]$ with rank k spans the signal subspace, and $\mathbf{U}_{xn} = [\mathbf{h}_{x(k+1)} \mathbf{h}_{x(k+2)} \ldots \mathbf{h}_{xm'}]$ spans the noise subspace. It is obvious that the signal subspace $\mathbf{U}_{xs}$ spans the column space of matrix $\mathbf{B}^H \mathbf{A}$ and the noise subspace $\mathbf{U}_{xn}$ is orthogonal to the $\mathbf{B}^H \mathbf{A}$. Based on the foregoing argument, it is natural to obtain the BS-MUSIC spectrum as:

$$P_{BMU}(\theta) = \frac{1}{\mathbf{v}^H(\theta)\mathbf{B}\mathbf{U}_{xn}\mathbf{U}_{xn}^H\mathbf{B}^H\mathbf{v}(\theta)} = \frac{1}{D_B(\theta)} \qquad (21.11)$$

21.3 Model for HF Radar

The radar signal processing problem is in essence narrowband, which means that the bandwidth of the impinging signal is much smaller than the reciprocal of the propagation time of the wavefront across the array. In HF radar, the narrowband echo is mixed down to baseband with in-phase and quadrature-phase (IQ) channels, where a wideband Doppler analysis (coherent integration) follows before detection. Here, an implied assumption is that the targets considered in the pulse (or time) domain produce discrete cisoidal Doppler signals, with energy concentrated in a limited number of Doppler bins after coherent integration, resulting in enhanced SNCR. A state space realization [29, 30] can take advantage of this characteristic of the target signals, and can be incorporated into the array model as follows. Assuming that the k signals impinging on the m-sensor array are all cisoids with distinct Doppler frequencies, which is a reasonable assumption with long coherent integration time, (21.1) can be rewritten into:

$$\mathbf{y}(t) = \sum_{i=1}^{k} \mathbf{a}(\boldsymbol{\theta}_i) \alpha_i e^{j\omega_i t} + \mathbf{n}(t) \qquad (21.12)$$

where α_i and ω_i are the complex amplitude and frequency of the ith signal source, respectively. A state space model for the array output is:

$$\mathbf{z}(t+1) = \boldsymbol{\Omega}\mathbf{z}(t) \qquad (21.13)$$
$$\mathbf{y}(t) = \mathbf{A}\mathbf{z}(t) + \mathbf{n}(t)$$

where t is the discrete time index, and the state vector $\mathbf{z}(t) \in \mathbb{C}^k$ and $\boldsymbol{\Omega}$ are defined as:

$$\mathbf{z}(t) = \left[\alpha_1 e^{j\omega_1 t}, \alpha_2 e^{j\omega_2 t}, \ldots, \alpha_k e^{j\omega_k t}\right]^T \qquad (21.14)$$
$$\boldsymbol{\Omega} = diag[e^{j\omega_1}, e^{j\omega_2}, \ldots, e^{j\omega_k}]$$

Note that the major distinction from conventional MUSIC is that, in the conventional MUSIC, the ith source signal is modeled simply as a white random process sample $s_i(t)$. Through (21.14), more specific signal knowledge is available. The additional knowledge incorporated into the signals' state equations can be combined with that of the beamspace spatial filtering, and the structured wavefront arrival delays across the array to enhance threshold and resolution performances.

21.4 Two-Dimensional Prefiltering-Based MUSIC

In this section, we introduce a new method that combines the subspace method with a two-stage two-dimensional (spatial and temporal) prefiltering. The usual white noise assumption of MUSIC can be relaxed, allowing for handling of both temporally and spatially colored clutter noise, which has a neat narrowband structure in either the Doppler or the spatial domain. Both our final derivation conclusions and experimental results verify this argument.

Under the state space model, the covariance matrix of the sensor snapshot $\mathbf{y}(t)$ has the following form, comparable to that in (21.3):

$$\mathbf{R_{yy}} = E\{\mathbf{y}(t)\mathbf{y}^H(t)\} = \mathbf{A R_{zz} A}^H + \sigma^2 \mathbf{I}_{m \times m} \qquad (21.15)$$

where $\mathbf{R_{zz}}$ will later be shown to be a diagonal matrix. We then form a new Nm-vector $\mathbf{y}_N(t)$ from the array signal vector $\mathbf{y}(t)$, as follows:

$$\mathbf{y}_N(t) = [\mathbf{y}^T(t) \quad \mathbf{y}^T(t+1) \quad \ldots \quad \mathbf{y}^T(t+N-1)]^T \qquad (21.16)$$

It is obvious that the state space model for this new-formed vector is:

$$\mathbf{y}_N(t) = \begin{bmatrix} \mathbf{A} \\ \mathbf{A\Omega} \\ \vdots \\ \mathbf{A\Omega}^{N-1} \end{bmatrix} \mathbf{z}(t) + \begin{bmatrix} \mathbf{n}(t) \\ \mathbf{n}(t+1) \\ \vdots \\ \mathbf{n}(t+N-1) \end{bmatrix} = \mathbf{\Omega}_N \mathbf{z}(t) + \mathbf{n}_N(t) \qquad (21.17)$$

where $\mathbf{\Omega}_N \in \mathbb{C}^{Nm \times k}$ and $\mathbf{n}_N(t) \in \mathbb{C}^{Nm}$. The covariance matrix $\mathbf{R}_{\mathbf{y}_N \mathbf{y}_N}$ of the vector $\mathbf{y}_N(t)$ is:

$$\mathbf{R}_{\mathbf{y}_N \mathbf{y}_N} = E\left[\mathbf{y}_N(t)\mathbf{y}_N(t)^H\right] = \mathbf{\Omega}_N \mathbf{R_{zz}} \mathbf{\Omega}_N^H + \sigma^2 \mathbf{I}_{Nm \times Nm} \qquad (21.18)$$

where $\mathbf{R_{zz}} \in \mathbb{C}^{k \times k}$ is the covariance matrix of $\mathbf{z}(t)$. Assuming that α_i are constants or uncorrelated random variables with zero mean, it is straightforward to show matrix $\mathbf{R_{zz}}$ to be:

$$\mathbf{R_{zz}} = E \begin{bmatrix} |\alpha_1|^2 & \alpha_1 \alpha_2^* e^{j(\omega_1-\omega_2)t} & \ldots & \alpha_1 \alpha_k^* e^{j(\omega_1-\omega_k)t} \\ \alpha_2 \alpha_1^* e^{j(\omega_2-\omega_1)t} & |\alpha_2|^2 & \ldots & \alpha_2 \alpha_k^* e^{j(\omega_2-\omega_k)t} \\ \vdots & \vdots & \ddots & \vdots \\ \alpha_k \alpha_1^* e^{j(\omega_k-\omega_1)t} & \alpha_k \alpha_2^* e^{j(\omega_k-\omega_2)t} & \ldots & |\alpha_k|^2 \end{bmatrix}$$

$$= \begin{bmatrix} \beta_1 & 0 & \ldots & 0 \\ 0 & \beta_2 & \ldots & 0 \\ \vdots & \vdots & \ddots & \vdots \\ 0 & 0 & \ldots & \beta_k \end{bmatrix} \qquad (21.19)$$

where $\beta_i = \sigma_{\alpha_i}^2$, when α_i are random variables, or $\beta_i = |\alpha_i|^2$, when α_i are deterministic constants. We should point out that the model proposed in (21.16) coincides with the standard STAP model [31]. However, the STAP assumes a single target and the availability of target-free training data to compute the optimal beamformer weights for target detection. The scenario here assumes that multiple targets are present in all the data, and MUSIC is used to estimate the DOA of the targets after a new rank-reduced covariance matrix is formed.

We let $\mathbf{B}$ be the same $m \times m'$ DPSS spatial beamforming matrix, as defined in (21.7), and define similar to $\mathbf{B}$ another $N \times N'$ temporal filtering matrix $\mathbf{C}$ [28]. The first-stage two-dimensional prefiltering matrix is defined as:

$$\mathbf{D} = \mathbf{C} \otimes \mathbf{B} \tag{21.20}$$

where $\otimes$ denotes Kronecker product, and $\mathbf{D} \in \mathbb{R}^{Nm \times N'm'}$ and $\mathbf{C}^H \mathbf{C} = \mathbf{I}_{N' \times N'}$. The two-dimensional preprocessor performs the following data transformation:

$$\mathbf{x}_N(t) = \mathbf{D}^H \mathbf{y}_N(t) = (\mathbf{C} \otimes \mathbf{B})^H \mathbf{\Omega}_N \mathbf{z}(t) + (\mathbf{C} \otimes \mathbf{B})^H \mathbf{n}_N(t) \tag{21.21}$$

where $\mathbf{x}_N(t) \in \mathbb{C}^{N'm'}$ is the transformed data vector. Under the assumption that the noise vector is uncorrelated with the signals, the covariance matrix $\mathbf{R}_{\mathbf{x}_N \mathbf{x}_N} \in \mathbb{C}^{N'm' \times N'm'}$ of vector $\mathbf{x}_N(t)$ is

$$\mathbf{R}_{\mathbf{x}_N \mathbf{x}_N} = E\left[\mathbf{x}_N(t)\mathbf{x}_N(t)^H\right] \tag{21.22}$$

$$= (\mathbf{C} \otimes \mathbf{B})^H \mathbf{\Omega}_N \mathbf{R}_{zz} \mathbf{\Omega}_N^H (\mathbf{C} \otimes \mathbf{B}) + \sigma^2 (\mathbf{C} \otimes \mathbf{B})^H \mathbf{I}_{Nm \times Nm} (\mathbf{C} \otimes \mathbf{B})$$

We first consider the second term on the right-hand side of (21.22), which can be readily shown to be an identity matrix:

$$\sigma^2 (\mathbf{C} \otimes \mathbf{B})^H \mathbf{I}_{Nm \times Nm} (\mathbf{C} \otimes \mathbf{B}) = \sigma^2 (\mathbf{C}^H \mathbf{C}) \otimes (\mathbf{B}^H \mathbf{B}) = \sigma^2 \mathbf{I}_{N'm' \times N'm'} \tag{21.23}$$

Next in Appendix A, we simplify the first part of (21.22), and obtain:

$$(\mathbf{C} \otimes \mathbf{B})^H \mathbf{\Omega}_N \mathbf{R}_{zz} \mathbf{\Omega}_N^H (\mathbf{C} \otimes \mathbf{B}) = \mathbf{G}\mathbf{G}^H \tag{21.24}$$

where

$$\mathbf{G} = \begin{bmatrix} \sqrt{\beta_1}\mathbf{B}^H \mathbf{a}(\theta_1)\mathbf{c}_1^H \boldsymbol{\omega}_1 & \sqrt{\beta_2}\mathbf{B}^H \mathbf{a}(\theta_2)\mathbf{c}_1^H \boldsymbol{\omega}_2 & \cdots & \sqrt{\beta_k}\mathbf{B}^H \mathbf{a}(\theta_k)\mathbf{c}_1^H \boldsymbol{\omega}_k \\ \sqrt{\beta_1}\mathbf{B}^H \mathbf{a}(\theta_1)\mathbf{c}_2^H \boldsymbol{\omega}_1 & \sqrt{\beta_2}\mathbf{B}^H \mathbf{a}(\theta_2)\mathbf{c}_2^H \boldsymbol{\omega}_2 & \cdots & \sqrt{\beta_k}\mathbf{B}^H \mathbf{a}(\theta_k)\mathbf{c}_2^H \boldsymbol{\omega}_k \\ \vdots & \vdots & \ddots & \vdots \\ \sqrt{\beta_1}\mathbf{B}^H \mathbf{a}(\theta_1)\mathbf{c}_{N'}^H \boldsymbol{\omega}_1 & \sqrt{\beta_2}\mathbf{B}^H \mathbf{a}(\theta_2)\mathbf{c}_{N'}^H \boldsymbol{\omega}_2 & \cdots & \sqrt{\beta_k}\mathbf{B}^H \mathbf{a}(\theta_k)\mathbf{c}_{N'}^H \boldsymbol{\omega}_k \end{bmatrix} \tag{21.25}$$

and $\mathbf{c}_i$ is the ith column of matrix $\mathbf{C}$, and $\boldsymbol{\omega}_i$ is defined as:

$$\boldsymbol{\omega}_i = [1 \quad e^{j\omega_i} \quad e^{j\omega_i(N-1)}]^T \qquad (21.26)$$

Note that in (21.25), $\mathbf{c}_i^H \boldsymbol{\omega}_j$ is the DFT of window i at the frequency of signal j. From another point of view, $\mathbf{c}_i^H \boldsymbol{\omega}_j$ is the convolution of the input with the filter response. Let us use symbol ψ_{ij} to represent $\mathbf{c}_i^H \boldsymbol{\omega}_j$, and find the matrix product result. Then, (21.24) becomes:

$$(\mathbf{C} \otimes \mathbf{B})^H \boldsymbol{\Omega}_N \mathbf{R}_{zz} \boldsymbol{\Omega}_N^H (\mathbf{C} \otimes \mathbf{B}) =$$

$$\begin{bmatrix} \mathbf{B}^H \left(\sum_{i=1}^{k} \beta_i |\psi_{1i}|^2 \mathbf{a}(\theta_i) \mathbf{a}(\theta_i)^H \right) \mathbf{B} & \mathbf{B}^H \left(\sum_{i=1}^{k} \beta_i \psi_{1i} \psi_{2i}^* \mathbf{a}(\theta_i) \mathbf{a}(\theta_i)^H \right) \mathbf{B} & \cdots & \mathbf{B}^H \left(\sum_{i=1}^{k} \beta_i \psi_{1i} \psi_{N'i}^* \mathbf{a}(\theta_i) \mathbf{a}(\theta_i)^H \right) \mathbf{B} \\ \mathbf{B}^H \left(\sum_{i=1}^{k} \beta_i \psi_{2i} \psi_{1i}^* \mathbf{a}(\theta_i) \mathbf{a}(\theta_i)^H \right) \mathbf{B} & \mathbf{B}^H \left(\sum_{i=1}^{k} \beta_i |\psi_{2i}|^2 \mathbf{a}(\theta_i) \mathbf{a}(\theta_i)^H \right) \mathbf{B} & \cdots & \mathbf{B}^H \left(\sum_{i=1}^{k} \beta_i \psi_{2i} \psi_{N'i}^* \mathbf{a}(\theta_i) \mathbf{a}(\theta_i)^H \right) \mathbf{B} \\ \vdots & \vdots & \ddots & \vdots \\ \mathbf{B}^H \left(\sum_{i=1}^{k} \beta_i \psi_{N'i} \psi_{1i}^* \mathbf{a}(\theta_i) \mathbf{a}(\theta_i)^H \right) \mathbf{B} & \mathbf{B}^H \left(\sum_{i=1}^{k} \beta_i \psi_{N'i} \psi_{2i}^* \mathbf{a}(\theta_i) \mathbf{a}(\theta_i)^H \right) \mathbf{B} & \cdots & \mathbf{B}^H \left(\sum_{i=1}^{k} \beta_i |\psi_{N'i}|^2 \mathbf{a}(\theta_i) \mathbf{a}(\theta_i)^H \right) \mathbf{B} \end{bmatrix}$$

$$(21.27)$$

Substituting (21.23) and (21.27) into (21.22), and taking the average of the N' diagonal block matrices from $\mathbf{R}_{x_N x_N}$ (second-stage lowpass filtering), gives the working covariance matrix $\mathbf{R}$:

$$\mathbf{R} = \frac{1}{N'} \left[\mathbf{B}^H \left(\sum_{i=1}^{k} \beta_i |\psi_{1i}|^2 \mathbf{a}(\theta_i) \mathbf{a}(\theta_i)^H \right) \mathbf{B} + \mathbf{B}^H \left(\sum_{i=1}^{k} \beta_i |\psi_{2i}|^2 \mathbf{a}(\theta_i) \mathbf{a}(\theta_i)^H \right) \mathbf{B} \right.$$

$$\left. + \ldots + \mathbf{B}^H \left(\sum_{i=1}^{k} \beta_i |\psi_{N'i}|^2 \mathbf{a}(\theta_i) \mathbf{a}(\theta_i)^H \right) \mathbf{B} \right] + \sigma^2 \mathbf{I}_{m' \times m'}$$

$$= \mathbf{B}^H \left[\sum_{i=1}^{k} \beta_i \left(\frac{1}{N'} \sum_{j=1}^{N'} |\psi_{ji}|^2 \right) \mathbf{a}(\theta_i) \mathbf{a}(\theta_i)^H \right] \mathbf{B} + \sigma^2 \mathbf{I}_{m' \times m'}$$

$$= \mathbf{B}^H \mathbf{A} \boldsymbol{\Omega}' \mathbf{A}^H \mathbf{B} + \sigma^2 \mathbf{I}_{m' \times m'}$$

$$(21.28)$$

where

$$\boldsymbol{\Omega}' = diag \left[\beta_1 \frac{1}{N'} \sum_{j=1}^{N'} |\psi_{j1}|^2 \quad \beta_2 \frac{1}{N'} \sum_{j=1}^{N'} |\psi_{j2}|^2 \quad \cdots \quad \beta_k \frac{1}{N'} \sum_{j=1}^{N'} |\psi_{jk}|^2 \right]$$

$$(21.29)$$

From the DPSS multitaper theory [27], we know that $\boldsymbol{\Omega}'$ is just the covariance matrix of a time domain DPSS filtering output. Hereafter, we realize the two-stage combined two-dimensional prefiltering in the working matrix $\mathbf{R}$. Assuming that the prefiltering passband will keep k' ($k' \leq k$) signals, the eigendecomposition of matrix $\mathbf{R}$ is:

$$\mathbf{R} = \sum_{i=1}^{m'} \lambda_{ri} \mathbf{h}_{ri} \mathbf{h}_{ri}^H = \mathbf{U}_{rs} \boldsymbol{\Sigma}_{rs} \mathbf{U}_{rs}^H + \mathbf{U}_{rn} \boldsymbol{\Sigma}_{rn} \mathbf{U}_{rn}^H \qquad (21.30)$$

where $\lambda_{r1} \geq \ldots \geq \lambda_{rk'} > \lambda_{r(k'+1)} = \ldots = \lambda_{rm'} = \sigma^2$ are the eigenvalues of $\mathbf{R}$, and $\mathbf{h}_{ri}$ are the corresponding orthonormal eigenvectors. $\mathbf{U}_{rs} = [\mathbf{h}_{r1} \mathbf{h}_{r2} \ldots \mathbf{h}_{rk'}]$ with rank k' spans the signal subspace, and $\mathbf{U}_{rn} = [\mathbf{h}_{r(k+1)} \mathbf{h}_{r(k+2)} \ldots \mathbf{h}_{rm'}]$ spans the noise subspace.

21.5 Algorithm Summary for 2DP-MUSIC

1. Form the new data vector $\mathbf{y}_N(t)$, according to (21.16).
2. Map data vector $\mathbf{y}_N(t)$ into $\mathbf{x}_N(t)$ by the first-stage two-dimensional prefiltering matrix (21.21).
3. Estimate the sample covariance matrix of $\mathbf{x}_N(t)$ by the time average

$$\hat{\mathbf{R}}_{\mathbf{x}_N \mathbf{x}_N} = \frac{1}{T-N+1} \sum_{i=1}^{T-N+1} \mathbf{x}_N(t) \mathbf{x}_N(t)^H$$

where T/N should be greater than $2N'm'$, according to the Reed-Mallett-Brennan rule.

4. Obtain the working matrix $\mathbf{R}$ by second-stage lowpass filtering along the main diagonal of $\hat{\mathbf{R}}_{\mathbf{x}_N \mathbf{x}_N}$, according to (21.28).
5. Estimate the signal subspace by SVD decomposition $\mathbf{R} = \mathbf{U}_{rs} \boldsymbol{\Sigma}_{rs} \mathbf{U}_{rs}^H + \mathbf{U}_{rn} \boldsymbol{\Sigma}_{rn} \mathbf{U}_{rn}^H$, where $\mathbf{U}_{rs}$ has rank k'.
6. Obtain the MUSIC spatial spectrum

$$P_{2\text{-dMU}}(\theta) = \frac{1}{\mathbf{v}^H(\theta) \mathbf{B} \mathbf{U}_{rn} \mathbf{U}_{rn}^H \mathbf{B}^H \mathbf{v}(\theta)} = \frac{1}{D_{2\text{-}d}(\theta)}$$

21.6 Performance Analysis

In this section, we analyze the asymptotic performance (i.e., number of snapshots T goes to infinity) of our proposed 2DP-MUSIC algorithm. We first give a simplified theoretical derivation of the resolution capability, which is characterized by the lowest possible threshold above which the two closely spaced emitters can be resolved. We then compare these asymptotic expressions to simulations in Section 21.7, in order to assess their accuracy for smaller values of T. In Section 21.7, we will also give the small error behavior of 2DP-MUSIC by Monte Carlo simulations when the SNR is well above the threshold, which means the probability of resolution is one. The elemental signal-to-noise ratio is defined by SNR $= P/\sigma^2$, and the array-signal-to-noise ratio (ASNR) is mP/σ^2, where P is target signal power.

In order to conduct the performance analysis on a fair basis, we assume that no interference and clutter exist, and that the background sensor noise is white Gaussian. There are only two closely spaced sources whose data is passing through our combined two-dimensional prefilter. Under such assumptions, 2DP-MUSIC will have the smallest performance gain over MUSIC and BS-MUSIC. Equation (21.28) can be expressed as

$$\mathbf{R} = \mathbf{B}^H \mathbf{A} \mathbf{\Omega}' \mathbf{A}^H \mathbf{B} + \sigma^2 \mathbf{I}_{m' \times m'} = \mathbf{B}^H \mathbf{R}' \mathbf{B} \tag{21.31}$$

where $\mathbf{R}'$ is equivalent to the covariance matrix of sensor domain MUSIC. Since both signals pass through the two-dimensional prefiltering, $\mathbf{R}'$ and $\mathbf{R}_{yy}$ are basically the same, to within a scalar. Our initial analysis can follow Kaveh and Barabell [7], who conducted the first comprehensive study on the resolution performance of the sensor domain MUSIC, and which is summarized in Appendix 21.B. Without loss of generality, we use $\mathbf{R}_{yy}$ instead of $\mathbf{R}'$. We must point out that our algorithm has the same form as BS-MUSIC (21.9) under the assumptions we have taken in this section. Thus, we will expect 2DP-MUSIC and BS-MUSIC to have similar resolution behavior. Lee and Wengrovitz [22] have already conducted a resolution analysis for BS-MUSIC, but that assumes arbitrary array geometry, general pre-beamforming, and unequal power sources, and this analysis is very difficult to follow. Since we adopt the specific DPSS beamformer and uniform linear array, we have a different and much simpler derivation. We will find that our conclusion is consistent with Lee's for the specific situation under consideration.

21.6.1 Formulation of Sample Null Spectra for MUSIC, BS-MUSIC, and 2DP-MUSIC

The reason for statistical performance analysis is that, in reality, the true covariance matrix is not available, and only a time-averaged sample covariance matrix is used:

$$\hat{\mathbf{R}}_{yy} = \frac{1}{T} \sum_{t=1}^{T} \mathbf{y}(t) \mathbf{y}^H(t)$$

$$\hat{\mathbf{R}}_{xx} = \frac{1}{T} \sum_{t=1}^{T} \mathbf{x}(t) \mathbf{x}^H(t) \tag{21.32}$$

$$\hat{\mathbf{R}}_{y_N y_N} = \frac{1}{T-N+1} \sum_{t=1}^{T-N+1} \mathbf{y}_N(t) \mathbf{y}_N(t)^H$$

where we use "^" to denote the estimate of the quantity over which it appears.

Based on the above three sample covariance matrices, we have the corresponding sample null spectra $\hat{D}(\theta)$, $\hat{D}_B(\theta)$, and $\hat{D}_{2\text{-}d}(\theta)$ for MUSIC, BS-MUSIC, and 2DP-MUSIC, respectively:

$$\hat{D}(\theta) = 1 - \mathbf{v}^H(\theta) \left(\sum_{i=1}^{k} \hat{\mathbf{h}}_i \hat{\mathbf{h}}_i^H \right) \mathbf{v}(\theta) \tag{21.33}$$

$$\hat{D}_B(\theta) = \mathbf{v}^H(\theta)\mathbf{B}\mathbf{B}^H\mathbf{v}(\theta) - \mathbf{v}^H(\theta)\mathbf{B}\left(\sum_{i=1}^{k}\hat{\mathbf{h}}_{xi}\hat{\mathbf{h}}_{xi}^H\right)\mathbf{B}^H\mathbf{v}(\theta) \quad (21.34)$$

$$\hat{D}_{2\text{-}d}(\theta) = \mathbf{v}^H(\theta)\mathbf{B}\mathbf{B}^H\mathbf{v}(\theta) - \mathbf{v}^H(\theta)\mathbf{B}\left(\sum_{i=1}^{k}\hat{\mathbf{h}}_{ri}\hat{\mathbf{h}}_{ri}^H\right)\mathbf{B}^H\mathbf{v}(\theta) \quad (21.35)$$

Note that the distinction of (21.34) and (21.35) is in the source of the eigenvectors $\hat{\mathbf{h}}_{xi}$ and $\hat{\mathbf{h}}_{ri}$, as compared to the $\hat{\mathbf{h}}_i$ in (21.33).

21.6.2 Expectation of 2DP-MUSIC's Null Spectrum

The approximate statistical behaviors of the spectra are studied, based on the mean and variance of those sample null spectra, since they will not exactly equal zero when the steering vector coincides with a source direction vector. The expected value of $\hat{D}_{2\text{-}d}(\theta)$ is a counterpart of (21.A.3):

$$E\{\hat{D}_{2\text{-}d}(\theta)\} = D_{2\text{-}d}(\theta) + \sigma^2 \mathbf{v}^H(\theta)\mathbf{B}\left[\frac{\lambda_{r1}(m'-2)}{(\lambda_{r1}-\sigma^2)^2}\mathbf{h}_{r1}\mathbf{h}_{r1}^H + \frac{\lambda_{r2}(m'-2)}{(\lambda_{r2}-\sigma^2)^2}\mathbf{h}_{r2}\mathbf{h}_{r2}^H\right]\frac{\mathbf{B}^H\mathbf{v}(\theta)}{T} \quad (21.36)$$

First, we find the relation between $\mathbf{h}_{ri}$ and $\mathbf{h}_i$. From (21.31), we have

$$\mathbf{R}\mathbf{B}^H\mathbf{h}_i = \mathbf{B}^H\mathbf{R}_{yy}\mathbf{B}\mathbf{B}^H\mathbf{h}_i \quad (21.37)$$

Since we adopt the DPSS beamforming matrix $\mathbf{B}$ here, the columns of $\mathbf{B}$ are orthonormal sets, since $\mathbf{B}$ is formed from eigendecompostion. In other words, the columns of $\mathbf{B}$ form the approximate orthonormal basis of the space of interest (i.e., passband in spatial domain), and $\mathbf{B}\mathbf{B}^H$ is a projection matrix onto this space. Since we assume that the two closely spaced sources pass through the spatial filtering, the subspace spanned by $\mathbf{B}$ incorporates signal subspace of $\mathbf{R}_{yy}$, which results in $\mathbf{B}\mathbf{B}^H\mathbf{h}_i = \mathbf{h}_i$. Then we have:

$$\mathbf{R}\mathbf{B}^H\mathbf{h}_i = \mathbf{B}^H\mathbf{R}_{yy}\mathbf{h}_i = \mathbf{B}^H\lambda_i\mathbf{h}_i = \lambda_i\mathbf{B}^H\mathbf{h}_i \quad (21.38)$$

which means that the normalized $\mathbf{B}^H\mathbf{h}_i$ are the eigenvectors of $\mathbf{R}$:

$$\mathbf{h}_{ri} = \frac{\mathbf{B}^H\mathbf{h}_i}{\left(\mathbf{h}_i^H\mathbf{B}\mathbf{B}^H\mathbf{h}_i\right)^{1/2}} = \mathbf{B}^H\mathbf{h}_i \quad (21.39)$$

and the eigenvalues of $\mathbf{R}$ are

$$\lambda_{ri} = \lambda_i \quad (21.40)$$

where $i = 1, 2$. Applying (21.39) and (21.40) into (21.36), we can obtain the following expression

$$E\{\hat{D}_{2\text{-}d}(\theta_i)\} = D_{2\text{-}d}(\theta_i) + \sigma^2 \mathbf{v}^H(\theta_i) \left[\frac{\lambda_1(m'-2)}{(\lambda_1 - \sigma^2)^2} \mathbf{h}_1 \mathbf{h}_1^H + \frac{\lambda_2(m'-2)}{(\lambda_2 - \sigma^2)^2} \mathbf{h}_2 \mathbf{h}_2^H \right] \frac{\mathbf{v}(\theta_i)}{T} \tag{21.41}$$

where $i = 1, 2, m, \theta_m = (\theta_1 + \theta_2)/2$. Note that $D_{2\text{-}d}(\theta_i)$ equals zero when $i = 1, 2$. The following also holds:

$$D_{2\text{-}d}(\theta_m) = 1 - \mathbf{v}^H(\theta_m) \mathbf{B} \left(\sum_{i=1}^{2} \mathbf{h}_{ri} \mathbf{h}_{ri}^H \right) \mathbf{B}^H \mathbf{v}(\theta_m)$$

$$= 1 - \mathbf{v}^H(\theta_m) \mathbf{B}\mathbf{B}^H \left(\sum_{i=1}^{2} \mathbf{h}_i \mathbf{h}_i^H \right) \mathbf{B}\mathbf{B}^H \mathbf{v}(\theta_m) \tag{21.42}$$

$$= D(\theta_m)$$

From (21.41), (21.B.1), and (21.B.2), we obtain the following equations:

$$E\{\hat{D}_{2\text{-}d}(\theta_i)\} = \frac{E\{\hat{D}(\theta_i)\}}{m-2}(m'-2) \approx \frac{(m'-2)}{T} \left[\frac{1}{2(\text{ASNR})} + \frac{1}{(\text{ASNR})^2 \Delta^2} \right] \quad i = 1, 2 \tag{21.43}$$

$$E\{\hat{D}_{2\text{-}d}(\theta_m)\} - D_{2\text{-}d}(\theta_m) = [E\{\hat{D}(\theta_m)\} - D(\theta_m)] \frac{(m'-2)}{m-2} \tag{21.44}$$

Considering (21.42), (21.44), and (21.B.3), it is straightforward to obtain:

$$E\{\hat{D}_{2\text{-}d}(\theta_m)\} \approx D_{2\text{-}d}(\theta_m) \approx \frac{1}{80} \Delta^4 \tag{21.45}$$

21.6.3 Resolution Threshold

We also consider the resolution threshold $\xi_{T_{2\text{-}d}}$ to be defined as the ASNR where

$$E\{\hat{D}_{2\text{-}d}(\theta_m)\} = E\{\hat{D}_{2\text{-}d}(\theta_1)\} = E\{\hat{D}_{2\text{-}d}(\theta_2)\} \tag{21.46}$$

From (21.43) to (21.46), we have:

$$\xi_{T_{2\text{-}d}} \approx \frac{1}{T} \left\{ 20(m'-2)\Delta^{-4} \left[1 + \sqrt{1 + \frac{T}{5(m'-2)} \Delta^2} \right] \right\} \tag{21.47}$$

The result in (21.47) is verified by, and consistent with, conclusions in [22], where a more general threshold expression is derived. Comparing (21.47) and (21.B.4), we find the only difference is that the factor of $m - 2$ in sensor MUSIC is replaced by $m' - 2$ in our algorithm. Since usually the number of beams m' is far smaller than the number of sensors m, a lower resolution threshold is achieved.

21.7 Simulations

In this section, we show comparative simulation results for probability of resolution, bias, and mean-square error of the DOA estimates under our three algorithms, and give an indication of their performance. In both simulation situations described below, a ULA with 16 sensors spaced at one-half radar carrier wavelength is adopted, which corresponds to a beamwidth 0.3927, or 7.2°. The parameters are set up as $N = 16$, $N' = 2$, and $m' = 5$.

In Simulation 1, we compare the algorithms when there is no interference and clutter. We verify that under this circumstance, the resolution threshold characteristic of 2dP-MUSIC is similar to that of BS-MUSIC, but with a gain of 1 to 2 dB. They both outperform MUSIC in the sense of resolution threshold.

In Simulation 2, we show the performance gain of 2DP-MUSIC over the other two algorithms when the interference is away from the interested space-frequency region.

Simulation 1

In this simulation, we inject two desired targets into a background of white Gaussian noise. Target 1 is at normalized frequency = 0.1 and DOA = 1°. Target 2 is at normalized frequency = 0.16 and DOA = –1°. Both signals have the same power, and thus the same SNR. The two-dimensional filter is set up to pass both target signals.

Figure 21.1 presents the probability of resolution of these two targets, based on 100 independent trials. In this simulation, the targets are considered to be

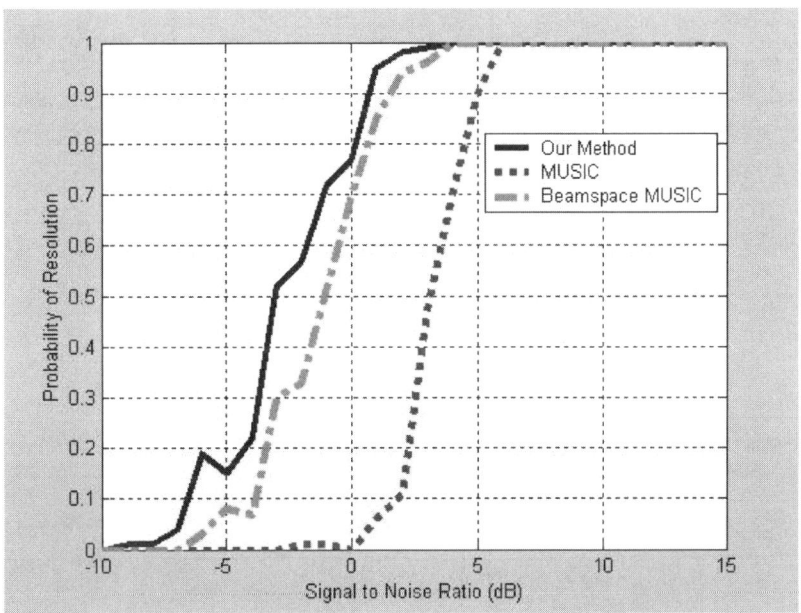

Figure 21.1 Probability of resolution versus SNR, with 100 independent trials. Solid line: our 2DP-MUSIC method; dash-dot line: BS-MUSIC; and dotted line: MUSIC.

resolved if the estimated peaks are in the range of 1° (one-half the spacing between the targets) around the true DOAs. The value of the angular separation parameter Δ is found to be 0.5065, and m' is chosen to be 5. The theoretical threshold (ASNR) predicted by (21.47) for 2DP-MUSIC is 8.66 dB, corresponding to SNR = −3.4 dB. From Figure 21.1, it is obvious that our simulation results for the threshold, corresponding to probability of resolution in the range from 0.33 to 0.5 [7], match the theory to within 0.3 dB, agreeing well with the interpretation. The theoretical threshold SNR for MUSIC is 1.5 dB, which matches the simulation result to within 1.5 dB.

Simulation 2

To illustrate the advantage of our proposed algorithm, we compare the estimate variances of 2DP-MUSIC with those of MUSIC and BS-MUSIC under two particularly difficult and simultaneous conditions:

1. An interference is within the same spatial beam as the desired target, but in a different frequency bin;
2. An interference is in the same frequency bin, but in a different spatial beam.

The desired target is simulated at DOA = 1° and normalized frequency = 0.1, with varying SNR. Two interferences with equal power are injected: one is at DOA = −1°, frequency = 0.9, and INR = 10 dB; the other one is at DOA = −40°, frequency = 0.16, and INR = 10 dB. The background noise is white Gaussian.

Figure 21.2 shows the spatial spectra of 2DP-MUSIC, MUSIC, and BS-MUSIC, when the SNR of the target is −5 dB. It is obvious that MUSIC and BS-MUSIC

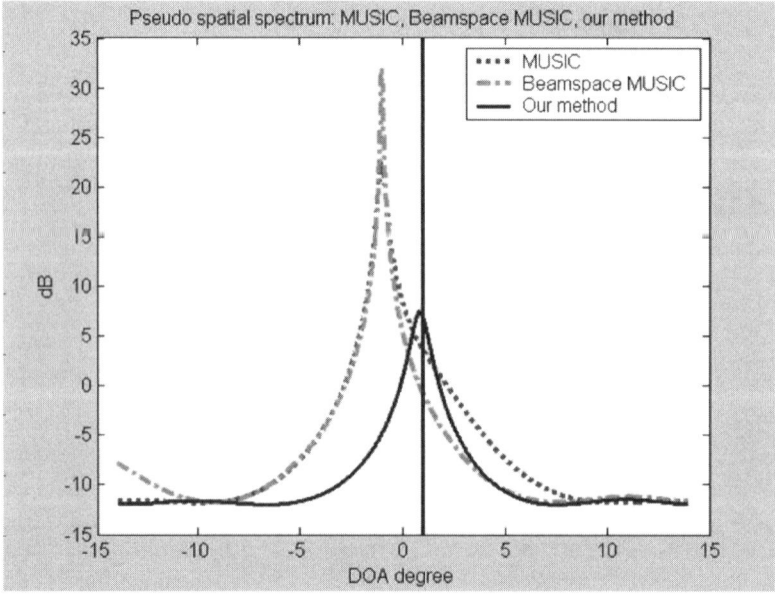

Figure 21.2 Pseudospatial spectrum from our 2DP-MUSIC method, MUSIC, and BS-MUSIC, at SNR = −5 dB. True DOA = 1°.

fail at this lower SNR. In Figure 21.3, it can be seen that MUSIC and BS-MUSIC have a threshold when the SNR is near 5 dB, while 2DP-MUSIC's SNR threshold is at −5 dB, which is a performance improvement of 10 dB.

The mean-squared error is plotted in Figure 21.4, where it is clear that 2DP-MUSIC has a consistently smaller estimation error than do the other two methods. Note that the performance curve for 2DP-MUSIC is very similar to the curve for BS-MUSIC in shape, which is consistent with our former theoretical conclusion in Section 21.6.

21.8 Experimental Result

In this section, we use real HF radar echo data, provided by Raytheon Canada, Ltd., to conduct an experiment to verify our algorithms. The radar utilizes a floodlight transmitting antenna and a 16-omnidirectional antenna ULA as receiver. The dataset is collected using a 3.2 MHz carrier frequency, and a dwell time of approximately 164 seconds, with and effective pulse repetition frequency of 1.6 Hz. The detection range is from 50 to 300 km, with a 1.5-km resolution.

A target signal is injected into the real sea clutter background with Doppler frequency of 0 Hz, and DOA of 0°, and with an approximate SNCR of 1.8 dB (see Figure 21.5). Dominating the Doppler power spectrum of the sea clutter are the first-order Bragg lines that appear around the specific Doppler positions, which are proportional to the square root of radar carrier frequency. The Bragg lines are first-order scatter from surface waves, which have a wavelength of one-half of the radar wavelength, and move directly away and towards the receiving array.

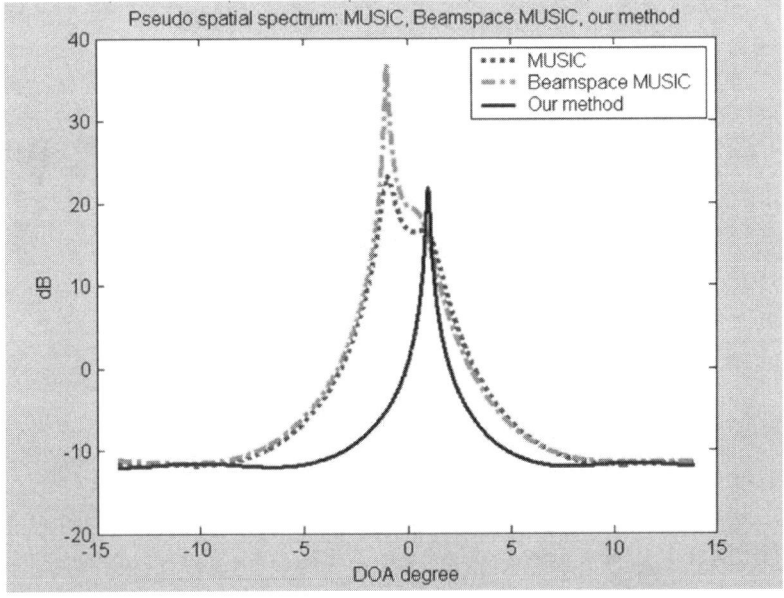

Figure 21.3 Pseudospatial spectrum from our 2DP-MUSIC method, MUSIC, and BS-MUSIC, at SNR = 5 dB. True DOA = 1°.

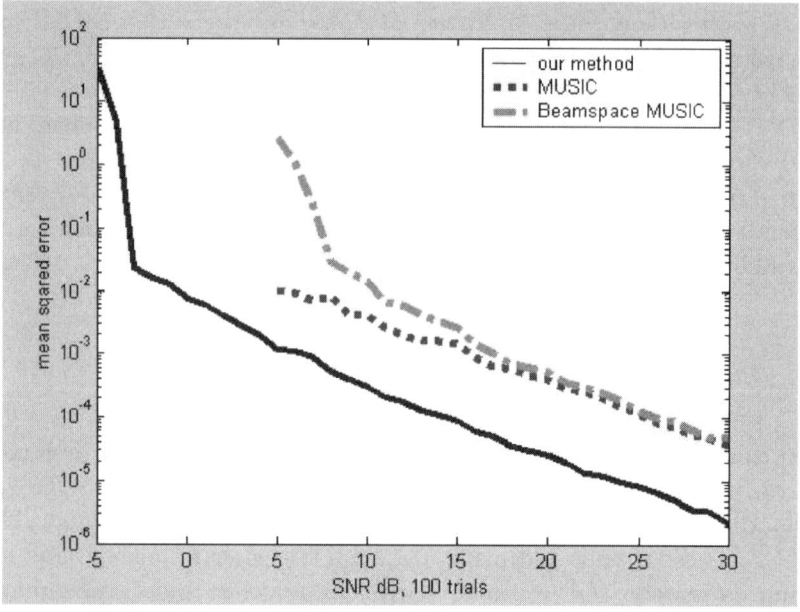

Figure 21.4 Mean-squared error, with 100 trials. Solid line: our 2DP-MUSIC method; dash-dot line: BS-MUSIC; dotted line: MUSIC.

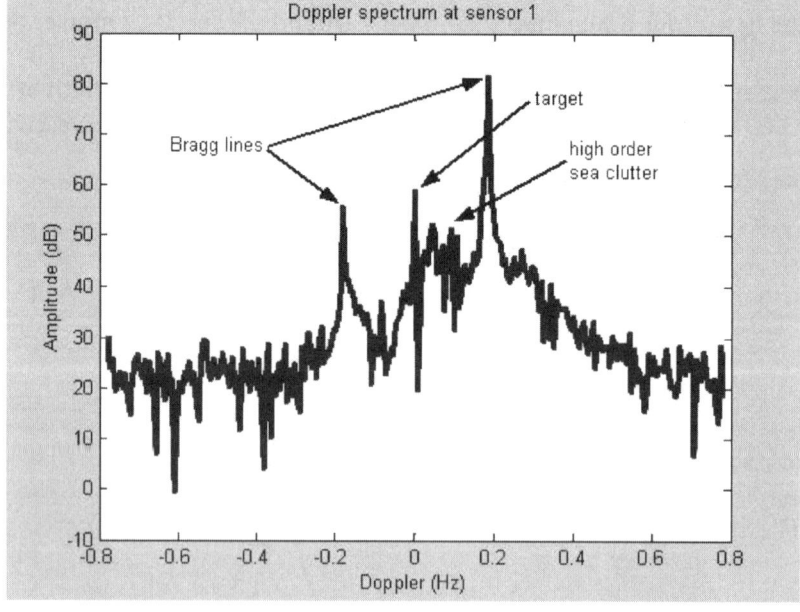

Figure 21.5 Doppler spectrum of sea clutter, with target at range 100 km and sensor 1.

Figure 21.6 shows the estimation performance from MUSIC, BS-MUSIC, and 2DP-MUSIC. Sea clutter has a well-defined Doppler structure. However, it has no directionality (i.e., it arrives from all directions), and both MUSIC and BS-MUSIC fail. However, 2DP-MUSIC takes advantage of sea clutter's Doppler structure, suppresses it from the temporal domain, and yields correct estimation.

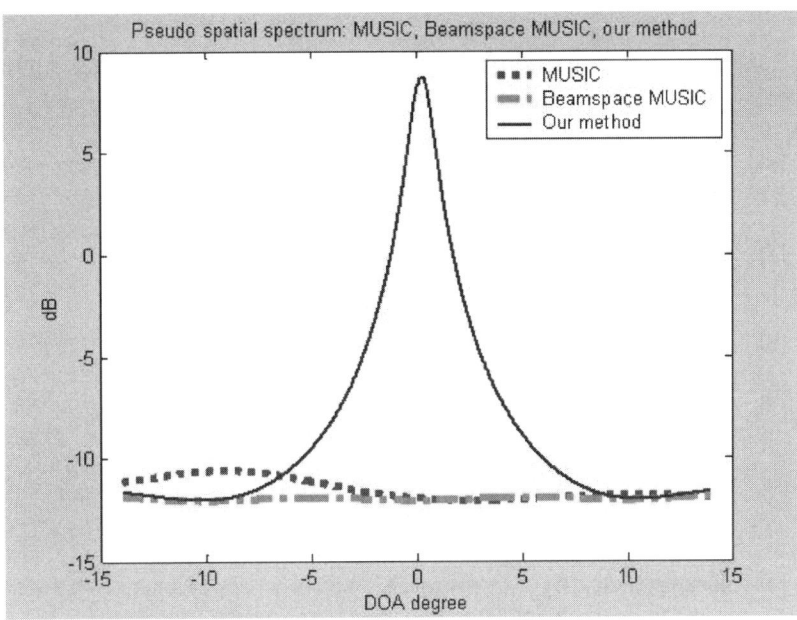

Figure 21.6 Pseudospatial spectrum from our 2DP-MUSIC method, MUSIC, and BS-MUSIC, at SCNR = 1.8 dB. True DOA = 0°.

21.9 Conclusions

We have proposed a new two-stage spatio-temporal prefilter that, when combined with the subsequent high resolution DOA estimation algorithm, provides a higher resolution, lower detection threshold, and lower estimation bias and variance than those of conventional high resolution methods that use only spatial prefiltering.

The algorithm utilizes DPSS to form the two-dimensional prefilter that applies to the output of a uniform linear array. The two-dimensional prefilter is formed to simultaneously filter out the unwanted noise, clutter, and interference from both the time and space domains, thereby improving the estimation performance compared to more restrictive algorithms. When the interference can only be suppressed in the frequency domain, the new algorithm produces a 10-dB lower-resolution threshold, along with lower estimation bias and variance, compared to conventional DPSS beam-space MUSIC [21] and conventional sensor-space MUSIC. The expected theoretical enhancement provided by the proposed algorithm has been verified by Monte Carlo simulation. The experimental result using real HF clutter data verifies that, when the interference or clutter has a well-defined structure in either the spatial or temporal domain, our algorithm will suppress it, and obtain an enhanced estimation result.

References

[1] Schmidt, R. O., "Multiple Emitter Location and Signal Parameter Estimation," *IEEE Trans. on Antennas and Propagation*, Vol. AP-34, No. 3, 1986, pp. 276–280.

[2] Kumaresan, R., and D. W. Tufts, "Estimating the Angles of Arrival of Multiple Plane Waves," *IEEE Trans. on Aerospace and Electronics Systems*, Vol. AES-19, January 1983.

[3] Paulraj, R. R., and T. Kailath, "A Subspace Rotation Approach to Signal Parameter Estimation," *IEEE Proc.*, July 1986, pp. 1044–1045.

[4] Roy, R., and T. Kailath, "ESPRIT-Estimation of Signal Parameters Via Rotational Invariance Techniques," *IEEE Trans. on Acoustics, Speech, and Signal Processing*, Vol. 37, No. 7, July 1989, pp. 984–995.

[5] Stoica, P., and K. C. Sharman, "Maximum Likelihood Methods for Direction-of-Arrival Estimation," *IEEE Trans. on Acoustics, Speech, and Signal Processing*, Vol. 38, No. 7, July 1990, pp. 1132–1143.

[6] Barabell, et al., *Performance Comparison of Superresolution Array Processing Algorithm*, Tech. Rep. TST-72, Lincoln Laboratory, M.I.T., 1984.

[7] Kaveh, M., and A. J. Barabell, "The Statistical Performance of the MUSIC and the Minimum-Norm Algorithms in Resolving Plane Waves in Noise," *IEEE Trans. on Acoustics, Speech, and Signal Processing*, Vol. ASSP-34, No. 2, April 1986, pp. 331–341.

[8] Stoica, P., and A. Nehorai, "MUSIC, Maximum Likelihood and Cramer-Rao Bound," *IEEE Trans. on Acoustics, Speech, and Signal Processing*, Vol. 37, No. 5, May 1989, pp. 720–741.

[9] Swindlehurst, A. L., and T. Kailath, "A Performance Analysis of Subspace-Based Methods in the Presence of Model Errors, Part I: The MUSIC Algorithm," *IEEE Trans. on Signal Processing*, Vol. 40, No. 7, July 1992, pp. 1758–1774.

[10] Zhou, C., F. Haber, and D. L. Jaggard, "A Resolution Measure for the MUSIC Algorithm and Its Application to Plane Wave Arrivals Contaminated by Coherent Interference," *IEEE Trans. on Signal Processing*, Vol. 39, No. 2, February 1991.

[11] Weiss, A. J., and B. Friedlander, "Effects of Modeling Errors on the Resolution Threshold of the MUSIC Algorithm," *IEEE Trans. on Signal Processing*, June 1994, pp. 1519–1526.

[12] Krim, H., and M. Viberg, "Two Decades of Array Signal Processing Research," *IEEE Signal Processing Magazine*, July 1996, pp. 67–94.

[13] Wiener, N., *Extrapolation, Interpolation and Smoothing of Stationary Time Series*, Cambridge, MA: MIT Press, 1949.

[14] Shan, T. J., M. Wax, and T. Kailath, "On Spatial Smoothing for Directions of Arrival Estimation of Coherent Signals," *IEEE Trans. on Acoustics, Speech, and Signal Processing*, ASSP-33, April 1985, pp. 806–811.

[15] Friedlander, B., and A. Weiss, "Direction Finding Using Spatial Smoothing with Interpolated Arrays," *IEEE Trans.*, AES. 28, April 1992, pp. 574–587.

[16] Kirlin, R. L., and W. Du, "Improvement on the Estimation of Covariance Matrices by Incorporating Crosscorrelations," *Radar and Signal Processing, IEE Proc. F*, Vol. 138, Issue 5, October 1991, pp. 479–482.

[17] Kirlin, R. L., and W. Du, "Enhancement of Covariance Matrix for Array Processing," *IEEE Trans. on Signal Processing*, Vol. 40, No. 10, October 1992, pp. 2602–2606.

[18] Kirlin, R. L., and W. Du, "Improved Spatial Smoothing Techniques for DOA Estimation of Coherent Signals," *IEEE Trans. on Signal Processing*, Vol. 39, No. 5, 1991, pp. 1208–1210.

[19] Bienvenu, G., and L. Kopp, "Decreasing High-Resolution Method Sensitivity by Conventional Beamformer Preprocessing," *ICASSP*, 1984, pp. 714–717.

[20] Buckley, K., and X. L. Xu, "Spatial-Spectrum Estimation in a Location Sector," *IEEE Trans. on Acoustics, Speech, and Signal Processing*, Vol. 38, No. 11, November 1990, pp. 1842–1852.

[21] Xu, X. L., and K. Buckley, "An Analysis of Beam-Space Source Localization," *IEEE Trans. on Signal Processing*, Vol. 41, No. 1, January 1993, pp. 501–504.

[22] Lee, H. B., and M. S. Wengrovitz, "Resolution Threshold of Beamspace MUSIC for Two Closely Spaced Emitters," *IEEE Trans. on Acoustics, Speech, and Signal Processing*, Vol. 38, No. 9, September 1990, pp. 1545–1559.

[23] Ralph, A., "Data Processing for a Groundwave HF Radar," *GEC J. Res.*, Vol. 6, No. 2, 1988, pp. 96–105.

[24] Sevgi, L., A. M. Ponsford, and H. C. Chan, "An Integrated Maritime Surveillance System Based on High-Frequency Surface-Wave Radars, 1. Theoretical Background and Numerical Simulations," *IEEE Antennas and Propagation Magazine*, Vol. 43, No. 4, 2001, pp. 28–43.

[25] Xie, J., Y. Yuan, and Y. Liu, "Super-Resolution Processing for HF Surface Wave Radar Based on Pre-Whitened MUSIC," *IEEE J. of Oceanic Engineering*, Vol. 23, No. 4, 1998, pp. 313–321.

[26] Wang, J., et al., "Small Ship Detection with High Frequency Radar Using an Adaptive Ocean Clutter Prewhitened Subspace Method," *IEEE Sensor Array and Multichannel Signal Processing Workshop*, Rosslyn, 2002, pp. 92–95.

[27] Percival, D. B., and A. T. Walden, "Spectral Analysis for Physical Applications: Multitaper and Conventional Univariate Techniques," Cambridge University Press, 1993.

[28] Bassias, A., and M. Kaveh, "Coherent Signal-Subspace Processing in a Sector," *IEEE Trans. on System, Man, and Cybernetics*, Vol. 21, No. 5, 1991, pp. 1088–1100.

[29] Ogata, K., *Modern Control Engineering*, 2nd ed., Upper Saddle River, NJ: Prentice Hall, 1990.

[30] Viberg, M., and P. Stoica, "A Computationally Efficient Method for Joint Direction Finding and Frequency estimation in Colored Noise," *Proc. 32nd Asilomar Conference on Signal, Systems, and Computers*, Pacific Cove, CA, November 1998, pp. 1547–1551.

[31] Ward, J., "Space-Time Adaptive Processing for Airborne Radar," MIT Lincoln Laboratory Technical Report 1015, December 1994.

Appendix 21.A   Derivation of (21.24)

$$(\mathbf{C} \otimes \mathbf{B})^H \mathbf{\Omega}_N \mathbf{R}_{zz} \mathbf{\Omega}_N^H (\mathbf{C} \otimes \mathbf{B})$$

$$= \begin{bmatrix} c(1,1)^* \mathbf{B}^H & c(2,1)^* \mathbf{B}^H & \cdots & c(N,1)^* \mathbf{B}^H \\ c(1,2)^* \mathbf{B}^H & c(2,2)^* \mathbf{B}^H & \cdots & c(N,2)^* \mathbf{B}^H \\ \vdots & \vdots & \ddots & \vdots \\ c(1,N')^* \mathbf{B}^H & c(2,N')^* \mathbf{B}^H & \cdots & c(N,N')^* \mathbf{B}^H \end{bmatrix} \begin{bmatrix} \mathbf{A} \\ \mathbf{A}\mathbf{\Omega} \\ \vdots \\ \mathbf{A}\mathbf{\Omega}^{N-1} \end{bmatrix} \mathbf{R}_{zz} \mathbf{\Omega}_N^H (\mathbf{C} \otimes \mathbf{B})$$

$$= \begin{bmatrix} \mathbf{B}^H \mathbf{A} \left(\sum_{i=1}^N c(i,1)^* \mathbf{\Omega}^{i-1} \right) \\ \mathbf{B}^H \mathbf{A} \left(\sum_{i=1}^N c(i,2)^* \mathbf{\Omega}^{i-1} \right) \\ \vdots \\ \mathbf{B}^H \mathbf{A} \left(\sum_{i=1}^N c(i,N')^* \mathbf{\Omega}^{i-1} \right) \end{bmatrix} \begin{bmatrix} \beta_1 & 0 & \cdots & 0 \\ 0 & \beta_2 & \cdots & 0 \\ \vdots & \vdots & \ddots & \vdots \\ 0 & 0 & \cdots & \beta_k \end{bmatrix} \begin{bmatrix} \mathbf{B}^H \mathbf{A} \left(\sum_{i=1}^N c(i,1)^* \mathbf{\Omega}^{i-1} \right) \\ \mathbf{B}^H \mathbf{A} \left(\sum_{i=1}^N c(i,2)^* \mathbf{\Omega}^{i-1} \right) \\ \vdots \\ \mathbf{B}^H \mathbf{A} \left(\sum_{i=1}^N c(i,N')^* \mathbf{\Omega}^{i-1} \right) \end{bmatrix}^H$$

$$= \mathbf{G}\mathbf{G}^H$$

(21.A.1)

where

$$G = \begin{bmatrix} \mathbf{B}^H\mathbf{A}\left(\sum_{i=1}^{N} c(i,1)^*\mathbf{\Omega}^{i-1}\right) \\ \mathbf{B}^H\mathbf{A}\left(\sum_{i=1}^{N} c(i,2)^*\mathbf{\Omega}^{i-1}\right) \\ \vdots \\ \mathbf{B}^H\mathbf{A}\left(\sum_{i=1}^{N} c(i,N')^*\mathbf{\Omega}^{i-1}\right) \end{bmatrix} \begin{bmatrix} \sqrt{\beta_1} & 0 & \cdots & 0 \\ 0 & \sqrt{\beta_2} & \cdots & 0 \\ \vdots & \vdots & \ddots & \vdots \\ 0 & 0 & \cdots & \sqrt{\beta_k} \end{bmatrix}$$

$$= \begin{bmatrix} \mathbf{B}^H\mathbf{a}(\theta_1)\left(\sum_{i=1}^{N} c(i,1)^*\sqrt{\beta_1}e^{j\omega_1(i-1)}\right) & \mathbf{B}^H\mathbf{a}(\theta_2)\left(\sum_{i=1}^{N} c(i,1)^*\sqrt{\beta_2}e^{j\omega_2(i-1)}\right) & \cdots & \mathbf{B}^H\mathbf{a}(\theta_k)\left(\sum_{i=1}^{N} c(i,1)^*\sqrt{\beta_k}e^{j\omega_k(i-1)}\right) \\ \mathbf{B}^H\mathbf{a}(\theta_1)\left(\sum_{i=1}^{N} c(i,2)^*\sqrt{\beta_1}e^{j\omega_1(i-1)}\right) & \mathbf{B}^H\mathbf{a}(\theta_2)\left(\sum_{i=1}^{N} c(i,2)^*\sqrt{\beta_2}e^{j\omega_2(i-1)}\right) & \cdots & \mathbf{B}^H\mathbf{a}(\theta_k)\left(\sum_{i=1}^{N} c(i,2)^*\sqrt{\beta_k}e^{j\omega_k(i-1)}\right) \\ \vdots & \vdots & \ddots & \vdots \\ \mathbf{B}^H\mathbf{a}(\theta_1)\left(\sum_{i=1}^{N} c(i,N')^*\sqrt{\beta_1}e^{j\omega_1(i-1)}\right) & \mathbf{B}^H\mathbf{a}(\theta_2)\left(\sum_{i=1}^{N} c(i,N')^*\sqrt{\beta_2}e^{j\omega_2(i-1)}\right) & \cdots & \mathbf{B}^H\mathbf{a}(\theta_k)\left(\sum_{i=1}^{N} c(i,N')^*\sqrt{\beta_k}e^{j\omega_k(i-1)}\right) \end{bmatrix}$$

$$= \begin{bmatrix} \sqrt{\beta_1}\mathbf{B}^H\mathbf{a}(\theta_1)\mathbf{c}_1^H\boldsymbol{\omega}_1 & \sqrt{\beta_2}\mathbf{B}^H\mathbf{a}(\theta_2)\mathbf{c}_1^H\boldsymbol{\omega}_2 & \cdots & \sqrt{\beta_k}\mathbf{B}^H\mathbf{a}(\theta_k)\mathbf{c}_1^H\boldsymbol{\omega}_k \\ \sqrt{\beta_1}\mathbf{B}^H\mathbf{a}(\theta_1)\mathbf{c}_2^H\boldsymbol{\omega}_1 & \sqrt{\beta_2}\mathbf{B}^H\mathbf{a}(\theta_2)\mathbf{c}_2^H\boldsymbol{\omega}_2 & \cdots & \sqrt{\beta_k}\mathbf{B}^H\mathbf{a}(\theta_k)\mathbf{c}_2^H\boldsymbol{\omega}_k \\ \vdots & \vdots & \ddots & \vdots \\ \sqrt{\beta_1}\mathbf{B}^H\mathbf{a}(\theta_1)\mathbf{c}_{N'}^H\boldsymbol{\omega}_1 & \sqrt{\beta_2}\mathbf{B}^H\mathbf{a}(\theta_2)\mathbf{c}_{N'}^H\boldsymbol{\omega}_2 & \cdots & \sqrt{\beta_k}\mathbf{B}^H\mathbf{a}(\theta_k)\mathbf{c}_{N'}^H\boldsymbol{\omega}_k \end{bmatrix} \quad (21.A.2)$$

Appendix 21.B Resolution of Spectral MUSIC

Assume within one beam there are only two closely spaced uncorrelated sources at θ_1 and θ_2, respectively [7]. We define $\hat{\mathbf{h}}_i = \mathbf{h}_i + \boldsymbol{\eta}_i$ and $\hat{\lambda}_i = \lambda_i + \beta_i$, where $\hat{\lambda}_i$ and $\hat{\mathbf{h}}_i$ are the ith eigenvalue and eigenvector of estimated covariance matrix $\hat{\mathbf{R}}_{yy}$, respectively. The following relationships hold:

$$E\{\hat{D}(\theta)\} = D(\theta) + \sigma^2 \mathbf{v}^H(\theta)\left[\frac{\lambda_1(m-2)}{(\lambda_1-\sigma^2)^2}\mathbf{h}_1\mathbf{h}_1^H + \frac{\lambda_2(m-2)}{(\lambda_2-\sigma^2)^2}\mathbf{h}_2\mathbf{h}_2^H\right]\frac{\mathbf{v}(\theta)}{T} \quad (21.B.1)$$

$$E\{\hat{D}(\theta_i)\} = \overline{D(\theta_i)} \approx \frac{(m-2)}{T}\left[\frac{1}{2(\text{ASNR})} + \frac{1}{(\text{ASNR})^2\Delta^2}\right] \quad (21.B.2)$$

$$E\{\hat{D}(\theta_m)\} \approx D(\theta_m) = \frac{1}{80}\Delta^4 \quad (21.B.3)$$

Appendix 21.B Resolution of Spectral MUSIC

where $i = 1, 2$, $\theta_m = (\theta_1 + \theta_2)/2$ and $\Delta \triangleq m2\pi d[\sin(\theta_1) - \sin(\theta_2)]/2\sqrt{3}\lambda$. The resolution threshold ξ_T is defined as the ASNR where $E\{\hat{D}(\theta_m)\} = E\{\hat{D}(\theta_1)\} = E\{\hat{D}(\theta_2)\}$. For $T \gg 1$, $\Delta \ll 1$, ASNR $\gg 1$

$$\xi_T \approx \frac{1}{T}\left\{20(m-2)\Delta^{-4}\left[1 + \sqrt{1 + \frac{T}{5(m-2)}\Delta^2}\right]\right\} \qquad (21.B.4)$$

List of Symbols

$()^T$	Transpose
$()^H$	Conjugate transpose
$\otimes$	Kronecker product
$E\{\}$	Expectation operator
θ_i	Directions of signal arrival
λ_i	Eigenvalues of covariance matrix
σ^2	Variance of noise
$\mathbf{a}(\theta)$	Signal direction vector
$\mathbf{A}$	Signal directions matrix
$\mathbf{B}$	Beamformer matrix
d	Distance between two sensors
$\mathbf{h}_i$	Eigenvectors of covariance matrix
$\mathbf{I}$	Identity matrix
k	Number of signals
$\mathbf{n}(t)$	Additive spatially white noise vector
m	Number of array sensors
m'	Dimension of beamspace
P_{MU}	Spatial spectrum of MUSIC
$\mathbf{R}$	Covariance matrix
$\mathbf{s}(t)$	Signal vector
$\mathbf{U}_s$	Signal subspace
$\mathbf{U}_n$	Noise subspace
$\mathbf{v}(\theta)$	Steering vector
$\mathbf{w}$	Beamformer weight vector
$\mathbf{x}(t)$	Vector in beamspace
$\mathbf{y}(t)$	Sensor array snapshot output at time instant
$\mathbf{y}_n(t)$	Nm-vector formed from N array signal vectors $\mathbf{y}(t)$

Acronyms

2DP-MUSIC	2-D spatial-temporal prefiltering-based MUSIC
AGS	ambiguity generator set
AIC	Akaike information criterion
ASNR	array signal-to-noise ratio
BICSSM	beamforming invariant CSSM
BS-MUSIC	beamspace MUSIC
CRB	Cramér-Rao bound
CSSM	coherent signal subspace method
DAA	direct augmentation approach
DDC	direct data covariance
DFT	discrete Fourier transform
DOA	direction of arrival
DPSS	discrete prolate spheroidal sequence
ESPAR	electronically steerable parasitic array radiator
ESPRIT	estimation of signal parameters via rotational invariance techniques
FFT	fast Fourier transform
FOV	field of view
GLRT	generalized likelihood ratio test
HF	high frequency
IMUSIC	incoherent MUSIC
ISSM	incoherent signal subspace method
ITC	information-theoretic criteria
LF	likelihood function
LR	likelihood ratio
MAP	maximum a posteriori probability
MDL	minimum description length
ML	maximum likelihood
MUSIC	multiple signal classification
NLA	nonuniform linear array
SNCR	signal-to-noise-and-clutter ratio
SNR	signal-to-noise ratio

STAP	space-time adaptive processing
SVD	singular value decomposition
TOPS	test of orthogonality of projected signal subspaces
UHF	ultra high frequency
ULA	uniform linear array
UWB	ultra wide band
VHF	very high frequency
WAVES	weighted average of signal subspace method
WSF	weighted subspace fitting

About the Authors

Habti Abeida received mastery engineering degrees in applied mathematics from Hassan II University, Casablanca, Morocco, in 2000, and from René Descartes University, Paris, France, in 2001, and a master's degree in statistics from Pierre et Marie Curie University, Paris, France, in 2002. He is currently pursuing a Ph.D. in applied mathematics and digital communications at the Institut National des Télécommunications, Evry, France. His research interests are in statistical signal processing.

Thushara D. Abhayapala obtained his undergraduate degree in interdisciplinary engineering with first class honors, and his doctoral degree in telecommunications engineering from the Australian National University (ANU). He worked in the industry for 2 years before his doctoral studies. Since 1999, he has worked for the ANU as an academic staff member at the Research School of Information Sciences and Engineering. Currently, he has an affiliation with the National ICT Australia (NICTA) as a senior researcher. His research interests include signal processing aspects of telecommunications systems, space time signal processing for wireless communication systems, spatio-temporal channel modeling, MIMO capacity analysis, space time receiver design, array signal processing, and acoustic signal processing. Dr. Abhayapala also has shared responsibility for the supervision of 12 Ph.D. students and 3 M.Phil. students and has coauthored approximately 75 publications in refereed journals and international conference proceedings. He is an editorial board member of the *EURASIP Journal on Wireless Communications and Networking* (EURASIP JWCN).

Yuri I. Abramovich received a Dipl. Eng. (Hons.) in radio electronics in 1967 and a Cand. Sci. degree (Ph.D. equivalent) in theoretical radio techniques in 1971, both from the Odessa Polytechnic University, Odessa, Ukraine. In 1981, he received a D.Sc. in radar and navigation from the Leningrad Institute for Avionics, Leningrad, Russia. From 1968 to 1994, he was a research fellow, professor, and vice-chancellor of science and research at the Odessa State Polytechnic University, Odessa, Ukraine. He has also previously worked for the Cooperative Research Centre for Sensor Signal and Information Processing (CSSIP), Adelaide, Australia. Since 2000, Dr. Abramovich has been a principal research scientist, seconded to CSSIP, at the Australian Defence Science and Technology Organisation (DSTO). His research interests are in signal processing (particularly spatio-temporal adaptive processing, beamforming, signal detection, and estimation), its application to radar (especially over-the-horizon radar), electronic warfare, and communication. He can be reached at Yuri.Abramovich@dsto.defence.gov.au.

Alon Amar received a B.Sc. in electrical engineering from the Technion-Israel Institute of Technology in 1997, and an M.Sc. in electrical engineering from Tel Aviv University, Tel Aviv, Israel, in 2003. He is currently pursuing a Ph.D. with the Department of Electrical Engineering Systems in Tel Aviv University, where he is also a teaching assistant. His main research interests are statistical and array signal processing for localization, communications, and estimation theory.

Moeness G. Amin received a Ph.D. in 1984 from the University of Colorado, Boulder. He has been on the faculty of Villanova University since 1985, where he is now a professor in the Department of Electrical and Computer Engineering and the director of the Center for Advanced Communications. Dr. Amin is a Fellow of the Institute of Electrical and Electronics Engineers (IEEE), the recipient of the IEEE Third Millennium Medal, Distinguished Lecturer of the IEEE Signal Processing Society for 2003–2004, Member of the Franklin Institute Committee of Science and Arts, Recipient of the 1997 Villanova University Outstanding Faculty Research Award, and Recipient of the 1997 IEEE Philadelphia Section Service Award. Dr. Amin has more than 280 publications in the areas of wireless communications, time-frequency analysis, smart antennas, interference cancellation in broadband communication platforms, digitized battlefield, direction finding, over-the-horizon radar, radar imaging, and channel equalizations.

Santana Burintramart received a B.S. from the Chulachomkloa Royal Military Academy, Nakhon-Nayok, Thailand, in 1998, and an M.S. from Syracuse University, Syracuse, New York, in 2004, where he is currently pursuing a Ph.D. in the Department of Electrical Engineering. He has been a member of the Royal Thai Army (RTA) since 1998. His current research interests include digital signal processing related to smart antenna problems, and space-time adaptive processing.

Christos N. Capsalis received a diploma in electrical and mechanical engineering from the National Technical University of Athens (NTUA) in 1979, and a B.S. in economics from the University of Athens in 1983. He obtained a Ph.D. in electrical engineering from NTUA in 1985. He is currently a professor and the director of the Laboratory of Long-Range Communications at the School of Electrical and Computer Engineering in NTUA. Professor Capsalis has published more than 70 papers in international journals. His current scientific activity concerns satellite and mobile communications, antenna theory and design, and electromagnetic compatibility.

Sathish Chandran obtained a Ph.D. in microwave telecommunication engineering from Loughborough University, United Kingdom, an M.Sc. in radio frequency communications engineering from the University of Bradford, United Kingdom, and a B.Tech. in electronics and communications engineering from the University of Kerala, India. Currently, he is a telecommunication consultant with various companies locally and internationally. He also serves as an adjunct faculty member at various educational establishments. Dr. Chandran has been a radio consultant to major telecommunication companies. He was employed as a postdoctoral fellow at the University of Nottingham, United Kingdom. He was also employed with Ericsson, Malaysia. Dr. Chandran was the president of the International Union of Radio Science (URSI), Malaysia. He has published works in his area of specialization and is the editor of *Adaptive Antenna Arrays: Trends and Applications* (Springer, 2004). He is also a member of the editorial board member of several international scientific and technical journals and serves on the technical review committee of various international conferences. Dr. Chandran is a Fellow of the IEE (United Kingdom).

Pascal Chargé received a Ph.D. in electrical engineering from the University of Nantes, France, in 2001. Until 2003, he was a research fellow with the electronic systems, telecomunications, and signal processing division at the IRCCyN Laboratory. Since September 2003, he has been an assistant professor at the National Institute of Applied Sciences (INSA) of Toulouse, and he is also a research fellow with the LESIA Laboratory. His research interests include sensors array processing, statistical signal processing, and signal processing for wireless communications.

Jiunn-Tsair Chen received a Ph.D. from Stanford University, Stanford, California, in 1998. Since 1999, he has been on the faculty of National Tsing-Hua University, Hsinchu, where he is an associate professor with the Department of Electrical Engineering. His current research interests include wireless communications, antenna array signal processing, adaptive signal processing, and power amplifier linearization.

Jun Cheng received a B.S. and an M.S. in telecommunications engineering from Xidian University, Xi'an, China, in 1984 and 1987, respectively, and a Ph.D. in electrical engineering from Doshisha University, Kyoto, Japan, in 2000. From 1987 to 1994, he was an assistant professor and lecturer in the Department of Information Engineering, Xidian University. From 1995 to 1996, he was an associate professor in the National Key Laboratory on Integrated Service Network, Xidian University. Dr. Cheng has previously worked as a visiting researcher at the ATR Adaptive Communications Research Laboratories in Kyoto, Japan; as a staff engineer at the R&D Center, Panasonic Mobile Communications Co., Ltd. (formerly Wireless Solution Labs, Matsushita Communication Industrial Co., Ltd.), Yokosuka, Japan; and as a staff engineer at Next-Generation Mobile Communications Development Center, Matsushita Electric Industrial Co., Ltd., Japan. Since April 2004, he has been an associate professor in the Department of Knowledge Engineering and Computer Science at Doshisha University, Kyotanabe-shi, Japan. His research interests are in the areas of communications theory, information theory, coding theory, array signal processing, and radio communication systems.

Jean Pierre Delmas received an engineering degree from Ecole Centrale de Lyon, France, in 1973, the Certificat d'études supérieures from the Ecole Nationale Supérieure des Télécommunications, Paris, France, in 1982, and the Habilitation à diriger des recherches (HDR) degree from the University of Paris, Orsay, France, in

2001. Since 1980, he has worked for the Institut National des Télécommunications, where he is presently a professor in the CTTI Department and in UMR-CNRS 5157. His teaching and research interests are in the areas of statistical signal processing, with application to communications and antenna array. He is currently an associate editor for the *IEEE Transactions on Signal Processing*.

Wen-Hsien Fang received a Ph.D. from University of Michigan, Ann Arbor, in 1991. In September 1991, he joined the faculty of National Taiwan University of Science and Technology, Taipei, where he is a professor with the Department of Electronic Engineering. His research interests include signal processing for wireless communications, fast algorithms and their VLSI implementations, and video coding.

Alfonso Farina received his doctorate in electronic engineering from the University of Rome (I) in 1973. In 1974, he joined Selenia, now Alenia Marconi Systems, where he has been a manager since May 1988. Since 2004, he has been a member of the Chief Technical Office and he was recently appointed scientific director. Dr. Farina has provided technical contributions to detection, signal, data and fusion, and image processing for radar systems, and he has provided leadership in many projects—also conducted in the international arena—in surveillance for ground and naval applications, in airborne early warning, and in imaging radar. Since 1979, he has also been *Professore Incaricato* of Radar Techniques at the University of Naples. In 1985, he was appointed to be an associate professor. He is the author of more than 330 peer-reviewed publications and the author of the following books and monographs: *Radar Data Processing* (Volumes 1 and 2) (translated in Russian and Chinese), 1985–1986; *Optimised Radar Processors*, 1987; and *Antenna Based Signal Processing Techniques for Radar Systems*, 1992. Dr. Farina has written Chapter 9 on "ECCM Techniques" in the *Radar Handbook, Second Edition* (1990), edited by Dr. M. I. Skolnik of the Naval Research Laboratory. He has been session chairman at many international radar conferences. He has lectured at universities and research centers in Italy and abroad. He also frequently gives tutorials at international radar conferences on signal, data and image processing for radar, in particular, on multisensor fusion, adaptive signal processing, and space-time adaptive processing (STAP) and detection. Dr. Farina is a referee of numerous publications submitted to several journals of IEEE, IEE, and Elsevier. He has also cooperated with the editorial board of *Electronics & Communication Engineering Journal* (ECEJ) of the IEE. Dr. Farina also serves as a member on the editorial board of *Signal Processing*. He has received several awards. Dr. Farina has been appointed as the NATO-SET (Sensor and Electronic Technology) Panel Member at Large. He is also a Fellow of the IEEE and a Fellow of the IEE.

Fulvio Gini received his Doctor Engineer (cum laude) and Research Doctor degrees in electronic engineering from the University of Pisa, Italy, in 1990 and 1995, respectively. In 1993 he joined the Department of "Ingegneria dell'Informazione" of the University of Pisa, where he has been an associate professor since October 2000. Dr. Gini has been a visiting researcher at the Department of Electrical Engineering, University of Virginia; a session chairman at international conferences; and the coauthor of two tutorials, entitled "Coherent Detection and Fusion in

High Resolution Radar Systems," presented at the 1999 International Conference on Radar, and "Advanced Radar Detection and Fusion," presented at the 2000 International Radar Conference. He is also an associate editor for the *IEEE Transactions on Signal Processing* and a member of the EURASIP JASP editorial board. He was corecipient of the 2001 IEEE Aerospace and Electronic Systems Society's Barry Carlton Award for Best Paper. He was recipient of the 2003 IEE Achievement Award for outstanding contribution in signal processing, and of the 2003 IEEE Aerospace and Electronic Systems Society Nathanson Award to the Young Engineer of the Year. He was a coguest editor of a special issue of the EURASIP *Signal Processing* journal on "New Trends and Findings in Antenna Array Processing for Radar," published in 2004, and is a coguest editor of a special issue of the *IEEE Signal Processing Magazine* on "Knowledge Based Systems for Adaptive Radar Detection, Tracking, and Classification," to be published in 2006. His general interests are in the areas of statistical signal processing, estimation, and detection theory, and, in particular, modeling and statistical analysis of recorded live sea and ground radar clutter data, non-Gaussian signal detection and estimation, parameter estimation and data extraction from multichannel interferometric SAR data, cyclostationary signal analysis, and estimation of nonstationary signals, with applications to radar signal processing. He authored or coauthored more than 70 journal papers and approximately 70 conference papers.

Giovanni Golino received his doctorate in electronic engineering from the University of Rome (II) in 1998. In 1999, he joined AMS, where he is a radar designer in the Radar and Technology & Operation Division, Systems Analysis Group. His actual areas of investigation are high-resolution radars, ECCM techniques, and systems of sensors.

Alexei Y. Gorokhov received an Ms.D. in 1993 from the Odessa Polytechnic University (OPU), Odessa, Ukraine, and a Ph.D. in 1997 from the École Nationale Supérieure des Télécommunications, Paris, France. From 1993 to 1994, he served as a research engineer with the research laboratory of the OPU. Since October 1997, he has been affiliated with the Centre National de la Recherche Scientifique (CNRS), Paris, France, where he holds a research position. From January 2000 to January 2004, he served as a senior scientist within the DSP group of Philips Research Laboratories, Eindhoven, the Netherlands, on long-term leave from CNRS. Since January 2004, he has been with QUALCOMM Inc., San Diego, California. His general interests cover wireless communications, spectral analysis, statistical signal processing, and information theory. His current research is mainly focused on error control coding, multiuser equalization, and antenna diversity for wireless communications. Dr. Gorokhov can be reached at gorokhov@qualcomm.com.

Maria Greco graduated in electronic engineering in 1993, and received a Ph.D. in telecommunication engineering in 1998 from University of Pisa, Italy. From December 1997 to May 1998, she worked at the Georgia Tech Research Institute, Atlanta, Georgia, as a visiting research scholar, where she carried on research activity in the field of radar detection in non-Gaussian background. In 1998, she joined the

Department of "Ingegneria dell'Informazione" of the University of Pisa, where she has been an assistant professor since April 2001. Her general interests are in the areas of statistical signal processing, estimation, and detection theory. In particular, her research interests include cyclostationarity signal analysis, clutter models, spectral analysis, DOA estimation, coherent and incoherent detection in non-Gaussian clutter, and CFAR techniques. She has been an IEEE Senior Member since June 2004, and a referee of many journals of IEEE, IEE, and Elsevier. She was corecipient, with P. Lombardo, F. Gini, A. Farina, and B. Billingsley, of the 2001 IEEE Aerospace and Electronic Systems Society's Barry Carlton Award for Best Paper.

Katsuyuki Haneda received a B.E. and an M.E. from the Tokyo Institute of Technology, Japan, in 2002 and 2004, respectively. His current interests are wireless propagation, channel modeling, array signal processing, MIMO systems, and ultrawideband radio. He won the Student Paper Award in the 7th International Symposium on Wireless Personal Multimedia Communications (WPMC 2004). He is a student member of the IEICE and IEEE.

K. V. S. Hari received a B.E. from Osmania University, Hyderabad, India, in 1983; an M.Tech. from the Indian Institute of Technology, New Delhi, India (IIT Delhi), in 1985; and a Ph.D. from the University of California, San Diego, La Jolla, in 1990. He has been an associate professor with the Department of Electrical Communication Engineering, Indian Institute of Science (IISc), Bangalore, India, since February 1998. His research interests are in statistical signal processing. He has worked on space-time signal processing algorithms for direction-of-arrival estimation, acoustic signal separation using microphone arrays, and MIMO wireless communication systems. He has also worked on MIMO wireless channel measurements and modeling, and is the coauthor of the IEEE 802.16 (WiMax) standard on wireless channel models for fixed broadband wireless communication systems. Dr. Hari has worked as a visiting member of the faculty at Helsinki University of Technology, Espoo, Finland; Stanford University, Stanford, California; and the Royal Institute of Technology, Stockholm, Sweden. He was an assistant professor with the Department of Electrical and Computer Engineering, IISc, and a scientist at Osmania University. He also worked for the Defence Electronics Research Lab, Hyderabad. He is currently a member of the editorial board of EURASIP's journal, *Signal Processing*.

Randy Haupt is an IEEE Fellow and senior scientist at the Penn State Applied Research Laboratory. He has a Ph.D. in electrical engineering from the University of Michigan, an M.S. in electrical engineering from Northeastern University, an M.S. in engineering management from Western New England College, and a B.S. in electrical engineering from the U.S.A.F. Academy. He was professor and department head of electrical and computer engineering at Utah State University from 1999 to 2003. He was a professor of electrical engineering at the U.S. Air Force Academy, and a professor and the chair of electrical engineering at the University of Nevada, Reno. In 1997, he retired as a lieutenant colonel in the U.S. Air Force. Dr. Haupt was a project engineer for the OTH-B radar, and a research antenna engineer for Rome Air Development Center. He was the Federal Engineer of the Year in 1993 and is a member of Tau Beta Pi, Eta Kappa Nu, URSI Commission B, and the Electromagnetics Academy. He has many journal articles, conference publications, and book chapters on antennas, radar cross-section, and numerical methods, and is the coauthor of the book *Practical Genetic Algorithms, Second Edition* (John Wiley & Sons, 2004). Dr. Haupt has eight patents in antenna technology.

Joby Joseph obtained a B.Tech in ECE in 1995 from Kerala University, India; and an M.E. in ECE in 1998, and a Ph.D. in 2004, both from the Indian Institute of Science, Bangalore, India. He is currently a researcher at the National Institute of Health, Maryland. During 1990, he obtained a DAAD fellowship, and visited the Control Systems and Signal Theory Group, University of Kaiserslautern, Germany, where he worked on automatic extraction of diagnostic parameters from ultrasound scan images of kidneys. His research interests are in MIMO signal processing, acoustic source separation using microphone arrays, modeling neural pathways, image processing, and adaptive signal processing. He has developed an open source audio analyzer tool (SAA, SSPLab Audio Analyzer), which is available at http://www.dsp.ece.iisc.ernet.in/~joby/saajoby/.

Lance M. Kaplan received a B.S. with distinction from Duke University, Durham, North Carolina, in 1989; and an M.S. and a Ph.D. from the University of Southern California, Los Angeles, in 1991 and 1994, respectively, all in electrical engineering. From 1987 to 1990, Dr. Kaplan worked as a technical assistant at the Georgia Tech Research Institute. He held a National Science Foundation Graduate Fellowship and a USC Dean's Merit Fellowship from 1990 to 1993, and worked as a research assistant in the Signal and Image Processing Institute at the University of Southern California from 1993 to 1994. He worked on the staff in the Reconnaissance Systems Department of the Hughes Aircraft Company from 1994 to 1996. From 1996 to 2004, he served on the faculty at Clark Atlanta University, in the Department of Engineering and the Center for Theoretical Studies of Physical Systems (CTSPS). Currently, he is a team leader in the EO/IR Image Processing Branch of the U.S. Army Research Laboratory. Dr. Kaplan serves as associate editor-in-chief and EO/IR systems editor for the *IEEE Transactions on Aerospace and Electronic Systems*. He was a three-time recipient of the Clark Atlanta University Electrical Engineering Instructional Excellence Award, from 1999 to 2001. His current research interests include signal and image processing, data fusion, automatic target recognition, modeling and synthesis, and multiresolution analysis.

R. Lynn Kirlin received a B.S.E.E. and an M.S.E.E. from the University of Wyoming in 1962 and 1963, and a Ph.D./E.E. from Utah State in 1968. His industrial experience includes communications systems at Martin-Marietta and Boeing, from 1963 to 1966; computer peripherals at Datel, Wyoming, 1969; applications software at Floating Point Systems, Oregon, 1979; and recent work at Rincon Research, Tucson. He was with the EE Department at the University of Wyoming from 1969 to 1986, and was with the ECE Department at the University of Victoria, Canada, from 1987 to 2002, where he is now Professor Emeritus. His major research and consulting associations have included U.S. and Canadian naval organizations and oil industries in the areas of statistical array signal processing. He was an associate editor for the *IEEE Transactions on Signal Processing*, from 1989 to 1991. He is a coauthor of the 1998 Best Paper in Geophysics and is the coeditor and major author of *Covariance Matrix Analysis for Seismic Signal Processing*. He is a Fellow of the IEEE.

Visa Koivunen received a D.Sc. (Tech) with honors from the University of Oulu, Department of Electrical Engineering, and was selected the primus doctor (best graduate) in 1990 to 1994. Dr. Kovunen was a visiting researcher at the University of Pennsylvania, Philadelphia, Pennsylvania; a faculty staff member at the Department of Electrical Engineering, University of Oulu; and an associate professor at the Signal Processing Laboratory, Tampere University of Technology. Since 1999, he has been a professor of signal processing in the Department of Electrical and Communications Engineering, Helsinki University of Technology (HUT), Finland. He is one of the principal investigators in SMARAD Center of Excellence in Radio and Communications Engineering nominated by the Academy of Finland. Since 2003, he has been an adjunct professor at the University of Pennsylvania. Dr. Koivunen's research interests include statistics, communications, and sensor array signal processing. He has published more than 160 papers in international scientific conferences and journals. He is a member of the editorial board of the *Signal Processing* journal. He is also a member of the IEEE Signal Processing for Communication Technical Committee (SPCOM-TC) and a senior member of the IEEE.

Xiaoli Lu received an M.S.E.E. from the University of Victoria in 2002 and is a Ph.D. candidate in the same university. Her major research and interests have included the areas of statistical signal processing, classification, and optimization. She is a holder of NSERC fellowship (national), Petch Research Scholarship, and other university awards. She is working with Raytheon Canada, Ltd. as an intern. She is a member of the IEEE.

Rodney A. Martin is an assistant staff engineer in the Communication and Navigation Office of the Penn State Applied Research Laboratory. While working at ARL, he has been involved in a variety of antenna and propagation projects, including the development of a radar-detecting phased array, a Java-based suite of antenna models, conformal antenna analysis software, urban RF propagation analysis, and fractal antenna engineering. Prior to college, Mr. Martin served 6 years as an electrician's mate in the Naval Nuclear Power Program—four of those years onboard a Sturgeon (637) class submarine. He received a B.S. and an M.S. in electrical engineering from the Pennsylvania State University, University Park, in 1999 and 2001, respectively. His research interests include antenna design, numerical methods, and computational electromagnetics. He is also a member of Eta Kappa Nu, and was the recipient of a College of Engineering Dean's Fellowship.

James H. McClellan received a B.S. in electrical engineering from Louisiana State University, Baton Rouge, in 1969, and M.S. and Ph.D. degrees from Rice University, Houston, Texas, in 1972 and 1973, respectively. From 1973 to 1982, he was a member of the research staff, and later a professor, with the Lincoln Laboratory, Massachusetts Institute of Technology, Lexington. From 1982 to 1987, he was with Schlumberger Well Services. Since 1987, he is a professor with the School of Electrical and Computer Engineering, Georgia Institute of Technology (Georgia Tech), Atlanta, where he presently holds the Byers Professorship in Signal Processing. He is a coauthor of the texts *Number Theory in Digital Signal Processing, Computer Exercises for Signal Processing, DSP First: A Multimedia Approach,* and *Signal Processing First,* which received the McGraw-Hill Jacob Millman Award for an innovative textbook in 2003. Professor McClellan received the W. Howard Ector Outstanding Teacher Award from Georgia Tech in 1998, and the Education Award from the IEEE Signal Processing Society in 2001. In 1987, he received the Technical Achievement Award for work on FIR filter design, and in 1996, the Society Award, both from the IEEE Signal Processing Society. In 2004, he was a co-recipient of the IEEE Kilby medal. He is a member of Tau Beta Pi and Eta Kappa Nu.

Stelios A. Mitilineos received his diploma in electrical and computer engineering from the National Technical University of Athens (NTUA) in October 2001. He is currently working toward a Ph.D. in electrical engineering at the same university. His main research interests are in the areas of antennas and propagation, smart antennas and mobile communications, and electromagnetic compatibility.

Baha Adnan Obeidat received a B.S. in electrical engineering from Temple University, Philadelphia, in 2002, and an M.S. in electrical engineering from Villanova University, Villanova, Pennsylvania, in 2004. He is currently pursuing a Ph.D. in electrical engineering at Villanova University. His research interests are in the areas of array processing, statistical signal processing, and time-frequency analysis.

Takashi Ohira received a D.E. in communication engineering from Osaka University, Japan, in 1983. He developed GaAs MMIC transponder modules and microwave beamforming networks aboard Japan national multibeam communication satellites at NTT Wireless Systems Laboratories, Yokosuka, Japan. From 1999, he was engaged in research on wireless ad hoc networks and microwave analog adaptive antennas for consumer electronic devices at ATR Adaptive Communications Research Laboratories, Kyoto, Japan. Currently, Dr. Ohira is the director for ATR Wave Engineering Laboratories. He serves as chair for URSI Commission C Japan Branch. He is an IEEE Fellow. He coauthored *Monolithic Microwave Integrated Circuits* (IEICE, 1997). Dr. Ohira was awarded the 1986 IEICE Shinohara Prize, the 1998 APMC Japan Microwave Prize, and the 2004 IEICE Electronics Society Prize.

Esa Ollila received a Ph.D. in statistics from the University of Jyväskylä in 2002. Currently, he is working as a postdoctoral fellow in SMARAD CoE, Helsinki University of Technology. His research interests are in robust and nonparametric multivariate statistics, array signal processing, and statistics of complex valued signals.

Stylianos C. Panagiotou received his diploma in electrical and computer engineering from the National Technical University of Athens (NTUA) in 2003. He is currently working toward a Ph.D. in electrical engineering at the same university. His main research interests are in the fields of multipath propagation, smart antennas, wireless MIMO systems, and antenna design.

Tapan K. Sarkar is a professor in the Department of Electrical and Computer Engineering at Syracuse University. His current research interests deal with numerical solutions of operator equations arising in electromagnetics and signal processing with application to system design. He has authored or coauthored several books, including *Iterative and Self Adaptive Finite-Elements in Electromagnetic Modeling* (Artech House, 1998), *Wavelet Applications in Electromagnetics and Signal Processing* (Artech House, 2002), *Smart Antennas* (John Wiley & Sons, 2003), and *History of Wireless* (John Wiley & Sons, 2005). He received Docteur Honoris Causa, both from the Universite Blaise Pascal, Clermont Ferrand, France, in 1998, and from the Politechnic University of Madrid, Madrid, Spain, in 2004. He received the medal of the Friend of the City of Clermont Ferrand, France, in 2000.

Daniel Segovia-Vargas received a telecommunication engineering degree at Polytechnic University of Madrid in 1993. In 1998, he received a Ph.D. from the Polytechnic University of Madrid. From 1993 to 1998, he worked at Valladolid University, teaching in the areas of radiocommunication and electronic design. Presently, he is professor at Carlos III University in Madrid, where he is in charge of the Antennas, Microwaves, and Radiocommunication courses. His research areas are devoted to the fields of printed antennas, and active and smart antennas. He is also interested in new materials for the design of microwave circuits and antennas. He has authored more than 40 papers in international journals and in international conferences. He has also been a member of the European Projects COST 260 and COST 284. He has taken part in several research projects in the area of active and smart antennas.

Manuel Sierra-Castañer received a degree of engineer of telecommunication in 1994, and a Ph.D. in 2000, both from the Polytechnic University of Madrid, Spain. In 1997, he was with the University "Alfonso X" as assistant, and since 1998, with the Polytechnic University of Madrid, as a research assistant, an assistant, and an associate professor. He is a member of the IEEE. His current research interests are planar antennas and antenna measurement systems.

Manuel Sierra-Pérez received an M.S. in 1975 and a Ph.D. in 1980, from the Universidad Politécnica de Madrid. He became full professor in the Department of Signals, Systems, and Radiocommunications at the same university in 1990. His current main interest is in passive and active array antennas, including design theory, measurement, and applications. He was an invited professor at the National Radio-Astronomy Observatory (NRAO) in Virginia, from 1981 to 1982, and at the University of Colorado, Boulder, from 1994 to 1995. He has been a member of IEEE since 1978. He was promoter and chairman of the IEEE Joint AP/MTT Spanish Chapter. He was the treasurer of the IEEE Spain Section, and is currently vice-chairman of the same section.

Apostolos I. Sotiriou graduated from Aristotle University of Thessaloniki (AUTH) in July 1999, receiving his diploma in electrical and computer engineering. In April 2001, he joined the transmission development team at COSMOTE MOBILE TELECOMMUNICATIONS S.A., where he currently works as a telecommunications engineer. In February 2002, he started working toward a Ph.D. in electrical engineering at the National Technical University of Athens (NTUA). His main research interests are in the areas of antennas and propagation, smart antennas, and cellular systems performance issues.

Nicholas K. Spencer received a B.Sc. (Hons.) in applied mathematics in 1985, and an M.Sc. in computational mathematics in 1992, both from the Australian National University, Canberra. He has worked with the Australian Department of Defence, Canberra; the Flinders University of South Australia, Adelaide; the University of Adelaide; and the Australian Centre for Remote Sensing, Canberra, in the areas of computational and mathematical sciences. He is currently a senior research fellow at the Cooperative Research Centre for Sensor Signal and Information Processing (CSSIP), Adelaide, Australia. His research interests include array signal processing, parallel and other high-performance computing, software best-practice, human-machine interfaces, multilevel numerical methods, modeling and simulation of physical systems, theoretical astrophysics, and cellular automata.

Jun-ichi Takada received a B.E., an M.E., and a D.E. from the Tokyo Institute of Technology, Japan, in 1987, 1989, and 1992, respectively. From 1992 to 1994, he was a research associate at Chiba University, Chiba, Japan. Since 1994, he has been an associate professor at Tokyo Institute of Technology, Tokyo, Japan. His previous research covered small antennas, radial line slot antennas, microwave hyperthemia, and numerical electromagnetics. His current research interests are wireless propagation and channel modeling, adaptive array and diversity antennas, ultrawideband radio, software defined radio, and applied radio instrumentation and measurements. He has been an active member of European COoperation in the field of Scientific and Technical research (COST) action 273 "Towards Mobile Broadband Multimedia Networks." Since 2003, he has been a part-time specialty researcher in the UWB Institute, National Institute of Communications and Information Technology (NICT), Japan, where he contributes to the propagation measurement and modeling, as well as the standardization of the UWB signal measurement in ITU-R. He is a member of IEEE, IEICE Japan, ACES, and ECTI Association, Thailand.

Luca Timmoneri received his doctorate in electronic engineering from the University of Rome, Italy, in 1989. In 1989, he joined Selenia S.p.A., now Alenia Marconi Systems (AMS), where he is currently the head of the Radar System Analysis Group of the Radar and Technology Division. His working interests span from the area of synthetic aperture radar (image formation and moving target detection and imaging), to space-time adaptive processing for AEW and ground-based radar, to parallel processing architectures with VLSI and COTS devices. He is presently involved in the areas of adaptive signal processing, detection and estimation with application to tridimensional ground and ship-based phased-array radar. He is the author of several peer-reviewed papers (also invited) on journals and conference proceedings. He is the coauthor of three tutorials on adaptive array and space-time adaptive processing, presented at the International IEEE Radar Conference in Washington, D.C., 1995, and in Boston, 1999 and 2003. Dr. Timmoneri received the 2002 AMS CEO Award, the 2003 AMS MD Award, the 2004 AMS CEO Award for Innovation Technology, and the first-prize award for Innovation Technology of Finmeccanica 2004.

Hiroaki Tsuchiya received a B.E. from Tokyo University of Agriculture and Technology, Japan, in 2003, and an M.E. from Tokyo Institute of Technology, Japan, in 2005. At present, he is studying towards a Ph.D. at Tokyo Institute of Technology. His research interests focus on the wireless propagation channel characterization and modeling of ultrawideband radio. He is currently a student member of IEICE, Japan.

Pantelis K. Varlamos received his diploma in electrical and computer engineering from the National Technical University of Athens (NTUA) in October 2000. He is currently working toward a Ph.D. in electrical engineering at the same university, supported by a scholarship from the Bodossakis Foundation. His main research interests are in the areas of antennas and propagation, smart antennas, and electromagnetic compatibility.

Jian Wang received an M.S.E.E. and a Ph.D./E.E. from the University of Victoria, in 2001 and 2004, respectively. His major research and interests have included the areas of statistical, radar, and array signal processing. He is working with Raytheon Canada, Ltd., since 2003. He is a member of the IEEE.

Yide Wang received a B.S. from Beijing University of Post and Telecommunication (BUPT), Beijing China, in 1984; and an M.S. and a Ph.D. in signal processing from the University of Rennes, France, in 1986 and 1989, respectively. He is now a professor at the polytechnic school of the University of Nantes, France. His research interests include array processing, spectral analysis, and mobile wireless communications systems.

Yung-Yi Wang received a Ph.D. from National Taiwan University of Science and Technology, Taipei, in 2000. Since 1994, he has been on the faculty of St. John's & St. Mary's Institute of Technology, Taipei, where he is an associate professor with the Department of Computer and Communications. Dr. Wang's research interests include statistical signal processing, wireless communication, and array signal processing.

Anthony J. Weiss received a B.Sc. from the Technion-Israel Institute of Technology, in 1973; and an M.Sc. and a Ph.D. from Tel Aviv University, Tel Aviv, Israel, in 1982 and 1985, respectively, all in electrical engineering. From 1973 to 1983, he was involved in the research and development of numerous projects in the fields of communications, command and control, and direction finding. In 1985, he joined the Department of Electrical Engineering-Systems, Tel Aviv University. From 1996 to 1999, he served as the department chairman, and as the chairman of IEEE Israel Section. In 1996, he cofounded Wireless Online Ltd., and served as the chief scientist for six years. Between 1998 and 2001, he also served as the chief scientist of SigmaOne Communications Ltd. His research interests include detection and estimation theory, signal processing, and sensor array processing, with applications to radar and sonar. Professor Weiss published more than 100 papers in professional magazines and conferences, and he holds nine U.S. patents. He was a recipient of the IEEE 1983 Acoustics, Speech, and Signal Processing Society's Senior Award, and the IEEE Third Millennium Medal. He has been an IEEE Fellow since 1997, and an IEE Fellow since 1999.

Yeo-Sun Yoon received a B.S. from Yonsei University, Seoul, Korea, in 1995, and an M.S. from the University of Michigan, Ann Arbor, Michigan, in 1998, both in electrical engineering. From 1999 to 2004, he was a graduate research assistant in the Center for Signal and Image Processing (CSIP) at the Georgia Institute of Technology, Atlanta, where he received a Ph.D. in 2004 in electrical engineering. From 2000 to 2002, he was a research assistant at the Center for Theoretical Studies of Physical Systems (CTSPS), Atlanta. He now works at the Samsung Thales Company, Korea, as a senior engineer. His research interests include array signal processing, signal parameter estimation, and radar signal processing for tracking.

Yimin Zhang received a Ph.D. from the University of Tsukuba, Tsukuba, Japan, in 1988. He joined the faculty of the Department of Radio Engineering, Southeast University, Nanjing, China, in 1988. He served as a technical manager at the Communication Laboratory Japan, Kawasaki, Japan, from 1995 to 1997, and was a visiting researcher at ATR Adaptive Communications Research Laboratories, Kyoto, Japan, from 1997 to 1998. Since 1998, he has been with Villanova University, where he is currently a research associate professor at the Center for Advanced Communications. Dr. Zhang's research interests are in the areas of array signal processing, space-time adaptive processing, multiuser detection, MIMO systems, cooperative networking, blind signal processing, digital mobile communications, and time-frequency analysis. Dr. Zhang is a Senior Member of the IEEE.

Index

2DP-MUSIC algorithm, 420, 423–27
  defined, 423
  null spectrum expectation, 429–30
  performance gain, 431
  pseudospatial spectrum, 432, 433, 435
  resolution behavior, 428
  resolution threshold characteristic, 431
  sample null spectra formulation, 428–29
  summary, 427
  See also MUSIC; Pulsed Doppler estimation

A

Acoustic source algorithm
  experimental results, 316–23
  parameter values, 316–18
  performance for sinusoidal sources, 318
  performance for synthesized array data, 317–18
  performance measure for listening tests, 321–22
  performance of sound source separation, 320–21
  performance with headlike structure, 322–23
  performance with increasing number of sources, 318–19
  performance with sensor spacing, 319–20
  resolvability, 320
  simulation conditions, 318
Adcock arrays, 9–10
  azimuth angle, 10
  defined, 9
  elevation angle estimate, 10
  illustrated, 9
  See also Arrays
Affine equivariance, 248
AML-RELAX algorithm, 327, 332–34
  defined, 332
  performance, 334
  RMSE, 337
  steps, 332–34
  See also Asymptotic ML (AML)
Amplitude
  comparison, 11
  complex, estimation, 92
  monopulse, 15
Amplitude modulated (AM) signals, 192
Angle-of-arrival. *See* AOAs
Angle-SINR Table (AST), 414
Angular coordinates
  CRLB estimation, 288–91
  MLE estimation, 286–88
Angular cross-correlation kernel, 193
Antennas
  canonical minimum scattering (CMSA), 269
  coupling in, 386
  DF, common, 6–7
  DF, spinning, 4–6
  directivity, 3
  directivity, deconvolution, 356
  ESPAR, 395–415
  farfield, calibration system, 388
  horns, 7
  mechanically rotating, 327–42
  minimum scattering (MSAs), 269
  receiving, modeling, 269
  reflectors, 6–7

459

Antennas (continued)
    single-port, correlation matrix, 402–3
    spirals, 7
    steering vector model, 260–64
    transmitting/receiving, configuration, 96
    See also Arrays
AOAs
    estimates, 213
    measurements, 214
    RMS error, 223, 224, 225, 226
Array factors
    circular, 15
    defined, 7–8
    four-element arrays, 11
    planar array, 17
    ULA, 8
    Wullenweber, 17
Array measurement, 381–93
    calibration process, 386–92
    conclusions, 392–93
    introduction, 381–82
    steering vector model, 382–86
Array model, 192–97
    coherently distributed, 193–95
    errors, 245
    incoherently distributed, 195
    See also Scattered sources
Array outputs, 7
    covariance for, 51
    frequency support, 50
    sequence of snapshots, 51
Arrays, 7–18
    Adcock, 9–10
    Butler matrix, 12–15
    circular, 15
    coupling model, 259–80
    DF algorithms and, 18
    dual-polarized, 106, 110
    elements, 7
    element spacing, 10–12
    four double-polarized patches, 277
    four-element, 11
    initial calibration, 259
    linear, 8–9
    on moving platforms, 18
    multiple beams, 11, 12
    as multipole network, 268
    performance, 7
    planar, 17–18
    processing examples, 254–55
    response, 7
    scattering matrix, 385
    snapshot in, 265
    steering vectors, 74
    switched parasitic (SPAs), 145, 146–48
    uniform circular (UCAs), 254, 255
    uniform linear (ULAs), 10, 50, 237–38, 243, 276
    Wullenweber, 15–17
Array signal model, 162–63
    covariance matrices, 162
    observation vectors, 162
    performance analysis, 163
Array-signal-to-noise ratio (ASNR), 427
Asymptotically minimum variance (AMV)
    algorithms, 161, 178, 179
    bounds, 161, 172, 188
    estimator, 172, 188
Asymptotic ML (AML), 327
    estimator, 331
    for known Doppler frequencies, 332
    See also AML-RELAX algorithm
Azimuth
    angle, 57, 297
    azimuth angle vs., 297
    elevation difference, 299
    pattern of difference, 287
    sum beam difference, 299, 300
    target, 301

B

Beamformers
    eigenspace, 272
    MVDR, 203, 265
Beamforming
    for cyclostationary signals, 374
    minimum variance, for cyclostationary signals, 374–75
    nonselective, 372–76
    variable, 409

Beamforming-invariant CSSM (BICSSM), 54–55
  defined, 54–55
  performance, 55
  transformation matrices, 55
Beamforming methods
  GC-1, 202, 209
  GC-2, 202–3, 209
  ID sources, 202–3
Beamspace MUSIC (BS-MUSIC), 420, 421–23
  defined, 420
  DPSS, 435
  resolution analysis, 428
  resolution behavior, 428
  review, 421–23
  sample null spectra formulation, 428–29
  spectrum, 422–23
  target direction and, 422
Bessel function, 75, 76
Binary phase shift keying (BPSK), 161, 364
  cyclic autocorrelation function, 368
  modulated signals, 192
  stochastic CRB for, 175–77
  stochastic sources for uniform white noise, 184
  unfiltered, 164, 177
Box-related transfer function (BRTF), 312, 321
Breakdown point, 248
Broadband sources
  DOA estimation, 305–23
  DOA estimation based on time delays/relative attenuation, 308–13
  experimental results, 316–23
  natural sound, 307–8
  recoverable, 307
  scenario and model, 306–7
  source separation algorithm, 314–16
  spectral components decomposition, 308
  spectrally weighted DOA estimation, 313–14

Butler matrix, 12–18
  beam locations, 14
  defined, 12
  for eight-element array, 15
  four-element, 13
  phase shifts, 14
  See also Arrays

C

Calibration, 386–92
  application example, 391–92
  coupling in RF subsystem, 390
  coupling through terminals, 387–88
  ESPAR antenna, 410–12
  extra phase measurement and, 391
  farfield antenna, 388
  matrix, 393
  process, 386–92
  radiation patterns measurement, 388–89
  radiation pattern with/without, 392
  for RF subsystem, 389–91
  self, 392–93
Canonical minimum scattering antennas (CMSA), 269
Capon method, 265, 375
Cauchy distributions, 248
CD sources
  data examples, 199–200
  DOA estimation, six sensors, 199
  DSPE algorithm, 197
  estimation techniques, 197–200
  See also Coherently distributed source model; Parameters estimation techniques
Circular polarization
  RMSE for estimated DOA, 278
  RMSE for estimated DOA (64 samples), 279
  RMSE for estimated DOA (256 samples), 279
Coarrays, 21
Coherently distributed source model, 193–95
  deterministic angular weighting function, 194
  generalized steering vector, 194–95

Coherent repeater interference (CRI), 296
Coherent signal subspace method
 (CSSM), 53–54
 beamforming-invariant (BICSSM),
  54–55
 defined, 53
 estimators, 54
 low SNR and, 54
Coherent wideband processing
 DOA estimation, 52–56
 focusing matrices for, 73–74
Compensated coupling estimation,
 273–78
Correlation matrix
 conventional antenna array, 402
 in reactance domain, 403
 for single-port antennas, 402–3
Coupling
 changes, 267
 coefficients, 269, 275
 correction schemes, 278
 in DOA estimation, 272–73
 EM matrix, 259
 input impedance change, 267
 levels, 273
 matrix, 271
 matrix representation, 386
 reception matrix, 386
 through terminals, 387–88
Coupling model, 266–72
 characterization, 266
 introduction, 259–60
 steering vector model, 260–64
Covariance fitting methods, 203–6
Covariance matching estimation
 technique extended invariance
 principle (COMET-EXIP)
 cost function, 204
 defined, 203
 estimates, 203
 estimator, 206
Covariance matching estimation
 techniques (COMET), 171
Covariance matrix, 51–52
 array measurements, 195
 array signal model, 162
 disturbance, 286, 288

emitted signal, 195
estimated, 62
fading vector, 126
focused data, 74
frequency averaged data, 73
modal, 79
noise-free, 196, 197
rank (RCM), 251
received signal, 127
spatial, 126
spatial sign, 250–51
tau (TCM), 251
time-varying polarizations, 113–14
zero-mean Gaussian distribution, 163,
 167, 169
C-PAR beams ratio, 297
Cramér-Rao bound (CRB), 21
 Gaussian stochastic matrix, 161
 inaccurate, 26
 for known waveforms, 228–30
 normalized Gaussian, 172
 RMS error, 223, 224, 225, 226
 root of lower bound (RCRLB), 334
 stochastic, for BPSK signals, 175–77
 stochastic, for noncircular Gaussian
  signals, 173–75
 stochastic Gaussian, 188
 stochastic sources for uniform white
  noise, 184
CRLB
 of angle accuracy, 290
 angular accuracy, 285
 of azimuth angle accuracy, 291, 292
 closed form expression, 285
 curves, 291
 estimation of target angular
  coordinates, 288–92
 of four-beam set, 291
 requirement, 290
 of three-beam set, 291
 values for standard deviation, 289
Cyclic autocorrelation function, 363, 368
Cyclic autocorrelation matrix, 365
 conjugate, 365
 extended, 366
Cyclic correlation matrix, 363
 ECM exploitation, 368
 estimation, 363

Cyclic MUSIC algorithm, 365
    Extended, 365–69
    Root, 369–72
Cyclostationarity, 363–64
    beamforming for, 374–75
    prior knowledge, 372

D

Data capture
    experimental setup, 296–97
    radar description, 295–96
Decentralized processing methods, 214
Decomposition
    into spectral components, 308
    linear array, 75–76
    modal, 71, 75–77
    singular value (SVD), 66
Delay-time-of-arrival (DTOA), 345
    definition, 348
    ray tracing and, 351
DF antennas
    common, 6–7
    horns, 7
    reflectors, 6–7
    spinning, 4–6
    spirals, 7
    See also Antennas
Dirac delta-function, 195
Direct augmentation approach (DAA), 21
Direct data covariance (DDC) matrix, 24
Direction estimation, 305–23
Direction finding (DF)
    accuracy, 3
    algorithms, 18
    defined, 3
    high-resolution methods, 241
    spinning, antenna, 4–6
    system block diagram, 3
Direction-of-arrival. *See* DOAs
Direction-of-departure (DOD)
    defined, 345
    ray tracing and, 351
    See also DOA/DOD/DTOA estimation
Direct position determination (DPD)
    advantages, 227
    benefits, 215
    cost, 215

known waveform signals, 219–22
mathematical preliminaries, 215–17
method, 214
of multiple radio transmitters, 213–27
numerical examples, 222–25
performance analysis, unknown
    waveforms, 233–38
position determination, 217–22
RMS error, 223, 224, 225, 226
unknown waveforms signals, 217–19
Discrete Fourier transform (DFT), 72,
    129–30
    filter bank, 313, 314
    sensor signal, 72
Dispersed signal parametric estimator
    (DISPARE), 200–201
Distance estimation method, 93–96
    defined, 93
    error prevention, 95
    pencil parameter, 94
    phase shifts, 94
    See also Target localization
Distributed source parameter estimator
    (DSPE), 197
    CD sources and, 197
    defined, 197
    performance evaluation, 199
Disturbance covariance matrix, 286, 288
DOA-delay estimation problem, 123
DOA/DOD/DTOA estimation, 345–60
    azimuth delay, 359, 360
    conclusions, 358–59
    elevation delay, 359, 360
    experiments, 356–58
    introduction, 345–46
    measurement environment, 357, 358
    measurement equipment specification,
        356–57
    ML-based parameter estimation,
        351–56
    parametric multipath modeling, 347–51
    results, 357–58
    UWB double directional channel
        sounding system, 346–47
DOA estimation
    algorithms, 10, 272
    ambiguous, 22

DOA estimation (continued)
   based on time delays/relative attenuation, 308–13
   of broadband sources, 305–23
   coherent broadband, 71–85
   for complex amplitude, 97
   coupling influence, 272–73
   ESPAR antenna signal processing for, 395–415
   high precision, 401–12
   incoherent wideband, 52
   with modal space processing, 71–85
   for noncircular signals, 161–88
   normal, 34
   performance, 114
   signal model, 242–44
   for signal of interest, 407
   for specified superior number of sources, 21
   spectrally weighted, 313–14
   systematic error (bias), 244
   time delays between sensors, 309
   TST-ESPRIT algorithm, 136–37
   under uncertainty, 241–56
   of wideband signals, 49–68
DOA estimation method, 87–93
   complex amplitude estimation, 92
   extraction of d_i, 91–92
   extraction of y_i, 90–91
   extraction of z_i, 91
   noisy signals, 92–93
   pairing y_i, z_i, and d_i, 92
   See also Target localization
DOAs
   antenna array measurement, 381–93
   detection system, 260
   in different beams, 151
   subspace-based algorithms, 172
   technique objectives, 73
   time-invariant, 114
   variation, 114
Double difference beam pattern, 288

E

Effective dimension, 200
Eigenspace beamformers, 272
Eigenvalue decomposition (EVD), 365
Eigenvalues
   of estimated covariance matrix, 438
   normalized noise-free covariance matrix, 201
Eigenvectors
   of estimated covariance matrix, 438
   noise subspace, 244
   signal subspace, 244
Electromagnetic (EM) coupling matrix, 259
Elementary vectors, 61
Element-by-element multiplication, 230
Elevation
   elevation angle vs., 298
   pattern of difference, 287
   sum beam difference, 299, 300
   target, 302
Elevation angle, 57, 298
   Adcock arrays, 10
   linear arrays, 50
   planar arrays, 18
Equivalent matrix, 267
Errors
   array model, 245
   compensation, mutual coupling, 273–78
   propagation model, 245
   signal model, 245
   steering vector, 264–66
   systematic, 244
ESPAR antenna
   angular diversity, 396
   applied to DOA estimation, 397
   calibration, 410–12
   computer simulations, 405–8
   conclusions, 414–15
   correlation matrix, 402–3
   cross-sectional view, 410
   cyclic symmetry of elements of, 399
   experimental verification, 408–12
   fabricated, 398
   fabricated 7-element, 409
   formulation, 397–400
   fundamentals, 397–401
   geometry, 400
   high precision DOA estimation, 401–12

introduction, 395–97
MUSIC spectrum for, 404
output formulation, 398
PC-based controller, 410
practical system applications, 412–14
reactance-domain signal processing, 403–5
satellite attitude/directional control, 412–13
signal model, 400–401
signal processing, 395–415
single-port output, 396
structure, 395, 397–400
structure/measurement configuration, 409–10
symmetric structure, 405
in transit mode, 398
wireless ad hoc network, 414
wireless locator and homing system, 413–14
ESPRIT, 51, 104, 420
algorithm, 123
implementation, 127
JADE algorithm, 123, 124, 137
one-dimensional, 141
S algorithm, 124, 127–28
T algorithm, 124, 128–31
TST algorithm, 124, 131–38
Expectation maximization (EM) algorithm, 352–53
Experimental setup, 296–97
Extended Cyclic MUSIC algorithm, 365–69
estimation performance, 367
extended autocorrelation matrix, 366
extended data model, 366
simulations, 367–69
SOI, handling, 369
spatial spectrum, 370
See also Cyclic MUSIC algorithm

F

Farfield plane wave, 262
Far-field system measurement, 388
Fast Fourier transform (FFT), 51, 96
Field of view (FOV), 55
Finite length observations, 238–39

Fisher information matrix (FIM), 229
associated blocks, 232
corresponding blocks, 231–32
inverting, 232
related to diagonal elements, 230
for zero-mean Gaussian signals, 230
Focusing angles, 53, 55
Focusing matrices, 77–78
Fourier coefficient, 216

G

Gaussian distributions, 248
Gauss-Newton procedure, 123
Generalized Capon-1 (GC-1) technique, 202, 209
Generalized Capon-2 (GC-2) technique, 202–3, 209
Generalized likelihood ratio test. *See* GLRT
General-purpose interface bus (GPIB), 346
Geometrically-based single bounce elliptical model (GBSBEM), 152
Global positioning systems (GPS), 145
GLRT
conventional case, 33–39
defined, 22
detection-estimation problem philosophy, 24–30
detection-estimation results, 30–39
framework, 26, 40
successful implementation, 26
superior case, 30–33

H

Heaviside function, 205
Hermitian matrices, 29, 233
High precision DOA estimation, 401–12
Horns, 7

I

ID sources, 200–209
beamforming methods, 202–3
covariance fitting methods, 203–6
data examples, 206–9
estimation techniques, 200–209
subspace-based methods, 200–202

ID sources (continued)
    See also Incoherently distributed source model; Parameters estimation techniques
IMUSIC, 64, 66
    likelihood surfaces, 66, 67
    modification, 66
Incoherently distributed source model, 195–97
Incoherent wideband DOA estimation, 52
Interference-to-noise ratio (INR), 289, 298
Interpolated array approach, 71
IQML algorithm, 123
Iterative linear programming (ILP), 30

J

Jacobi-Anger expansion, 75
JADE-ESPRIT algorithm, 123, 124, 137
    estimates distribution, 140
    RMSE of estimated delays, 139
    RMSE of estimated DOAs, 139
    See also ESPRIT
JADE-MUSIC
    estimated distribution, 140
    RMSE of estimated delays, 139
    RMSE of estimated DOAs, 139
    See also MUSIC

K

Known waveform signals, 219–22
    CRB for, 228–30
    defined, 219
Kronecker delta function, 143, 196

L

Lagrange multipliers, 265
Legendre function, 75
Likelihood function (LF)
    optimization, 41
    in signal subspace projection, 62–63
Likelihood ratio (LR)
    analysis, 32
    distributions, 31, 32
    distributions for fully augmentable array, 31

function, 23
lower-bound analysis, 30
maximization, 25, 34
"nested" property, 25
optimization, 31
Linear arrays, 8–9
    defined, 8
    modal decomposition, 75
    uniform (ULA), 8
    See also Arrays
Linear time-invariant system, 125
Listening tests, 321–22
    defined, 321–22
    fidelity, 322
    intelligibility, 322
    results, 322
    suppression, 322
    See also Acoustic source algorithm
Localization
    auditory, 305–23
    of scattered sources, 191–209
    source, problems, 211–12
    target, 87–101
Location-based services (LBS), 145
Log likelihood function, 219, 228, 230

M

Main beam interference (MBI), 285
Manifold ambiguity, 22
Maximum a posteriori (MAP) estimate, 214
Maximum likelihood estimate (MLE), 214
    algorithms for TDOA estimation, 296
    algorithm simulation results, 292–94
    for estimation of target angular coordinates, 286–88
    on recorded live data, 301
Maximum-likelihood (ML)
    asymptotic (AML), 327
    estimates, 35, 40
    estimator, 22
    locally optimal, 32
    optimization, 37, 38
    parameter estimation, 351–56
    stochastic (SML), 252–53
Mean square errors (MSE), 182

Meijer's G-functions, 29
M-estimation, 251–52
Method of direction estimation (MODE) algorithm, 214
Minimum scattering antennas (MSAs), 269, 382
Minimum shift keying (MSK), 364
Minimum variance distortionless response (MVDR), 202
  beamformer, 265
  beamformer generalization, 203
  filter, 265
Min-Norm, 419, 420
Mixing condition, 328
MMEE Algorithm, 34
Modal covariance matrix, 79
Modal decomposition, 71, 75–77
  application of, 72
  linear arrays, 75–76
  truncation, 76–77
Modal space, 72
  algorithm, 79–80
  modal data vector in, 79
Modal space processing (MSP), 77–79
  defined, 72
  DOA estimation with, 71–85
  estimated spectral spectrum, 81, 82, 83, 84
  matrix focusing, 77–78
  spatial resampling matrices, 78–79
Monopulse, 285
  amplitude, 15
  defined, 15
  system with sum/difference beam, 16
Monte Carlo simulations, 26, 27, 30
  performance comparison with, 65
  "strict" inequality, 30
Moore Penrose inverse, 172
Multichannel digital data recorder (MDDR), 295
Multidimensional minimization problem, 198
Multipath model, 347–48
Multiple-input, multiple-output (MIMO) channels, 75
Multiple radio transmitters, 213–27
Multiple signal classification. *See* MUSIC

Multiple speech sources, DOA estimation
  generation approach, 309
  relative attenuation between sensors, 311–13
  subspace methods, 310–11
  time delay between sensors, 309–10
Multiple target estimation, 327–42
  2-D maximizations, 331
  AML-RELAX estimator, 332–34
  data model, 328–29
  estimation algorithm, 329–34
  introduction, 327–28
  numerical performance analysis, 334–42
  problem statement, 328–29
Multipole network arrays, 270
MUSIC, 22, 51
  2DP, 420
  asymptotic efficiency, 23
  beamspace (BS), 420, 421–23
  classical, 153
  column weighting, 168
  cost function, 168
  Cyclic, 365
  DOA estimated produced by, 27
  empirical variances, 182
  estimates, 35
  experimental spectrum, 411, 412
  Extended Cyclic, 365–69
  ITC+, 23
  modified algorithm for SPAs, 153–57
  optimal weighting, 169, 170, 180
  performance, 34, 36
  performance breakdown, 33, 42
  polarimetric (P-MUSIC), 112, 118, 119–20
  polarimetric time-frequency, 110–11
  PTF algorithm, 104
  reactance-domain estimation, 405
  resolvability, 419
  review, 421–23
  Root Cyclic, 369–72
  sample null spectra formulation, 428–29
  SPA-based, 153–57
  spatial smoothing, 256
  spectrum, resolution of, 438–39

MUSIC (continued)
  spectrum calculation, 411
  spectrum for SNR, 407
  spectrum with estimated DOA, 406, 408
  t-f-based algorithms, 111
  theoretical variances, 182
  weighted cost function, 168
  white noise assumption, 424
MUSIC-like algorithms, 163–70
  built from $R_{y,T}$ and $R'_{y,T}$, 164–70
  built only from $R'_{y,T}$, 163–64
  defined, 163
  root, 166
Mutual coupling, 259
  changes due to, 264
  error compensation, 273–78
  matrix, estimation, 274
  measurement, 275
  noise vector and, 272
  See also Coupling

N

Narrowband DOA estimator, 51
Narrowband signals, 49
Natural sound forces, characteristics, 307–8
Newton-Raphson algorithm, 288
Noise
  model, uncertainty, 247–53
  subspace, eigenvectors, 244
  variance, 243
  vector, 272
Noise-like interferences (NLI), 296
Noisy signals, 92–93
Noncircular complex Gaussian (NCG) sources, 173
Noncircular distributed source parameter estimator (NC-DSPE) method, 197
  defined, 197
  performances, 199
Noncircular signals
  array signal model, 162–63
  DOA estimation for, 161–88
  illustrative examples, 177–87
  MUSIC-like algorithms, 163–70

  second-order algorithm, 175
  stochastic CRB (BPSK signals), 175–77
  stochastic CRB (noncircular Gaussian signals), 173–75
Nonparametric statistics, 250–51
Nonselective beamforming techniques, 372–76
  for cyclostationary signals, 374
  data model, 373–74
  discussion, 375
  minimum variance beamforming for cyclostationary signals, 374–75
  simulations, 375–76
Nonuniform linear arrays (NLAs)
  defined, 21
  partially augmentable, 31
  See also Linear arrays
Nyquist spacing, 308

O

Offset quadrature phase shift keying (OQPSK), 161, 364
  modulated sources, 375
  unfiltered, 164
Optimal weighted MUSIC algorithm
  DOA estimate variance, 169
  implementation, 180
  performance, 180
  reduction to standard MUSIC algorithm, 170
  See also MUSIC
Optimum weighting matrix, 247
Orthogonal vectors, 262

P

Parameters estimation techniques, 197–209
  for CD sources, 197–200
  for ID sources, 200–209
Parametric multipath modeling, 346, 347–51
  directional measurements, pairing, and clusterization, 350–51
  multipath model, 347–48
  spherical wave model, 349–50
  subband model, 349
  for UWB, 347–51

Pencil parameter, 94
Phase shifts, 94, 301
Planar arrays
  array factor, 17
  defined, 17
  elevation angle, 18
  See also Arrays
Polarimetric modeling, 106–8
Polarimetric MUSIC (P-MUSIC), 112
  performance, 118
  tracking performance illustration, 119–20
  See also MUSIC
Polarimetric signatures, 113
Polarimetric time-frequency distributions (PTFDs), 108
  defined, 108
  generalization, 109
Polarimetric time-frequency MUSIC (PTF-MUSIC), 110–11
  algorithm, 104
  instantaneous polarization characteristics, 116
  performance vs. P-MUSIC, 118
  polarization diversity, 114–16
  tracking performance illustration, 119–20
  See also MUSIC
Polarization diversity, 113–14
  in PTF-MUSIC, 114–16
  utilization, 113
Polarizations
  circular, 278, 279
  independent modulations for, 262
  orthogonal linear, 275
  source information autoterms, 115
  time-varying, 112–16
  vertical, 277
Polarization signatures, 108, 115
  spatio, 109
  time-varying, 117
  vector, 109
Polarization waves, 262
Position determination, 217–22
  known waveform signals, 219–22
  unknown waveform signals, 217–19

See also Direct position determination (DPD)
Probability density function (PDF), 192, 329, 330
Probability distribution function (PDF), 28, 29
Propagation environment, 241
Propagation model errors, 245
Pseudo Wigner-Ville distribution (PWVD), 117, 118
Pulsed Doppler estimation, 419–35
  2DP-MUSIC, 423–27
  experimental result, 433–35
  HF radar model, 423
  introduction, 419–20
  performance analysis, 427–30
  simulations, 431–33
Pulse repetition frequency (PRF), 328
Pulse repetition time (PRT), 328

Q

Qualitative robustness, 249–50
  defined, 249
  infinitesimal contamination, 249
  influence function, 250
  See also Robustness
Quantitative robustness, 248–49
  breakdown point, 248
  dissimilarity measure, 249
  See also Robustness

R

Radar
  for data capture, 295–96
  HF, model, 423
  multiple target estimation, 327–42
  passive phased array C-band, 295
  pulsed Doppler, 419–35
Radiation patterns measurement, 388–89
Rank covariance matrix (RCM), 251
  defined, 251
  robust, 255
RARE, 245–46
Reactance domain
  correlation matrix in, 403
  MUSIC estimation, 405
  signal processing, 403–5

Reception coupling matrix, 386
Reflectors, 6–7
Relative attenuation, 308–13
   measurements, 311
   between sensors, 308, 311–13
Relative delay, 315
RELAXation method, 327
Reverberant room, subspace methods in, 310–11
RF system
   calibration process for, 389–91
   calibration scheme for coupling, 390, 391
   coupling in, 386
Robustness, 248–50
   improving, 245–47
   qualitative, 249–50
   quantitative, 248–49
   UCA and, 254
Robust procedures, 250–53
   M-estimation, 251–52
   nonparametric statistics, 250–51
   stochastic ML (SML), 252–53
Root Cyclic MUSIC algorithm, 369–72
   behavior, 371
   defined, 369
   roots calculation, 369
   selected polynomial roots by, 372
   simulations, 371–72
   See also Cyclic MUSIC algorithm
Root-mean-square-error (RMSE), 63, 138, 139
   of AML-RELAX algorithm, 337
   of DOA estimator, 336, 337
   of DOA estimator vs. number of integrated pulses, 339
   of DOA estimator vs. SDR, 340
   of DOA estimator vs. SDR_2, 341
   of Doppler frequency estimator, 335, 336
   of Doppler frequency estimator vs. number of integrated pulses, 339
   of Doppler frequency estimator vs. SDR, 340
   of Doppler frequency estimator vs. SDR_2, 341
   of estimated delays, 139
   estimated DOA for circular polarization, 278
   estimated DOA for circular polarization (64 samples), 279
   estimated DOA for circular polarization (256 samples), 279
   estimated DOA for vertical polarization, 277
   estimated DOAs, 139
Root of Cramér-Rao lower bound (RCRLB), 334
   of DOA estimator, 336, 337
   of DOA estimator vs. number of integrated pulses, 339
   of DOA estimator vs. SDR, 340
   of DOA estimator vs. SDR_2, 341
   of Doppler frequency estimator, 335, 336
   of Doppler frequency estimator vs. number of integrated pulses, 339
   of Doppler frequency estimator vs. SDR, 340
   of Doppler frequency estimator vs. SDR_2, 341
   See also Cramér-Rao bound (CRB)
Rotational signal subspace (RSS), 54

S
SAGE algorithm, 353–54
   defined, 353
   SIC-type procedure of, 355
Satellite attitude/directional control, 412–13
Scattered sources
   array model, 192–97
   localization of, 191–209
   parameters estimation techniques, 197–209
Scattering matrix, 267, 381
   for antenna array, 385
   influence function, 250
   parameters, 267
Self-calibration process, 392–93
Sensor-CLEAN algorithm, 346
S-ESPRIT algorithm, 124, 127–28
   defined, 127
   DOAs with diagonal elements, 128
   See also ESPRIT

Sidelobe blanker (SLB), 295
Sidelobe canceller (SLC), 295
Signal model, 105–6
   errors, 245
   spatial time-frequency distributions, 105–6
   time-frequency distributions, 105
   uncertainties, 244–47
Signal properties exploitation, 363–77
   cyclic MUSIC algorithm, 365
   cyclostationarity, 363–64
   data model, 364–65
   high-resolution techniques, 365–72
   nonselective beamforming techniques, 372–76
Signal subspace
   eigenvectors, 244
   projection, 62–63
Signal-to-interference ratio (SIR), 309
SI-JADE algorithm, 124
Simulations, 80–84
   ESPAR antenna, 405–8
   Extended Cyclic MUSIC algorithm, 367–69
   group of two sources, 80–82
   Monte Carlo, 26, 27, 30, 65
   nonselective beamforming, 375–76
   pulsed Doppler estimation, 431–33
   Root Cyclic MUSIC algorithm, 371–72
   target localization, 96–101
   three groups of five sources, 82–84
Singular value decomposition (SVD), 56, 365
   computation, 366
   of pseudodata matrix, 56
Smart antenna algorithms, 250
Snapshot vector, 262–63
SNR
   array (ASNR), 427
   high, 266
   low, 54, 266
   threshold, 33
Sound source separation performance, 320–21
Source localization problems, 211–80

Source models, 193–97
   coherently distributed, 193–95
   incoherently distributed, 195–97
Source separation algorithm, 314–16
SPA-based MUSIC algorithm, 153–57
   method reliability, 156–57
   MUSIC spectrum computation, 154
   performance analysis, 155–57
   steering matrix, 153, 154
   summary, 154
   See also Switched parasitic arrays (SPAs)
Space division multiple access (SDMA), 145
Space reference algorithm, 266
Space-time adaptive processing (STAP) algorithms, 297, 425
Spatial beamforming, 137
Spatial covariance matrix, 126
Spatial polarimetric time-frequency distributions (SPTFDs), 108–10
   extended mixing matrix, 109
   matrices, 110, 112
   matrix formulation, 110
Spatial rank function, 250
Spatial resampling method, 71, 74–75
   average covariance matrix, 75
   defined, 74
   matrices, 78–79
Spatial sign function, 250
Spatial TFDs (STFDs), 103
   of data vector, 105
   matrices, 104, 106, 110
   noise-free, 105–6
Spatial time-frequency distributions, 105–6
Spatio-polarimetric correlations, 111–12
   coefficient, 112
   product of, 112
Spectrally weighted DOA estimation, 313–14
   illustration, 314
   steps, 313
Spheres
   three, results, 99, 100
   two, results, 98, 99
   two, with 50m spacing, 100

Spherical wave model, 349–50
   double directional, 350
   illustrated, 349
   parameters, 349
Spinning DF antennas, 4–6
Spirals, 7
Steered covariance method (STCM), 58–59
Steering errors, 264–66
Steering vectors, 74
   base station, 237
   coherently distributed source model, 194–95
   computation, 260
   derivatives, 238
   errors, 264–66
   extended, 166
   as matrix product, 271
   model, 260–64, 382–86
   phase factor formula, 264
   position, 271
   representation, 263
   true, finding, 200
   ULA replacing, 166
   uniform circular array, 243
Stochastic CRB, 173–77
   for BPSK signals, 175–77
   for noncircular Gaussian signals, 173–75
   See also Cramér-Rao bound (CRB)
Stochastic ML (SML), 252–53
   estimation, 253
   noise/signal distribution, 252
   See also Maximum likelihood (ML)
Subband model, 349
Subdiag method, 205
   Subdiag-1, 206
   Subdiag-2, 206
Subspace-based algorithms, 123, 166, 172
Subspace-based methods
   for ID sources, 200–202
   in reverberant room, 310–11
Successive interference cancellation (SIC)-type procedure, 354–55
   defined, 354
   drawback, 354
   illustrated, 355

Sum beams
   azimuth angle vs., 197
   azimuth difference, 299, 300
   elevation angle vs., 298
   elevation difference, 299, 300
   pattern, 286
Superior scenarios GLRT, 22
Switched parasitic arrays (SPAs), 145, 146–48
   circularly symmetric radiation patterns, 146
   circular (SPCA), 147
   electronically controllable radiation pattern, 146
   for electronic beam steering, 146
   method reliability for, 151
   modified MUSIC algorithm, 153–57
   nonsymmetric radiation patterns, 146
   normalized radiation patterns, 148
   seven-element, 147, 151
Sylvester's inequality, 61
Systematic error (bias), 244
System model, 72–73, 124–27
System parameter uncertainties, 244–47

T

Target direction-of-arrival (TDOA), 285, 296
Target localization, 87–101
   conclusion, 101
   distance estimation method, 93–96
   DOA estimation method, 87–93
   estimation/simulation results, 96–101
Targets
   angular coordinates estimation (CRLB), 288–92
   angular coordinates estimation (MLE), 286–88
   azimuth, 301
   elevation, 302
   multiple, estimation, 327–42
   perfect conducting (PEC) spheres, 97
Tau covariance matrix (TCM), 251
Taylor series
   expansion, 115
   formulas, 245
TDMA wireless system, 124, 126

Temporal array vector, 135
Temporal filtering, 133, 136
Terminals, coupling through, 387–88
T-ESPRIT algorithm, 124, 128–31
    band selection, 130–31
    channel estimation, 129
    deconvolution, 131
    defined, 128
    DFT, 129–30
    preprocess, 129
    procedures illustration, 131
    temporal samples use, 128
    See also ESPRIT
Test of Orthogonality of Projected Signal (TOPS), 52, 56–57
    algorithm, 62
    defined, 56–57
    frequency bins, 63
    likelihood estimator, 63, 64
    likelihood surfaces, 66, 67
    modification for localization, 66–67
    theory, 57
Time delays, 308–13
Time-frequency distributions (TFDs), 105
    cross-term, 105
    polarimetric (PTFDs), 108, 109
    sensor data, 103
    source waveforms, 103
    spatial polarimetric, 106–10
    spatial (STFDs), 103, 105–6
    use, 103
Time-frequency MUSIC (TF-MUSIC)
    defined, 110
    polarimetric (PTF-MUSIC), 110–11
Time-of-arrivals (TOAs)
    estimates, 213
    measurements, 214
Time-varying polarizations
    covariance matrix, 113–14
    moving sources with, 112–16
    signal effect, 114
    signatures, 117
    See also Polarizations
Toeplitz matrix, 31, 125–26, 197
TST-ESPRIT algorithm, 124, 131–38
    accuracy, 138
    computational complexity, 137, 138

delay estimation, 137
DOA estimation, 136–37
estimated distribution, 140–41
grouping, 136
postmultiplication, 133
premultiplication, 134
rationale, 131
rays, 132
rays resolution, 138
RMSE of estimated delays, 139
RMSE of estimated DOAs, 139
spatial beamforming, 137
steps, 136–38
structure illustration, 133
temporal filtering, 133, 136
tree structure, 132
See also ESPRIT
Two-dimensional spatial-temporal prefiltering-based MUSIC. *See* 2DP-MUSIC algorithm
Two-path model, 148–53

U

Uncertainty
    DOA estimation under, 241–56
    in noise model, 241, 247–53
    in signal model, 241, 244–47
    sources, 245
    in system parameters, 244–47
Uniform circular arrays (UCAs)
    robustness and, 254
    root-MUSIC algorithm for, 255
    See also Arrays
Uniform linear arrays (ULAs), 8, 237–38
    DOA estimation, 276
    of M sensors, 243
    number of elements/spacing, 10
    output, 50
    See also Arrays
Unknown waveform signals, 217–19
    cost function, 226
    CRB for, 230–32
    DPD performance analysis, 233–38
    parameters, 217
UWB
    directional measurements, pairing, clusterization, 350–51

UWB (continued)
    multipath model, 347–48
    parametric multipath modeling, 346, 347–51
    spherical wave model, 349–50
    subband model, 349
UWB double directional channel
    ML-based parameter estimation, 351–56
    modeling, 345–60
    sounding system, 346–47

V

Vandermonde form, 50, 243
Vector network analyzer (VNA), 346, 347
Vertical polarization, RMSE for estimated DOA, 277

W

Waveform signals, 217–22
    CRB for, 228–33
    known, 219–22
    unknown, 217–19
Weighted average of signal subspace (WAVES), 53, 55–56
    covariance matrices, 56
    defined, 55–56
    drawback, 56
    noise subspace determination, 56
Weighted pseudosubspace fitting (WPSF), 201

Weighted subspace fitting (WSF) method, 55
Weighting matrix, 168
Weight vectors
    Capon, 266
    optimum, 265
White-noise power, 25
Wideband direction-finding method, 148–53
    description, 149–51
    performance analysis, 151–53
    two-path model, 148–53
    wideband signaling, 149
Wideband DOA estimation, 52–56
    coherent, 52–56
    incoherent, 52
    performance assessment, 63–66
Wideband signals
    array output, 50
    DOA estimation, 49–68
    in multipath environments, 149
    narrowband decomposition, 68
Wireless ad hoc network, 414
Wireless local area networks (WLANs), 145
Wireless locator/homing system, 413–14
Wullenweber array, 15–17
    array factor, 16, 17
    defined, 15
    illustrated, 16
    See also Arrays

Recent Titles in the Artech House Radar Library

David K. Barton, Series Editor

Advanced Techniques for Digital Receivers, Phillip E. Pace

Advances in Direction-of-Arrival Estimation, Sathish Chandran, editor

Airborne Pulsed Doppler Radar, Second Edition, Guy V. Morris and Linda Harkness, editors

Bayesian Multiple Target Tracking, Lawrence D. Stone, Carl A. Barlow, and Thomas L. Corwin

Beyond the Kalman Filter: Particle Filters for Tracking Applications, Branko Ristic, Sanjeev Arulampalam, and Neil Gordon

Computer Simulation of Aerial Target Radar Scattering, Recognition, Detection, and Tracking, Yakov D. Shirman, editor

Design and Analysis of Modern Tracking Systems, Samuel Blackman and Robert Popoli

Detecting and Classifying Low Probability of Intercept Radar, Phillip E. Pace

Digital Techniques for Wideband Receivers, Second Edition, James Tsui

Electronic Intelligence: The Analysis of Radar Signals, Second Edition, Richard G. Wiley

Electronic Warfare in the Information Age, D. Curtis Schleher

EW 101: A First Course in Electronic Warfare, David Adamy

EW 102: A Second Course in Electronic Warfare, David L. Adamy

Fourier Transforms in Radar and Signal Processing, David Brandwood

Fundamentals of Electronic Warfare, Sergei A. Vakin, Lev N. Shustov, and Robert H. Dunwell

Fundamentals of Short-Range FM Radar, Igor V. Komarov and Sergey M. Smolskiy

Handbook of Computer Simulation in Radio Engineering, Communications, and Radar, Sergey A. Leonov and Alexander I. Leonov

High-Resolution Radar, Second Edition, Donald R. Wehner

Introduction to Electronic Defense Systems, Second Edition, Filippo Neri

Introduction to Electronic Warfare, D. Curtis Schleher

Introduction to Electronic Warfare Modeling and Simulation, David L. Adamy

Introduction to RF Equipment and System Design, Pekka Eskelinen

Microwave Radar: Imaging and Advanced Concepts, Roger J. Sullivan

Millimeter-Wave Radar Targets and Clutter, Gennadiy P. Kulemin

Modern Radar System Analysis, David K. Barton

Multitarget-Multisensor Tracking: Applications and Advances Volume III, Yaakov Bar-Shalom and William Dale Blair, editors

Principles of High-Resolution Radar, August W. Rihaczek

Principles of Radar and Sonar Signal Processing, François Le Chevalier

Radar Cross Section, Second Edition, Eugene F. Knott et al.

Radar Evaluation Handbook, David K. Barton et al.

Radar Meteorology, Henri Sauvageot

Radar Reflectivity of Land and Sea, Third Edition, Maurice W. Long

Radar Resolution and Complex-Image Analysis, August W. Rihaczek and Stephen J. Hershkowitz

Radar Signal Processing and Adaptive Systems, Ramon Nitzberg

Radar System Analysis and Modeling, David K. Barton

Radar System Performance Modeling, Second Edition, G. Richard Curry

Radar Technology Encyclopedia, David K. Barton and Sergey A. Leonov, editors

Range-Doppler Radar Imaging and Motion Compensation, Jae Sok Son et al.

Signal Detection and Estimation, Second Edition, Mourad Barkat

Space-Time Adaptive Processing for Radar, J. R. Guerci

Theory and Practice of Radar Target Identification, August W. Rihaczek and Stephen J. Hershkowitz

Time-Frequency Transforms for Radar Imaging and Signal Analysis, Victor C. Chen and Hao Ling

For further information on these and other Artech House titles, including previously considered out-of-print books now available through our In-Print-Forever® (IPF®) program, contact:

Artech House
685 Canton Street
Norwood, MA 02062
Phone: 781-769-9750
Fax: 781-769-6334
e-mail: artech@artechhouse.com

Artech House
46 Gillingham Street
London SW1V 1AH UK
Phone: +44 (0)20 7596-8750
Fax: +44 (0)20 7630-0166
e-mail: artech-uk@artechhouse.com

Find us on the World Wide Web at: www.artechhouse.com